MW00845178

PROBABILITY

PROBABILITY

With Applications and R

Second Edition

AMY S. WAGAMAN
Department of Mathematics and Statistics
Amherst College
Amherst, MA

ROBERT P. DOBROW
Department of Mathematics
Carleton College
Northfield, MN

This edition first published 2021

Edition History
John Wiley & Sons, Inc. (1e, 2014)

Registered Office
John Wiley & Sons, Inc., 111 River Street, Hoboken, NJ 07030, USA

Editorial Office
111 River Street, Hoboken, NJ 07030, USA

For details of our global editorial offices, customer services, and more information about Wiley products visit us at www.wiley.com.

Wiley also publishes its books in a variety of electronic formats and by print-on-demand. Some content that appears in standard print versions of this book may not be available in other formats.

Library of Congress Cataloging-in-Publication Data

Names: Wagaman, Amy Shepherd, 1982- author. | Dobrow, Robert P., author.
Title: Probability : with applications and R / Amy S. Wagaman, Department of Mathematics and Statistics, Amherst College, Amherst, MA, Robert P. Dobrow, Department of Mathematics, Carleton College, Northfield, MN.
Description: Second edition. | Hoboken, NJ : Wiley, [2021] | Includes bibliographical references and index.
Identifiers: LCCN 2021007900 (print) | LCCN 2021007901 (ebook) | ISBN 9781119692386 (cloth) | ISBN 9781119692348 (adobe pdf) | ISBN 9781119692416 (epub)
Subjects: LCSH: Probabilities–Data processing. | R (Computer program language)
Classification: LCC QA276.45.R3 D63 2021 (print) | LCC QA276.45.R3 (ebook) | DDC 519.20285/5133–dc23
LC record available at https://lccn.loc.gov/2021007900
LC ebook record available at https://lccn.loc.gov/2021007901

Cover Design: Wiley
Cover Image: © D3Damon/Getty Images, NicoElNino/Shutterstock

Set in 10/12pt TimesLTStd by Straive, Chennai, India

10 9 8 7 6 5 4 3 2 1

Amy: To my fantastic, supportive fiancé, Stephen,
my beloved parents (rest in peace, Mom), and my Aunt Pat

Bob: To my wonderful family
Angel, Joe, Danny, Tom

CONTENTS

PREFACE

*Probability: With Applications and **R*** is a probability textbook for undergraduates. The second edition contains modest changes from the first, including some reorganization of material. It assumes knowledge of differential and integral calculus (two semesters of calculus, rather than three semesters). Double integrals are introduced to work with joint distributions in the continuous case, with instruction in working with them provided in an appendix. While the material in this book stands on its own as a "terminal" course, it also prepares students planning to take upper level courses in statistics, stochastic processes, and actuarial sciences.

There are several excellent probability textbooks available at the undergraduate level, and we are indebted to many, starting with the classic *Introduction to Probability Theory and Its Applications* by William Feller.

Our approach is tailored to our students and based on the experience of teaching probability at a liberal arts college. Our students are not only math majors but come from disciplines throughout the natural and social sciences, especially biology, physics, computer science, and economics. Sometimes we will even get a philosophy, English, or arts history major. They tend to be sophomores and juniors. These students love to see connections with "real-life" problems, with applications that are "cool" and compelling. They are fairly computer literate. Their mathematical coursework may not be extensive, but they like problem solving and they respond well to the many games, simulations, paradoxes, and challenges that the subject offers.

Several features of our textbook set it apart from others. First is the emphasis on simulation. We find that the use of simulation, both with "hands-on" activities in the classroom and with the computer, is an invaluable tool for teaching probability.

We use the free software **R** and provide supplemental resources (on the text website) for getting students up to speed in using and understanding the language. We recommend that students work through the introductory **R** supplement, and encourage use of the other supplements that enhance the code and discussion from the textbook with additional practice. The book is not meant to be an instruction manual in **R**; we do not teach programming. But the book does have numerous examples where a theoretical concept or exact calculation is reinforced by a computer simulation. The **R** language offers simple commands for generating samples from probability distributions. The book references numerous **R** script files, that are available for download, and are contained in the **R** supplements, also available for download from the text website. It also includes many short **R** "one-liners" that are easily shown in the classroom and that students can quickly and easily duplicate on their computer. Throughout the book are numerous "**R**" display boxes that contain these code and scripts. Students and instructors may use the supplements and scripts to run the book code without having to retype it themselves. The supplements also include more detail on some examples and questions for further practice.

In addition to simulation, another emphasis of the book is on applications. We try to motivate the use of probability throughout the sciences and find examples from subjects as diverse as homelessness, genetics, meteorology, and cryptography. At the same time, the book does not forget its roots, and there are many classical chestnuts like the problem of points, Buffon's needle, coupon collecting, and Montmort's problem of coincidences. Within the context of the examples, when male and female are referred to (such as in the example on colorblindness affecting males more than females), we note that this refers to biological sex, not gender identity. As such, we use the term "sex" not "gender" in the text.

Following is a synopsis of the book's 11 chapters.

Chapter 1 begins with basics and general principles: random experiment, sample space, and event. Probability functions are defined and important properties derived. Counting, including the multiplication principle, permutations, and combinations (binomial coefficients) are introduced in the context of equally likely outcomes. A first look at simulation gives accessible examples of simulating several of the probability calculations from the chapter.

Chapter 2 emphasizes conditional probability, along with the law of total probability and Bayes formula. There is substantial discussion of the birthday problem. It closes with a discussion of independence.

Random variables are the focus of Chapter 3. The most important discrete distributions—binomial, Poisson, and uniform—are introduced early and serve as a regular source of examples for the concepts to come.

Chapter 4 contains extensive material on discrete random variables, including expectation, functions of random variables, and variance. Joint discrete distributions are introduced. Properties of expectation, such as linearity, are presented, as well as the method of indicator functions. Covariance and correlation are first introduced here.

Chapter 5 highlights several families of discrete distributions: geometric, negative binomial, hypergeometric, multinomial, and Benford's law. Moment-generating functions are introduced to explore relationships between some distributions.

Continuous probability begins with Chapter 6. Expectation, variance, and joint distributions are explored in the continuous setting. The chapter introduces the uniform and exponential distributions.

Chapter 7 highlights several important continuous distributions starting with the normal distribution. There is substantial material on the Poisson process, constructing the process by means of probabilistic arguments from i.i.d. exponential inter-arrival times. The gamma and beta distributions are presented. There is also a section on the Pareto distribution with discussion of power law and scale invariant distributions. Moment-generating functions are used again to illustrate relationships between some distributions.

Chapter 8 examines methods for finding densities of functions of random variables. This includes maximums, minimums, and sums of independent random variables (via the convolution formula). Transformations of two or more random variables are presented next. Finally, there is material on geometric probability.

Chapter 9 is devoted to conditional distributions, both in the discrete and continuous settings. Conditional expectation and variance are emphasized as well as computing probabilities by conditioning. The bivariate normal is introduced here to illustrate many of the conditional properties.

The important limit theorems of probability—law of large numbers and central limit theorem—are the topics of Chapter 10. Applications of the strong law of large numbers are included via the method of moments and Monte Carlo integration. Moment-generating functions are used to prove the central limit theorem.

Chapter 11 has optional material for supplementary discussion and/or projects. These three sections center on random walks on graphs and Markov chains, culminating in an introduction to Markov chain Monte Carlo. The treatment does not assume linear algebra and is meant as a broad strokes introduction.

There is more than enough material in this book for a one-semester course. The range of topics allows much latitude for the instructor. We feel that essential material for a first course would include Chapters 1–4, 6, and parts of Chapters 7, 9, and 10.

The second edition adds learning outcomes for each chapter, the **R** supplements, and many of the chapter review exercises, as well as fixes many typos from the first edition (in both the text and the solutions).

Additional features of the book include the following:

- Over 200 examples throughout the text and some 800 end-of-chapter exercises. Includes short numerical solutions for most odd-numbered exercises.

- Learning outcomes at the start of each chapter provide information for instructors and students. The learning outcome with a (C) is a computational learning outcome.
- End-of-chapter summaries highlight the main ideas and results from each chapter for easy access.
- Chapter review exercises, which are provided online, offer a good source of additional problems for students preparing for midterm and/or final exams.
- Starred subsections are optional and contain more challenging material and may assume a higher mathematical level.
- The **R** supplements (available online) contain the book code and scripts with enhanced discussion, additional examples, and questions for practice for interested students and instructors.
- The introductory **R** supplement introduces students to the basics of **R**. (Enhanced version of first edition Appendix A, available online as part of the **R** supplements.)
- A website containing relevant material (including the **R** supplements, script files, and chapter review exercises) and errata has been established. The URL is `www.wiley.com/go/wagaman/probability2e`.
- An instructor's solutions manual with detailed solutions to all the exercises is available for instructors who teach from this book.

Amy
Amherst, MA
September 2020

ACKNOWLEDGMENTS

From Amy for the second edition:

We are indebted to many individuals who supported the work of creating a second edition of this text. First, we thank Bob, for his thoughts and encouragement when we inquired about revising the text. We also thank our student interns, especially Sabir and Tyler, for their hard work reviewing the text, working on the new supplements, and typing solutions for exercises. The students were generously supported by funding from Amherst College. We also thank our colleagues Nick Horton and Tanya Leise for helpful discussions of the first edition.

Wiley's staff supported us well during the revision, especially Kimberly Monroe-Hill, who had very helpful suggestions. We would also like to thank Mindy Okura-Marszycki, Kathleen Santoloci, and Linda Christina, for their support getting the project off the ground.

From Bob for the first edition:

We are indebted to friends and colleagues who encouraged and supported this project. The students of my Fall 2012 Probability class were real troopers for using an early manuscript that had an embarrassing number of typos and mistakes and offering a volume of excellent advice. We also thank Marty Erickson, Jack Goldfeather, Matthew Rathkey, WenliRui, and Zach Wood-Doughty. Professor Laura Chihara field-tested an early version of the text in her class and has made many helpful suggestions. Thank you to Jack O'Brien at Bowdoin College for a detailed reading of the manuscript and for many suggestions that led to numerous improvements.

Carleton College and the Department of Mathematics were enormously supportive, and I am grateful for a college grant and additional funding that supported this work. Thank you to Mike Tie, the Department's Technical Director, and Sue Jandro, the Department's Administrative Assistant, for help throughout the past year.

The staff at Wiley, including Steve Quigley, Amy Hendrickson, and Sari Friedman, provided encouragement and valuable assistance in preparing this book.

ABOUT THE COMPANION WEBSITE

This book is accompanied by a companion website:

www.wiley.com/go/wagaman/probability2e

The book companion site is split into:

- The student companion site includes chapter reviews and is open to all.
- The instructor companion site includes the instructor solutions manual.

INTRODUCTION

All theory, dear friend, is gray, but the golden tree of life springs ever green.

—Johann Wolfgang von Goethe

Probability began by first considering games of chance. But today, it has practical applications in areas as diverse as astronomy, economics, social networks, and zoology that enrich the theory and give the subject its unique appeal.

In this book, we will flip coins, roll dice, and pick balls from urns, all the standard fare of a probability course. But we have also tried to make connections with real-life applications and illustrate the theory with examples that are current and engaging.

You will see some of the following case studies again throughout the text. They are meant to whet your appetite for what is to come.

I.1 WALKING THE WEB

There are about one trillion websites on the Internet. When you google a phrase like "Can Chuck Norris divide by zero?," a remarkable algorithm called PageRank searches these sites and returns a list ranked by importance and relevance, all in the blink of an eye. PageRank is the heart of the Google search engine. The algorithm assigns an "importance value" to each web page and gives it a rank to determine how useful it is.

PageRank is a significant accomplishment of mathematics and linear algebra. It can be understood using probability. Of use are probability concepts called Markov chains and random walks, explored in Chapter 11. Imagine a web surfer who starts at some web page and clicks on a link at random to find a new site. At each page, the surfer chooses from one of the available hypertext links equally at random. If there are two links, it is a coin toss, heads or tails, to decide which one to pick. If there are 100 links, each one has a 1% chance of being chosen. As the web surfer moves from page to random page, they are performing a random walk on the web.

What is the PageRank of site x? Suppose the web surfer has been randomly walking the web for a very long time (infinitely long in theory). The probability that they visit site x is precisely the PageRank of that site. Sites that have lots of incoming links will have a higher PageRank value than sites with fewer links.

The PageRank algorithm is actually best understood as an assignment of *probabilities* to each site on the web. Such a list of numbers is called a *probability distribution*. And since it comes as the result of a theoretically infinitely long random walk, it is known as the *limiting distribution* of the random walk. Remarkably, the PageRank values for billions of websites can be computed quickly and in real time.

I.2 BENFORD'S LAW

Turn to a random page in this book. Look in the middle of the page and point to the first number you see. Write down the first digit of that number.

You might think that such first digits are equally likely to be any integer from 1 to 9. But a remarkable probability rule known as Benford's law predicts that most of your first digits will be 1 or 2; the chances are almost 50%. The probabilities go down as the numbers get bigger, with the chance that the first digit is 9 being less than 5% (Fig. I.1).

Benford's law, also known as the "first-digit phenomenon," was discovered over 100 years ago, but it has generated new interest in recent years. There are a huge number of datasets that exhibit Benford's law, including street addresses, populations of cities, stock prices, mathematical constants, birth rates, heights of mountains, and line items on tax returns. The last example, in particular, caught the eye

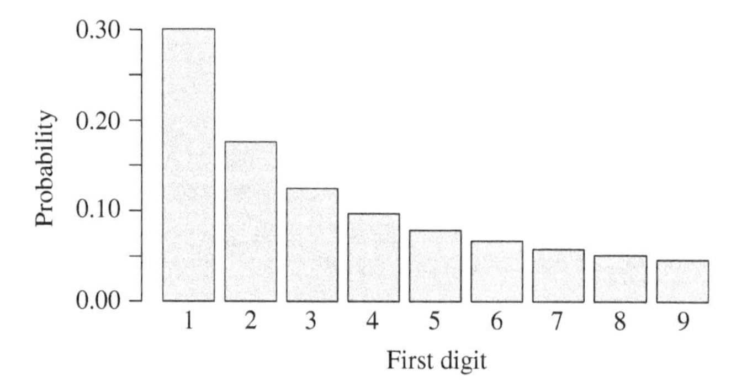

FIGURE I.1: Benford's law describes the frequencies of first digits for many real-life datasets.

of business Professor Mark Nigrini who showed that Benford's law can be used in forensic accounting and auditing as an indicator of fraud [2012].

Durtschi et al. [2004] describe an investigation of a large medical center in the western United States. The distribution of first digits of check amounts differed significantly from Benford's law. A subsequent investigation uncovered that the financial officer had created bogus shell insurance companies in her own name and was writing large refund checks to those companies. Applications to international trade were investigated in Cerioli et al. [2019].

I.3 SEARCHING THE GENOME

Few areas of modern science employ probability more than biology and genetics. A strand of DNA, with its four nucleotide bases adenine, cytosine, guanine, and thymine, abbreviated by their first letters, presents itself as a sequence of outcomes of a four-sided die. The enormity of the data—about three billion "letters" per strand of human DNA—makes randomized methods relevant and viable.

Restriction sites are locations on the DNA that contain a specific sequence of nucleotides, such as G-A-A-T-T-C. Such sites are important to identify because they are locations where the DNA can be cut and studied. Finding all these locations is akin to finding patterns of heads and tails in a long sequence of coin tosses. Theoretical limit theorems for idealized sequences of coin tosses become practically relevant for exploring the genome. The locations for such restriction sites are well described by the Poisson process, a fundamental class of random processes that model locations of restriction sites on a chromosome, as well as car accidents on the highway, service times at a fast food chain, and when you get your text messages.

On the macrolevel, random processes are used to study the evolution of DNA over time in order to construct evolutionary trees showing the divergence of species. DNA sequences change over time as a result of mutation and natural selection. Models for sequence evolution, called Markov processes, are continuous time analogues of the type of random walk models introduced earlier.

Miller et al. [2012] analyze the sequenced polar bear genome and give evidence that the size of the bear population fluctuated with key climactic events over the past million years, growing in periods of cooling and shrinking in periods of warming. Their paper, published in the *Proceedings of the National Academy of Sciences*, is all biology and genetics. But the appendix of supporting information is all probability and statistics. Similar analyses, rooted in probability theory, continue to be performed investigating relationships between species, as described in Mather et al. [2020].

I.4 BIG DATA

The search for the Higgs boson, the so-called "God particle," at the Large Hadron Collider in Geneva, Switzerland, generated 200 petabytes of data (1 petabyte = 10^{15}

bytes). That is as much data as the total amount of printed material in the world at the time! In physics, genomics, climate science, marketing, even online gaming and film, the sizes of datasets being generated are staggering. How to store, transmit, visualize, and process such data is one the great challenges of science.

Probability is being used in a central way for such problems in a methodology called *compressed sensing*.

In the average hospital, many terabytes (1 terabyte = 10^{12} bytes) of digital magnetic resonance imaging (MRI) data are generated each year. A half-hour MRI scan might collect 100 Mb of data. These data are then compressed to a smaller image, say 5 Mb, with little loss of clarity or detail. Medical and most natural images are compressible since lots of pixels have similar values. Compression algorithms work by essentially representing the image as a sum of simple functions (such as sine waves) and then discarding those terms that have low information content. This is a fundamental idea in signal processing, and essentially what is done when you take a picture on your cell phone and then convert it to a JPEG file for sending to a friend or uploading to the web.

Compressed sensing asks: If the data are ultimately compressible, is it really necessary to acquire all the data in the first place? Can just the final compressed data be what is initially gathered? And the startling answer is that by *randomly* sampling the object of interest, the final image can be reconstructed with similar results as if the object had been fully sampled. Random sampling of MRI scans produces an image of similar quality as when the entire object is scanned. The new technique has reduced MRI scan time to one-seventh the original time, from about half an hour to less than 5 minutes, and shows enormous promise for many other applied areas. For more information on this topic, the reader is directed to Mackenzie [2009].

I.5 FROM APPLICATION TO THEORY

Having sung the praises of applications and case studies, we come back to the importance of theory.

Probability has been called the science of uncertainty. "Mathematical probability" may seem an oxymoron like jumbo shrimp or civil war. If any discipline can profess a claim of "certainty," surely it is mathematics with its adherence to rigorous proof and timeless results.

One of the great achievements of modern mathematics was putting the study of probability on a solid scientific foundation. This was done in the 1930s, when the Russian mathematician Andrey Nikolaevich Kolmogorov built up probability theory in a rigorous way similarly to how Euclid built up geometry. Much of his work is the material of a graduate-level course, but the basic framework of axiom, definition, theory, and proof sets the framework for the modern treatment of the subject.

One of the joys of learning probability is the compelling imagery we can exploit. Geometers draw circles and squares; probabilists toss coins and roll dice. There is no perfect circle in the physical universe. And the "fair coin" is an idealized model. Yet when you take *real* pennies and toss them repeatedly, the results conform so beautifully to the theory.

In this book, we use the computer program **R**. **R** is free software and an interactive computing environment available for download at `http://www.r-project.org/`. If you have never used **R** before, we encourage you to work through the introductory **R** supplement to familiarize yourself with the language. As you work through the text, the associated supplements support working with the code and script files. The script files only require **R**. For working with the supplements, you can read the pdf versions, or if you want to run the code yourself, we recommend using RStudio to open these RMarkdown files. RStudio has a free version, and it provides a useful user interface for **R**. RMarkdown files allow **R** code to be interwoven with text in a reproducible fashion.

Simulation plays a significant role in this book. Simulation is the use of random numbers to generate samples from a random experiment. Today, it is a bedrock tool in the sciences and data analysis. Many problems that were for all practical purposes impossible to solve before the computer age are now easily handled with simulation.

There are many compelling reasons for including simulation in a probability course. Simulation helps build invaluable intuition for how random phenomena behave. It will also give you a flexible platform to test how changes in assumptions and parameters can affect outcomes. And the exercise of translating theoretical models into usable simulation code (easy to do in **R**) will make the subject more concrete and hopefully easier to understand.

And, most importantly, it is fun! Students enjoy the hands-on approach to the subject that simulation offers. It is thrilling to see some complex theoretical calculation "magically" verified by a simulation.

To succeed in this subject, read carefully, work through the examples, and do as many problems as you can. But most of all, enjoy the ride!

> The results concerning fluctuations in coin tossing show that widely held beliefs . . . are fallacious. They are so amazing and so at variance with common intuition that even sophisticated colleagues doubted that coins actually misbehave as theory predicts. The record of a simulated experiment is therefore included . . . .
>
> —William Feller, *An Introduction to Probability Theory and Its Applications*, Vol. 1, Third Edition (1968), page xi.

1

FIRST PRINCIPLES

The beginning is the most important part of the work.

—Plato

Learning Outcomes

1. Define basic probability and set theory terms.
2. Give examples of sample spaces, events, and probability models.
3. Apply properties of probability functions.
4. Solve problems involving equally likely outcomes and using counting methods.
5. (C) Explore simulation basics in **R** with a focus on reproducibility.

1.1 RANDOM EXPERIMENT, SAMPLE SPACE, EVENT

Probability begins with some activity, process, or experiment whose outcome is uncertain. This can be as simple as throwing dice or as complicated as tomorrow's weather.

Given such a "random experiment," the set of all possible outcomes is called the *sample space*. We will use the Greek capital letter Ω (omega) to represent the sample space.

Perhaps the quintessential random experiment is flipping a coin. Suppose a coin is tossed three times. Let H represent heads and T represent tails. The sample space is

$$\Omega = \{\text{HHH}, \text{HHT}, \text{HTH}, \text{HTT}, \text{THH}, \text{THT}, \text{TTH}, \text{TTT}\},$$

Probability: With Applications and R, Second Edition. Amy S. Wagaman and Robert P. Dobrow.

Companion Website: www.wiley.com/go/wagaman/probability2e

consisting of eight outcomes. The Greek lowercase omega ω will be used to denote these outcomes, the elements of Ω.

An *event* is a set of outcomes, and as such is a subset of the sample space Ω. Often, we refer to events by assigning them a capital letter near the beginning of the alphabet, such as event A. The event of getting all heads in three coin tosses can be written as

$$A = \{\text{Three heads}\} = \{HHH\}.$$

Event A contains a single outcome, and clearly, $A \subseteq \Omega$. More commonly, events include multiple outcomes. The event of getting at least two tails is

$$B = \{\text{At least two tails}\} = \{HTT, THT, TTH, TTT\}.$$

We often desire probabilities of events. But before learning how to find these probabilities, we first learn to identify the sample space and relevant event for a given problem.

Example 1.1 The weather forecast for tomorrow says rain. The number of umbrellas students bring to class can be considered an outcome of a random experiment. If at most each of n students brings one umbrella, then the sample space is the set $\Omega = \{0, 1, \ldots, n\}$. The event that between 2 and 4 umbrellas are brought to class is $A = \{2, 3, 4\}$. ■

Dice are often used to illustrate probability concepts. Unless stated otherwise, in this text, rolling a die refers to rolling a fair six-sided die with the usual numeric labels of the numbers 1 through 6.

Example 1.2 Roll a pair of dice. Find the sample space and identify the event that the sum of the two dice is equal to 7.

The random experiment is rolling two dice. Keeping track of the roll of each die gives the sample space

$$\Omega = \{(1,1), (1,2), (1,3), (1,4), (1,5), (1,6), (2,1), (2,2), \ldots, (6,5), (6,6)\}.$$

The event is $A = \{\text{Sum is } 7\} = \{(1,6), (2,5), (3,4), (4,3), (5,2), (6,1)\}$.

The sample space can also be presented using an array format, where the rows denote the first roll and the columns denote the second roll. The cell entries are the sum of the row and column numbers. All 36 outcomes will be represented in the resulting cells. The event A can then be identified by finding the cells that correspond to the desired criteria. ■

Example 1.3 Yolanda and Zach are running for president of the student association. One thousand students will be voting, and each voter will pick one of the two candidates. We will eventually ask questions like, What is the probability that Yolanda wins the election over Zach by at least 100 votes? But before actually finding this probability, first identify (i) the sample space and (ii) the event that Yolanda beats Zach by at least 100 votes.

(i) The outcome of the vote can be denoted as $(x, 1000 - x)$, where x is the number of votes for Yolanda, and $1000 - x$ is the number of votes for Zach. Then the sample space of all voting outcomes is

$$\Omega = \{(0, 1000), (1, 999), (2, 998), \ldots, (999, 1), (1000, 0)\}.$$

(ii) Let A be the event that Yolanda beats Zach by at least 100 votes. The event A consists of all outcomes in which $x - (1000 - x) \geq 100$, or $550 \leq x \leq 1000$. That is, $A = \{(550, 450), (551, 449), \ldots, (999, 1), (1000, 0)\}$. ■

Example 1.4 Diego will continue to flip a coin until heads appears. Identify the sample space and the event that it will take Diego at least three coin flips to get a head.

The sample space is the set of all sequences of coin flips with one head preceded by some number of tails. That is,

$$\Omega = \{H, TH, TTH, TTTH, TTTTH, TTTTTH, \ldots\}.$$

The desired event is $A = \{TTH, TTTH, TTTTH, \ldots\}$. Note that in this case both the sample space and the event A are infinite, meaning they contain an infinite number of outcomes. ■

1.2 WHAT IS A PROBABILITY?

What does it mean to say that *the probability that A occurs* or *the probability of A* is equal to x?

From a formal, purely mathematical point of view, a probability is a number between 0 and 1 that satisfies certain properties, which we will describe later. From a practical, empirical point of view, a probability matches up with our intuition of the likelihood or "chance" that an event occurs. An event that has probability 0 "never" happens. An event that has probability 1 is "certain" to happen. In repeated coin flips, a fair coin comes up heads about half the time, and the probability of heads is equal to one-half.

Let A be an event associated with some random experiment. One way to understand the probability of A is to perform the following thought exercise: imagine conducting the experiment over and over, infinitely often, keeping track of how often A occurs. Each experiment is called a *trial*. If the event A occurs when the experiment is performed, that is a *success*. The proportion of successes is the probability of A, written $P(A)$.

This is the *relative frequency* interpretation of probability, which says that the probability of an event is equal to its relative frequency in a large number of trials.

When the weather forecaster tells us that tomorrow there is a 20% chance of rain, we understand that to mean that if we could repeat today's conditions—the air pressure, temperature, wind speed, etc.—over and over again, then 20% of the resulting "tomorrows" will result in rain. Closer to what weather forecasters actually do in coming up with that 20% number, together with using satellite and radar information along with sophisticated computational models, is to go back in the historical record and find other days that match up closely with today's conditions and see what proportion of those days resulted in rain on the following day.

There are definite limitations to constructing a rigorous mathematical theory out of this intuitive and empirical view of probability. One cannot actually repeat an experiment infinitely many times. To define probability carefully, we need to take a formal, axiomatic, mathematical approach. Nevertheless, the relative frequency viewpoint will still be useful in order to gain intuitive understanding. And by the end of the book, we will actually derive the relative frequency viewpoint as a consequence of the mathematical theory.

1.3 PROBABILITY FUNCTION

We assume for the next several chapters that the sample space is *discrete*. This means that the sample space is either finite or countably infinite.

A set is *countably infinite* if the elements of the set can be arranged as a sequence. The natural numbers $1, 2, 3, \ldots$ is the classic example of a countably infinite set. And all countably infinite sets can be put in one-to-one correspondence with the natural numbers.

If the sample space is finite, it can be written as $\Omega = \{\omega_1, \ldots, \omega_k\}$. If the sample space is countably infinite, it can be written as $\Omega = \{\omega_1, \omega_2, \ldots\}$.

The set of all real numbers is an infinite set that is not countably infinite. It is called *uncountable*. An interval of real numbers, such as (0,1), the numbers between 0 and 1, is also uncountable. Probability on uncountable spaces will require differential and integral calculus and will be discussed in the second half of this book.

A *probability function* assigns numbers between 0 and 1 to events according to three defining properties.

PROBABILITY FUNCTION

Given a random experiment with discrete sample space Ω, a *probability function* P is a function on Ω with the following properties:

1.

$$P(\omega) \geq 0, \text{ for all } \omega \in \Omega.$$

2.

$$\sum_{\omega \in \Omega} P(\omega) = 1. \quad (1.1)$$

3. For all events $A \subseteq \Omega$,

$$P(A) = \sum_{\omega \in A} P(\omega). \quad (1.2)$$

You may not be familiar with some of the notation in this definition. The symbol $\in$ means "is an element of." So $\omega \in \Omega$ means ω is an element of Ω. We are also using a generalized Σ-notation in Equations 1.1 and 1.2, writing a condition under the Σ to specify the summation. The notation $\sum_{\omega \in \Omega}$ means that the sum is over all ω that are elements of the sample space, Ω, that is, all outcomes in the sample space.

In the case of a finite sample space $\Omega = \{\omega_1, \ldots, \omega_k\}$, Equation 1.1 becomes

$$\sum_{\omega \in \Omega} P(\omega) = P(\omega_1) + \cdots + P(\omega_k) = 1.$$

And in the case of a countably infinite sample space $\Omega = \{\omega_1, \omega_2, \ldots\}$, this gives

$$\sum_{\omega \in \Omega} P(\omega) = P(\omega_1) + P(\omega_2) + \cdots = \sum_{i=1}^{\infty} P(\omega_i) = 1.$$

In simple language, probabilities sum to 1. The third defining property of a probability function says that the probability of an event is the sum of the probabilities of all the outcomes contained in that event. We might describe a probability function with a table, function, graph, or qualitative description. Multiple representations are possible, as shown in the next example.

Example 1.5 A type of candy comes in red, yellow, orange, green, and purple colors. Choose a piece of candy at random. What color is it? The sample space is $\Omega = \{R, Y, O, G, P\}$. Assuming the candy colors are equally likely outcomes, here are three equivalent ways of describing the probability function:

1.

R	Y	O	G	P
0.20	0.20	0.20	0.20	0.20

2. $P(\omega) = 1/5$, for all $\omega \in \Omega$.
3. The five colors are equally likely. ■

In the discrete setting, we will often use *probability model* and *probability distribution* interchangeably with probability function. In all cases, to specify a probability function requires identifying (i) the outcomes of the sample space and (ii) the probabilities associated with those outcomes.

Letting H denote heads and T denote tails, an obvious model for a simple coin toss is $P(\text{H}) = P(\text{T}) = 0.50$.

Actually, there is some extremely small, but nonzero, probability that a coin will land on its side. So perhaps a better model would be

$$P(\text{H}) = P(\text{T}) = 0.49999999995 \quad \text{and} \quad P(\text{Side}) = 0.0000000001.$$

Ignoring the possibility of the coin landing on its side, a more general model is

$$P(\text{H}) = p \quad \text{and} \quad P(\text{T}) = 1 - p,$$

where $0 \leq p \leq 1$. If $p = 1/2$, we say the coin is *fair*. If $p \neq 1/2$, we say that the coin is *biased*. In this text, assume coins are fair unless otherwise specified.

In a mathematical sense, all of these coin tossing models are "correct" in that they are consistent with the definition of what a probability is. However, we might debate which model most accurately reflects reality and which is most useful for modeling actual coin tosses.

■ **Example 1.6** Suppose that a college has six majors: biology, geology, physics, dance, art, and music. The percentage of students taking these majors are 20, 20, 5, 10, 10, and 35, respectively, with double majors not allowed. Choose a random student. What is the probability they are a science major?

The random experiment is choosing a student. The sample space is

$$\Omega = \{\text{Bio, Geo, Phy, Dan, Art, Mus}\}.$$

The probability model is given in Table 1.1. The event in question is

$$A = \{\text{Science major}\} = \{\text{Bio, Geo, Phy}\}.$$

Finally,

$$\begin{aligned} P(A) &= P(\{\text{Bio, Geo, Phy}\}) = P(\text{Bio}) + P(\text{Geo}) + P(\text{Phy}) \\ &= 0.20 + 0.20 + 0.05 = 0.45. \end{aligned}$$

■

TABLE 1.1. Probability model for majors.

Bio	Geo	Phy	Dan	Art	Mus
0.20	0.20	0.05	0.10	0.10	0.35

This example is probably fairly clear and may seem like a lot of work for a simple result. However, when starting out, it is good preparation for the more complicated problems to come to clearly identify the sample space, event, and probability model before actually computing the final probability.

Example 1.7 In three coin tosses, what is the probability of getting at least two tails?

Although the probability model here is not explicitly stated, the simplest and most intuitive model for fair coin tosses is that every outcome is equally likely. As the sample space

$$\Omega = \{HHH, HHT, HTH, THH, HTT, THT, TTH, TTT\}$$

has eight outcomes, the model assigns to each outcome the probability $1/8$.

The event of getting at least two tails can be written as $A = \{HTT, THT, TTH, TTT\}$. This gives

$$\begin{aligned} P(A) &= P(\{HTT, THT, TTH, TTT\}) \\ &= P(HTT) + P(THT) + P(TTH) + P(TTT) \\ &= \frac{1}{8} + \frac{1}{8} + \frac{1}{8} + \frac{1}{8} = \frac{1}{2}. \end{aligned}$$

■

1.4 PROPERTIES OF PROBABILITIES

Events can be combined together to create new events using the connectives "or," "and," and "not." These correspond to the set operations union, intersection, and complement.

For sets $A, B \subseteq \Omega$, the *union* $A \cup B$ is the set of all elements of Ω that are in either A or B or both. The *intersection* AB is the set of all elements of Ω that are in both A and B. (Another common notation for the intersection of two events is $A \cap B$.) The *complement* A^c is the set of all elements of Ω that are not in A.

In probability word problems, descriptive phrases are typically used rather than set notation. See Table 1.2 for some equivalences.

A Venn diagram is a useful tool for working with events and subsets. A rectangular box denotes the sample space Ω, and circles are used to denote events. See Figure 1.1 for examples of Venn diagrams for the most common combined events obtained from two events A and B.

One of the most basic, and important, properties of a probability function is the simple addition rule for mutually exclusive events. We say that two events are *mutually exclusive*, or *disjoint*, if they have no outcomes in common. That is, A and B are mutually exclusive if $AB = \emptyset$, the empty set.

TABLE 1.2. Events and sets.

Description	Set notation
Either A or B or both occur	$A \cup B$
A and B	AB
Not A	A^c
A implies B; A is a subset of B	$A \subseteq B$
A but not B	AB^c
Neither A nor B	A^cB^c
At least one of the two events occurs	$A \cup B$
At most one of the two events occurs	$(AB)^c = A^c \cup B^c$

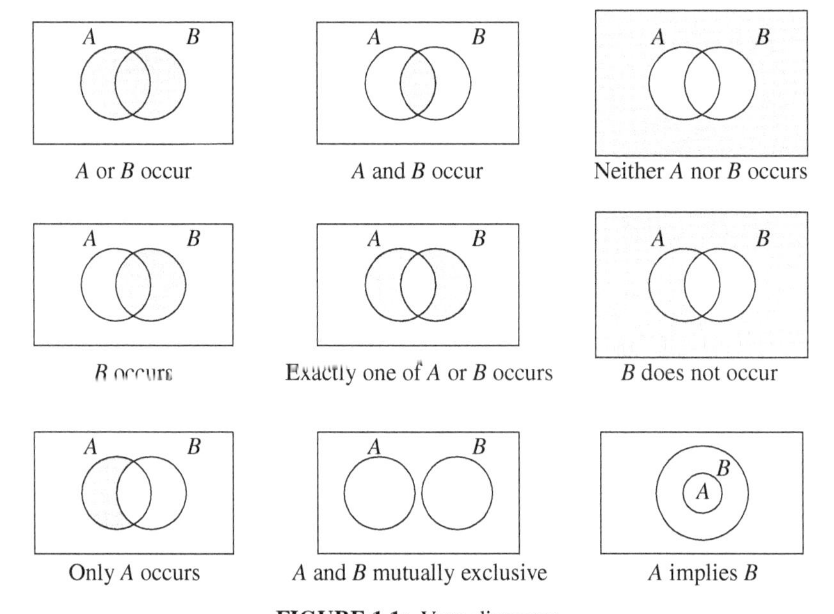

FIGURE 1.1: Venn diagrams.

> **ADDITION RULE FOR MUTUALLY EXCLUSIVE EVENTS**
>
> If A and B are mutually exclusive events, then
>
> $$P(A \text{ or } B) = P(A \cup B) = P(A) + P(B).$$

The addition rule is a consequence of the third defining property of a probability function. We have that

$$P(A \text{ or } B) = P(A \cup B) = \sum_{\omega \in A \cup B} P(\omega)$$

$$= \sum_{\omega \in A} P(\omega) + \sum_{\omega \in B} P(\omega)$$
$$= P(A) + P(B),$$

where the third equality follows because the events are disjoint, so no outcome ω will be counted twice. The addition rule for mutually exclusive events extends to more than two events.

EXTENSION OF ADDITION RULE FOR MUTUALLY EXCLUSIVE EVENTS

Suppose $A_1, A_2, \ldots$ is a sequence of pairwise mutually exclusive events. That is, A_i and A_j are mutually exclusive for all $i \neq j$. Then

$$P(\text{at least one of the } A_i\text{'s occurs}) = P\left(\bigcup_{i=1}^{\infty} A_i\right) = \sum_{i=1}^{\infty} P(A_i).$$

Next, we highlight other key properties that are consequences of the defining properties of a probability function and the addition rule for disjoint events.

PROPERTIES OF PROBABILITIES

1. If A implies B, that is, if $A \subseteq B$, then $P(A) \leq P(B)$.
2. $P(A \text{ does not occur}) = P(A^c) = 1 - P(A)$.
3. For all events A and B,

$$P(A \text{ or } B) = P(A \cup B) = P(A) + P(B) - P(AB). \quad (1.3)$$

Each property is derived next.

1. As $A \subseteq B$, write B as the disjoint union of A and BA^c. By the addition rule for disjoint events,

$$P(B) = P(A \cup BA^c) = P(A) + P(BA^c) \geq P(A),$$

because probabilities are nonnegative.

2. The sample space Ω can be written as the disjoint union of any event A and its complement A^c. Thus,

$$1 = P(\Omega) = P(A \cup A^c) = P(A) + P(A^c).$$

Rearranging gives the result.

3. Write $A \cup B$ as the disjoint union of A and A^cB. Also write B as the disjoint union of AB and A^cB. Then $P(B) = P(AB) + P(A^cB)$ and thus,

$$P(A \cup B) = P(A) + P(A^cB) = P(A) + P(B) - P(AB).$$

Observe that the addition rule for mutually exclusive events follows from Property 3 because if A and B are disjoint, then $P(AB) = P(\emptyset) = 0$.

Example 1.8 In a city, suppose 75% of the population have brown hair, 40% have brown eyes, and 25% have both brown hair and brown eyes. A person is chosen at random from the city. What is the probability that they

1. Have brown eyes or brown hair?
2. Have neither brown eyes nor brown hair?

To gain intuition, draw a Venn diagram, as in Figure 1.2. Let H be the event of having brown hair; let E denote brown eyes.

1. The probability of having brown eyes or brown hair is

$$P(E \text{ or } H) = P(E) + P(H) - P(EH) = 0.75 + 0.40 - 0.25 = 0.90.$$

Notice that E and H are not mutually exclusive. If we made a mistake and used the simple addition rule $P(E \text{ or } H) = P(E) + P(H)$, we would mistakenly get $0.75 + 0.40 = 1.15 > 1$.

2. The complement of having neither brown eyes nor brown hair is having brown eyes or brown hair. Thus,

$$P(E^cH^c) = P((E \text{ or } H)^c) = 1 - P(E \text{ or } H) = 1 - 0.90 = 0.10.$$

■

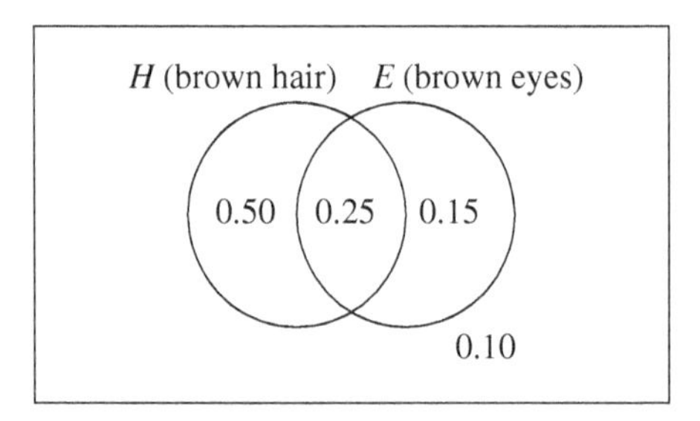

FIGURE 1.2: Venn diagram.

1.5 EQUALLY LIKELY OUTCOMES

The simplest probability model for a finite sample space is that all outcomes are equally likely. If Ω has k elements, then the probability of each outcome is $1/k$, as probabilities sum to 1. That is, $P(\omega) = 1/k$, for all $\omega \in \Omega$.

Computing probabilities for equally likely outcomes takes a fairly simple form. Suppose A is an event with s elements, with $s \leq k$. As $P(A)$ is the sum of the probabilities of all the outcomes contained in A,

$$P(A) = \sum_{\omega \in A} P(\omega) = \sum_{\omega \in A} \frac{1}{k} = \frac{s}{k} = \frac{\text{Number of elements of } A}{\text{Number of elements of } \Omega}.$$

In other words, probability with equally likely outcomes reduces to *counting* elements in A and Ω.

Example 1.9 A *palindrome* is a word that reads the same forward or backward. Examples include mom, civic, and rotator. Pick a three-letter "word" at random choosing from D, O, or G for each letter. What is the probability that the resulting word is a palindrome? (Words in this context do not need to be real words in English, e.g., *OGO* is a palindrome.)

There are 27 possible words (three possibilities for each of the three letters). List and count the palindromes: DDD, OOO, GGG, DOD, DGD, ODO, OGO, GDG, and GOG. The probability of getting a palindrome is $9/27 = 1/3$. ■

Example 1.10 A bowl has r red balls and b blue balls. A ball is drawn randomly from the bowl. What is the probability of selecting a red ball?

The sample space consists of $r + b$ balls. The event $A = \{\text{Red ball}\}$ has r elements. Therefore, $P(A) = r/(r + b)$. ■

A model for equally likely outcomes assumes a finite sample space. Interestingly, it is impossible to have a probability model of equally likely outcomes on an infinite sample space. To see why, suppose $\Omega = \{\omega_1, \omega_2, \dots\}$ and $P(\omega_i) = c$ for all i, where c is a nonzero constant. Then summing the probabilities gives

$$\sum_{i=1}^{\infty} P(\omega_i) = \sum_{i=1}^{\infty} c = \infty \neq 1.$$

While equally likely outcomes are not possible in the infinite case, there are many ways to assign probabilities for an infinite sample space where outcomes are not equally likely. For instance, let $\Omega = \{\omega_1, \omega_2, \dots\}$ with $P(\omega_i) = (1/2)^i$, for $i = 1, 2, \dots$. Then, using results for geometric series,

$$\sum_{i=1}^{\infty} P(\omega_i) = \sum_{i=1}^{\infty} \left(\frac{1}{2}\right)^i = \left(\frac{1}{2}\right) \frac{1}{1 - (1/2)} = 1.$$

We introduce some basic counting principles in the next two sections because counting plays a fundamental role in probability when outcomes are equally likely.

1.6 COUNTING I

Counting sets is sometimes not as easy as $1, 2, 3, \dots$. But a basic counting principle known as the *multiplication principle* allows for tackling a wide range of problems.

MULTIPLICATION PRINCIPLE

If there are m ways for one thing to happen, and n ways for a second thing to happen, there are $m \times n$ ways for both things to happen.

More generally—and more formally—consider an n-element sequence $(a_1, a_2, \dots, a_n)$. If there are k_1 possible values for the first element, k_2 possible values for the second element, $\dots$, and k_n possible values for the nth element, there are $k_1 \times k_2 \times \cdots \times k_n$ possible sequences.

For instance, in tossing a coin three times, there are $2 \times 2 \times 2 = 2^3 = 8$ possible outcomes. Rolling a die four times gives $6 \times 6 \times 6 \times 6 = 6^4 = 1296$ possible rolls.

Example 1.11 License plates in Minnesota are issued with three letters from A to Z followed by three digits from 0 to 9. If each license plate is equally likely, what is the probability that a random license plate starts with G-Z-N?

The solution will be equal to the number of license plates that start with G-Z-N divided by the total number of license plates. By the multiplication principle, there are $26 \times 26 \times 26 \times 10 \times 10 \times 10 = 26^3 \times 10^3 = 17{,}576{,}000$ possible license plates.

For the number of plates that start with G-Z-N, think of a six-element plate of the form G-Z-N-_-_-_. For the three blanks, there are $10 \times 10 \times 10$ possibilities. Thus, the desired probability is $10^3/(26^3 \times 10^3) = 1/26^3 = 0.0000569$. ■

Example 1.12 A DNA strand is a long polymer string made up of four nucleotides—adenine, cytosine, guanine, and thymine. It can be thought of as a sequence of As, Cs, Gs, and Ts. DNA is structured as a double helix with two paired strands running in opposite directions on the chromosome. Nucleotides always pair the same way: A with T and C with G. A *palindromic sequence* is equal to its "reverse complement." For instance, the sequences CACGTG and TAGCTA are palindromic sequences (with reverse complements GTGCAC and ATCGAT, respectively), but TACCAT is not (reverse complement is ATGGTA). Such sequences play a significant role in molecular biology.

Suppose the nucleotides on a DNA strand of length six are generated in such a way so that all strands are equally likely. What is the probability that the DNA sequence is a palindromic sequence?

By the multiplication principle, the number of DNA strands is 4^6 because there are four possibilities for each site. A palindromic sequence of length six is completely determined by the first three sites. There are 4^3 palindromic sequences. The desired probability is $4^3/4^6 = 1/64$. ■

Example 1.13 Logan is taking four final exams next week. His studying was erratic and all scores A, B, C, D, and F are equally likely for each exam. What is the probability that Logan will get at least one A?

Take complements (often an effective strategy for "at least" problems). The complementary event of getting at least one A is getting no A's. As outcomes are equally likely, by the multiplication principle there are 4^4 exam outcomes with no A's (four grade choices for each of four exams). And there are 5^4 possible outcomes in all. The desired probability is $1 - 4^4/5^4 = 0.5904$. ■

1.6.1 Permutations

Given a set of distinct objects, a *permutation* is an ordering of the elements of the set. For the set $\{a, b, c\}$, there are six permutations:

$$(a, b, c), (a, c, b), (b, a, c), (b, c, a), (c, a, b), \text{ and } (c, b, a).$$

How many permutations are there of an n-element set? There are n possibilities for the first element of the permutation, $n - 1$ for the second, and so on. The result follows by the multiplication principle.

COUNTING PERMUTATIONS

There are $n \times (n - 1) \times \cdots \times 1 = n!$ permutations of an n-element set.

The factorial function $n!$ grows very large very fast. In a classroom of 10 people with 10 chairs, there are $10! = 3{,}628{,}800$ ways to seat the students. There are $52! \approx 8 \times 10^{67}$ orderings of a standard deck of cards, which is "almost" as big as the number of atoms in the observable universe, which is estimated to be about 10^{80}.

Functions of the form c^n, where c is a constant, are said to exhibit *exponential* growth. The factorial function $n!$ grows like n^n, which is sometimes called *super-exponential* growth.

The factorial function $n!$ is pervasive in discrete probability. A good approximation when n is large is given by Stirling's approximation

$$n! \approx n^n e^{-n} \sqrt{2\pi n}. \tag{1.4}$$

More precisely,

$$\lim_{n\to\infty} \frac{n!}{n^n e^{-n} \sqrt{2\pi n}} = 1.$$

We say that $n!$ "is asymptotic to" the function $n^n e^{-n} \sqrt{2\pi n}$.

For first impressions, it looks like the right-hand side of Equation 1.4 is more complicated than the left. However, the right-hand side is made up of relatively simple, elementary functions, which makes it possible to obtain useful approximations of factorials. Modern computational methods swap to computing logarithms of factorials to handle large computations, so in practice, you will likely not need to employ this formula.

How do we use permutations to solve problems? The following examples illustrate some applications.

Example 1.14 Maria has three bookshelves in her dorm room and 15 books—5 are math books and 10 are novels. If each shelf holds exactly five books and books are placed randomly on the shelves (all orderings are equally likely), what is the probability that the bottom shelf contains all the math books?

There are $15!$ ways to permute all the books on the shelves. There are $5!$ ways to put the math books on the bottom shelf and $10!$ ways to put the remaining novels on the other two shelves. Thus, by the multiplication principle, the desired probability is $(5!10!)/15! = 1/3003 = 0.000333$. ■

Example 1.15 A bag contains six Scrabble tiles with the letters A-D-M-N-O-R. You reach into the bag and take out tiles one at a time. What is the probability that you will spell the word R-A-N-D-O-M?

How many possible words can be formed? All the letters are distinct and a "word" is a permutation of the set of six letters. There are $6! = 720$ possible words. Only one of them spells R-A-N-D-O-M, so the desired probability is $1/720 = 0.001389$. ■

Example 1.16 Scrabble continued. Change the previous example. After you pick a tile from the bag, write down that letter and then return the tile to the bag. So every time you reach into the bag, it contains the six original letters. What is the probability that you spell R-A-N-D-O-M now when drawing six tiles?

With the change, there are $6 \times \cdots \times 6 = 6^6 = 46{,}656$ possible words, and only one still spells R-A-N-D-O-M, so the desired probability is $1/46{,}656 = 0.0000214$. ■

SAMPLING WITH AND WITHOUT REPLACEMENT

The last examples highlight two different sampling methods called *sampling without replacement* and *sampling with replacement*. When sampling with replacement, a unit that is selected from a population is returned to the population before another unit is selected. When sampling without replacement, the unit is not returned to the population after being selected. When solving a probability problem involving sampling (such as selecting cards or picking balls from urns), make sure you know the sampling method before computing the related probability.

Example 1.17 When national polling organizations conduct nationwide surveys, they often select about 1000 people sampling without replacement. If N is the number of people in a target population, then by the multiplication principle there are $N \times (N-1) \times (N-2) \times \cdots \times (N-999)$ possible ordered samples. For national polls in the United States, where N, the number of people age 18 or over, is about 250 million, that gives about $(250{,}000{,}000)^{1000}$ possible ordered samples, which is a mind-boggling 2.5 with 8000 zeros after it. ■

As defined, permutations tell us the number of possible arrangements of n objects where all objects are distinguishable and order matters when sampling without replacement. As in the last example, perhaps we only want to arrange k of the n objects, $k \leq n$, sampling without replacement where order matters. What changes when we want to know the number of permutations of k objects out of n?

The first object can still be any of the n objects, while the second can be any of the remaining $n-1$ objects, etc. However, rather than continuing to the last of the n objects, we stop at the kth object. Considering the pattern, the kth object can be any of the remaining $n-k+1$ objects at that point. Thus, the number of possible arrangements is $n \times (n-1) \times \cdots \times (n-k+1)$ which is more easily written as $n!/(n-k)!$ We illustrate this in the following examples.

Example 1.18 A club with seven members needs to elect three officers (president, vice president, and secretary/treasurer) for the upcoming year. Members can only hold one position. How many sets of officers are possible?

Thinking through the problem, the president can be any of the seven members. Once the president is in place, the vice president can be any of the remaining six members, and finally, the secretary/treasurer can be any of the remaining five members. By the multiplication rule, the number of possible sets of officers is $7 \times 6 \times 5 = 210$. We obtain the same result with $n = 7$ and $k = 3$, as $7!/4! = 210$. ■

Example 1.19 A company decides to create inventory stickers for their product. Each sticker will consist of three digits (0–9) which cannot repeat among themselves, two capital letters which cannot repeat, and another three digits that cannot repeat amongst themselves. For example, valid stickers include 203AZ348 and 091BE289, but 307JM449 is not valid. How many different possible stickers can the company make?

We use both permutations and the multiplication rule to solve the problem. For the three digits, there are $n = 10$ possible options and we need $k = 3$ in order. This occurs twice. For each set of digits, there are thus $n!/(n-k)! = 10!/7! = 10 \times 9 \times 8 = 720$ arrangements possible. For the capital letters, there are $n = 26$ options with $k = 2$. Thus, the number of possible arrangements is $26 \times 25 = 650$. Combining this we find the number of possible stickers is $720 \times 650 \times 720 = 336{,}960{,}000$. The company can keep inventory on up to almost 337 million objects with this scheme for stickers. ■

1.7 COUNTING II

In the last section, you learned how to count ordered lists and permutations. Here we count unordered sets and subsets. For example, given a set of n distinct objects, how many subsets of size k can we select when sampling without replacement? To proceed, we first show a simple yet powerful correspondence between subsets of a set and binary sequences, or lists. This correspondence will allow us to relate counting results for sets to those for lists, and vice versa.

A *binary sequence* is a list, each of whose elements can take one of two values, which we generically take to be zeros and ones. Our questions about subsets of size k drawn from n distinct objects is equivalent to asking: how many binary sequences of length n contain exactly k ones?

To illustrate, consider a group of n people lined up in a row and numbered 1 to n. Each person holds a card. On one side of the card is a 0; on the other side is a 1. Initially, all the cards are turned to 0. The n cards from left to right form a binary list.

Select a subset of the n people. Each person in the subset turns over his/her card, from 0 to 1. The cards taken from left to right form a new binary list. For instance, if $n = 6$ and the first and third persons are selected, the corresponding list is $(1, 0, 1, 0, 0, 0)$.

Conversely, given a list of zeros and ones, we select those people corresponding to the ones in the list. That is, if a one is in the kth position in the list, then person k is selected. If the list is $(1, 0, 1, 1, 1, 1)$, then all but the second person are selected.

This establishes a one-to-one correspondence between subsets of $\{1, \ldots, n\}$ and binary lists of length n. Table 1.3 shows the correspondence for the case $n = 3$.

A one-to-one correspondence between two finite sets means that both sets have the same number of elements. Our one-to-one correspondence shows that the number of subsets of an n-element set is equal to the number of binary lists of length n. The number of binary lists of length n is easily counted by the multiplication principle. As there are two choices for each element of the list, there are 2^n binary lists. The number of subsets of an n-element set immediately follows as 2^n.

TABLE 1.3. Correspondence between subsets and binary lists.

Subset	List
$\emptyset$	$(0,0,0)$
$\{1\}$	$(1,0,0)$
$\{2\}$	$(0,1,0)$
$\{3\}$	$(0,0,1)$
$\{1,2\}$	$(1,1,0)$
$\{1,3\}$	$(1,0,1)$
$\{2,3\}$	$(0,1,1)$
$\{1,2,3\}$	$(1,1,1)$

1.7.1 Combinations and Binomial Coefficients

Our goal is to count the number of binary lists of length n with exactly k ones. We will do so by first counting the number of k-element subsets of an n-element set. In the subset-list correspondence, observe that every k-element subset of $\{1, \dots, n\}$ corresponds to a binary list with k ones. And conversely, every binary list with exactly k ones corresponds to a k-element subset. This is true for each $k = 0, 1, \dots, n$. For instance, in the case $n = 4$ and $k = 2$, the subsets are

$$\{1,2\}, \{1,3\}, \{1,4\}, \{2,3\}, \{2,4\}, \{3,4\}$$

with corresponding lists

$$(1,1,0,0), (1,0,1,0), (1,0,0,1), (0,1,1,0), (0,1,0,1), (0,0,1,1).$$

Given a specific k-element subset, there are $k!$ ordered lists that can be formed by permuting the elements of that subset. For instance, the three-element subset $\{1,3,4\}$ yields the $3! = 6$ lists: (1,3,4), (1,4,3), (3,1,4), (3,4,1), (4,1,3), and (4,3,1).

It follows that the number of lists of length k made up of the elements $\{1, \dots, n\}$ is equal to $k!$ times the number of k-element subsets of $\{1, \dots, n\}$. By the multiplication principle, there are $n \times (n-1) \times \cdots \times (n-k+1)$ such lists. Thus, the number of k-element subsets of $\{1, \dots, n\}$ is equal to

$$\frac{n \times (n-1) \times \cdots \times (n-k+1)}{k!} = \frac{n!}{k!(n-k)!}.$$

This quantity is so important it gets its own name. It is known as a *binomial coefficient*, written

$$\binom{n}{k} = \frac{n!}{k!(n-k)!},$$

and read as "n choose k." On calculators, the option may be shown as "nCr."

TABLE 1.4. Common values of binomial coefficients.

Binomial coefficients
$\binom{n}{0} = \binom{n}{n} = 1$
$\binom{n}{1} = \binom{n}{n-1} = n$
$\binom{n}{2} = \binom{n}{n-2} = n(n-1)/2$
$\binom{n}{k} = \binom{n}{n-k}$

The binomial coefficient is often referred to as a way to count *combinations*, numbers of arrangements where order does not matter, as opposed to permutations, where order does matter. From the equation, one can see that the number of combinations of obtaining k objects from n is equal to the number of permutations of k objects out of n objects divided by the factorial of k, the size of the subset desired (the number of permutations of the k objects). This is accounting for the fact that in combinations, we do not care about the order of the objects in the subset.

By the one-to-one correspondence between k-element subsets and binary lists with exactly k ones, we have the following.

COUNTING k-ELEMENT SUBSETS AND LISTS WITH k ONES

There are $\binom{n}{k}$ k-element subsets of $\{1, \ldots, n\}$.

There are $\binom{n}{k}$ binary lists of length n with exactly k ones.

There are $\binom{n}{k}$ ways to select a subset of k objects from a set of n objects when the order the objects are selected in does not matter.

Binomial coefficients are defined for nonnegative integers n and k, where $0 \leq k \leq n$. For $k < 0$ or $k > n$, set $\binom{n}{k} = 0$. Common values of binomial coefficients are given in Table 1.4.

Example 1.20 A classroom of ten students has six females and four males. (i) What are the number of ways to pick five students for a project? (ii) How many ways can we pick a group of two females and three males?

(i) There are $\binom{10}{5} = 252$ ways to pick five students.

(ii) There are $\binom{6}{2} = 15$ ways to pick the females, and $\binom{4}{3} = 4$ ways to pick the males. By the multiplication principle, there are $15 \times 4 = 60$ ways to pick the group.

■

Example 1.21 In a poker game, players are dealt five cards from a standard deck of 52 cards as their starting hand. The best hand in the game of poker is a royal straight flush consisting of 10-Jack-Queen-King-Ace, all of the same suit. What is the probability of getting dealt a royal straight flush?

There are four possible royal straight flushes, one for each suit. A five-card hand in poker is a five-element subset of a 52-element set. Thus,

$$P(\text{Royal straight flush}) = \frac{4}{\binom{52}{5}} = 1.539 \times 10^{-6},$$

or about 1.5 in a million. ■

Example 1.22 Texas hold ’em. In Texas hold ’em poker, players are initially dealt two cards from a standard deck. What is the probability of being dealt at least one ace?

Consider the complementary event of being dealt no aces. There are $\binom{48}{2}$ ways of being dealt two cards neither of which are aces. The desired probability is

$$1 - \binom{48}{2} \Big/ \binom{52}{2} = 1 - \frac{188}{221} = \frac{33}{221} = 0.149321.$$

■

Example 1.23 Twenty-five people will participate in a clinical trial, where 15 people receive the treatment and 10 people receive the placebo. In a group of six people who participated in the trial, what is the probability that four received the treatment and two received the placebo?

There are $\binom{15}{4}$ ways to pick the four who received the treatment, and $\binom{10}{2}$ ways to pick the two who received the placebo. There are $\binom{25}{6}$ possible subgroups of six people. The desired probability is

$$\binom{15}{4}\binom{10}{2} \Big/ \binom{25}{6} = \frac{351}{1012} = 0.347.$$

■

Example 1.24 Powerball lottery. In the Powerball lottery, the player picks five numbers between 1 and 59 and then a single “powerball” number between 1 and 35. To win the jackpot, you need to match all six numbers. What is the probability of winning the jackpot?

There are $\binom{35}{1}\binom{59}{5}$ possible plays. Of these, one is the jackpot winner. The desired probability is

$$P(\text{Jackpot}) = 1 \Big/ 35\binom{59}{5} = 5.707 \times 10^{-9},$$

or almost 1 out of 200 million.

You can win \$10,000 in the Powerball lottery if you match the powerball and exactly four of the other five numbers. The number of ways to make such a match is $\binom{5}{4}\binom{54}{1}$. This is accomplished by selecting the matching powerball $\binom{1}{1}$, four of the five selected nonpowerball numbers $\binom{5}{4}$, and one of the remaining nonselected nonpowerball numbers $\binom{54}{1}$. Note that the coefficient $\binom{1}{1}$ for selecting the powerball winner is not shown in the computation. This is a common convention for coefficients that evaluate to 1 or n, as shown below with $\binom{35}{1} = 35$. The probability of winning \$10,000 in Powerball is thus

$$P(\$10{,}000) = \binom{5}{4}\binom{54}{1} \Big/ 35\binom{59}{5} = 1.54089 \times 10^{-6},$$

about the same as the probability of being dealt a royal straight flush in poker. ■

Example 1.25 Bridge. In the game of bridge, all 52 cards are dealt out to four players. Each player gets 13 cards. A *perfect* bridge hand is getting all cards of the same suit. (i) What is the probability of being dealt a perfect hand? (ii) What is the probability that all four players will be dealt perfect hands?

(i) There are $\binom{52}{13}$ possible bridge hands. Of those, four contain all the same suit. Thus, the probability of being dealt a perfect hand is

$$\frac{4}{\binom{52}{13}} = \frac{1}{158{,}753{,}389{,}900} = 6.29908 \times 10^{-12}.$$

(ii) There are $\binom{52}{13}$ ways for the first player to be dealt 13 cards. Then $\binom{39}{13}$ ways to deal the remaining 39 cards to the second player, and so on. There are

$$\binom{52}{13}\binom{39}{13}\binom{26}{13}\binom{13}{13} = \frac{52!}{13!13!13!13!}$$

possible ways to deal out the deck. Perfect hands differ by the suit. And there are $4! = 24$ ways to permute the four suits among the four players. The desired probability is

$$\frac{4!13!13!13!13!}{52!} = 4.47388 \times 10^{-28}.$$

■

Perfect Bridge Hands

According to the BBC News [2003], on January 27, 1998, in a whist club in Bucklesham, England, four card players at a table were dealt perfect bridge hands.

"Witnesses in the village hall where the game was being played," reported the BBC, "have confirmed what happened. Eighty-seven-year-old Hilda Golding was the first to pick up her hand. She was dealt 13 clubs in the pack.... Hazel Ruffles had all the diamonds. Alison Chivers held the hearts...."

"Chivers insists that the cards were shuffled properly. 'It was an ordinary pack of cards,' she said."

Another report of a perfect bridge hand appeared in *The Herald* of Sharon, Pa., on August 16, 2010.

An article in 2011 (https://aperiodical.com/2011/12/four-perfect-hands-an-event-never-seen-before-right/) reported several instances of perfect bridge hands being dealt starting in 1929.

Do you think these reports are plausible?

Example 1.26 In 20 coin tosses, what is the probability of getting exactly 10 heads?

Here is a purely counting approach. There are 2^{20} possible coin tosses. Of those, there are $\binom{20}{10}$ sequences of H's and T's with exactly 10 H's. The desired probability is

$$\frac{\binom{20}{10}}{2^{20}} = 0.176197.$$

A slightly different approach first counts the number of possible outcomes and then computes the probability of each. There are $\binom{20}{10}$ possible outcomes. By independence, each outcome occurs with probability $(1/2)^{20}$. This gives the same result. ■

Example 1.27 A DNA strand can be considered a sequence of As, Cs, Gs, and Ts. Positions on the DNA sequence are called *sites*. Assume that the letters at each site on a DNA strand are equally likely and independent of other sites (a simplifying assumption that is not true with actual DNA). Suppose we want to find (i) the probability that a DNA strand of length 20 is made up of 4 As and 16 Gs and (ii) the probability that a DNA strand of length 20 is made up of 4 As, 5 Gs, 3 Ts, and 8 Cs.

(i) There are $\binom{20}{4}$ binary sequences of length 20 with 4 As and 16 Gs. By independence, each sequence occurs with probability $1/4^{20}$. The desired probability is

$$\frac{\binom{20}{4}}{4^{20}} = 4.4065 \times 10^{-9}.$$

(ii) Consider the *positions* of the letters. There are $\binom{20}{4}$ choices for the positions of the As. This leaves 16 positions for the Gs. There are $\binom{16}{5}$ choices for those. Of the remaining 11 positions, there are $\binom{11}{3}$ positions for the Ts. And the last eight positions are fixed for the Cs. This gives

$$\binom{20}{4}\binom{16}{5}\binom{11}{3}\binom{8}{8} \bigg/ 4^{20} = \frac{20!}{4!5!3!8!} \bigg/ 4^{20} = 0.00317.$$

This last expression is an example of a *multinomial* probability, discussed in Chapter 5.

■

Binomial theorem. The classic binomial theorem describes the algebraic expansion of powers of a polynomial with two terms. The algebraic proof uses induction and is somewhat technical. Here is a *combinatorial* proof.

BINOMIAL THEOREM

For nonnegative integer n, and real numbers x and y,

$$(x + y)^n = \sum_{k=0}^{n} \binom{n}{k} x^k y^{n-k}.$$

Proof: It will help the reader in following the proof to choose a small n, say $n = 3$, and expand $(x + y)^3$ by hand.

Observe that all the terms of the expansion have the form $x^k y^{n-k}$, for $k = 0, 1, \ldots, n$. Fix k and consider the coefficient of $x^k y^{n-k}$ in the expansion. The product $(x + y)^n = (x + y) \cdots (x + y)$ consists of n factors. There are k of these factors to choose an x from, and the remaining $n - k$ factors to choose a y from. There are $\binom{n}{k}$ ways to do this. Thus, the coefficient of $x^k y^{n-k}$ is $\binom{n}{k}$, which gives the result. □

$$
\begin{array}{c}
1\\
1 \quad 1\\
1 \quad 2 \quad 1\\
1 \quad 3 \quad 3 \quad 1\\
1 \quad 4 \quad 6 \quad 4 \quad 1\\
1 \quad 5 \quad 10 \quad 10 \quad 5 \quad 1\\
1 \quad 6 \quad 15 \quad 20 \quad 15 \quad 6 \quad 1\\
1 \quad 7 \quad 21 \quad 35 \quad 35 \quad 21 \quad 7 \quad 1\\
1 \quad 8 \quad 28 \quad 56 \quad 70 \quad 56 \quad 28 \quad 8 \quad 1\\
1 \quad 9 \quad 36 \quad 84 \quad 126 \quad 126 \quad 84 \quad 36 \quad 9 \quad 1
\end{array}
$$

FIGURE 1.3: Pascal's triangle.

Let $x = y = 1$ in the binomial theorem. This gives

$$2^n = \sum_{k=0}^{n} \binom{n}{k} 1^k 1^{n-k} = \sum_{k=0}^{n} \binom{n}{k}.$$

There is a combinatorial interpretation of this identity. The left-hand side counts the number of sets of size n. The right-hand side counts the number of such sets by summing the number of subsets of size 0, size 1, …, and size n.

Binomial coefficients appear in the famous Pascal's triangle, shown in Figure 1.3. Each entry of the triangle is the sum of the two numbers above it. The entries are all binomial coefficients. Enumerate the rows starting at $n = 0$ at the top. The entries of each row are numbered from the left starting at $k = 0$. The kth number on the nth row is $\binom{n}{k}$. The fact that each entry is the sum of the two entries above it gives the identity

$$\binom{n}{k} = \binom{n-1}{k-1} + \binom{n-1}{k}. \tag{1.5}$$

The algebraic proof of this identity is an exercise in working with factorials.

$$
\begin{aligned}
\binom{n-1}{k-1} + \binom{n-1}{k} &= \frac{(n-1)!}{(k-1)!(n-k)!} + \frac{(n-1)!}{k!(n-1-k)!}\\
&= \frac{(n-1)!k}{k!(n-k)!} + \frac{(n-k)(n-1)!}{k!(n-k)!}\\
&= \frac{(n-1)!(k+(n-k)}{k!(n-k)!}\\
&= \frac{(n-1)!n}{k!(n-k)!}\\
&= \frac{n!}{k!(n-k)!} = \binom{n}{k}.
\end{aligned}
$$

Here is a combinatorial proof:

Question: There are n students in the room, including Addison. How many ways are there to pick a group of k students?

Answer #1: Choose k students from the set of n students in $\binom{n}{k}$ ways.

Answer #2: Pick k students that include Addison. Then pick k students that do not include Addison. If Addison is included, there are $\binom{n-1}{k-1}$ ways to pick the remaining $k-1$ students in the group. If Addison is not included, there are $\binom{n-1}{k}$ ways to pick the group (choosing from everyone except Addison). Thus, there are $\binom{n-1}{k-1} + \binom{n-1}{k}$ ways to pick a group of k students.

The two solutions answer the same question, proving the desired identity.

Example 1.28 Ballot problem. This classic problem introduced by Joseph Louis François Bertrand in 1887 asks, "In an election where candidate A receives p votes and candidate B receives q votes with $p > q$, what is the probability that A will be strictly ahead of B throughout the count?" The problem assumes that votes for A and B are equally likely.

For instance, if A receives $p = 3$ votes and B receives $q = 2$ votes, the possible vote counts are given in Table 1.5. Of the 10 possible voting outcomes, only the first two show A always ahead throughout the count. The desired probability is $2/10 = 1/5$.

We show that the solution to the ballot problem is $(p-q)/(p+q)$.

A voting outcome can be thought of as a list of length $p+q$ with p As and q Bs. Thus, there are $\binom{p+q}{p}$ possible voting outcomes.

TABLE 1.5. Voting outcomes for the ballot problem.

Voting pattern	Net votes for A during the count
AAABB	1,2,3,2,1
AABAB	1,2,1,2,1
AABBA	1,2,1,0,1
ABABA	1,0,1,0,1
ABAAB	1,0,1,2,1
ABBAA	1,0,−1,0,1
BAAAB	−1,0,1,2,1
BAABA	−1,0,1,0,1
BABAA	−1,0,−1,0,1
BBAAA	−1,−2,0,1,2

A receives three votes and B receives two votes.

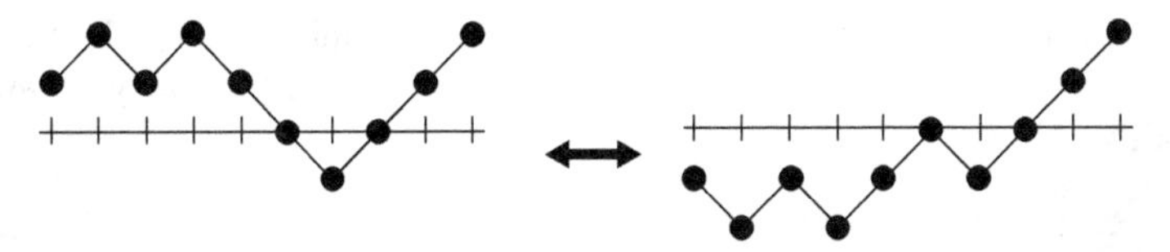

FIGURE 1.4: Illustrating the correspondence between "bad" lists that start with A and lists that start with B.

Consider the number of outcomes in which A is always ahead. Clearly, such an outcome must begin with a vote for A. The number of outcomes that begin with A is $\binom{p+q-1}{p-1}$, because the first element of the list is fixed and there are $p + q - 1$ positions to fill with $p - 1$ A's. Some of these outcomes are "good" (A stays ahead throughout) and some are "bad". We need to subtract off the number of "bad" lists.

To count such lists, we give a one-to-one correspondence between "bad" lists that start with A and general lists that start with B. To do so, we represent a voting outcome by a path where the vertical axis represents the number of votes for A. Thus, when a path crosses the horizontal axis, it represents a tie.

See the example in Figure 1.4. The left diagram corresponds to the voting outcome AABABBBAAA. The outcome is "bad" in that there is eventually a tie and the path crosses the horizontal axis. For such a path, "reflect" the portion of the path up until the tie across the x axis, giving the outcome in the right diagram. The reflection results in a path that starts with B.

Conversely, consider a path that starts with B. As there are more As than Bs, at some point in the count, there must be a tie and the path crosses the x-axis. Reflecting the portion of the path up until the tie across the x-axis gives a "bad" path that starts with A.

Having established a one-to-one correspondence we see that the number of "bad" lists that start with A is equal to the number of lists that start with B. There are $\binom{p+q-1}{q-1}$ lists that start with B. This gives the desired probability

$$
\begin{aligned}
&P(A \text{ is ahead throughout the count}) \\
&= \frac{\text{Number of good lists that start with } A}{\text{Number of voting outcomes}} \\
&= \frac{\text{Number of lists that start with } A - \text{Number of bad lists that start with } A}{\text{Number of voting outcomes}} \\
&= \frac{\left[\binom{p+q-1}{p-1} - \binom{p+q-1}{q-1}\right]}{\binom{p+q}{p}}.
\end{aligned}
$$

We leave it to the reader to check that this last expression simplifies to $(p - q)/(p + q)$. ■

To round out this chapter, in the next sections, we examine some problem-solving strategies and take a first look at simulation as tools to help in your study of probability.

1.8 PROBLEM-SOLVING STRATEGIES: COMPLEMENTS AND INCLUSION–EXCLUSION

Consider a sequence of events $A_1, A_2, \ldots$. In this section, we consider strategies to find the probability that *at least* one of the events occurs, which is the probability of the union $\cup_i A_i$, i.e., $A_1 \cup A_2 \cup A_3 \cup \cdots$.

Sometimes the complement of an event can be easier to work with than the event itself. The complement of the event that at least one of the A_is occurs is the event that none of the A_is occur, which is the intersection $\cap_i A_i^c$.

Check with a Venn diagram (and if you are comfortable working with sets prove it yourself) that

$$(A \cup B)^c = A^c B^c \quad \text{and} \quad (AB)^c = A^c \cup B^c.$$

Complements turn unions into intersections, and vice versa. These set-theoretic results are known as DeMorgan's laws. The results extend to infinite sequences. Given events $A_1, A_2, \ldots,$

$$\left(\bigcup_{i=1}^{\infty} A_i \right)^c = \bigcap_{i=1}^{\infty} A_i^c \quad \text{and} \quad \left(\bigcap_{i=1}^{\infty} A_i \right)^c = \bigcup_{i=1}^{\infty} A_i^c.$$

Example 1.29 Four dice are rolled. Find the probability of getting at least one 6.

The sample space is the set of all outcomes of four dice rolls

$$\Omega = \{(1,1,1,1), (1,1,1,2), \ldots, (6,6,6,6)\}.$$

By the multiplication principle, there are $6^4 = 1296$ elements. If the dice are fair, each of these outcomes is equally likely. It is not obvious, without some new tools, how to count the number of outcomes that have at least one 6.

Let A be the event of getting at least one 6. Then the complement A^c is the event of getting no sixes in four rolls. An outcome has no sixes if the dice rolls a 1, 2, 3, 4, or 5 on every roll. By the multiplication principle, there are $5^4 = 625$ possibilities. Thus, $P(A^c) = 5^4/6^4 = 625/1296$ and

$$P(A) = 1 - P(A^c) = 1 - \frac{625}{1296} = 0.5177. \qquad \blacksquare$$

Recall the formula in Equation 1.3 for the probability of a union of two events. We generalize for three or more events using the principle of inclusion–exclusion.

For events A, B, and C,

$$P(A \cup B \cup C) = P(A) + P(B) + P(C) - P(AB) - P(AC) - P(BC) + P(ABC).$$

As we first *include* the sets, then *exclude* the pairwise intersections, then *include* the triple intersection, this is called the *inclusion–exclusion principle*. The proof is intuitive with the help of a Venn diagram, which we leave to the reader. Write

$$A \cup B \cup C = [A \cup B] \cup [C(AC \cup BC)^c].$$

The bracketed sets $A \cup B$ and $C(AC \cup BC)^c$ are disjoint. Thus,

$$\begin{aligned} P(A \cup B \cup C) &= P(A \cup B) + P(C(AC \cup BC)^c) \\ &= P(A) + P(B) - P(AB) + P(C(AC \cup BC)^c). \end{aligned} \tag{1.6}$$

Write C as the disjoint union

$$C = [C(AC \cup BC)] \cup [C(AC \cup BC)^c] = [AC \cup BC] \cup [C(AC \cup BC)^c].$$

Rearranging gives

$$P(C(AC \cup BC)^c) = P(C) - P(AC \cup BC).$$

Together with Equation 1.6, we find

$$\begin{aligned} P(A \cup B \cup C) &= P(A) + P(B) - P(AB) \\ &\quad + P(C) - [P(AC) + P(BC) - P(ABC)]. \end{aligned}$$

Extending further to more than three events gives the general principle of inclusion–exclusion. We will not prove it, but if you know how to use mathematical induction, give it a try.

INCLUSION–EXCLUSION

Given events $A_1, \ldots, A_n$, the probability that at least one event occurs is

$$\begin{aligned} P(A_1 \cup \cdots \cup A_n) &= \sum_i P(A_i) - \sum_{i<j} P(A_i A_j) \\ &\quad + \sum_{i<j<k} P(A_i A_j A_k) - \cdots + (-1)^{n+1} P(A_1, \ldots, A_n). \end{aligned}$$

Example 1.30 An integer is drawn uniformly at random from $\{1, \ldots, 1000\}$ such that each number is equally likely. What is the probability that the number drawn is divisible by 3, 5, or 6?

Let D_3, D_5, and D_6 denote the events that the number drawn is divisible by 3, 5, and 6, respectively. The problem asks for $P(D_3 \cup D_5 \cup D_6)$. By inclusion–exclusion,

$$P(D_3 \cup D_5 \cup D_6) = P(D_3) + P(D_5) + P(D_6) - P(D_3D_5) - P(D_3D_6)$$

Let $\lfloor x \rfloor$ denote the integer part of x. There are $\lfloor 1000/x \rfloor$ numbers from 1 to 1000 that are divisible by x. Because all selections are equally likely,

$$P(D_3) = \lfloor 1000/3 \rfloor / 1000 = 0.333.$$
$$P(D_5) = \lfloor 1000/5 \rfloor / 1000 = 0.20.$$
$$P(D_6) = \lfloor 1000/6 \rfloor / 1000 = 0.166.$$

A number is divisible by 3 and 5 if and only if it is divisible by 15. Thus, $D_3D_5 = D_{15}$. If a number is divisible by 6, it is also divisible by 3, so $D_3D_6 = D_6$. Also, $D_5D_6 = D_{30}$. And $D_3D_5D_6 = D_{30}$. This gives

$$P(D_3D_5) = \lfloor 1000/15 \rfloor / 1000 = 0.066.$$
$$P(D_3D_6) = 0.166.$$
$$P(D_5D_6) = \lfloor 1000/30 \rfloor / 1000 = 0.033.$$
$$P(D_3D_5D_6) = 0.033.$$

Putting it all together gives us that $P(D_3 \cup D_5 \cup D_6)$ is equal to

$$0.333 + 0.2 + 0.166 - 0.066 - 0.166 - 0.033 + 0.033 = 0.467.$$ ■

We have presented two different ways of computing the probability that at least one of several events occurs: (i) a "back-door" approach of taking complements and working with the resulting "and" probabilities and (ii) a direct "frontal-attack" by inclusion–exclusion. Here is a third way, which illustrates decomposing an event into a union of mutually exclusive subsets.

Example 1.31 Consider a random experiment that has k equally likely outcomes, one of which we call *success*. Repeat the experiment n times. Let A be the event that at least one of the n outcomes is a success. We want to find $P(A)$.

For instance, consider rolling a die 10 times, where success means rolling a three. Here $n = 10$, $k = 6$, and A is the event of rolling at least one 3.

Define a sequence of events $A_1, \ldots, A_n$, where A_i is the event that the ith trial is a success. Then $A = A_1 \cup \cdots \cup A_n$ and $P(A) = P(A_1 \cup \cdots \cup A_n)$. We cannot use the addition rule on this probability as the A_is are not mutually exclusive.

To define a sequence of mutually exclusive events, let B_i be the event that the *first* success occurs on the ith trial. Then the B_is are mutually exclusive. Furthermore,

$$B_1 \cup \cdots \cup B_n = A_1 \cup \cdots \cup A_n = A.$$

Thus, $P(A) = P(B_1 \cup \cdots \cup B_n) = P(B_1) + \cdots + P(B_n)$.

To find $P(B_i)$, observe that if the first success occurs on the ith trial, then the first $i - 1$ trials are necessarily not successes and the ith trial is a success. There are $k - 1$ possible outcomes for each of the first $i - 1$ trials, one outcome for the ith trial, and k possible outcomes for each of the remaining $n - i$ trials. By the multiplication principle, there are $(k - 1)^{i-1}k^{n-i}$ outcomes where the first success occurs on the ith trial, and there are k^n possible outcomes in all. Thus,

$$P(B_i) = \frac{(k-1)^{i-1}k^{n-i}}{k^n} = \frac{1}{k}\left(\frac{k-1}{k}\right)^{i-1} = \frac{1}{k}\left(1 - \frac{1}{k}\right)^{i-1},$$

for $i = 1, \ldots, n$. The desired probability is

$$\begin{aligned} P(A) = P(B_1) + \cdots + P(B_n) &= \sum_{i=1}^{n} \frac{1}{k}\left(1 - \frac{1}{k}\right)^{i-1} \\ &= \frac{1}{k}\left(\frac{1-(1-1/k)^n}{1-(1-1/k)}\right) \\ &= 1 - \left(1 - \frac{1}{k}\right)^n. \end{aligned}$$

For instance, the probability of rolling at least one 3 in 10 rolls of a die is

$$1 - \left(1 - \frac{1}{6}\right)^{10} = 1 - \left(\frac{5}{6}\right)^{10} = 0.8385.$$

■

1.9 A FIRST LOOK AT SIMULATION

Using random numbers on a computer to simulate probabilities is called the Monte Carlo method. Today, Monte Carlo tools are used extensively in statistics, physics, engineering, and across many disciplines. The name was coined in the 1940s by mathematicians John von Neumann and Stanislaw Ulam working on the Manhattan Project. It was named after the famous Monte Carlo casino in Monaco.

Ulam's description of his inspiration to use random numbers to simulate complicated problems in physics is quoted in Eckhardt [1987]:

> The first thoughts and attempts I made to practice [the Monte Carlo method] were suggested by a question which occurred to me in 1946 as I was convalescing from an illness and playing solitaires.

> The question was what are the chances that a Canfield solitaire laid out with 52 cards will come out successfully? After spending a lot of time trying to estimate them by pure combinatorial calculations, I wondered whether a more practical method than "abstract thinking" might not be to lay it out say one hundred times and simply observe and count the number of successful plays. This was already possible to envisage with the beginning of the new era of fast computers, and I immediately thought of problems of neutron diffusion and other questions of mathematical physics, and more generally how to change processes described by certain differential equations into an equivalent form interpretable as a succession of random operations. Later [in 1946], I described the idea to John von Neumann, and we began to plan actual calculations.

The Monte Carlo simulation approach is based on the relative frequency model for probabilities. Given a random experiment and some event A, the probability $P(A)$ is estimated by repeating the random experiment many times and computing the proportion of times that A occurs.

More formally, define a sequence $X_1, X_2, \ldots,$ where

$$X_k = \begin{cases} 1, & \text{if } A \text{ occurs on the } k\text{th trial} \\ 0, & \text{if } A \text{ does not occur on the } k\text{th trial,} \end{cases}$$

for $k = 1, 2, \ldots.$ Then

$$\frac{X_1 + \cdots + X_n}{n}$$

is the proportion of times in which A occurs in n trials. For large n, the Monte Carlo method estimates $P(A)$ by

$$P(A) \approx \frac{X_1 + \cdots + X_n}{n}. \tag{1.7}$$

MONTE CARLO SIMULATION

Implementing a Monte Carlo simulation of $P(A)$ requires three steps:

1. **Simulate a trial:** Model, or translate, the random experiment using random numbers on the computer. One iteration of the experiment is called a "trial."
2. **Determine success:** Based on the outcome of the trial, determine whether or not the event A occurs. If yes, call that a "success."
3. **Replication:** Repeat the aforementioned two steps many times. The proportion of successful trials is the simulated estimate of $P(A)$.

Monte Carlo simulation is intuitive and matches up with our sense of how probabilities "should" behave. We give a theoretical justification for the method in Chapter 10, where we study limit theorems and the law of large numbers.

Here is a most simple, even trivial, starting example.

Example 1.32 Consider simulating the probability that an ideal fair coin comes up heads. One could do a *physical* simulation by just flipping a coin many times and taking the proportion of heads to estimate $P(\text{Heads})$.

Using a computer, choose the number of trials n (the larger the better) and type the **R** command

```
> sample(0:1, n, replace = T)
```

The command samples with replacement from $\{0, 1\}$ n times such that outcomes are equally likely. Let 0 represent tails and 1 represent heads. The output is a sequence of n ones and zeros corresponding to heads and tails. The average, or mean, of the list is precisely the proportion of ones. To simulate $P(\text{Heads})$, type

```
> mean(sample(0:1, n, replace = T))
```

Repeat the command several times (use the up arrow key). These give repeated Monte Carlo estimates of the desired probability. Observe the accuracy in the estimate with one million trials:

```
> mean(sample(0:1, 1000000, replace = T))
[1] 0.500376
> mean(sample(0:1, 1000000, replace = T))
[1] 0.499869
> mean(sample(0:1, 1000000, replace = T))
[1] 0.498946
> mean(sample(0:1, 1000000, replace = T))
[1] 0.500115
```

The **R** script **CoinFlip.R** simulates a familiar probability—the probability of getting three heads in three coin tosses.

R: SIMULATING THE PROBABILITY OF THREE HEADS IN THREE COIN TOSSES

```
# CoinFlip.R
# Trial
> trial <- sample(0:1, 3, replace = TRUE)
# Success
> if (sum(trial) == 3) 1 else 0
# Replication
> n <- 10000    # Number of repetitions
> simlist <- numeric(n) # Initialize vector
```

```
> for (i in 1:n) {
    trial <- sample(0:1, 3, replace = TRUE)
    success <- if (sum(trial) == 3) 1 else 0
    simlist[i] <- success }
> mean(simlist) # Proportion of trials with 3 heads
[1] 0.1293
```

The script is divided into three parts to illustrate (i) coding the trial, (ii) determining success, and (iii) implementing the replication.

To simulate three coin flips, use the `sample` command. Again letting 1 represent heads and 0 represent tails, the command

```
> trial <- sample(0:1, 3, replace = TRUE)
```

chooses a head or tails three times. The three results are stored as a three-element list (called a vector in **R**) in the variable `trial`.

After flipping three coins, the routine must decide whether or not they are all heads. This is done by summing the outcomes. The sum will equal three if and only if all flips are heads. This is checked with the command

```
> if (sum(trial) == 3) 1 else 0
```

which returns a 1 for success, and 0, otherwise.

For the actual simulation, the commands are repeated n times in a loop. The output from each trial is stored in the vector `simlist`. This vector will consist of n ones and zeros corresponding to success or failure for each trial, where success is flipping three heads.

Finally, after repeating n trials, we find the proportion of successes in all the trials, which is the proportion of ones in `simlist`. Given a list of zeros and ones, the average, or mean, of the list is precisely the proportion of ones in the list. The command `mean(simlist)` finds this average giving the simulated probability of getting three heads.

Run the script via the script file or **R** supplement to see that the resulting estimate is fairly close to the exact solution $1/8 = 0.125$. Increase n to 100,000 or even a million to get more precise estimates. ■

Reproducibility. If you and a classmate both ran the code above exactly as presented, you would get different estimates of the probability via the simulation. This is due to having different random numbers in your simulations. Some readers may know that computer random number generators are really only pseudorandom. The computer is using an algorithm that generates numbers which behave like random numbers would. It turns out that this is actually a plus when writing code or doing simulations in the sense that you can make the computer regenerate the same random numbers by "setting a seed." You can think of a seed as telling the computer where to start the algorithm for generating the random numbers. Then, if anyone else runs the code (including the seed), they would get the same result that you did with that seed (provided they have the same software and version). In **R** this is accomplished with the command

```
> set.seed(360)
```

where any number can be used as the input. We will use a seed of 360 unless otherwise specified and will not show this command in the text (though it is shown in the supplements and scripts), but it is used before every script where random numbers are generated. Have fun and set seeds using numbers you enjoy!

The reason that setting seeds is important is that this makes the work reproducible. Reproducibility means that others can take the code and obtain the same results, lending credibility to the work. While writing simulations, you should aim to make sure all your code is reproducible.

Example 1.33 The script **Divisible356.R** simulates the divisibility problem in Example 1.30 that a random integer from $\{1, \ldots, 1000\}$ is divisible by 3, 5, or 6. The problem, and the resulting code, is more complex.

The function `simdivis()` simulates one trial. Inside the function, the expression `num%%x==0` checks whether `num` is divisible by x. The `if` statement checks whether `num` is divisible by 3, 5, or 6, returning 1 if it is, and 0, otherwise.

After defining the function, typing `simdivis()` will simulate one trial. By repeatedly typing `simdivis()` on your computer, you get a feel for how this random experiment behaves over repeated trials.

In this script, instead of writing a loop, we use the `replicate` command. This powerful **R** command is an alternative to writing loops for simple expressions. The syntax is `replicate(n,expr)`. The expression `expr` is repeated n times creating an n-element vector. Thus, the result of typing

```
> simlist <- replicate(1000, simdivis())
```

is a vector of 1000 ones and zeros stored in the variable `simlist` corresponding to success or failure in the divisibility experiment. The average `mean(simlist)` gives the simulated probability.

Play with this script. Based on 1000 trials, you might guess that the true probability is between 0.45 and 0.49. Increase the number of trials to 10,000 and the estimates are roughly between 0.46 and 0.48. At 100,000, the estimates become even more precise between 0.465 and 0.468.

We can actually quantify this increase in precision in Monte Carlo simulation as n gets large. But that is a topic that will have to wait until Chapter 11.

R: SIMULATING THE DIVISIBILITY PROBABILITY

```
# Divisible356.R
# simdivis() simulates one trial
> simdivis <- function()  {
   num <- sample(1:1000,1)
   if (num%%3==0 || num%%5==0 || num%%6==0) 1 else 0
   }
> simlist <- replicate(10000, simdivis())
mean(simlist)
[1] 0.4707
```

■

1.10 SUMMARY

In this chapter, the first principles of probability were introduced: from random experiment and sample space to the properties of probability functions. We start with discrete sample spaces—sets are either finite or countably infinite. The simplest probability model is when outcomes in a finite sample space are equally likely. In that case, probability reduces to "counting." Counting principles are presented for both permutations and combinations. Binomial coefficients count: (i) the number of k-element subsets of an n-element set and (ii) the number of n-element binary sequences with a k ones. General properties of probabilities are derived from the three defining properties of a probability function. The chapter ends with problem-solving strategies and a first look at simulation.

- **Random experiment:** An activity, process, or experiment in which the outcome is uncertain.
- **Sample space Ω:** Set of all possible outcomes of a random experiment.
- **Outcome ω:** The elements of a sample space.
- **Event:** A subset of the sample space; a collection of outcomes.
- **Probability function:** A function P that assigns numbers to the elements $\omega \in \Omega$ such that
  1. $P(\omega) \geq 0$
  2. $\sum_{\omega} P(\omega) = 1$
  3. For events A, $P(A) = \sum_{\omega \in A} P(\omega)$.
- **Equally likely outcomes:** Probability model for a finite sample space in which all elements have the same probability.

- **Counting**
  1. **Multiplication principle:** If there are m ways for one thing to happen, and n ways for a second thing to happen, then there are mn ways for both things to happen.
  2. **Permutations:** A permutation of $\{1, \ldots, n\}$ is an n-element ordering of the n numbers. There are $n!$ permutations of an n-element set.
  3. **Binomial coefficient:** The binomial coefficient $\binom{n}{k} = n!/(k!(n-k)!)$ or "n choose k" counts: (i) the number of k-element subsets of $\{1, \ldots, n\}$ and (ii) the number of n element $0-1$ sequences with exactly k ones. Each subset is also referred to as a **combination**.
- **Stirling's approximation:** For large n,

$$n! \approx n^n e^{-n} \sqrt{2\pi n}.$$

- **Sampling:** When sampling from a population, *sampling with replacement* is when objects are returned to the population after they are sampled; *sampling without replacement* is when objects are not returned to the population after they are sampled.
- **Properties of probabilities:**
  1. **Simple addition rule:** If A and B are mutually exclusive, that is, disjoint, then $P(A \text{ or } B) = P(A \cup B) = P(A) + P(B)$.
  2. **Implication:** If A implies B, that is, if $A \subseteq B$, then $P(A) \leq P(B)$.
  3. **Complement:** The probability that A does not occur $P(A^c) = 1 - P(A)$.
  4. **General addition rule:** For all events A and B, $P(A \text{ or } B) = P(A \cup B) = P(A) + P(B) - P(AB)$.
- **Monte Carlo simulation** is based on the relative frequency interpretation of probability. Given a random experiment and an event A, $P(A)$ is approximately the fraction of times in which A occurs in n repetitions of the random experiment. A Monte Carlo simulation of $P(A)$ is based on three principles:
  1. **Trials:** Simulate the random experiment, typically on a computer using the computer's random numbers.
  2. **Success:** Based on the outcome of each trial, determine whether or not A occurs. Save the result.
  3. **Replication:** Repeat the aforementioned steps n times. The proportion of successful trials is the simulated estimate of $P(A)$.
- **Setting seeds** for reproducibility is vital when generating random numbers.
- **Problem-solving strategies:**
  1. **Taking complements:** Finding $P(A^c)$, the probability of the complement of an event, might be easier in some cases than finding $P(A)$, the probability of the event. This arises in "at least" problems. For instance, the complement of the event that "at least one of several things occur" is

the event that "none of those things occur." In the former case, the event involves a union. In the latter case, the event involves an intersection.

2. **Inclusion–exclusion:** This is another method for tackling "at least" problems. For three events, inclusion–exclusion gives $P(A \cup B \cup C)$ equals

$$P(A) + P(B) + P(C) - P(AB) - P(AC) - P(BC) + P(ABC).$$

EXERCISES

Understanding Sample Spaces and Events

1.1 Your friend was sick and unable to make today's class. Explain to your friend, using your own words, the meaning of the terms (i) random experiment, (ii) sample space, and (iii) event.

For the following problems 1.2–1.5, identify (i) the random experiment, (ii) the sample space, and (iii) the event of interest.

1.2 Roll four dice. Consider the probability of getting all fives.

1.3 A pizza shop offers three toppings: pineapple, peppers, and pepperoni. A pizza can have 0, 1, 2, or 3 toppings. Consider the probability that a random customer asks for two toppings.

1.4 Bored one day, you decide to play the video game Angry Birds until you win. Every time you lose, you start over. Consider the probability that you win in less than 1000 tries.

1.5 In Julia's garden, there is a 3% chance that a tomato will be bad. Julia harvests 100 tomatoes and wants to know the probability that at most five tomatoes are bad.

1.6 In two dice rolls, let X be the outcome of the first die, and Y the outcome of the second die. Then $X + Y$ is the sum of the two dice. Describe the following events in terms of simple outcomes of the random experiment:

(a) $\{X + Y = 4\}$. (Example solution: $\{(1, 3), (2, 2), (3, 1)\}$.)

(b) $\{X + Y = 9\}$.

(c) $\{Y = 3\}$.

(d) $\{X = Y\}$.

(e) $\{X > 2Y\}$.

1.7 A bag contains r red and b blue balls. You reach into the bag and take k balls. Let R be the number of red balls you take. Let B be the number of blue balls. Express the following events in terms of R and B, assuming valid values for r, b, and k:

(a) You pick no red balls. (Example solution: $\{R = 0\}$.)

(b) You pick one red and two blue balls.

(c) You pick four balls.

(d) You pick twice as many red balls as blue balls.

(e) You pick at least two red balls.

1.8 A couple plans to continue having children until they have a girl or until they have six children, whichever comes first. Describe an appropriate sample space for this random experiment.

Probability Functions

1.9 A sample space has four elements $\omega_1, \ldots, \omega_4$ such that ω_1 is twice as likely as ω_2, which is three times as likely as ω_3, which is four times as likely as ω_4. Find the probability function.

1.10 A sample space has four elements $\omega_1, \ldots, \omega_4$ such that ω_1 is ten times as likely as ω_3, which is four times as likely as ω_4. Finally, ω_2 is as likely as ω_3 and ω_4 combined. Find the probability function.

1.11 A sample space has four elements. For the potential probability functions for the sample space below, state whether they are valid or not. Provide support for your response.

(a) $P(\omega_1) = 0.6, P(\omega_2) = 0.05, P(\omega_3) = 0.4, P(\omega_4) = 0.2$.

(b) $P(\omega_1) = 0.5, P(\omega_2) = 0.2, P(\omega_3) = 0.2, P(\omega_4) = 0.1$.

(c) $P(\omega_1) = 0.15, P(\omega_2) = 0.3, P(\omega_3) = 0.1, P(\omega_4) = 0.45$.

(d) $P(\omega_1) = 0.3, P(\omega_2) = 0.3, P(\omega_3) = -0.2, P(\omega_4) = 0.6$.

1.12 A random experiment has three possible outcomes a, b, and c, with

$$P(a) = p, \; P(b) = p^2, \text{ and } \; P(c) = p.$$

What choice(s) of p makes this a valid probability model?

1.13 Let P_1 and P_2 be two probability functions on Ω. Define a new function P such that $P(A) = (P_1(A) + P_2(A))/2$. Show that P is a probability function.

1.14 Suppose $P_1, \ldots, P_k$ are probability functions on Ω. Let $a_1, \ldots, a_k$ be a sequence of numbers. Under what conditions on the a_i's will

$$P = a_1 P_1 + \cdots + a_k P_k$$

be a probability function?

1.15 Let P be a probability function on $\Omega = \{a, b\}$ such that $P(a) = p$ and $P(b) = 1 - p$ for $0 \leq p \leq 1$. Let Q be a function on Ω defined by $Q(\omega) = [P(\omega)]^2$. For what value(s) of p will Q be a valid probability function?

Equally Likely Outcomes and Counting

1.16 A club has 10 members including Nasir, Rose, and Devin, and is choosing a president, vice-president, and treasurer. All selections are equally likely.

(a) What is the probability that Nasir is selected president?

(b) What is the probability that Rose is chosen president and Devin is chosen treasurer?

(c) What is the probability that neither Nasir, Rose, or Devin obtain a position?

1.17 A fair coin is flipped six times. What is the probability that the first two flips are heads and the last two flips are tails? Use the multiplication principle.

1.18 Suppose that license plates can be two, three, four, or five letters long, taken from the alphabets A to Z. All letters are possible, including repeats. A license plate is chosen at random in such a way so that all plates are equally likely.

(a) What is the probability that the plate is "A-R-R?"

(b) What is the probability that the plate is four letters long?

(c) What is the probability that the plate is a palindrome?

(d) What is the probability that the plate has at least one "R?"

1.19 Suppose you throw five dice and all outcomes are equally likely.

(a) What is the probability that all dice are the same? (In the game of Yahtzee, this is known as a *yahtzee*.)

(b) What is the probability of getting at least one 4?

(c) What is the probability that all the dice are different?

1.20 Tori is picking her fall term classes. She needs to fill three time slots, and there are 20 distinct courses to choose from, including probability 101, 102, and 103. She will pick her classes at random so that all outcomes are equally likely.

(a) What is the probability that she will get probability 101?

(b) What is the probability that she will get probability 101 and probability 102?

(c) What is the probability she will get all three probability courses?

1.21 Suppose k numbers are chosen from $\{1, \ldots, n\}$, where $k < n$, sampling without replacement. All outcomes are equally likely. What is the probability that the numbers chosen are in increasing order?

1.22 There are 40 pairs of shoes in Bill's closet. They are all mixed up.

(a) If 20 shoes are picked, what is the chance that Bill's favorite sneakers will be in the group?

(b) If 20 shoes are picked, what is the chance that at most one shoe from each of the 40 pairs will be picked? (Remember, a left shoe is different than a right shoe.)

1.23 Many bridge players believe that the most likely distribution of the four suits (spades, hearts, diamonds, and clubs) in a bridge hand is 4-3-3-3 (four cards in one suit, and three cards of the other three).

(a) Show that the suit distribution 4-4-3-2 is more likely than 4-3-3-3.

(b) In fact, besides the 4-4-3-2 distribution, there are three other patterns of suit distributions that are more likely than 4-3-3-3. Can you find them?

1.24 Find the probability that a bridge hand contains a nine-card suit. That is, the number of cards of the longest suit is nine.

1.25 A chessboard is an eight-by-eight arrangement of 64 squares. Suppose eight chess pieces are placed on a chessboard at random so that each square can receive at most one piece. What is the probability that there will be exactly one piece in each row and in each column?

1.26 Find the probabilities for being dealt the following poker hands. They are arranged in increasing order of probability.

(a) Straight flush. (Five cards in a sequence and of the same suit.)

(b) Four of a kind. (Four cards of one face value and one other card.)

(c) Full house. (Three cards of one face value and two of another face value.)

(d) Flush. (Five cards of the same suit. Does not include a straight flush.)

(e) Straight. (Five cards in a sequence. Does not include a straight flush. Ace can be high or low.)

(f) Three of a kind. (Three cards of one face value. Does not include four of a kind or full house.)

(g) Two pair. (Does not include four of a kind or full house.)

(h) One pair. (Does not include any of the aforementioned conditions.)

1.27 A walk in the positive quadrant of the plane consists of a sequence of moves, each one from a point (a, b) to either $(a + 1, b)$ or $(a, b + 1)$.

(a) Show that the number of walks from the origin $(0, 0)$ to (x, y) is $\binom{x+y}{x}$.

(b) Suppose a walker starts at the origin $(0, 0)$ and at each discrete unit of time moves either up one unit or to the right one unit each with probability 1/2. If $x > y$, find the probability that a walk from (0,0) to (x, y) always stays above the main diagonal.

1.28 See Example 1.24 for a description of the Powerball lottery. A \$100 prize is won by either (i) matching exactly three of the five balls and the powerball or (ii) matching exactly four of the five balls and not the powerball. Find the probability of winning \$100.

1.29 Give a combinatorial argument (not an algebraic one) for why

$$\binom{n}{k} = \binom{n}{n-k}.$$

1.30 Give a combinatorial proof that

$$\sum_{i=0}^{k} \binom{m}{i}\binom{n}{k-i} = \binom{m+n}{k}. \tag{1.8}$$

Hint: How many ways can you choose k people from a group of m men and n women? From Equation 1.8 show that

$$\sum_{k=0}^{n} \binom{n}{k}^2 = \binom{2n}{n}. \tag{1.9}$$

Properties of Probabilities

1.31 Suppose $P(A) = 0.50$, $P(AB) = 0.20$, and $P(A \text{ or } B) = 0.70$. Find

(a) $P(B)$.

(b) P(exactly one of the two events occurs).

(c) P(neither event occurs).

1.32 Suppose $P(A) = 0.40$, $P(B) = 0.60$, and $P(A \text{ or } B) = 0.80$. Find

(a) P(neither A nor B occur).

(b) $P(AB)$.

(c) P(one of the two events occurs, and the other does not).

1.33 Suppose A and B are mutually exclusive, with $P(A) = 0.30$ and $P(B) = 0.60$. Find the probability that

(a) At least one of the two events occurs.

(b) Both of the events occur.

(c) Neither event occurs.

(d) Exactly one of the two events occur.

1.34 Suppose $P(A \cup B) = 0.6$ and $P(A \cup B^c) = 0.8$. Find $P(A)$.

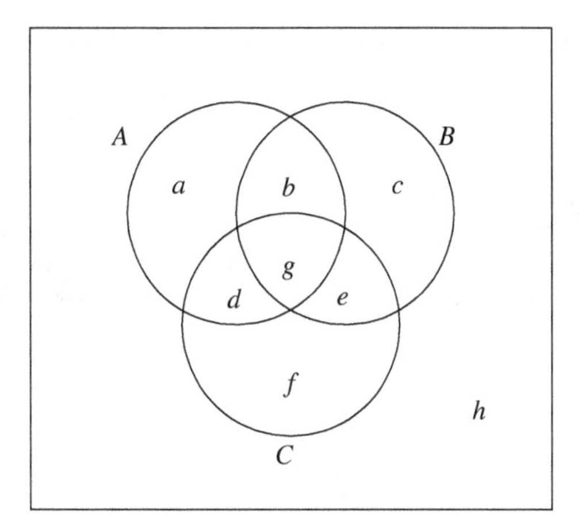

FIGURE 1.5: Venn diagram.

1.35 Let A, B, C, be three events. At least one event always occurs. But it never happens that exactly one event occurs. Nor does it ever happen that all three events occur. If $P(AB) = 0.10$ and $P(AC) = 0.20$, find $P(B)$.

1.36 See the assignment of probabilities to the Venn diagram in Figure 1.5. Find the following:

(a) P(No events occur).

(b) P(Exactly one event occurs).

(c) P(Exactly two events occur).

(d) P(Exactly three events occur).

(e) P(At least one event occurs).

(f) P(At least two events occur).

(g) P(At most one event occurs).

(h) P(At most two events occur).

1.37 Suppose that probabilities have been assigned to the Venn Diagram in Figure 1.5 as follows: $a + c = f = h$, $b = d = e = g = 0.1$, and $c = 3a$. Find the following:

(a) P(No events occur).

(b) P(Exactly two events occur).

(c) P(At most one event occurs).

(d) P(At most two events occur).

1.38 For three events A, B, and C, the following is known: $P(A) = P(B) = P(C)$, $P(BC) = 0.2P(B \cup C)$, $P(AB) = P(AC) = 0$, and the probability of no events occurring is 0.1.

(a) Sketch a Venn diagram that matches the information provided about the three events.

(b) Find $P(B)$.

(c) Find $P(ABC)$.

(d) Find $P[(A \cup B) \cup C^c]$.

1.39 Four coins are tossed. Let A be the event that the first two coins both come up heads. Let B be the event that the number of heads is odd. Assume that all 16 elements of the sample space are equally likely. Describe and find the probabilities of (i) AB, (ii) $A \cup B$, and (iii) AB^c.

1.40 Two dice are rolled. Let X be the maximum number obtained. (Thus, if 1 and 2 are rolled, $X = 2$; if 5 and 5 are rolled, $X = 5$.) Assume that all 36 elements of the sample space are equally likely. Find the probability function for X. That is, find $P(X = x)$, for $x = 1, 2, 3, 4, 5, 6$.

1.41 Judith has a penny, nickel, dime, and quarter in her pocket. So does Kory. They both reach into their pockets and choose a coin. Let X be the greater (in cents) of the two.

(a) Construct a sample space and describe the events $\{X = k\}$ for $k = 1, 5, 10, 25$.

(b) Assume that coin selections are equally likely. Find the probabilities for each of the aforementioned four events.

(c) What is the probability that Judith's coin is worth more than Kory's? (It is not $1/2$.)

1.42 A tetrahedral dice is four-sided and labeled with 1, 2, 3, and 4. When rolled it lands on the base of a pyramid and the number rolled is the number on the base. In five rolls, what is the probability of rolling at least one 2?

1.43 Let

$$Q(k) = \frac{2}{3^{k+1}}, \quad \text{for } k = 0, 1, 2, \ldots.$$

(a) Show that Q is a probability function. That is, show that the terms are nonnegative and sum to 1.

(b) Let X be defined such that $P(X = k) = Q(k)$, for $k = 0, 1, 2, \ldots$. Find $P(X > 2)$ without summing an infinite series.

1.44 The function

$$P(k) = c\frac{3^k}{k!}, \quad \text{for } k = 0, 1, 2, \ldots,$$

is a probability function for some choice of c. Find c.

1.45 Let A, B, C be three events. Find expressions for the events:

(a) At least one of the events occurs.

(b) Only B occurs.

(c) At most one of the events occurs.

(d) All of the events occur.

(e) None of the events occur.

1.46 The *odds in favor* of an event is the ratio of the probability that the event occurs to the probability that it will not occur. For example, the odds that you were born on a Friday, assuming birthdays are equally likely, is 1 to 6, often written 1 : 6 or 1 to 6, obtained from (1/7)/(6/7).

(a) In Texas Hold'em Poker, the odds of being dealt a pair (two cards of the same denomination) is 1 : 16. What is the chance of not being dealt a pair?

(b) For sporting events, bookies usually quote odds as odds against, as opposed to odds in favor. In the Kentucky Derby horse race, our horse Daddy Long Legs was given 9 to 2 odds. What is the chance that Daddy Long Legs wins the race?

1.47 An exam had three questions. One-fifth of the students answered the first question correctly; one-fourth answered the second question correctly; and one-third answered the third question correctly. For each pair of questions, one-tenth of the students got that pair correct. No one got all three questions right. Find the probability that a randomly chosen student did not get any of the questions correct.

1.48 Suppose $P(ABC) = 0.05$, $P(AB) = 0.15$, $P(AC) = 0.2$, $P(BC) = 0.25$, $P(A) = P(B) = P(C) = 0.5$. For each of the events given next, write the event using set notation in terms of A, B, and C, and compute the corresponding probability.

(a) At least one of the three events A, B, C occur.

(b) At most one of the three events occurs.

(c) All of the three events occurs.

(d) None of the three events occurs.

(e) At least two of the three events occurs.

(f) At most two of the three events occurs.

1.49 Find the probability that a random integer between 1 and 5000 is divisible by 4, 7, or 10.

1.50 Each of the four squares of a two-by-two checkerboard is randomly colored red or black. Find the probability that at least one of the two columns of the checkerboard is all red.

1.51 Given events A and B, show that the probability that exactly one of the events occurs equals

$$P(A) + P(B) - 2P(AB).$$

1.52 Given events A, B, C, show that the probability that exactly one of the events occurs equals

$$P(A) + P(B) + P(C) - 2P(AB) - 2P(AC) - 2P(BC) + 3P(ABC).$$

Simulation and R

1.53 Modify the code in the **R** script **CoinFlip.R** to simulate the probability of getting exactly one head in four coin tosses.

1.54 Modify the code in the **R** script **Divisible356.R** to simulate the probability that a random integer between 1 and 5000 is divisible by 4, 7, or 10. Compare with your answer in Exercise 1.49.

1.55 Use **R** to simulate the probability of two rolled dice having values that sum to 8.

1.56 Explain what reproducibility means to you in a few sentences.

1.57 Use **R** to simulate the probability in Exercise 1.41 part c.

1.58 Use **R** to simulate the probability in Exercise 1.42.

1.59 Make up your own random experiment and write an **R** script to simulate it. Be sure your results are reproducible.

1.60 See the help file for the `sample` command (type `?sample`). Write a function `dice(`k`)` for generating k throws of a fair die. Use your function and **R**'s `sum` function to generate the sum of two dice throws.

Chapter Review

Chapter review exercises are available through the text website. The URL is `www.wiley.com/go/wagaman/probability2e`.

2

CONDITIONAL PROBABILITY AND INDEPENDENCE

In the fields of observation chance favors only the prepared mind.

—Louis Pasteur

Learning Outcomes

1. Define key conditional probability terms and recognize notation.
2. Solve conditional probability problems using the general multiplication rule and techniques such as trees.
3. Identify and solve Bayes' rule problems.
4. Give examples of independence and independent events.
5. (C) Explore simulations of conditional probability problems with a focus on common functions and loops.

2.1 CONDITIONAL PROBABILITY

Sixty students were asked, "Would you rather be attacked by a big bear or swarming bees?" Their answers, along with their sex, are collected in the following *contingency table*, a common way to present data for two variables, in this case sex and attack preference. The table includes row and column totals, called marginals or marginal totals, and the overall total surveyed.

Probability: With Applications and R, Second Edition. Amy S. Wagaman and Robert P. Dobrow.

Companion Website: www.wiley.com/go/wagaman/probability2e

	Big bear (B)	**Swarming bees** (S)	**Total**
Female (F)	27	9	36
Male (M)	10	14	24
Total	37	23	60

The table of counts is the basis for creating a probability model for selecting a student at random and asking their sex and attack preference. The sample space consists of the four possible responses

$$\Omega = \{(F, B), (F, S), (M, B), (M, S)\},$$

where M is male, F is female, B is big bear, and S is swarming bees. The probability function is constructed from the contingency table so that the probability of each outcome is the corresponding proportion of responses from the sample. That is,

	Big bear (B)	**Swarming bees** (S)
Female (F)	$27/60 = 0.450$	$9/60 = 0.150$
Male (M)	$10/60 = 0.167$	$14/60 = 0.233$

Some questions of interest are

1. What is the probability that a student is female and would rather be attacked by a big bear?
2. What is the probability that a female student would rather be attacked by a big bear?

These questions are worded similarly but ask different things. Let us answer them to see the differences. For the first question, the proportion of students who are female and prefer a big bear attack is $27/60 = 0.450$. That is, $P(F \text{ and } B) = 0.450$. For the second question, the proportion of female students who prefer a big bear attack is $27/36 = 0.75$ because there are 36 females and 27 of them prefer a big bear attack to a swarming bees attack.

The second probability is an example of a *conditional probability*. In a conditional probability, some information about the outcome of the random experiment is known—in this case that the selected student is female. The probability is *conditional* on that knowledge.

For events A and B, the conditional probability of A given that B occurs is written $P(A|B)$. We also read this as "the probability of A conditional on B." Hence, the probability the student would rather be attacked by a big bear conditional on being female is $P(B|F) = 0.75$.

The probability of preferring a big bear attack conditional on being female is computed from the table by taking the number of students who are both female and prefer a big bear attack as a proportion of the total number of females. That is,

$$P(B|F) = \frac{\text{Number of females who prefer a big bear attack}}{\text{Number of females}}.$$

Dividing numerator and denominator by the total number of students, this is equivalent to

$$P(B|F) = \frac{P(B \text{ and } F)}{P(F)}.$$

This suggests how to define the general conditional probability $P(A|B)$.

CONDITIONAL PROBABILITY

For events A and B such that $P(B) > 0$, the *conditional probability of A given B* is

$$P(A|B) = \frac{P(AB)}{P(B)}. \tag{2.1}$$

Example 2.1 In a population, suppose 60% of the people have brown hair (H), 40% have brown eyes (E), and 30% have both (H and E). The probability that someone has brown eyes given that they have brown hair is

$$P(E|H) = \frac{P(EH)}{P(H)} = \frac{0.30}{0.60} = 0.50.$$ ■

Example 2.2 Consider the Venn diagram in Figure 2.1. If each outcome x is equally likely, then $P(A) = 5/14$, $P(B) = 7/14$, $P(AB) = 2/14$, $P(A|B) = 2/7$, and $P(B|A) = 2/5$. ■

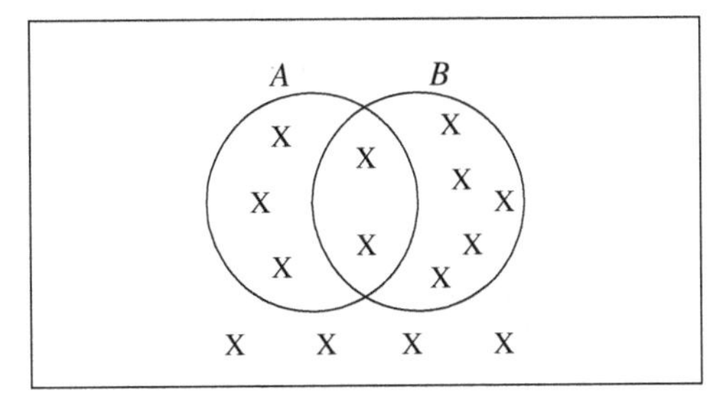

FIGURE 2.1:

Example 2.3 Two dice are rolled. What is the probability that the first die is a 2 given that the sum of the dice is 7?

We use variables to notate the problem. Let X_1 and X_2 be the outcomes of the first and second die, respectively. Then the sum of the dice is $X_1 + X_2$. The problem asks for

$$\begin{aligned} P(X_1 = 2|X_1 + X_2 = 7) &= \frac{P(X_1 = 2 \text{ and } X_1 + X_2 = 7)}{P(X_1 + X_2 = 7)} \\ &= \frac{P(X_1 = 2 \text{ and } 2 + X_2 = 7)}{P(X_1 + X_2 = 7)} \\ &= \frac{P(X_1 = 2 \text{ and } X_2 = 5)}{P(X_1 + X_2 = 7)} \\ &= \frac{P(\{(2,5)\})}{P(\{(1,6),(2,5),(3,4),(4,3),(5,2),(6,1)\})} \\ &= \frac{1/36}{6/36} = \frac{1}{6}. \end{aligned}$$

Observe that for the second equality the variable X_1 is replaced with its stated given value 2.

It is interesting to observe that the *unconditional* probability $P(X_1 = 2)$ that the first die is 2 is also equal to 1/6. In other words, the information that the sum of the dice is 7 did not affect the probability that the first die is 2.

On the other hand, if we are given that the sum is 6, then

$$P(X_1 = 2|X_1 + X_2 = 6) = \frac{P(\{(2,4)\})}{P(\{(1,5),(2,4),(3,3),(4,2),(5,1)\})} = \frac{1}{5} > \frac{1}{6}.$$

Information that the sum of the dice is 6 makes the probability that first die is 2 a little more likely than if nothing was known about the sum of the two dice. ■

R: SIMULATING A CONDITIONAL PROBABILITY

Simulating the conditional probability $P(A|B)$ requires repeated simulations of the underlying random experiment, but restricting to trials in which B occurs. To simulate the conditional probability in the last example requires simulating repeated pairs of dice tosses, but the only data that are relevant are those pairs that result in a sum of 7.

See **ConditionalDice.R**. Every time a pair of dice is rolled, the routine checks whether the sum is 7 or not. If not, the dice are rolled again until a 7 occurs. Once a 7 is rolled, success is recorded if the first die is 2, and failure if it is not. The proportion of successes is taken just for those pairs of dice rolls that sum to 7.

A counter `ctr` is used that is iterated every time a 7 is rolled. This keeps track of the actual trials. The while statement is a loop that continues to run while the condition that the `ctr` is less than n holds. Be careful with while loops! If the condition is always met, the loop just continues. Here, the condition will fail once we reach the desired number of trials because then `ctr` is equal to n, which exits the loop. We set the number of trials at n =10,000. However, more than 10,000 trials will be attempted. As the probability of getting a 7 is $1/6$, the 10,000 trials are about one-sixth of the total number of trials attempted, which we expect to be about 60,000. In each trial, two random numbers are generated, one for each die roll. So we are likely generating about 120,000 random numbers but only looking at results involving 20,000 of them.

```
# ConditionalDice.R
> n <- 10000
> ctr <- 0
> simlist <- numeric(n)
> while (ctr < n) {
    trial <- sample(1:6, 2, replace = TRUE)
    if (sum(trial) == 7)    { # Check if sum is 7
    success <- if (trial[1] == 2) 1 else 0
    ctr <- ctr + 1
    simlist[ctr] <- success }        }
> mean(simlist)
[1] 0.1706
```

Example 2.4 Jayden flips three fair coins. The probability of getting all heads is $(1/2)^3 = 1/8$. Suppose Aimee peeks and sees that the first coin came up heads. For Aimee, what is the probability that Jayden gets all heads?

Aimee's probability is conditional on the first coin coming up heads. Given the first coin is heads, we find

$$\begin{aligned} P(\text{HHH}|\text{ First coin is H}) &= \frac{P(\text{HHH and First coin is H})}{P(\text{First coin is H})} \\ &= \frac{P(\text{HHH})}{P(\text{First coin is H})} = \frac{1/8}{1/2} = \frac{1}{4}. \end{aligned}$$

■

Warning: A common mistake when first working with conditional probability is to write $P(A|B) = P(A)/P(B)$. In general, this is just wrong. However, in the special case when A implies B, that is, $A \subseteq B$, then it is correct because $A \cap B = A$ and thus,

$$P(A|B) = \frac{P(A \cap B)}{P(B)} = \frac{P(A)}{P(B)}.$$

That is what happens in the last example, because getting all heads implies that the first coin is heads.

2.2 NEW INFORMATION CHANGES THE SAMPLE SPACE

In Example 2.4, the fact that Aimee's probability of getting three heads is different from Jayden's highlights the fact that probability is not a static property of a random experiment. It changes based on information and context. When we ask what is the probability of getting all heads in three coin tosses, implicit in that question is that you have not seen the outcome of the experiment. If you see the outcome, then you know that either all heads came up or they did not, so the probability is either 1 or 0. On the other hand, when some part of the experiment is observed, then that partial information becomes relevant in the probability calculation.

Partial information about the outcome of a random experiment actually changes the set of possible outcomes, that is, it changes the sample space of the original experiment and reduces it based on new information. So you can think of conditioning as asking questions about probabilities on a reduced sample space determined by the given event. For the three coin tosses, before Aimee peeks, the sample space is

$$\Omega = \{\text{HHH}, \text{HHT}, \text{HTH}, \text{THH}, \text{HTT}, \text{THT}, \text{TTH}, \text{TTT}\}.$$

But after she looks and sees that the first coin is heads, the sample space reduces to

$$\Omega' = \{\text{HHH}, \text{HHT}, \text{HTH}, \text{HTT}\}.$$

The resulting conditional probability is a probability function computed on the restricted sample space.

Conditional probability is a probability function. In Chapter 1, a probability function is defined based on a function satisfying three properties—taking only nonnegative values and satisfying Equations 1.1 and 1.2. Here we show that for a fixed event B, the conditional probability $P(A|B)$—as a function of A—is itself a similarly defined probability function, but on a reduced sample space.

Starting from a random experiment and sample space Ω, let $B \subseteq \Omega$ be an event such that $P(B) > 0$. We use lowercase letters b to denote the elements of B. Consider $P(A|B)$ as a function of A. It is better to write this as $P(\cdot|B)$ to emphasize that the conditional probability is a function of its first argument. This function is itself a probability function on the restricted sample space B, and as a result, satisfies the three conditions for a probability function, listed below. That is,

1. $P(b|B) \geq 0$, for all $b \in B$.
2. $\sum_{b \in B} P(b|B) = 1$.
3. For all $A \subseteq B$, $P(A|B) = \sum_{b \in A} P(b|B)$.

The properties are verified subsequently:

1. This is true because $P(b|B)$ is defined as a ratio of two probabilities, which are both nonnegative. (Recall $P(B) > 0$.)
2.
$$\sum_{b\in B} P(b|B) = \sum_{b\in B} \frac{P(\{b\} \text{ and } B)}{P(B)} = \sum_{b\in B} \frac{P(b)}{P(B)}$$
$$= \frac{1}{P(B)} \sum_{b\in B} P(b) = \frac{P(B)}{P(B)} = 1.$$
3. Let $A \subseteq \Omega$. Then
$$P(A|B) = \frac{P(A \cap B)}{P(B)} = \sum_{b\in A\cap B} \frac{P(b)}{P(B)}$$
$$= \sum_{b\in A} \frac{P(\{b\} \cap B)}{P(B)} = \sum_{b\in A} P(b|B).$$

In summary, conditional probabilities are themselves probability functions defined on the restricted sample space of the conditioning event. To demonstrate, consider the next example.

Example 2.5 Let X be a random integer picked uniformly from 0 to 6. The probability function for X is

$$P(X = 0) = \cdots = P(X = 6) = \frac{1}{7}.$$

We are told that X is odd. Then for $k = 1, 3, 5$,

$$P(X = k|X \text{ is odd}) = \frac{P(X = k \text{ and } X \text{ is odd})}{P(X \text{ is odd})} = \frac{P(X = k)}{P(X \text{ is odd})} = \frac{1/7}{3/7} = \frac{1}{3}.$$

This gives the new probability function, $\tilde{P}$, where

$$\tilde{P}(X = 1) = \tilde{P}(X = 3) = \tilde{P}(X = 5) = \frac{1}{3},$$

with reduced sample space $\{1, 3, 5\}$. This relationship can be summarized as

$$\tilde{P}(X = k) = P(X = k|X \text{ is odd}), \quad \text{for } k = 1, 3, 5.$$ ■

2.3 FINDING $P(A$ AND $B)$

So far we have focused on finding the conditional probability $P(A|B)$, which requires knowledge of $P(A \text{ and } B)$. But sometimes what is unknown is precisely

$P(A \text{ and } B)$. Rearranging the formula $P(A|B) = P(A \text{ and } B)/P(B)$ gives the following, commonly referred to as the *general multiplication rule.*

GENERAL FORMULA FOR $P(A$ AND $B)$

$$P(A \text{ and } B) = P(AB) = P(A|B)P(B), \tag{2.2}$$

This is a general formula for working with "and" probabilities. Observe that by switching the roles of A and B, this also gives

$$P(A \text{ and } B) = P(B|A)P(A).$$

Example 2.6 Draw two cards from a standard deck. What is the probability of getting two aces?

The probability of drawing an ace is $4/52 = 1/13$. If we have already drawn an ace, the probability of drawing a second ace is $3/51$, as there are three aces left in a reduced deck of 51. Let A_1 be the event of drawing an ace on the first card, and A_2 the event of drawing an ace on the second card. Then

$$P(A_1A_2) = P(A_2|A_1)P(A_1) = \left(\frac{3}{51}\right)\left(\frac{1}{13}\right) = \frac{1}{221} = 0.0045. \quad \blacksquare$$

Draw three cards from a standard deck. What is the probability of getting three aces? Intuitively, you might guess the answer $(4/52)(3/51)(2/50) = 0.000178$. The three factors in the probability are (i) the probability of getting an ace on the first card, (ii) the probability of getting an ace on the second card given that the first card is an ace, and (iii) the probability of getting an ace on the third card given that the first two cards are aces.

This intuition is correct and suggests the extension of Equation 2.2 for more than two events. Given events A_1, A_2, and A_3,

$$\boxed{P(A_1A_2A_3) = P(A_3|A_1A_2)P(A_2|A_1)P(A_1).} \tag{2.3}$$

To see this, let $A = A_1$ and $B = A_2A_3$. Then

$$\begin{aligned} P(A_1A_2A_3) &= P(AB) = P(A|B)P(B) \\ &= P(A_1|A_2A_3)P(A_2A_3) \\ &= P(A_1|A_2A_3)P(A_2|A_3)P(A_3). \end{aligned}$$

More generally, for k events $A_1, \ldots, A_k$, the general multiplication rule is

$$\boxed{P(A_1 \cdots A_k) = P(A_k|A_1 \cdots A_{k-1})P(A_{k-1}|A_1 \cdots A_{k-2}) \cdots P(A_2|A_1)P(A_1).} \tag{2.4}$$

The general result is proven using mathematical induction in a similar fashion.

Example 2.7 A subject in an experiment is given three tries to complete a task. On the first try, the probability of success is 0.30. If they fail, the chance of success on the second attempt is 0.50. And if they fail that, the chance of success on the third try is 0.65. What is the probability that they complete the task, $P(S)$?

Let S_1, S_2, S_3 denote the events that the task is completed on the first, second, and third tries, respectively. Then S can be expressed as $S_1 \cup S_2 \cup S_3$. The desired probability $P(S)$ is

$$\begin{aligned} P(S) = P(S_1 \cup S_2 \cup S_3) &= 1 - P(S_1^c S_2^c S_3^c) \\ &= 1 - P(S_1^c)P(S_2^c | S_1^c)P(S_3^c | S_1^c S_2^c) \\ &= 1 - (0.70)(0.50)(0.35) = 0.8775. \end{aligned}$$ ■

Tree diagrams. Tree diagrams are useful tools for computing probabilities. They often arise when events can be ordered sequentially (first one thing happens, then the next). They are also great visual aids that decompose a problem into smaller logical units. Probabilities are written on the branches of the tree, and outcomes are written at the end of each branch. The outermost branch has unconditional probabilities, while the inner branches have probabilities conditional on transversing that branch of the tree.

Figure 2.2 illustrates the random experiment of picking two balls from a bag containing two red and three blue balls. The outcome of picking two red balls is described by the top branch of the tree. First, we select a red ball (with probability 2/5), and then we select a second red ball given that the first ball was red (with probability 1/4). The probability of the final outcome is obtained by multiplying along the branch ($1/10 = 2/5 \times 1/4$). Observe that the branches of the tree are labeled with conditional probabilities.

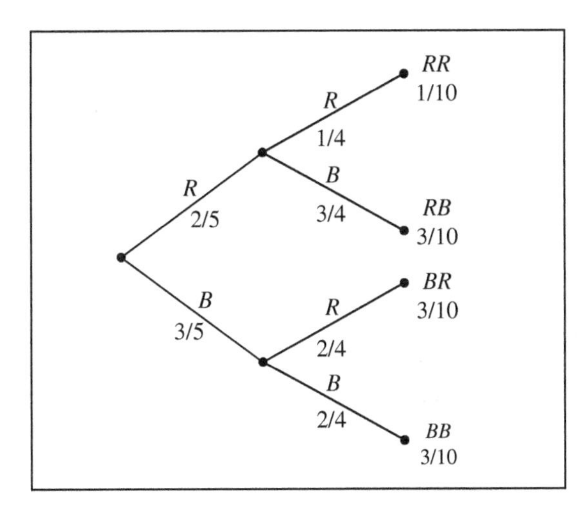

FIGURE 2.2: Tree diagram for picking two balls from a bag of two red and three blue balls.

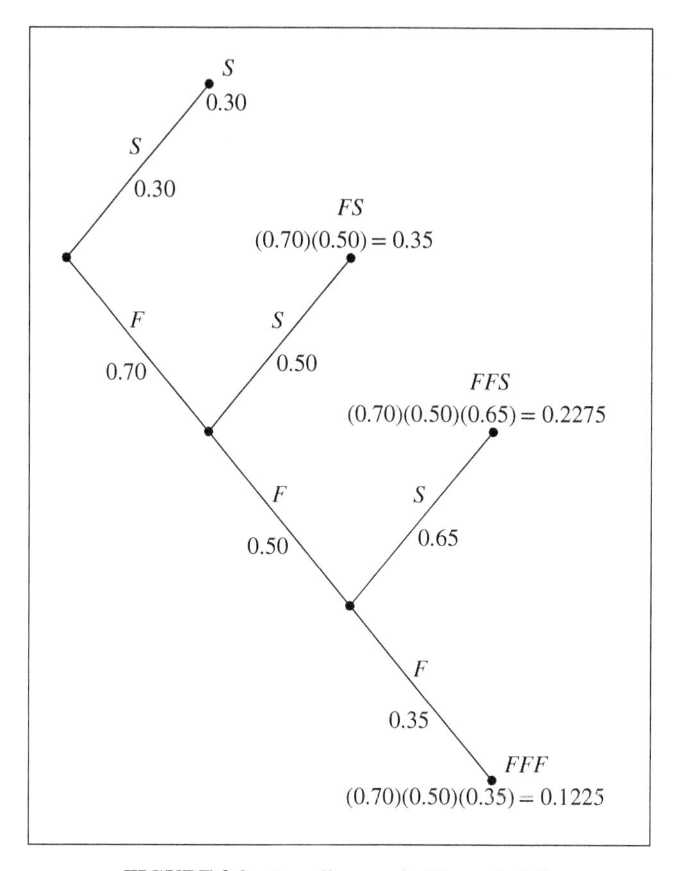

FIGURE 2.3: Tree diagram for Example 2.7.

Example 2.7 lends itself naturally to a tree diagram analysis because of the sequential nature of the random experiment. The tree is presented in Figure 2.3. Recall that the subject is given three tries to complete a task. The event that the subject eventually completes their task is the disjoint union of the events that the subject completes their task on the first, second, or third try, respectively. These are the paths that end with an S in the tree. Thus, the probability that the subject eventually completes their task is the sum

$$P(S) + P(FS) + P(FFS) = 0.30 + 0.35 + 0.2275 = 0.8775.$$

Alternatively, using complements and finding $1 - P(FFF)$ also yields the probability 0.8775.

Example 2.8 Blackjack, or twenty-one, is a popular casino game. (For more details about the game, the reader can look it up online.) To start a game, the player

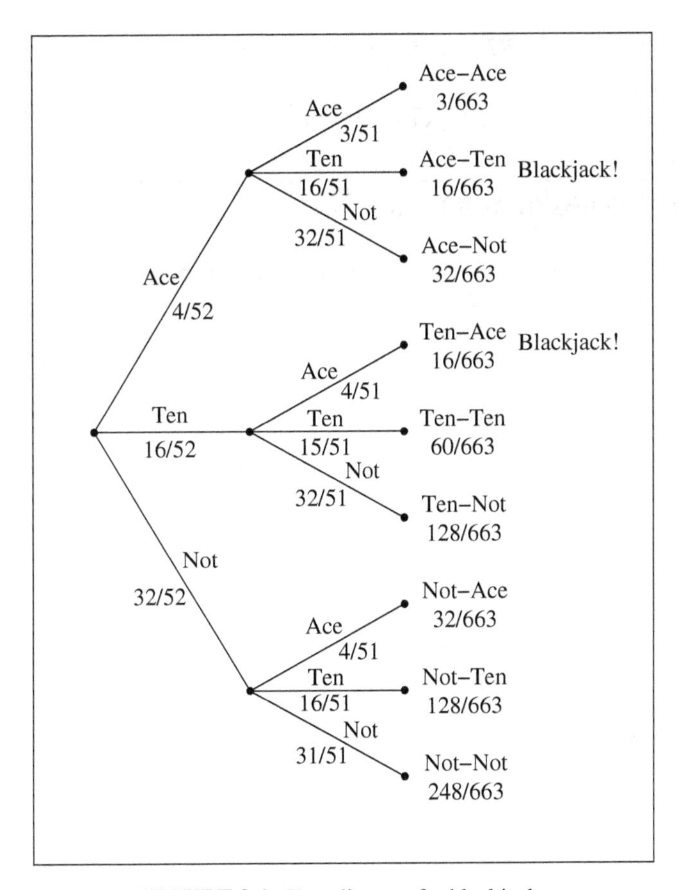

FIGURE 2.4: Tree diagram for blackjack.

is dealt two cards. A *blackjack* is an ace and a ten card (10, jack, queen, or king). What is the probability of being dealt a blackjack?

We illustrate the solution with the tree diagram in Figure 2.4. Blackjack is obtained by either getting an ace on the first card and then a ten card, or a ten card first and then an ace. We use "Not" to denote any of the 32 cards that are neither an ace nor a ten card.

There are two outcomes that correspond to a blackjack—being dealt an ace on the first card and then a ten card on the second, or vice versa. Let A_1 and A_2 denote the events of getting an ace on the first and second cards, respectively. Similarly, define T_1 and T_2 for getting a ten card on the first and second cards. Then the events A_1T_2 and A_2T_1 are mutually exclusive and we find that

$$\begin{aligned} P(\text{Blackjack}) &= P(A_1T_2 \text{ or } T_1A_2) \\ &= P(A_1T_2) + P(T_1A_2) \end{aligned}$$

$$= P(T_2|A_1)P(A_1) + P(A_2|T_1)P(T_1)$$

$$= \left(\frac{16}{51}\right)\left(\frac{4}{52}\right) + \left(\frac{4}{51}\right)\left(\frac{16}{52}\right) = \frac{2 \times 4 \times 16}{51 \times 52} = 0.048.$$

■

R: SIMULATING BLACKJACK

The script **Blackjack.R** simulates the blackjack probability. The numbers 1–52 represent a deck of cards. We assign the four aces to the numbers 1, 2, 3, 4, and the 16 ten cards to the numbers 37–52. The command

```
> trial <- sample(1:52, 2, replace=FALSE)
```

chooses two cards from 1 to 52, sampling without replacement, and assigns them to the variable `trial`. The command

```
> success <- if (trial[1] <= 4 && trial[2] >= 37
|| trial[1] >= 37 && trial[2] <= 4) 1 else 0
```

uses the logical operators || (or) and && (and) to determine if blackjack occurred, returning a 1 if yes, and 0, otherwise.

```
# Blackjack.R
> n <- 50000
> simlist <- replicate(n, 0)
> for (i in 1:n) {
   trial <- sample(1:52, 2, replace = FALSE)
   success <- if (trial[1] <= 4 && trial[2] >= 37 ||
   trial[1] >= 37 && trial[2] <= 4) 1 else 0
   simlist[i] <- success  }
> mean(simlist)
[1] 0.0462
```

2.3.1 Birthday Problem

The birthday problem is a classic probability delight first introduced by the mathematician Richard Von Mises in 1939. Von Mises asked, "How many people must be in a room before the probability that some share a birthday, ignoring the year and ignoring leap days, becomes at least 50%?"

For a group of k people, let B be the event that at least two people have the same birthday. We find $P(B)$. Remember the problem-solving strategy of taking complements for "at least" probabilities. The complement B^c is the event that none of the k people have the same birthday. We compute that probability with a tree diagram.

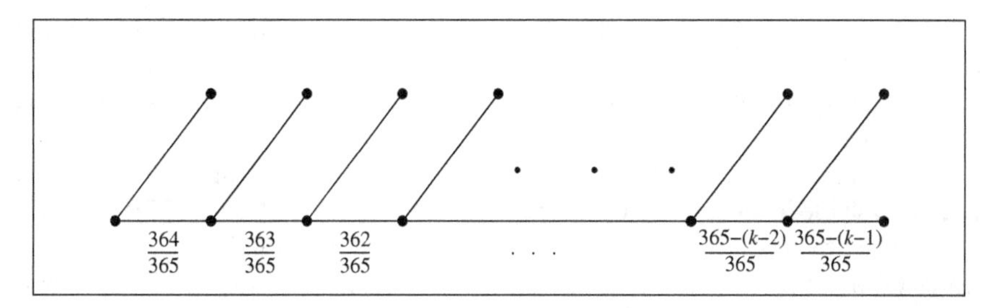

FIGURE 2.5: Solving the birthday problem with a tree diagram.

Consider asking people one by one their birthday and checking whether their birthday is different from the birthdays of those previously asked. The first person's birthday is fixed. The second person's birthday either matches the first birthday, which occurs with probability 1/365, or does not, with probability 364/365. Two branches grow out of the first node labeled with these probabilities as in Figure 2.5.

The full tree will have a lot of branches. But we are only interested in one path of the tree, where everyone's birthday is different. So it is not necessary to draw the entire tree, just that relevant path.

From the second node, the probability that the third person has a birthday different from the previous two, given that the previous two birthdays are different, is 363/365 (because two birthdays have been picked and there are 363 available ones left). Continuing in this way, we see that the ith branch of the tree gives the probability that the $(i + 1)$st person's birthday is different from the previous i birthdays, given that the previous i birthdays are all different, which occurs with probability $(365 - i)/365$. This gives

$$P(B^c) = \left(\frac{364}{365}\right)\left(\frac{363}{365}\right)\cdots\left(\frac{365-(k-1)}{365}\right) = \prod_{i=1}^{k-1}\left(1 - \frac{i}{365}\right). \tag{2.5}$$

And thus the birthday probability that at least two people have the same birthday is

$$P(B) = 1 - P(B^c) = 1 - \prod_{i=1}^{k-1}\left(1 - \frac{i}{365}\right). \tag{2.6}$$

At $k = 22$, $P(B) = 0.476$, and at $k = 23$, $P(B) = 0.507$. So the answer to Von Mises' question is, remarkably, 23 people. The number is much smaller than most people think. Table 2.5 gives birthday probabilities for different group sizes. With just $k = 15$ people there is a 25% chance of at least one birthday match. And with $k = 50$ people the likelihood of at least one match is virtually certain with $P(B) = 0.970$.

To explain the seemingly paradoxical result, intuitively observe that in a group of 23 people there are actually 253 ways for people to be paired. And we just need one of those pairs to have a common birthday for the desired event to occur.

There are many fun ways to illustrate the birthday problem. Consider the birthdays of the first 45 US presidents. With $k = 45$, the probability of a match is 0.941. And we find that Warren G. Harding and James K. Polk were both born on November 2.

The (planned for in 2020) 26-player active roster of a major league baseball team is a nice vehicle for checking the birthday problem. Each team's roster gives a nearly 60% chance that two players on the team have the same birthday. Among the 30 teams of major league baseball, we estimate that about 18 teams will have at least one birthday match with two players on the active roster with the same birthday. And we invite a baseball fan to check our conjecture.

The birthday problem can be cast in a very general framework. Suppose we distribute k balls into m boxes in such a way so that each of the m boxes is equally likely to receive any ball. Think of the boxes as birthdays and the balls as people. The probability that some box contains two or more balls is equivalent to the birthday probability that among k people there are two birthdays in common. This probability is equal to

$$P(\text{Some box contains at least two balls}) = 1 - \prod_{i=1}^{k-1}\left(1 - \frac{i}{m}\right). \tag{2.7}$$

If k is large, this probability does not lend itself to easy calculation or interpretation. A simpler closed form expression can be gotten with the help of calculus, in particular Taylor series (see Appendix C). We give the derivation thinking of the birthday problem with $m = 365$.

Let $p = \prod_{i=1}^{k-1}(1 - i/365)$. Then

$$\ln p = \ln \prod_{i=1}^{k-1}\left(1 - \frac{i}{365}\right) = \sum_{i=1}^{k-1} \ln\left(1 - \frac{i}{365}\right).$$

The Taylor series expansion for $\ln(1 - x)$ is

$$\ln(1 - x) = -x - \frac{x^2}{2} - \frac{x^3}{3} - \cdots,$$

which converges for $-1 < x < 1$. Truncate off all but the first term of the series to obtain the approximation $\ln(1 - x) \approx -x$. The approximation is good for small values of x close to 0.

For $1 \leq i < k < 365$, $i/365$ will be "small," justifying the use of the approximation. This gives

$$\ln p = \sum_{i=1}^{k-1} \ln\left(1 - \frac{i}{365}\right) \approx \sum_{i=1}^{k-1} -\frac{i}{365} = -\frac{k(k-1)}{2 \times 365},$$

TABLE 2.1. Birthday probabilities.

k	Exact	Approximate
15	0.253	0.250
23	0.507	0.500
30	0.706	0.696
40	0.891	0.882
50	0.970	0.965
60	0.994	0.992

using the fact that the sum of the first $k - 1$ integers is $k(k - 1)/2$. Exponentiating both sides gives $p \approx e^{-k(k-1)/(2\times365)}$. And the birthday probability is

$$P(B) \approx 1 - e^{-k(k-1)/(2\times365)} \approx 1 - e^{-k^2/(2\times365)}. \tag{2.8}$$

In Table 2.1, we compare the approximation with the exact probabilities for select values of k.

More generally, in the balls-and-boxes setting with k balls and m boxes, the probability that some box contains two or more balls is approximately equal to $1 - e^{-k^2/(2m)}$.

There are many applied settings that fit the balls-into-boxes framework of the birthday problem as shown in the next examples.

Example 2.9 In 2001, the Arizona Department of Public Safety reported, in response to a court order, that a search of its state offender database of 65,493 DNA profiles found a "nine-locus" DNA match, where the DNA of two samples agreed at nine positions on the chromosome. See Troyer et al. [2001]. The estimated probability of such a match is about 1 in 754 million. At the time, the DNA match was said to be so unlikely as to call into question the reliability of the state's database and even of the use of DNA evidence in court. What is the probability of finding two such matching profiles in the database?

If we assume that all 754 million DNA outcomes are equally likely, then this gives an application of the birthday problem with $k = 65{,}493$ balls and $m = 7.54 \times 10^8$ boxes. The probability of a DNA match is

$$1 - \prod_{i=1}^{65,492} \left(1 - \frac{i}{7.54 \times 10^8}\right) \approx 1 - e^{-65,493^2/(2\times7.54\times10^8)} = 0.942.$$

What was originally thought to be an extremely rare coincidence actually has a very high probability of occurrence.

An intuitive explanation for this high probability is that in a database of 65,493 DNA profiles there are about two billion different *pairs* of profiles. An event which has a one in a billion chance of occurring, if the experiment is repeated two billion times, will likely occur twice! ■

Coincidences, like having the same birthday as your roommate, or having gone to the same high school as the person sitting next to you on the plane, or always seeing license plates that seem to start with the same three letters as your own, seem to defy logic. Yet they can often be explained by the laws of probability.

If an event has a one-in-a-million chance of occurring, then in a country of 300 million you would expect about 300 occurrences.

In "Methods for Studying Coincidences," Diaconis and Mosteller [1989] assert the *Law of Truly Large Numbers*: "With a large enough sample, any outrageous thing is likely to happen." Their highly readable and entertaining paper is part human psychology and gives a guide to the probabilistic and statistical techniques for studying coincidences.

Many variations and extensions of the birthday problem have been proposed and studied. In our discussion, we assume that the 365 days of nonleap years are equally likely. This is not the case in the United States, where months such as July, August, and September see more births than January. We encourage the reader to investigate further if truly interested with references such as Borja and Haigh [2007].

The birthday problem does *not* ask for the probability that among k people there will be a match of any one particular birthday, but rather that some pair of people will have the same birthday. If you survey your classmates in your classes of about 25 students, you would expect to find that two students in a class have the same birthday more than 50% of the time. But you might never find a student whose birthday matches your own.

2.4 CONDITIONING AND THE LAW OF TOTAL PROBABILITY

According to the Howard Hughes Medical Institute, about 7% of men and 0.4% of women are colorblind, meaning they either cannot distinguish red from green or see red and green differently from most people. In the United States, about 49% of the population is male and 51% is female. A person is selected at random. What is the probability they are colorblind?

As you contemplate answering this question, you might find yourself saying, "Well, it depends—on whether you are male or female." The problem provides conditional information based on sex but the question asks for an *unconditional* probability.

In this section, we introduce a powerful technique for using conditional probability for solving "unconditional" problems.

The event $C = \{\text{Colorblind}\}$ can be decomposed into the disjoint union

$$\{\text{Colorblind}\} = \{\text{Colorblind and Male}\} \cup \{\text{Colorblind and Female}\}.$$

We then obtain

$$\begin{aligned} P(C) &= P(CM \cup CF) = P(CM) + P(CF) \\ &= P(C|M)P(M) + P(C|F)P(F) \\ &= (0.07)(0.49) + (0.004)(0.51) = 0.03634. \end{aligned}$$

The approach to solving this problem is known as *conditioning*. In this case, we are conditioning on sex because the conditional probabilities $P(C|M)$ and $P(C|F)$ are easier and more "natural" to solve than the unconditional probability $P(C)$.

More generally, say that a collection of events $\{B_1, \ldots, B_k\}$ is a *partition* of the sample space Ω if (i) the events have no outcomes in common and (ii) their union is equal to Ω. Given an event A, the *law of total probability* shows how to find the unconditional probability $P(A)$ by conditioning on the B_i's.

LAW OF TOTAL PROBABILITY

Suppose $B_1, \ldots, B_k$ is a partition of the sample space. Then

$$P(A) = \sum_{i=1}^{k} P(A|B_i)P(B_i). \tag{2.9}$$

Observe that we can decompose A into the disjoint union

$$A = AB_1 \cup \cdots \cup AB_k$$

as illustrated in Figure 2.6. The law of total probability follows by taking probabilities and applying the conditional probability formula to each term of the resulting sum because

$$P(A) = P\left(\bigcup_{i=1}^{k} AB_i\right) = \sum_{i=1}^{k} P(AB_i) = \sum_{i=1}^{k} P(A|B_i)P(B_i).$$

A common special case of the law of total probability occurs when $k = 2$. For any event B, the sets B and B^c partition the sample space. This gives

$$P(A) = P(A|B)P(B) + P(A|B^c)P(B^c). \tag{2.10}$$

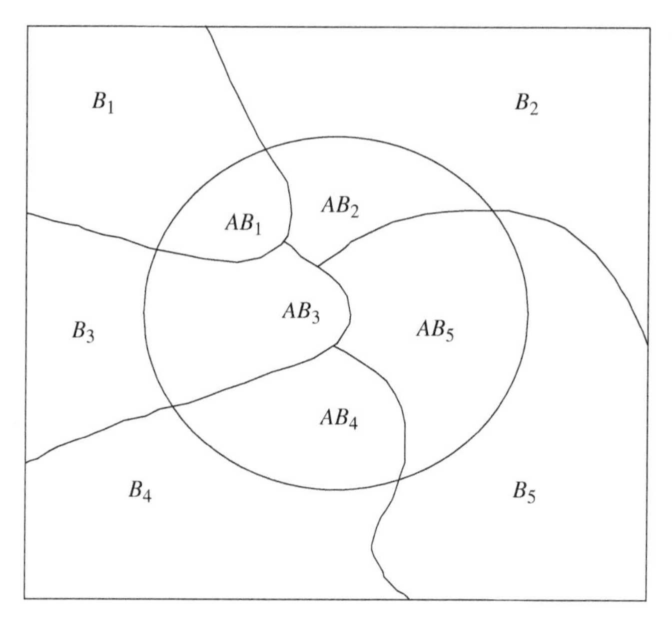

FIGURE 2.6: The events $B_1, \ldots, B_5$ partition the sample space. The circle represents event A, which is decomposed into the disjoint union $A = AB_1 \cup \cdots \cup AB_5$.

Solving a general probability $P(A)$ by conditioning refers to introducing "supplemental" disjoint events $B_1, \ldots, B_k$. We find the "partial" probabilities $P(A|B_1), \ldots, P(A|B_k)$ in order to find the "total" probability $P(A)$.

Example 2.10 An insurance company predicts the likelihood that a person in a particular age group will have an auto accident during the next year. For individuals under the age of 25, 11% are likely to have an accident, for those between 25 and 39, 3% are likely to have an accident, and for those over 40, only 2% are likely to have an accident. The company's policyholders are 20% under the age of 25, 30% between 25 and 39, and 50% over the age of 40. What is the probability that a random policyholder will have an auto accident next year?

Denote the three age groups by G_1, G_2, and G_3, respectively. Let A be the event of having an auto accident. Conditioning on age group, the law of total probability gives

$$\begin{aligned} P(A) &= P(A|G_1)P(G_1) + P(A|G_2)P(G_2) + P(A|G_3)P(G_3) \\ &= (0.11)(0.20) + (0.03)(0.30) + (0.02)(0.50) \\ &= 0.041. \end{aligned}$$

■

Example 2.11 How to ask a sensitive question? Statisticians are sometimes confronted with how to obtain information on sensitive issues. What proportion of people use illegal drugs? How many students ever cheated on an exam? Surveying people directly and asking these types of sensitive questions is not likely to get honest responses and useful data.

Using probabilistic methods, statisticians have developed interesting ways to ask sensitive questions that protect confidentiality. Here is one example.

Respondents are given a coin and told to flip it in private, not letting anyone see the outcome. If it lands heads, they answer the sensitive question of interest (e.g., "Have you ever taken illegal drugs?") (hopefully, truthfully). If tails, they answer an innocuous question such as "Were you born in the first half of the year—January through June?") The respondent reports a yes or no, but does not say which question they actually answered. And from a sample of such yes—no responses, statisticians can estimate the parameter of interest, such as the proportion of people who have ever taken illegal drugs. How can this be done?

Let Y and N denote responses of yes and no, respectively. Let Q_S denote the sensitive question and Q_I the innocuous question. The unknown parameter that surveyors want to estimate is $p = P(Y|Q_S)$, the probability that someone answers yes given that they were asked the sensitive question. We assume that the innocuous question is (i) easy to answer and (ii) has a known probability of yes and no, in this case 50% each.

Consider the unconditional probability $P(Y)$. By the law of total probability,

$$\begin{aligned} P(Y) &= P(Y|Q_S)P(Q_S) + P(Y|Q_I)P(Q_I) \\ &= p\left(\frac{1}{2}\right) + \left(\frac{1}{2}\right)\left(\frac{1}{2}\right) \\ &= \frac{p}{2} + \frac{1}{4}. \end{aligned}$$

When this survey is given to n people, the final data will consist of n yes's and no's. The proportion of yes responses is a simulated estimate of the unknown $P(Y)$. And thus

$$\frac{p}{2} + \frac{1}{4} = P(Y) \approx \frac{\text{Number of yes's in the sample}}{n}.$$

Solving for p gives

$$p \approx 2\left(\frac{\text{Number of yes's in the sample}}{n} - \frac{1}{4}\right),$$

which is the final estimate of the parameter of interest.

Thirty students participated in such a classroom experiment where the "sensitive" question was "Are you wearing running shoes?" The instructor gave each student a coin with instructions for doing the anonymous survey. Sixteen students responded yes and 14 students responded no. The instructor estimated that $p = P(Y|Q_S) \approx 2((16/30) - 1/4) = 0.567$. As there were 30 students in the class, the instructor guessed that $30 \times (0.567) = 17$ students were wearing running shoes. In fact, there were 15. ■

Example 2.12 Finding the largest number. The following problem, known originally as the secretary problem, was first introduced by Martin Gardner in his *Mathematical Games* column in *Scientific American*. It was originally cast in terms of a manager trying to hire the most qualified secretary from a group of n applicants. Versions include finding the best lottery prize among n prizes, and the following game to find the highest number in a list of n numbers (Ferguson [1989]).

On n pieces of paper are written n distinct numbers. Let z be the largest number in the group. You will be shown the pieces of paper one at a time and you must decide after each number whether to choose that number—and the game stops—or reject it and move on to the next number. Your goal is to find z. When shown the kth number, the only information you are given is the relative rank of that number compared to the previous $k - 1$ numbers.

What should your strategy be for deciding which number to choose? And using such a strategy, what is the probability of getting the top number?

It seems at first reading that any strategy will produce a very small probability of getting the top number. Remarkably, one can do reasonably well.

Consider the following strategy: For some r between 0 and n, reject the first r numbers and then choose the first number that is better than the first r. We find the probability of choosing the top number z for a fixed r. We will then find the choice of r which does best.

Let A be the event of choosing z. Let R be the relative rank of z. Assume all arrangements of the numbers are equally likely. (Also, we make no assumptions about the size of the numbers.) By conditioning on R,

$$P(A) = \sum_{i=1}^{n} P(A|R = i)P(R = i) = \frac{1}{n} \sum_{i=1}^{n} P(A|R = i).$$

Consider $P(A|R = i)$, the probability of choosing the top number z given that it is in position i.

Suppose z is at position i, where $i \leq r$. Then z will be rejected, and you lose the game.

Suppose z is at position i, where $i > r$. Then z will be chosen if and only if the largest of the first $i - 1$ numbers is among the first r numbers. (Otherwise, the largest of the first $i - 1$ numbers will be chosen.) The largest of the first $i - 1$ numbers can be in one of $i - 1$ equally likely positions. The probability that it is among the first r numbers is thus $r/(i - 1)$. This gives

$$P(A) = \frac{1}{n} \sum_{i=1}^{n} P(A|R = i) = \frac{1}{n} \sum_{i=r+1}^{n} \frac{r}{i - 1} = \frac{r}{n} \sum_{i=r+1}^{n} \frac{1}{i - 1} = \frac{r}{n} \sum_{i=r}^{n-1} \frac{1}{i}.$$

The harmonic series $\sum_{i=1}^{n} 1/i$ diverges as $n \to \infty$. For large n, the sum of the series is approximately equal to $\log n$. This gives

$$P(A) = \frac{r}{n} \sum_{i=r}^{n-1} \frac{1}{i} \approx \frac{r}{n}(\log n - \log r) = \frac{r}{n} \log \left(\frac{n}{r}\right).$$

To find the r that does best, maximize the function $f(x) = (r/x)\log(x/r)$. Taking the derivative with respect to x and setting it equal to 0 finds that the maximum is achieved at $x = n/e$. For $r = n/e$, the probability that you will choose z is

$$P(A) \approx \frac{n/e}{n} \log \left(\frac{n}{n/e}\right) = \frac{1}{e} = 0.368.$$

In the following simulation, we choose from a list of $n = 100$ numbers, rejecting the first $n/e \approx 37$. The simulation is repeated 10,000 times.

R: FINDING THE LARGEST NUMBER

```
# TopNumber.R
# Numbers are 1, ..., n
> ntrials <- 10000
> n <- 100
> r <- round(n/exp(1))    # r = n/e = 37
> simlist <- numeric(ntrials)
> for (j in 1:ntrials) {
   numbers <- sample(1:n, n)
   best <- which(numbers==n) # position of top number
   prob <- 0
   firstmax <- max(numbers[1:r]) # largest of first r
   for (i in (r+1):n)  { # look after r-th number
      if (numbers[i] > firstmax)
         {if (numbers[i] == n) prob <- 1
        break}
    else {prob<-0}  }
     simlist[j] <- prob  }
 >  mean(simlist)
[1] 0.3749
```

■

Example 2.13 Random permutations. There are many settings where one wants to generate a uniformly random permutation. Random permutations are used in many computer algorithms. A common application is shuffling a deck of cards, which can be considered a permutation of $\{1, \ldots, 52\}$.

The following extremely fast method of generating a uniformly random permutation is known as the *Knuth shuffle*, named after the computer scientist Donald Knuth. Start with the list $(1, 2, \ldots, n)$. Move down the list from the first to the $(n-1)$st position. At each position i, swap the element in that position with a randomly chosen element from positions i to n. After $n-1$ such swaps, the resulting list will have the desired distribution.

We show that a permutation produced by the Knuth shuffle has the desired probability distribution. Let $(R_1, R_2, \ldots, R_n)$ denote the final output of the Knuth algorithm. Suppose $(r_1, r_2, \ldots, r_n)$ is a permutation of $\{1, 2, \ldots, n\}$. We need to show that $P(R_1 = r_1, R_2 = r_2, \ldots, R_n = r_n) = 1/n!$.

Using the general formula for the intersection of n events,

$$\begin{aligned} &P(R_1 = r_1, R_2 = r_2, \ldots, R_n = r_n) \\ &\quad = P(R_1 = r_1)P(R_2 = r_2|R_1 = r_1)\cdots P(R_n = r_n|R_1 = r_1, \ldots, R_{n-1} = r_{n-1}). \end{aligned}$$

We have that $P(R_1 = r_1) = 1/n$, because R_1 can take any of n values, all of which are equally likely. Observe that $P(R_2 = r_2|R_1 = r_1) = 1/(n-1)$, because if $R_1 = r_1$, then R_2 can take any value except r_1, all of which are equally likely. Similarly, for each $i = 2, \ldots, n-1$,

$$P(R_i = r_i|R_1 = r_1, \ldots, R_{i-1} = r_{i-1}) = \frac{1}{n-(i-1)}.$$

Finally, $P(R_n = r_n|R_1 = r_1, \ldots, R_{n-1} = r_{n-1}) = 1$, because if $n-1$ values have been assigned to the first $n-1$ positions of the list, the last remaining value must be assigned to the last position of the list. We have that

$$P(R_1 = r_1, R_2 = r_2, \ldots, R_n = r_n) = \prod_{i=1}^{n-1} \frac{1}{n-(i-1)} = \frac{1}{n!},$$

giving the result. ■

R: SIMULATING RANDOM PERMUTATIONS

The following code implements the Knuth shuffle (**KnuthShuffle.R**) to generate a uniformly random permutation. A permutation of size $n = 12$ is output.

```
> n <- 12
> perm <- 1:n
> for (i in 1:(n-1)) {
  x <- sample(i:n,1)
  old <- perm[i]
  perm[i] <- perm[x]
```

```
  perm[x] <- old }
> perm
[1] 10  3  5 11  4  9  2  7  8  1 12  6
```

2.5 BAYES FORMULA AND INVERTING A CONDITIONAL PROBABILITY

It should be clear from many previous examples that in general $P(A|B) \neq P(B|A)$. The probability that someone uses hard drugs given that they smoke marijuana (fairly low) is not equal to the probability that they smoke marijuana given that they use hard drugs (fairly high—no pun intended). When conditional probabilities arise in real-world problems, they can be confusing and subject to misinterpretation. Data may often be given in the form $P(A|B)$, but what is really desired is the "inverse probability" $P(B|A)$.

Bayes formula, also known as Bayes theorem, is a simple but remarkably powerful result for tackling such conditional probability problems.

BAYES FORMULA

For events A and B,

$$P(B|A) = \frac{P(A|B)P(B)}{P(A|B)P(B) + P(A|B^c)P(B^c)}.$$

The result is a consequence of two applications of the basic conditional probability formula and the law of total probability, as follows

$$\begin{aligned} P(B|A) &= \frac{P(BA)}{P(A)} = \frac{P(AB)}{P(A)} \\ &= \frac{P(AB)}{P(AB) + P(AB^c)} \\ &= \frac{P(A|B)P(B)}{P(A|B)P(B) + P(A|B^c)P(B^c)}. \end{aligned}$$

Here is a more general form of the formula: Given event A and a sequence of events $B_1, \ldots, B_k$ that partition the sample space, then for each $j = 1, \ldots, k$,

$$P(B_j|A) = \frac{P(A|B_j)P(B_j)}{\sum_{i=1}^{k} P(A|B_i)P(B_i)}.$$

Example 2.14 Diagnostic tests. Diagnostic tests are commonly used to determine the likelihood of disease. Results are never certain due to the possibility of false positives and false negatives. Confusion about conditional probability can lead to erroneous conclusions about the efficacy of a particular test.

Suppose a rare disease affects 1% of the population. A hypothetical blood test to detect the disease seems to be relatively accurate. On the one hand, the test has a 99% *sensitivity*, which means that if someone has the disease the chance that the test result is "positive" is 0.99. This also means that there is a 1% chance of error, called the false-negative rate. On the other hand, the test has a 90% *specificity*, which means that if someone does not have the disease the test will be "negative" 9 times out of 10. That is, there is a 10% false-positive rate.

The terms "sensitivity" and "specificity" are used by epidemiologists, public health workers who study the distribution patterns of disease and health events. A major tool in their arsenal is probability.

Suppose a random person gets tested, and the test comes back positive. What is the probability that the person actually has the disease?

Before proceeding, you might want to test your intuition and guess the answer without doing any computations. Is the probability of having the disease close to 10, 50, or 90%?

Many people, even experienced doctors, when asked this question assume that the test is fairly accurate and give a high estimate for the probability of disease. Let us solve the problem and find out what the probability is.

Let D be the event that a person has the disease. Let S be the event that the test comes back positive. The problem is asking for $P(D|S)$.

The probabilities provided, however, are of the form $P(S|D)$ and $P(S|D^c)$ The 99% sensitivity rate means $P(S|D) = 0.99$. And the false-negative rate gives $P(S^c|D) = 0.01$. The 90% specificity rate means that $P(S^c|D^c) = 0.90$. And the false-positive rate gives $P(S|D^c) = 0.10$. We are also told that $P(D) = 0.01$.

The information provided in the problem is probabilities that are conditional on having or not having the disease. But the problem is asking for a conditional probability given the outcome of the test. In order to solve the problem, we need to *invert* the conditional probability $P(D|S)$ to use the available information. By Bayes formula,

$$\begin{aligned} P(D|S) &= \frac{P(S|D)P(D)}{P(S|D)P(D) + P(S|D^c)P(D^c)} \\ &= \frac{0.99(0.01)}{0.99(0.01) + 0.10(0.99)} = 0.091. \end{aligned}$$

The chance of actually having the disease after testing positive is less than 10%!

While the final result might be perplexing, even paradoxical, the key to understanding it is the very low 1% probability of having the disease. Most people do not have the disease. Even though the diagnostic test has a low false-positive rate, the low rate applied to a large population of people who do not have the disease results in a lot of people with false positives due to the low disease rate.

Imagine a hypothetical town of 10,000 individuals. About 100 people (1%) will have the disease. Assume the entire town is tested. Out of those who have the disease, about 99 people would test positive and one person would test negative. On the other hand, about 9900 people do not have the disease (99%). And if everyone takes the test, 8910 of them (90%) will test negative. But 990 (10%) will test positive.

This means that about $99 + 990 = 1089$ people test positive in total. And of those, only 99 have the disease, so the probability of having the disease given you test positive is $99/1089 = 0.091$. ■

Hypothetical 10,000. It may not surprise you to know that tree diagrams can be used to help solve problems where Bayes formula is used. You should be able to make a tree to illustrate what happens in the last example. In case you are not a fan of trees however, there is another strategy that can help provide a visual aid for such problems. In the last example, we imagined what would happen in a hypothetical town of 10,000 individuals. This could be done with another n, such as 1000, or one more convenient based on the probabilities you have, and the results displayed in a two-way table that you might be more comfortable working with. The probability information provided in the problem determines the values in each cell. Here, we demonstrate what our hypothetical 10,000 table would look like for the last example as shown in Table 2.2.

With the table, we can easily see that 1089 individuals would test positive, but only 99 actually had the disease. Thus, $P(D|S) = 99/1089$, as above.

Example 2.15 Color blindness continued. Given the color-blind rates for males and females presented at the beginning of Section 2.4, we found the probability that a random person is color-blind. Even though color blindness is fairly unusual, it is much more common among men than women. Suppose a person is color-blind (C). What is the probability they are male (M)? The problem asks for $P(M|C)$, and again, we must invert the conditional probability in order to use the given data, which is conditional on sex, not color blindness. We show two solutions to this problem, one using the hypothetical table with 100,000 individuals, and the other using Bayes formula.

The information about the United States being about 49% male and 51% female sets the marginal distribution for Sex. Then, we use the conditional probabilities

TABLE 2.2. Hypothetical 10,000 table.

Test versus Disease	D	D^c	Total
S (+)	99	990	1089
S^c (−)	1	8910	8911
Total	100	9900	10,000

TABLE 2.3. Hypothetical 1000 table for color blindness.

Colorblind versus sex	Male	Female	Total
Colorblind	3430	204	3634
Not colorblind	45,570	50,796	96,366
Total	49,000	51,000	100,000

that about 7% of men and 0.4% of women are colorblind to fill in the interior cells, as shown in Table 2.3. Note that we chose a large n here to make sure our counts were integers. Then, in order to find $P(M|C)$, we find that out of 3634 color-blind individuals, 3430 are male, so $P(M|C) = 3430/3634 = 0.944$.

By Bayes formula, we simply plug the probabilities in to find that

$$\begin{aligned} P(M|C) &= \frac{P(C|M)P(M)}{P(C|M)P(M) + P(C|F)P(F)} \\ &= \frac{(0.07)(0.49)}{(0.07)(0.49) + (0.004)(0.51)} = 0.944. \end{aligned}$$

■

Example 2.16 Auto accidents continued. Based on insurance company data in Example 2.10, we found the probability that a random policyholder will have a car accident next year. The data show that adults under 25 years old are more likely to have an accident than older people. Suppose a policyholder has an accident. What is the probability they are under 25?

By Bayes formula,

$$\begin{aligned} P(G_1|A) &= \frac{P(A|G_1)P(G_1)}{P(A|G_1)P(G_1) + P(A|G_2)P(G_2) + P(A|G_3)P(G_3)} \\ &= \frac{(0.11)(0.20)}{(0.11)(0.20) + (0.03)(0.30) + (0.02)(0.50)} = 0.537. \end{aligned}$$

■

We leave it to the reader to find this probability via a tree or table for practice.

Bayesian statistics. Bayes formula is intimately connected to the field of Bayesian statistics. Statistical inference uses data to infer knowledge about an unknown parameter in a population. For instance, 100 fish are caught and measured to estimate the mean length of all the fish in a lake. The 100 fish measurements are the sampled data, and the mean length of all the fish in the lake is the unknown parameter.

In Bayesian statistics, the unknown population parameter is itself considered random and the tools of probability are used to make probabilistic estimates of the parameter. One conditions on the data in order to compute the $P(\text{Parameter}|\text{Data})$.

In other words, how likely are possible parameter values given the observed data? Let us consider an example.

Suppose your friend has three coins: one is fair, one is two-headed, and one is two-tailed. A coin is picked uniformly at random. It is tossed and comes up heads. Which coin is it?

In a Bayesian context, the type of coin is the unknown parameter. The outcome of the coin toss—heads in this case—is the observed data.

Let $C = 1, 2,$ or 3, depending upon whether the coin is fair, two-headed, or two-tailed, respectively. Let H denote heads. For $c = 1, 2, 3$, Bayes formula gives

$$P(\text{Parameter}|\text{Data}) = P(C = c|\text{H}) = \frac{P(\text{H}|C = c)P(C = c)}{P(\text{H})} = \frac{P(\text{H}|C = c)}{3P(\text{H})}.$$

By the law of total probability,

$$\begin{aligned} P(\text{H}) &= P(\text{H}|C = 1)P(C = 1) + P(\text{H}|C = 2)P(C = 2) + P(\text{H}|C = 3)P(C = 3) \\ &= \left(\frac{1}{2}\right)\left(\frac{1}{3}\right) + (1)\left(\frac{1}{3}\right) + (0)\left(\frac{1}{3}\right) = \frac{1}{2}. \end{aligned}$$

This gives

$$P(C = c|\text{H}) = 2P(\text{H}|C = c)/3 = \begin{cases} 1/3, & \text{if the coin is fair } (c = 1), \\ 2/3, & \text{if the coin is two-headed } (c = 2), \\ 0, & \text{if the coin is two-tailed } (c = 3). \end{cases}$$

In Bayesian statistics, this probability distribution is called the *posterior distribution* of the parameter (coin) given the data. A "best guess" of your friend's coin is that it is two-headed. It is twice as likely to be two-headed than it is to be fair. See the simulation in **Bayes.R**.

Example 2.17 Bertrand's box paradox. The French mathematician Joseph Louis François Bertrand posed the following problem in 1889. There are three boxes. One box contains two gold coins; one box contains two silver coins; and one box contains one gold and one silver coin. A box is picked uniformly at random. A coin is picked from the box and it is gold. What is the probability that the other coin in the box is also gold?

The correct answer is 2/3. Many people feel the answer should be 1/2, according to the following logic: The gold coin must have come from one of two boxes that are equally likely, either the gold–gold box or the gold–silver box. Thus, the gold–gold box is chosen half the time.

The fallacy is that once we know the coin is gold, the two boxes are not equally likely. There are *three* gold coins. Two of them come from the gold–gold box, and one from the gold–silver box. If the second coin is gold, it must have come from the gold–gold box and the resulting probability is two out of three.

Here is a conditional probability analysis. Let G_1 and G_2 denote that the first coin and second coin chosen are gold, respectively. Then, $P(G_2|G_1) = P(G_2G_1)/P(G_1)$. The numerator is equal to the probability of picking the gold–gold box, which is 1/3. By conditioning on which box was chosen,

$$\begin{aligned} P(G_1) &= \frac{1}{3}(P(G_1|\text{gold–gold}) + P(G_1|\text{silver–silver}) + P(G_1|\text{gold–silver})) \\ &= \frac{1}{3}\left(1 + 0 + \frac{1}{2}\right) = \frac{1}{2}. \end{aligned}$$

The desired probability is $P(G_2|G_1) = (1/3)/(1/2) = 2/3$. ■

Example 2.18 Perhaps the most well-known probability paradox of the past two decades is the infamous Monty Hall problem. This was popularized in Marilyn vos Savant's *Ask Marilyn* column in Parade magazine in 1990 (vos Savant 2013). She wrote,

> Suppose you are on a game show, and you are given the choice of three doors: Behind one door is a car; behind the others, goats. You pick a door, say No. 1, and the host, who knows what is behind the doors, opens another door, say No. 3, which has a goat. He then says to you, "Do you want to pick door No. 2?" Is it to your advantage to switch your choice of doors?

Savant answered—correctly—that it is beneficial to switch. Without switching, the chance of picking the right door is 1/3. If you switch, the probability increases to 2/3.

One way to see that 2/3 is correct is to observe that with the switching strategy you always win if you initially pick a goat, which happens with probability 2/3. If you picked a goat to start, by revealing the other goat, the host is showing you where the car is. The only way you would lose is if you chose the car to start, which would happen with probability 1/3.

Some 10,000 people wrote to Parade magazine, including many with PhDs and even some mathematicians, insisting that Savant was wrong. However, one is easily convinced after simulating the problem, either on a computer or by "playing the game show" in class. We invite the reader to search the web for the many articles, applets, simulations, and discussion of this intriguing problem. ■

2.6 INDEPENDENCE AND DEPENDENCE

In the sections above, you were introduced to conditional probability, and learned that probabilities of events can change depending on whether other events occur. The probability that you get an *A* in your math class is probably dependent on how much you study. But it probably is not dependent on the color of your roommate's hair. Intuitively, your grade and your roommate's hair color are independent events.

On the other hand, you are more likely to get an A in math class if you study hard. Most likely,

$$P(A \text{ in math class}|\text{Roommate is a red head}) = P(A \text{ in math class})$$

while

$$P(A \text{ in math class}|\text{Study hard}) > P(A \text{ in math class}).$$

This suggests the definition of independent events.

INDEPENDENT EVENTS

Events A and B are *independent* if

$$P(A|B) = P(A). \tag{2.11}$$

Events that are not independent are said to be *dependent*.

Example 2.19 A card is drawn from a standard deck. Let A be the event that it is a spade. Let B be the event that it is an ace. Then

$$P(A|B) = \frac{P(AB)}{P(B)} = \frac{P(\text{Ace of Spades})}{P(\text{Ace})} = \frac{1/52}{1/13} = \frac{1}{4} = P(A).$$

The two events are independent. ■

This example illustrates that independence is *not* the same as mutually exclusive. These terms are sometimes confused. The events A and B are not mutually exclusive since $AB = \{\text{Ace of Spades}\} \neq \emptyset$. One can tell if two events are mutually exclusive by looking at the Venn diagram or examining their included outcomes. Independence is more subtle. A Venn diagram alone will not identify independence. Knowledge of probabilities is required. In particular, mutually exclusive events are dependent because if one occurs, the other cannot occur.

If A and B are independent, then B and A are independent. And thus $P(B|A) = P(B)$. This follows from the defining formula Equation 2.11 as

$$P(B|A) = \frac{P(BA)}{P(A)} = \frac{P(AB)}{P(A)} = \frac{P(A|B)P(B)}{P(A)} = \frac{P(A)P(B)}{P(A)} = P(B).$$

If A and B are independent events, then rearranging the conditional probability formula Equation 2.1 gives

$$P(AB) = P(A|B)P(B) = P(A)P(B).$$

The equation

$$P(AB) = P(A)P(B) \tag{2.12}$$

is sometimes used as the primary definition of independent events as it is referred to as the (simple) *multiplication rule for independent events.*

The advantage of the definition of independent events and Equation 2.11 is that it is intuitive—events are independent if knowledge of whether or not one event occurs does not affect the probability of the other. On the other hand, Equation 2.12 highlights an important computational advantage of working with independent events—the probability that two independent events both occur is the *product* of their individual probabilities.

From a practical modeling perspective, we often decide *a priori* that events are independent based on assumptions that seem plausible in a real-world context. For instance, successive coin flips are modeled as independent events. The model may be useful and a reasonable approximation of reality, but no model is 100% true.

COIN TOSSING IN THE REAL WORLD

In a fascinating investigation of coin tossing, Diaconis [2007] tries to analyze the natural process of flipping a coin that is caught in the hand. They use high-speed slow motion cameras to record data of actual coin tosses. Their paper, which contains a lot of physics, shows that vigorously flipped coins are slightly biased to come up the same way they started. They show that

$$P(\text{Heads}|\text{Start with Heads}) = 0.508 \neq 0.50.$$

■ **Example 2.20** What is the probability of getting "snake-eyes"—two ones—when rolling two dice?

In Chapter 1, we enumerated the sample space for two dice rolls. By the multiplication principle, there are $6 \times 6 = 36$ outcomes. Assuming each outcome is equally likely, $P(\text{Snake-eyes}) = 1/36$.

Now an alternate derivation can be given using independence. Let A_1 and A_2 denote getting a one on the first and second rolls, respectively. Then

$$P(\text{Snake-eyes}) = P(A_1A_2) = P(A_1)P(A_2) = \left(\frac{1}{6}\right)\left(\frac{1}{6}\right) = \frac{1}{36}. \qquad ■$$

Independence of two events means that knowledge of whether or not one event occurs does not affect the probability of the other event occurring. Thus, if A and B are independent events, then intuitively the pairs (A, B^c), (A^c, B), and (A^c, B^c) are also independent events.

To show independence of A and B^c, write $A = AB \cup AB^c$. Then $P(A) = P(AB) + P(AB^c)$. Hence,

$$\begin{aligned} P(AB^c) &= P(A) - P(AB) \\ &= P(A) - P(A)P(B) \\ &= P(A)[1 - P(B)] = P(A)P(B^c). \end{aligned}$$

Switching the roles of A and B shows independence of A^c and B. See Exercise 2.33 for showing that A^c and B^c are independent.

Mutual and pairwise independence. How should independence be defined for more than two events? Surely if three events A, B, C are independent, then each pair (A, B), (A, C), and (B, C) should also be independent. Generalizing the simple multiplication rule Equation 2.12, you might also guess that independence is equivalent to the identity $P(ABC) = P(A)P(B)P(C)$. This, however, is not enough. Examples can be found where this equation holds for A, B, and C, but no two of the three events are independent (see Exercise 2.36).

For three events A, B, C, independence requires the multiplication rule to hold for the collection of three events *and* for all subgroups of two events. That is, we need

$$P(ABC) = P(A)P(B)P(C),$$

$$P(AB) = P(A)P(B),\ P(AC) = P(A)P(C), \quad \text{and} \quad P(BC) = P(B)P(C),$$

to all hold.

For larger collections of events—including infinite collections—independence requires the multiplication rule to hold for *all* finite subgroups of events in the collection. This gives the general definition.

INDEPENDENCE FOR A COLLECTION OF EVENTS

A collection of events is independent if for every finite subgroup $A_1, \ldots, A_k$,

$$P(A_1, \ldots, A_k) = P(A_1), \ldots, P(A_k). \tag{2.13}$$

This definition of independence is also called *mutual independence*. That is, mutual independence is a synonym for independence.

If we restrict to the case $k = 2$ and only require pairs of events to satisfy Equation 2.13, we say that the collection is *pairwise independent*. That is, a collection of events is pairwise independent if the simple multiplication rule holds for every pair of events. Clearly, mutual independence implies pairwise independence. However, the converse is not always true, as shown in the next example.

Example 2.21 Flip two coins. Let A be the event that the first coin comes up heads; B the event that the second comes up heads; and C the event that both coins come up the same, either heads or tails. Check that $P(A) = P(B) = P(C) = 1/2$. Also,

$$P(AB) = P(A)P(B) = \frac{1}{4}$$
$$= P(AC) = P(A)P(C) = P(BC) = P(B)P(C).$$

Thus, the three events are pairwise independent. But $P(ABC) = P(\text{Two heads}) = 1/4$ and

$$P(A)P(B)P(C) = (1/2)(1/2)(1/2) = 1/8 \neq 1/4.$$

So the three events are not mutually independent. The fact that the three events are not independent can be seen without any calculation because if A and B both occur, then so does C. ■

Example 2.22 Data from the Red Cross on the distribution of blood type in the United States are given in Table 2.4.

In the case of needed blood plasma, people with blood group O (both O+ and O−) can donate plasma to anyone. Suppose three people are selected at random. What is the probability they are all blood group O?

Let O_1, O_2, O_3 be the events that the first, second, and third persons selected are from blood group O, respectively. By independence,

$$P(O_1O_2O_3) = P(O_1)P(O_2)P(O_3) = P(O_1)^3$$
$$= (0.374 + 0.066)^3 = (0.44)^3 = 0.085.$$ ■

TABLE 2.4. Distribution of blood type in the United States.

O+	A+	B+	AB+	O−	A−	B−	AB−
0.374	0.357	0.085	0.034	0.066	0.063	0.015	0.006

How might independence be violated in this last example? If we chose three people from the same family, or the same nationality or ethnicity, this would violate independence, as people from such groups might be more likely to have similar blood types. Note that without the property of independence, it would not be possible to answer the probability question without additional information.

Sampling with and without replacement. Independence is often associated with sampling with replacement. For instance, suppose a bowl contains 10 balls of different colors, including red and green. Pick two balls. Let R_1 be the event that the first ball is red. Let G_2 be the event that the second ball is green. If we sample with replacement, then $P(G_2|R_1) = 1/10 = P(G_2)$, and the events are independent. After the first ball is picked, it is returned to the bowl and the second selection is made as if nothing changed.

On the other hand, sampling without replacement gives $P(G_2|R_1) = 1/9$, as once the first red ball is picked there are nine balls remaining. Now consider $P(G_2)$. Not knowing anything about the first ball, the second ball is equally likely to be any of the 10 colors. Thus, $P(G_2) = 1/10$, and the events are not independent.

Typically, when sampling with replacement, successive outcomes are independent events. When sampling without replacement, they are not independent. However, when the population size is very big (when the number of balls in the bowl is large), the actual numerical probabilities resulting from the two sampling schemes are practically the same. As a mind stretch, you can think of sampling without replacement from a bowl of size n, and then let $n \to \infty$. Sampling without replacement from an "infinite bowl" gives sampling with replacement!

In statistical surveying, while many practical sampling schemes from large populations are done without replacement, the analysis is often done with replacement to exploit the computational advantages of working with independence.

Example 2.23 Coincidences and the birthday problem. It is the first week on campus. Six new students are sitting together in the cafeteria. They start asking about each other. What dorm floor are you on? (There are 30 possibilities.) What frosh seminar are you in (with 40 to choose from)? What is your Zodiac sign (there are 12)?

Remarkably, two students have matching replies to all three questions. What a coincidence! But is it really that remarkable?

Generalizing the birthday problem (see Section 2.3.1), the probability that none of the six students are on the same dorm floor is $(29 \cdot 28 \cdot 27 \cdot 26 \cdot 25)/30^5$. Similar calculations are done for frosh seminar and Zodiac sign. If we assume that the three categories dorm floor, seminar, and Zodiac sign are independent, then the probability that there is no match for any category among the six students at the table is the product of the probabilities of no match in each category, which is

$$\left(\frac{29}{30} \cdots \frac{25}{30}\right)\left(\frac{39}{40} \cdots \frac{35}{40}\right)\left(\frac{11}{12} \cdots \frac{7}{12}\right) = 0.0882.$$

Therefore, the probability of a least one match among six people is

$$P(\text{Match}) = 1 - 0.0882 = 0.9118.$$

Not such a coincidence after all. ■

Example 2.24 *A* before *B*. The following scenario is very general. A random experiment is performed repeatedly with independent trials. Events A and B are mutually exclusive, but not complements. What is the probability that A occurs before B?

We assumed that $B \neq A^c$. Why? The solution is not very interesting if $B = A^c$, as then either A or B occurs and the solution is just $P(A)$.

For ease of notation, let $p = P(A)$, $q = P(B)$, with $p + q < 1$. We give two solutions to this problem—one analytical, the other probabilistic.

Solution 1: Let E be the event that A occurs before B. For E to happen, either A occurs right away or neither A nor B occurs for one or more trials and then A occurs. That is, for some $k \geq 0$, neither A nor B occur for k trials, and then A occurs on trial $k + 1$. For each k, let E_k denote the event that (i) A first occurs on trial $k + 1$ and (ii) neither A nor B occur on the first k trials. Observe that the E_k's are mutually exclusive and $E = \bigcup_{k=0}^{\infty} E_k$.

The probability that on a particular trial neither A nor B occur is $1 - (p + q) = 1 - p - q$. By independence, $P(E_k) = (1 - p - q)^k p$. This gives,

$$\begin{aligned} P(E) &= P\left(\bigcup_{k=0}^{\infty} E_k\right) = \sum_{k=0}^{\infty} P(E_k) \\ &= \sum_{k=0}^{\infty} (1 - p - q)^k p \\ &= p\left(\frac{1}{1 - (1 - p - q)}\right) = \frac{p}{p + q}. \end{aligned}$$

The geometric series converges as $0 < p + q < 1$ and thus $0 < 1 - p - q < 1$.

Solution 2: Condition on the first trial. There are three possibilities:

(i) If A occurs on the first trial, then E occurs.

(ii) If B occurs on the first trial, then E does not occur.

(iii) If neither A nor B occur on the first trial, then we "start over again" to determine whether or not A occurs first. It is as if the first trial did not happen, and the problem begins anew at the second trial. This is a consequence of independence. The event that neither A nor B occur on the first trial is independent of the event that A occurs before B. Letting C be the event that neither A nor B occurs on the first trial, this gives that $P(E|C) = P(E)$.

By the law of total probability,

$$\begin{aligned} P(E) &= P(E|A)P(A) + P(E|B)P(B) + P(E|C)P(C) \\ &= 1(p) + 0(q) + P(E)(1 - p - q) \\ &= p + P(E)(1 - p - q). \end{aligned}$$

The equation contains $P(E)$ on both sides. Solving for $P(E)$ gives $P(E) = p/(p + q)$. We restate this useful result.

A BEFORE B

In repeated independent trials, if A and B are mutually exclusive events, the probability that A occurs before B is

$$\frac{P(A)}{P(A)+P(B)} \tag{2.14}$$

For example, when repeatedly rolling pairs of dice, what is the probability that a sum of 9 appears before a sum of 7?

The probability of getting nine is $P(\{(3,6),(4,5),(5,4),(6,3)\}) = 4/36$. The probability of getting 7 is 6/36. The desired probability is

$$P(9 \text{ before } 7) = \frac{4/36}{4/36+6/36} = \frac{2}{5}.$$

■

We make use of these ideas in the analysis of a popular casino game.

Example 2.25 Craps. The dice game craps is fast-paced, exciting, and typically offers the best odds at the casino. The player rolls two dice. If you get a 7 or 11, you win immediately. If you get a 2, 3, or 12, you lose. For any other outcome (4, 5, 6, 8, 9, 10), the number rolled is your *point*. You now roll again and keep rolling until you either roll your point again or roll a 7. If the 7 comes before the point, you lose. If the point comes before the 7, you win. What is the probability of winning at craps?

Let W denote winning. Conditioning on the first outcome,

$$\begin{aligned}
P(W) &= \sum_{k=2}^{12} P(W|k)P(k) \\
&= (0)[P(2)+P(3)+P(12)] + (1)[P(7)+P(11)] \\
&\quad + \sum_{k\in\{4,5,6,8,9,10\}} P(W|k)P(k) \\
&= P(7)+P(11) + \sum_{k\in\{4,5,6,8,9,10\}} P(k \text{ before } 7)P(k),
\end{aligned}$$

where $\{k \text{ before } 7\}$ denotes the event that the point k comes up before 7 in repeated rolls. The probability of winning at craps is thus

$$P(W) = P(7)+P(11) + \sum_{k\in\{4,5,6,8,9,10\}} P(k \text{ before } 7)P(k)$$

$$
\begin{aligned}
&= \frac{6}{36} + \frac{2}{36} + \left(\frac{3/36}{3/36+6/36}\right)\frac{3}{36} + \left(\frac{4/36}{4/36+6/36}\right)\frac{4}{36} \\
&\quad + \left(\frac{5/36}{5/36+6/36}\right)\frac{5}{36} + \left(\frac{5/36}{5/36+6/36}\right)\frac{5}{36} \\
&\quad + \left(\frac{4/36}{4/36+6/36}\right)\frac{4}{36} + \left(\frac{3/36}{3/36+6/36}\right)\frac{3}{36} \\
&= 0.4929.
\end{aligned}
$$

See the script **Craps.R** to simulate the game. ■

2.7 PRODUCT SPACES*

In this section, we treat some technical issues associated with modeling independent events.

When we repeat a random experiment several times, or combine the results of two or more different random experiments, we are in effect creating a new sample space.

For instance, in rolling one die, the sample space is $\{1, 2, 3, 4, 5, 6\}$. But in rolling two dice, a larger sample space is used of all $6 \times 6 = 36$ pairs of dice rolls

$$\{(1, 1), (1, 2), \ldots, (6, 5), (6, 6)\}.$$

If the two dice rolls are independent, then for each ordered pair (x, y) the probability of rolling an x on the first die and a y on the second die $P((x, y))$ is equal to $P(x)P(y)$, the product of the individual probabilities. The larger sample space makes it possible to consider probabilities involving two dice. To consider three dice rolls, the underlying sample space would be lists of length three

$$\{(1, 1, 1), (1, 1, 2), \ldots, (6, 6, 5), (6, 6, 6)\}$$

and so on.

Consider two random experiments with respective sample spaces Ω and Ω'. (The sample spaces may be the same, as in the dice example, or different.) The *product space* $\Omega \times \Omega'$ is the set of all ordered pairs (ω, ω') such that $\omega \in \Omega$ and $\omega' \in \Omega'$.

To define a probability function on that product space, let

$$P((\omega, \omega')) = P(\omega)P(\omega')$$

for all $\omega \in \Omega$ and $\omega' \in \Omega'$. This gives a valid probability function because

$$\sum_{(\omega,\omega') \in \Omega \times \Omega'} P((\omega, \omega')) = \sum_{\omega \in \Omega} \sum_{\omega' \in \Omega'} P((\omega, \omega'))$$

$$
\begin{aligned}
&= \sum_{\omega \in \Omega} \sum_{\omega' \in \Omega'} P(\omega)P(\omega') \\
&= \sum_{\omega \in \Omega} P(\omega) \sum_{\omega' \in \Omega'} P(\omega') \\
&= (1)(1) = 1.
\end{aligned}
$$

With this construction, the outcomes of the Ω random experiment are independent of the outcomes of the Ω' random experiment.

If $A \subseteq \Omega$ and $B \subseteq \Omega'$, the event $A \cap B$ is not necessarily defined on Ω or Ω', but it is defined on the product space $\Omega \times \Omega'$. And with the probability function just defined,

$$
\begin{aligned}
P(AB) &= \sum_{(\omega,\omega') \in AB} P((\omega, \omega')) \\
&= \sum_{\omega \in A} \sum_{\omega' \in B} P((\omega, \omega')) \\
&= \sum_{\omega \in A} \sum_{\omega' \in B} P(\omega)P(\omega') \\
&= \sum_{\omega \in A} P(\omega) \sum_{\omega' \in B} P(\omega') \\
&= P(A)P(B).
\end{aligned}
$$

(We use the same capital letter P for what is in effect three different probability functions—one on Ω, one on Ω', and one on $\Omega \times \Omega'$—corresponding to three different random experiments.)

Here is an example. Roll a four-sided tetrahedral die. The sample space is $\Omega = \{1, 2, 3, 4\}$ with each outcome equally likely. Draw a letter at random from a bag that contains letters a, b, and c. The sample space is $\Omega' = \{a, b, c\}$, with each outcome equally likely. To model the random experiment of both rolling the die and drawing a letter so that outcomes of the first experiment are independent of the second, consider the 12-element product space

$$
\Omega \times \Omega' = \{(1, a), (1, b), \ldots, (4, b), (4, c)\}
$$

and define

$$
P(\omega, \omega') = P(\omega)P(\omega') = \left(\frac{1}{4}\right)\left(\frac{1}{3}\right) = \frac{1}{12},
$$

for all $\omega \in \Omega$ and $\omega' \in \Omega'$.

Suppose A is the event of rolling a 1 or 2 on the die. And B is the event of drawing the letter b from the bag. Then $P(A) = 1/2$ and $P(B) = 1/3$. And $P(AB) = P(\{(1, b), (2, b)\}) = P(1, b) + P(2, b) = 2/12 = 1/6$. The events A and B are independent.

Product spaces are a natural construction for working with repeated random experiments. They extend in an obvious way to more than two sample spaces. We can even define product spaces with infinite products for modeling, say, an infinite sequence of coin flips, although there are technical issues to be resolved when working with infinite sequences and infinite products that require more advanced analytical tools than what is presented in this first course.

2.8 SUMMARY

Conditional probability is introduced. Conditional probability can often be useful for finding $P(A \text{ and } B)$ via the multiplication rule. Tree diagrams are useful devices for finding probabilities, especially when a sequence of events occur in succession. The classic birthday problem is discussed. The law of total probability is a powerful tool for computing probabilities by *conditioning*. Sometimes a problem asks for $P(A|B)$, but the information we are given is of the form $P(B|A)$. This is a natural setting for Bayes formula, which can be thought of as a way to "invert" a conditional probability. The concept of independence is introduced via independent events. Finally, technical issues about product spaces are addressed in an optional section.

- **Conditional probability formula:** $P(A|B) = P(AB)/P(B)$.
- **Conditional probability as a probability function:** For fixed B, the conditional probability $P(A|B)$ as a function of its first argument *is* a probability function that satisfies the three defining properties.
- **General multiplication rule:** $P(A \text{ and } B) = P(A|B)P(B) = P(B|A)P(A)$.
- **Law of total probability:** If $B_1, \ldots, B_k$ partition Ω, then

  $$P(A) = P(A|B_1)P(B_1) + \cdots + P(A|B_k)P(B_k).$$

  For $k = 2$, this gives

  $$P(A) = P(A|B)P(B) + P(A|B^c)P(B^c).$$

  We say that we are *conditioning* on B.
- **Bayes formula:**

  $$P(B|A) = \frac{P(A|B)P(B)}{P(A|B)P(B) + P(A|B^c)P(B^c)}.$$

- **Problem-solving strategies**
  1. **Tree diagrams:** Tree diagrams are intuitive and useful tools for finding probabilities of events that can be ordered sequentially.

2. **Conditioning:** Given an event A for which we want to find $P(A)$, introducing disjoint events $B_1, \dots, B_k$ and applying the law of total probability, whereby the conditional probabilities $P(A|B_i)$ are easier and more natural to solve than $P(A)$.
3. **Hypothetical tables:** Hypothetical tables can be constructed for many scenarios involving probabilities and can be used in many of the same situations as tree diagrams.

- **Independent events:** Events A and B are independent if $P(A|B) = P(A)$. Equivalently, $P(AB) = P(A)P(B)$.
- **Mutual independence:** For general collections of events, independence means that for every finite subcollection $A_1, \dots, A_k$,

$$P(A_1, \dots, A_k) = P(A_1), \dots, P(A_k).$$

  Mutual independence is a synonym for independence.
- **Pairwise independence:** A collection of events is pairwise independent if $P(A_iA_j) = P(A_i)P(A_j)$ for all pairs of events.
- **Independence notes:** (i) A collection of independent events is pairwise independent. But events that are pairwise independent are not necessarily independent. (ii) Do not confuse independence with mutually exclusive.
- **Product spaces:** Product spaces are useful when modeling two or more independent events.

EXERCISES

Basics of Conditional Probability

2.1 Your friend missed probability class today. Explain to your friend, in simple language, the meaning of *conditioning*.

2.2 Your friend missed probability class today. Explain to your friend, in simple language, what the *Law of Total Probability* says.

2.3 A survey of residential college students at a large university revealed the following breakdown of whether their major is in STEM or not and whether the students were living in a single room or not.

	STEM	Not STEM
Single room	75	62
Not single room	113	207

Find the probability that a randomly selected student from the survey:

(a) Has a major in STEM.

(b) Has a major in STEM and a single room.

(c) Has a major in STEM given that they have a single room.

(d) Does not have a single room given they have a major not in STEM.

2.4 Suppose $P(A) = P(B) = 0.3$ and $P(A|B) = 0.5$. Find $P(A \cup B)$.

2.5 Suppose $P(A) = P(B) = p_1$ and $P(A \cup B) = p_2$. Find $P(A|B)$.

2.6 A survey of households in a city explored access to fresh fruit (defined as access within 15 minutes walking distance or not) and whether the home had fresh fruit in the home that day. The following table shows the results of the survey.

	Access	**No access**
Fresh fruit	54	11
No fresh fruit	35	16

Find the probability that a randomly selected household from the survey:

(a) Had access and fresh fruit in the home.

(b) Had no access given that they had fresh fruit in the home.

(c) Had no fresh fruit in the home given that they had access to fresh fruit.

2.7 A total of 108 students filled out a survey for a psychology class project. A total of 36 students indicated they were athletes. Of those students, 21 said they preferred to work out in the morning as opposed to the afternoon. For the nonathletes, 25 said they preferred to work out in the morning. Find the following probabilities for a randomly selected student who took the survey:

(a) P(Athlete given prefer morning workout).

(b) P(Prefer morning workout given nonathlete).

(c) P(Nonathlete given prefer nonmorning workout).

2.8 **A paradox?** Jayden flips three pennies.

(a) Aimee peeks and sees that the first coin lands heads. What is the probability of getting all heads?

(b) Robbie peeks and sees that one of the coins lands heads. What is the probability of getting all heads? (The two probabilities are different.)

2.9 Find a simple expression for $P(A|B)$ under the following conditions:

(a) A and B are disjoint.

(b) $A = B$.

(c) A implies B.

(d) B implies A.

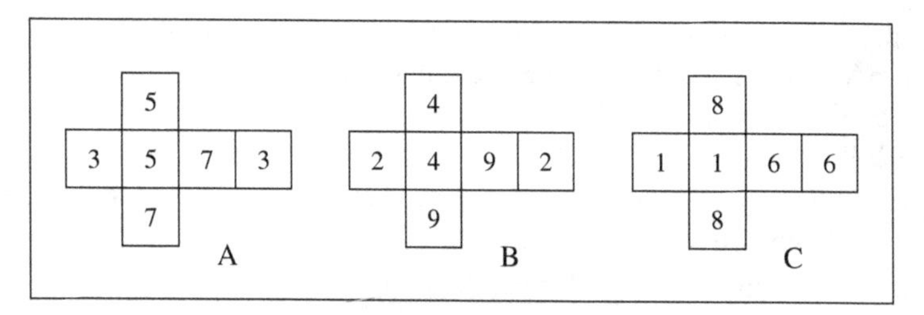

FIGURE 2.7: Nontransitive dice.

2.10 **Nontransitive dice:** Consider three nonstandard dice. Instead of the numbers 1 through 6, die A has two 3s, two 5s, and two 7s; die B has two 2s, two 4s, and two 9s; and die C has two 1s, two 6s, and two 8s, as in Figure 2.7. Suppose dice A and B are rolled. (i) Show that A is more likely to get the higher number. That is, $P(A > B) > 0.50$, where $\{A > B\}$ denotes the event that A beats B. Hint: Condition on the outcome of die A.

(ii) Now show that if B and C are rolled, B is more likely to get the higher number. And, remarkably (take our word for it), if C and A are rolled, C is more likely to get the higher number.

Many relationships in life are *transitive*. For instance, if Amy is taller than Blane and Blane is taller than Chloe, then Amy is taller than Chloe. But these dice show that the relation "more likely to roll a higher number" is not transitive.

The dice are the basis of a magic trick. You pick any die. Then I can always pick a die that is more likely to beat yours. If you pick A, I pick C. If you pick B, I pick A. And if you pick C, I pick B.

2.11 True or False. Either show the statement to be true for any event A and B or exhibit a counterexample.

(a) $P(A|B) + P(A|B^c) = 1$.

(b) $P(A|B) + P(A^c|B) = 1$.

2.12 A bag of 15 Scrabble tiles contains three each of the letters A, C, E, H, and N. If you pick six letters one at a time, what is the chance that you spell C-H-A-N-C-E?

2.13 In the game of poker, a *flush* is five cards of the same suit. Use conditional probability to find the probability of being dealt a flush.

2.14 Box A contains one white ball and two red balls. Box B contains one white ball and three red balls. A ball is picked at random from box A and put into box B. A ball is then picked at random from box B. Draw a tree diagram for this problem and use it to find the probability that the final ball picked is white.

2.15 Eduardo is taking a test. There are two questions he is stumped on and he decides to guess. Let A be the event that he gets the first question right; let B be the event he gets the second question right (adapted from Blom et al. [1991]).

(a) Obtain an expression for p_1, the probability that he gets both questions right conditional on getting the first question right.

(b) Obtain an expression for p_2, the probability that he gets both questions right conditional on getting either of the two questions right (A or B).

(c) Show that $p_2 \leq p_1$. This may seem paradoxical. Knowledge that A or B has taken place makes the conditional probability that A and B happens smaller than when we know that A has happened. Can you untangle the paradox?

2.16 Suppose $P(A) = 1/2$, $P(B^c|AC) = 1/3$, and $P(C|A) = 1/4$. Find $P(ABC)$.

2.17 Prove the addition rule for conditional probabilities. That is, show that for events A, B, and C,

$$P(A \cup B|C) = P(A|C) + P(B|C) - P(AB|C).$$

Conditioning, Law of Total Probability, and Bayes Formula

2.18 The planet Mars revolves around the sun in 687 days. Answer Von Mises' birthday question for Martians. That is, how many Martians must be in a room before the probability that some share a birthday becomes at least 50%?

2.19 Jimi has 5000 songs on his iPod shuffle, which picks songs uniformly at random. Jimi plans to listen to 100 songs today. What is the chance he will hear at least one song more than once?

2.20 A standard deck of cards has one card missing. A card is then picked from the deck. What is the chance that it is a heart? Solve this problem in two ways:

(a) Condition on the missing card.

(b) Appeal to symmetry. That is, make a *qualitative* argument for why the answer should not depend on the heart suit.

2.21 Maya has two bags of candy. The first bag contains two packs of M&Ms and three packs of Gummi Bears. The second bag contains four packs of M&Ms and two packs of Gummi Bears. Maya chooses a bag uniformly at random and then picks a pack of candy. What is the probability that the pack chosen is Gummi Bears? Solve (i) by using a tree diagram and (ii) by another method.

2.22 In a roll of two tetrahedral dice, each labeled one to four, let X be the sum of the dice. Let $A = \{X \text{ is prime}\}$ and

$$B_1 = \{X = 2\}, B_2 = \{3 \leq X \leq 5\}, B_3 = \{6 \leq X \leq 7\}, \text{and}$$
$$B_4 = \{X = 8\}.$$

Observe that the B_is partition the sample space.

(a) Draw a diagram as in Figure 2.6. Label all events in the diagram.

(b) Illustrate the law of total probability by writing out formula 2.9 and finding the probabilities for each term in the equation.

2.23 Give a formula for $P(A|B^c)$ in terms of $P(A)$, $P(B)$, and $P(AB)$ only.

2.24 **Lewis Carroll's pillow problem #5.** Lewis Carroll, author of *Alice's Adventures in Wonderland*, is the pen name of Charles Lutwidge Dodgson, who was an Oxford mathematician and logician. Lewis Carroll's *Pillow Problems* [1958], is a collection of 72 challenging, and sometimes amusing, mathematical problems, several of which involve probability. Here is Problem #5.

A bag contains one counter, known to be either white or black. A white counter is put in, the bag shaken, and a counter drawn out, which proves to be white. What is now the chance of drawing a white counter?

2.25 **Pillow problem #72.** Here is Lewis Carroll's last pillow problem [1958].

A bag contains two counters, as to which nothing is known except that each is either black or white. Ascertain their colors without taking them out of the bag.

Carroll's answer is

One is black, and the other is white.

The rationale provided for his solution is as follows:

We know that, if a bag contained three counters, two being black and one white, the chance of drawing a black one would be 2/3; and that any *other* state of things would *not* give this chance.

Now the chances, that the given bag contains (a) BB, (b) BW, (c) WW, are respectively, 1/4, 1/2, 1/4. Add a black counter. Then the chances that it contains (a) BBB, (b) BWB, (c) WWB are, as before, 1/4, 1/2, 1/4.

Hence, the chance of now drawing the black one is

$$\frac{1}{4}(1) + \frac{1}{2}\left(\frac{2}{3}\right) + \frac{1}{4}\left(\frac{1}{3}\right) = \frac{2}{3}.$$

Hence, the bag now contains BBW (since any *other* state of things would *not* give this chance). Hence, before the black counter was added, it contained BW, i.e., one black counter and one white.

(a) What result or concept is being applied to compute the probability of drawing a black counter?

(b) Critique Carroll's solution.

2.26 Consider flipping coins until either two heads HH or heads then tails HT first occurs. By conditioning on the first coin toss, find the probability that HT occurs before HH.

2.27 In a certain population of youth, the probability of being a smoker is 20%. The probability that at least one parent is a smoker is 30%. And if at least one parent is a smoker, the probability of being a smoker is 35%. Find the probability of being a smoker if neither parent is a smoker.

2.28 According to the National Cancer Institute, for women aged 50, there is a 2.38% risk (probability) of being diagnosed with breast cancer. Screening mammography has a sensitivity of about 85% for women aged 50, and a 95% specificity. That is, the false-negative rate is 15% and the false-positive rate is 5%. If a woman aged 50 has a mammogram, and it comes back positive for breast cancer, what is the probability that she has the disease?

2.29 A polygraph (lie detector) is said to be 90% *reliable* in the following sense: There is a 90% chance that a person who is telling the truth will pass the polygraph test; and there is a 90% chance that a person telling a lie will fail the polygraph test.

(a) Suppose a population consists of 5% liars. A random person takes a polygraph test, which concludes that they are lying. What is the probability that they are actually lying?

(b) Consider the probability that a person is actually lying given that the polygraph says that they are. Using the definition of reliability, how reliable must the polygraph test be in order that this probability is at least 80%?

2.30 An eyewitness observes a hit-and-run accident in New York City, where 95% of the cabs are yellow and 5% are blue. The witness asserts the cab was blue. Police experts believe that eyewitnesses are 80% reliable. That is, an eyewitness will correctly identify the color of a cab 80% of the time. What is the probability that the cab actually was blue?

2.31 Your friend has three dice. One die is fair. One die has fives on all six sides. One die has fives on three sides and fours on three sides. A die is chosen at random. It comes up five. Find the probability that the chosen die is the fair one.

Independence

2.32 Your friend missed probability class today. Explain to your friend, in simple language, what the difference between disjoint and independent events is.

2.33 Suppose A and B are independent events. Show that A^c and B^c are independent events.

2.34 Suppose A, B, and C are independent events with respective probabilities $1/3$, $1/4$, and $1/5$. Find

(a) $P(ABC)$.

(b) $P(A \text{ or } B \text{ or } C)$.

(c) $P(AB|C)$.

(d) $P(B|AC)$.

(e) P(At most one of the three events occurs).

2.35 There is a 70% chance that a tree is infected with either root rot or bark disease. The chance that it does not have bark disease is 0.4. Whether or not a tree has root rot is independent of whether it has bark disease. Find the probability that a tree has root rot.

2.36 Toss two dice. Let A be the event that the first die rolls 1, 2, or 3. Let B be the event that the first die rolls 3, 4, or 5. Let C be the event that the sum of the dice is 9. Show that $P(ABC) = P(A)P(B)P(C)$, but no pair of events is independent.

2.37 **The first probability problem.** A gambler's dispute in 1654 is said to have led to the creation of mathematical probability. Two French mathematicians, Blaise Pascal and Pierre de Fermat, considered the probability that in 24 throws of a pair of dice at least one "double six" occurs. It was commonly believed by gamblers at the time that betting on double sixes in 24 throws would be a profitable bet (i.e., greater than 50% chance of occurring). But Pascal and Fermat showed otherwise. Find this probability.

2.38 A lottery will be held. From 1000 numbers, one will be chosen as the winner. Each lottery ticket is a number between 1 and 1000. How many tickets do you need to buy in order for the probability of winning to be at least 50%?

2.39 A local church is holding a bake sale which includes a cake raffle. One cake will be raffled off based on raffle ticket numbers from 1 to 30. How many tickets do you need to buy in order for the probability of winning the cake to be at least 25%?

2.40 **Coincidences.** Diaconis and Mosteller [1989]. See Section 2.3.1 on the birthday problem. Some categories (like birthdays) are equally likely to occur, with c possible values.

(a) Let k be the number of people needed so that the probability of at least one match is 95%. Show $k \approx 2.45\sqrt{c}$. (Hint: Use Equation 2.5.)

(b) Suppose there are m categories, all of which are independent and take c possible values. Let k be the number of people needed so that the probability of at least one match in any category is 95%. Show $k \approx 2.45\sqrt{c/m}$.

(c) A group of k people is comparing (i) their birthdays, (ii) the last two digits on their social security card, and (iii) the two-digit ticket number on their movie stubs. How big should k be so that there is a 50% chance of at least one match? A 95% chance?

Simulation and R

2.41 Make up your own random experiment involving conditional probability. Write an **R** script to simulate your problem and compare the simulation to your exact solution.

2.42 The **R** command

```
> sample(1:365, 23, replace = T)
```

simulates birthdays from a group of 23 people. The expression

```
> 2 %in% table(sample(1:365, 23, replace = T))
```

can be used to simulate the birthday problem. It creates a frequency table showing how many people have each birthday, and then determines if two is in that table; that is, whether two people have the same birthday. Use and suitably modify the expression for the following problems.

(a) Simulate the probability that two people have the same birthday in a room of 23 people.

(b) Estimate the number of people needed so that the probability of a match is 95%.

(c) Find the approximate probability that three people have the same birthday in a room of 50 people.

(d) Estimate the number of people needed so that the probability that three people have the same birthday is 50%.

2.43 Simulate the nontransitive dice probabilities in Exercise 2.10.

2.44 The following problem appeared in the news column "Ask Marilyn" on September 19, 2010.

> Four identical sealed envelopes are on a table. One of them contains a $100 bill. You select an envelope at random and hold it in your hand without opening it. Two of the

three remaining envelopes are then removed and set aside, unopened. You are told that they are empty. You are given the choice of keeping the envelope you chose or exchanging it for the one on the table.

What should you do? (a) Keep your envelope. (b) Switch it. (c) It does not matter.

Write a simulation to find the probability of selecting the $100 bill when you switch. Confirm the results of your simulation with an exact analysis.

2.45 Modify the **Blackjack.R** script to simulate the probability of being dealt two cards of the same suit. Compare with the exact answer.

2.46 See Example 2.13 for generating a random permutation. Implement the algorithm in **R** for shuffling a standard deck of cards. Use it to simulate the probability that in a randomly shuffled deck the top and bottom cards are the same suit.

2.47 Simulate Bertrand's box paradox (Example 2.17).

2.48 Simulate the gambler's dispute from 1654 in Exercise 2.37.

Chapter Review

Chapter review exercises are available through the text website. The URL is `www.wiley.com/go/wagaman/probability2e`.

3

INTRODUCTION TO DISCRETE RANDOM VARIABLES

It is utterly implausible that a mathematical formula should make the future known to us, and those who think it can would once have believed in witchcraft.

—Jacob Bernoulli, Ars Conjectandi

Learning Outcomes

1. Define the term "random variable."
2. Give examples of discrete random variables and independent random variables.
3. Solve problems using the discrete uniform distribution and Bernoulli sequences.
4. Distinguish between the binomial and Poisson distributions; understand their relationship and common applications.
5. (C) Evaluate probabilities in **R** for the binomial and Poisson distributions.

3.1 RANDOM VARIABLES

Often, the outcomes of a random experiment take on numerical values. For instance, we might be interested in how many heads occur in three coin tosses. Let X be the number of heads. Then X is equal to 0, 1, 2, or 3, depending on the outcome of the coin tosses. The object X is called a *random variable*. The possible *values* of X are 0, 1, 2, and 3. Each outcome has an associated value of the random variable.

Probability: With Applications and R, Second Edition. Amy S. Wagaman and Robert P. Dobrow.

Companion Website: www.wiley.com/go/wagaman/probability2e

RANDOM VARIABLE

A random variable assigns numerical values to the outcomes of a random experiment.

Random variables are enormously useful and allow us to use algebraic expressions, equalities, and inequalities when manipulating events. In many of the previous examples, we have been working with random variables without using the name, for example, the number of threes in rolls of a die, the number of votes received, the number of palindromes, the number of heads in repeated coin tosses. In particular, we have been primarily working with *discrete random variables*, where the number of possible values of the random variable is countable, most often associated with a discrete sample space for the underlying experiment.

Example 3.1 In tossing three coins, let X be the number of heads. Then the event of getting two heads can be written as $\{X = 2\}$. The probability of getting two heads is thus

$$P(X = 2) = P(\{HHT, HTH, THH\}) = \frac{3}{8}.$$

■

We write $\{X = 2\}$ for the event that the random variable takes the value 2. More generally, we write $\{X = x\}$ for the event that the random variable X takes the value x, where x is a specific number. The difference between the uppercase X (a random variable) and the lowercase x (a number) can be confusing but is extremely important to clarify.

Random variables as functions. Random variables are central objects in probability. However, they are really neither "random" nor a "variable" in the way that that word is used in algebra or calculus. A random variable is actually a *function*, a function whose domain is the sample space of the experiment.

A random variable assigns every outcome of the sample space a real number. Consider the three coins example, letting X be the number of heads in three coin tosses. Depending upon the outcome of the experiment, X takes on different values. To emphasize the dependency of X on the outcome ω, we can write $X(\omega)$, rather than just X. In particular, for the three coins example,

$$X(\omega) = \begin{cases} 0, & \text{if } \omega = TTT, \\ 1, & \text{if } \omega = HTT, THT, \text{or } TTH, \\ 2, & \text{if } \omega = HHT, HTH, \text{or } THH, \\ 3, & \text{if } \omega = HHH. \end{cases}$$

The probability of getting exactly two heads is written as $P(X = 2)$, which is shorthand for $P(\{\omega : X(\omega) = 2\})$.

You may be unfamiliar with this last notation, used for describing sets. The notation $\{\omega : \text{Property}\}$ describes the set of all ω that satisfies some property; so $\{\omega : X(\omega) = 2\}$ is the set of all ω with the property that $X(\omega) = 2$. That is, the set of all outcomes that result in exactly two heads, which is $\{HHT, HTH, THH\}$.

Similarly, the probability of getting at most one head in three coin tosses is

$$P(X \leq 1) = P(\{\omega : X(\omega) \leq 1\}) = P(\{TTT, HTT, THT, TTH\}).$$

Because of simplicity and ease of notation, authors (including us) typically use the shorthand X in writing random variables instead of the more verbose $X(\omega)$.

We can approach many problems with random variables that we once solved with direct counting.

Example 3.2 If we throw two dice, what is the probability that the sum of the dice is greater than four?

Let Y be the sum of two dice rolls. Then Y is a random variable whose possible values are $2, 3, \ldots, 12$. The event that the sum is greater than 4 can be written as $\{Y > 4\}$. The complementary event is $\{Y \leq 4\}$, with

$$\begin{aligned} P(Y \leq 4) &= P(Y = 2 \text{ or } Y = 3 \text{ or } Y = 4) = P(Y = 2) + P(Y = 3) + P(Y = 4) \\ &= P(\{(1, 1)\}) + P(\{(1, 2), (2, 1)\}) + P(\{(1, 3), (2, 2), (3, 1)\}) \\ &= \frac{1}{36} + \frac{2}{36} + \frac{3}{36} = \frac{1}{6}. \end{aligned}$$

The desired probability is $P(Y > 4) = 1 - P(Y \leq 4) = 1 - (1/6) = 5/6$. ■

Example 3.3 Recall Example 1.3. One thousand students are voting. Suppose the number of votes that Yolanda receives is equally likely to be any number from 0 to 1000. What is the probability that Yolanda beats Zach by at least 100 votes?

We approach the problem using random variables. Let Y be the number of votes for Yolanda. Let Z be the number of votes for Zach. Then the total number of votes is $Y + Z = 1000$. Thus, $Z = 1000 - Y$. The event that Yolanda beats Zach by at least 100 votes is $\{Y - Z \geq 100\} = \{Y - (1000 - Y) \geq 100\} = \{2Y \geq 1100\} = \{Y \geq 550\}$. The desired probability is

$$P(Y - Z \geq 100) = P(Y \geq 550) = 451/1001,$$

as there are 1001 possible values of the number of votes for Yolanda and 451 of them are greater than or equal to 550. ■

Discrete uniform distribution. If a random variable X takes values in a finite set, all of whose elements are equally likely, we say that X is *uniformly distributed* on that set.

UNIFORM RANDOM VARIABLE

Let $S = \{s_1, \ldots, s_k\}$ be a finite set. A random variable X is *uniformly distributed on* S if

$$P(X = s_i) = \frac{1}{k}, \quad \text{for } i = 1, \ldots, k.$$

We write $X \sim \text{Unif}(S)$. The symbol $\sim$ stands for "is distributed as."

Example 3.4 Rachel picks an integer "at random" between 1 and 50. (i) Find the probability that she picks 13. (ii) Find the probability that her number is between 10 and 20. (iii) Find the probability that her number is prime.

Before we find the probabilities desired, we outline a strategy to assist in solving this and similar problems. First, define any necessary random variables or events. Then, state what information is known about those variables or events. After that, one can proceed to solving the problem, or laying out further steps needed. In this particular case, we start by defining a random variable to work with.

Let X be Rachel's number. Then X is uniformly distributed on $\{1, \ldots, 50\}$.

(i) The probability that Rachel picks 13 is $P(X = 13) = 1/50 = 0.02$.

(ii) There are 11 numbers between 10 and 20 (including both 10 and 20). The desired probability is

$$P(10 \leq X \leq 20) = \frac{11}{50} = 0.22.$$

(iii) There are 15 prime numbers between 1 and 50. Thus,

$$P(X \text{ is prime}) = \frac{15}{50} = 0.3.$$

■

Example 3.5 Roll a pair of dice. What is the probability of getting a sum of 4? Of getting each number between 2 and 12 as the sum of the rolls?

Assuming the dice are fair, each die number is equally likely. There are six possibilities for the first roll, six possibilities for the second roll, so $6 \times 6 = 36$ possible rolls. We thus assign the probability of 1/36 to each possible roll. Let X be the sum of the two dice. Then

$$\begin{aligned} P(X = 4) &= P(\{(1,3),(3,1),(2,2)\}) \\ &= P((1,3)) + P((3,1)) + P((2,2)) = 3\left(\frac{1}{36}\right) = \frac{1}{12}. \end{aligned}$$

Consider $P(X = x)$ for $x = 2, 3, \ldots, 12$. By counting all the possible combinations, verify the probabilities in Table 3.1.

Observe that while the outcomes of each individual die are equally likely, the values of the *sum* of two dice are not. ■

TABLE 3.1. Probability distribution for the sum of two dice.

x	2	3	4	5	6	7	8	9	10	11	12
$P(X = x)$	$\frac{1}{36}$	$\frac{2}{36}$	$\frac{3}{36}$	$\frac{4}{36}$	$\frac{5}{36}$	$\frac{6}{36}$	$\frac{5}{36}$	$\frac{4}{36}$	$\frac{3}{36}$	$\frac{2}{36}$	$\frac{1}{36}$

3.2 INDEPENDENT RANDOM VARIABLES

The intuitive notion of independence introduced in Chapter 2 extends to random variables. We say that random variables X and Y are independent to mean that knowledge of the outcome of X does not affect the probability of the outcome of Y. That is, independence for discrete random variables means that for all x and y, the events $\{X = x\}$ and $\{Y = y\}$ are independent events. For notation, $X \perp Y$ may be used to indicate that X and Y are independent.

INDEPENDENCE OF RANDOM VARIABLES

Discrete random variables X and Y are said to be *independent* if

$$P(X = x|Y = y) = P(X = x), \text{ for all } x, y. \tag{3.1}$$

Equivalently,

$$P(X = x, Y = y) = P(X = x)P(Y = y), \text{ for all } x, y. \tag{3.2}$$

A collection of discrete random variables, such as an infinite sequence, is independent if for all finite subgroups $X_1, \ldots, X_k$ of the collection,

$$P(X_1 = x_1, \ldots, X_k = x_k) = P(X_1 = x_1) \cdots P(X_k = x_k),$$

for all $x_1, \ldots, x_k$.

Example 3.6 The Current Population Survey of 2010 provides data on the proportion of family households in the United States by number of children under 18 years old (see Table 3.2). In a sample of four households, what is the probability that no household has children?

TABLE 3.2. Distribution of number of children in US households.

0	1	2	3 or more
0.5533	0.1922	0.1642	0.0903

Although the actual sampling was done without replacement, because the population is so large, we assume sampling with replacement and independence, as the difference is negligible. This is common practice when working with a small sample from a much larger population.

For $k = 1, 2, 3, 4$, let X_k be the number of children for the kth family household in the sample. Then the desired probability is

$$\begin{aligned} &P(X_1 = 0, X_2 = 0, X_3 = 0, X_4 = 0) \\ &\quad = P(X_1 = 0)P(X_2 = 0)P(X_3 = 0)P(X_4 = 0) = (0.5533)^4 = 0.0938. \quad \blacksquare \end{aligned}$$

For a subset A of real numbers, the event $\{X \in A\}$ is the event that X takes values in A. For instance, if A is the interval $(1, 5)$, then $P(X \in A) = P(1 < X < 5)$. If A is the set of positive even numbers, then

$$P(X \in A) = P(X \text{ is even}) = P\left(\bigcup_{k=1}^{\infty}\{X = 2k\}\right).$$

With this notation, independence of random variables can be expressed more generally as follows:

INDEPENDENT RANDOM VARIABLES

Random variables X and Y are independent if for all $A, B \subseteq \mathfrak{R}$,

$$P(X \in A, Y \in B) = P(X \in A)P(Y \in B).$$

It should be clear that this definition implies Equation 3.2. Simply take $A = \{x\}$ and $B = \{y\}$. Conversely, suppose Equation 3.2 holds. Then

$$\begin{aligned} P(X \in A, Y \in B) &= \sum_{x\in A}\sum_{y\in B} P(X = x, Y = y) \\ &= \sum_{x\in A}\sum_{y\in B} P(X = x)P(Y = y) \\ &= \left(\sum_{x\in A} P(X = x)\right)\left(\sum_{y\in B} P(Y = y)\right) \\ &= P(X \in A)P(Y \in B). \end{aligned}$$

3.3 BERNOULLI SEQUENCES

A random variable that takes only two values 0 and 1 is called a *Bernoulli random variable*.

There are a vast range of applications that can be modeled as sequences of Bernoulli random variables, starting with coin flips. We often refer to the states of a Bernoulli variable as "success" and "failure," with the parameter p identified as the *success parameter* or *probability of success*.

BERNOULLI DISTRIBUTION

A random variable X has a *Bernoulli distribution with parameter p* if

$$P(X = 1) = p \text{ and } P(X = 0) = 1 - p,$$

for $0 < p < 1$. We write $X \sim \text{Ber}(p)$.

Example 3.7 A manufacturing process produces electronic components that occasionally are defective. There is a one-in-a-thousand chance that a component is defective. Furthermore, whether or not a component is defective is independent of any other component's status. If n components are produced in a day, find the probability that at least one is defective.

Let

$$X_k = \begin{cases} 1, & \text{if the } k\text{th component is defective,} \\ 0, & \text{otherwise,} \end{cases}$$

for $k = 1, \ldots, n$. Each X_k has a Bernoulli distribution with parameter $p = 0.001$. The probability that at least one component is defective is

$$\begin{aligned} &P(\text{At least one component defective}) \\ &\quad = 1 - P(\text{No defective components}) \\ &\quad = 1 - P(X_1 = 0, \ldots, X_n = 0) \\ &\quad = 1 - P(X_1 = 0) \cdots P(X_n = 0) \\ &\quad = 1 - (1 - 0.001) \cdots (1 - 0.001) = 1 - (0.999)^n. \end{aligned}$$

If the manufacturer produces 500 components per day, then the probability that at least one component will be defective is $1 - 0.999^{500} = 0.3936$. Thus, we would not be surprised to find at least one defective component. This would happen about four times out of every 10 days of production of 500 components! ■

In this last example, the random variables $X_1, \ldots, X_n$ were all independent and had the same Bernoulli distribution. In this case, we say that $X_1, \ldots, X_n$ is an *independent and identically distributed (i.i.d.) sequence* of random variables.

INDEPENDENT AND IDENTICALLY DISTRIBUTED (i.i.d.) SEQUENCES OF RANDOM VARIABLES

A sequence of random variables is said to be i.i.d. if the random variables are independent and have the same probability distribution (including all distribution parameters).

An i.i.d. sequence consisting of Bernoulli random variables is called a *Bernoulli sequence*. Random samples in statistics are often modeled as i.i.d. sequences of random variables.

Example 3.8 A national pollster wants to estimate the president's approval rating. Let p be the unknown proportion of US adults who approve of how the president is handling their job. A random sample of 1000 adults is taken. Each person is asked, "Do you approve or disapprove with the president's handling of their job?" The responses are modeled as an i.i.d. Bernoulli sequence $X_1, \ldots, X_{1000}$, where

$$X_k = \begin{cases} 1, & \text{if the } k\text{th person in the sample approves,} \\ 0, & \text{if the } k\text{th person in the sample does not approve or did not reply.} \end{cases}$$

The goal is to use these data to estimate the unknown proportion p. The proportion of people in the sample who approve of how the president is handling their job is $(X_1 + \cdots + X_{1000})/1000$. This *sample proportion* is an estimate of the *population proportion p*. ■

Example 3.9 A four-sided tetrahedral (a triangular prism) die labeled 1, 2, 3, 4 is thrown four times. What is the probability of rolling a three exactly twice?

The four die rolls are modeled by four independent Bernoulli trials with common success probability $p = 1/4$. Let 3 denote rolling a three and N denote not rolling a 3. A not-so-elegant solution is to enumerate all the possible ways to get a three exactly twice: 33*NN*, 3*N*3*N*, 3*NN*3, *N*33*N*, *N*3*N*3, *NN*33. There are six possibilities. By independence, each outcome occurs with probability $(1/4)^2(3/4)^2 = 9/256$. Thus, the probability of rolling a three exactly twice in four rolls is $6(9/256) = 27/128 = 0.211$.

Recall that you can also count the number of arrangements with combinations, as introduced in Chapter 1. We will generalize this problem and simplify its solution when we introduce the binomial distribution next. ■

3.4 BINOMIAL DISTRIBUTION

Let $X_1, \ldots, X_n$ be a Bernoulli sequence with success parameter p. Let 1 denote "success," and 0 "failure." The sum $X = X_1 + \cdots + X_n$ counts the number of successes in n trials.

For $k = 0, \ldots, n$, consider $P(X = k)$, the probability of obtaining exactly k successes in n trials. Each outcome of k successes and $n - k$ failures can be represented by a binary list of length n with exactly k ones. We learned to count the number of these arrangements in Chapter 1. There are $\binom{n}{k}$ such lists. The probability of each such outcome, by independence of the trials, is $p^k(1-p)^{n-k}$. Combining these gives the probability function of the binomial distribution.

BINOMIAL DISTRIBUTION

A random variable X is said to have a *binomial distribution with parameters n and p* if

$$P(X = k) = \binom{n}{k} p^k (1-p)^{n-k}, \quad \text{for } k = 0, 1, \ldots, n. \tag{3.3}$$

We write $X \sim \text{Binom}(n, p)$ or $\text{Bin}(n, p)$.

The probability function for the binomial distribution is indeed a valid probability function as the terms are "nonnegative" and sum to 1. By the binomial theorem,

$$\sum_{k=0}^{n} P(X = k) = \sum_{k=0}^{n} \binom{n}{k} p^k (1-p)^{n-k} = (p + (1-p))^n = 1^n = 1.$$

The most common setting in which the binomial distribution arises is modeling the number of successes in n independent Bernoulli trials with constant probability of success, p. Examples include the following:

- The number of heads in n fair coin tosses has a binomial distribution with parameters n and $p = 1/2$.
- Suppose 500 bits of data are sent through a digital transmission such that there is a 1% chance that any bit is received in error. If bit errors are independent of each other, then the number of errors has a binomial distribution with $n = 500$ and $p = 0.01$.
- Mutations on a DNA strand can be modeled as a sequence of independent Bernoulli trials. The total number of mutations has a binomial distribution. The parameters are the length of the strand n and the mutation rate p.

Probabilities that are modeled with the binomial distribution assume an underlying sequence of Bernoulli trials, which are i.i.d. Without independence, a fixed number of trials n, or a constant probability p for each trial, the binomial model is not valid.

The following situations are *not* appropriate for a binomial model:

1. **Ask 50 people their height**: In a binomial setting, each trial has *two* possible outcomes. Heights have many values.
   However, if people are asked whether or not they are at least 6 feet tall, then there are two outcomes, and the binomial model could be appropriate, assuming independence.
2. **The chance that a manuscript has a typo on any page is 1%. We will read the manuscript until we find 10 typos:** In a binomial setting, the number of trials n is fixed. In this case, the number of pages read is not fixed.
   However, if we read 100 pages and consider the probability of getting 10 typos, then the binomial setting is appropriate.
3. **Five cards are dealt from a standard deck of cards:** What is the probability you were dealt one ace? In dealing cards, we are sampling without replacement. Successive draws are neither independent nor have the same probability of success. That is, successive draws are not Bernoulli trials.
   However, if we replaced the card after each draw so that the sampling was with replacement, then the binomial setting would work.

VISUALIZING THE BINOMIAL DISTRIBUTION

To visualize the binomial distribution using **R**, type

```
> n <- 8
> p <- 0.15
> barplot(dbinom(0:n, n, p), names.arg = 0:n)
```

with your own choices of n and p. See Figure 3.1 for four examples. The script **Binom.R** has the example code for these four distributions.

Example 3.10 A multiple choice exam has 10 questions with four choices for each question. If a student guesses on each question from the available choices, what is the chance they will get exactly two questions right?

Briefly, we revisit problem-solving strategy. Before computing anything, we define a random variable and identify a suitable distribution to work with. Then, we work to find the probability requested.

Let X be the number of questions the student gets correct. Then X has a binomial distribution with parameters $n = 10$ and $p = 1/4$. This gives

$$P(X = 2) = \binom{10}{2}\left(\frac{1}{4}\right)^2\left(\frac{3}{4}\right)^8 = 0.281.$$

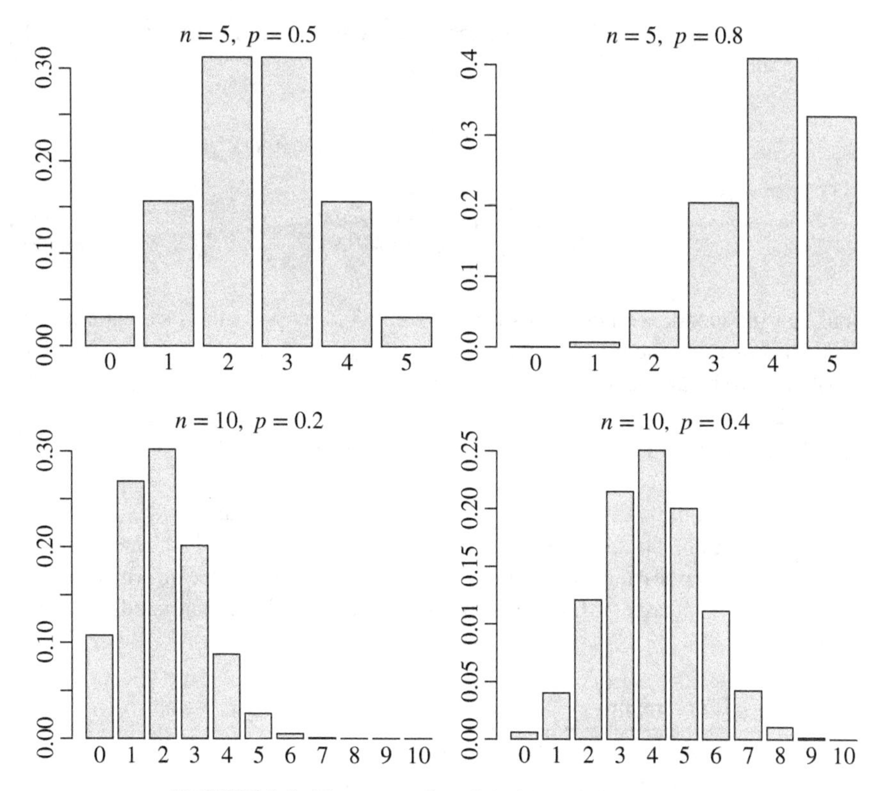

FIGURE 3.1: Four examples of the binomial distribution.

If the question for this problem had asked, "What is the chance that the student will get two questions right?" without the word "exactly," the question could be considered vague. Does it mean *exactly* two questions right or *at least* two questions right? In the latter case,

$$P(X \geq 2) = 1 - P(X \leq 1) = 1 - P(X = 0) - P(X = 1)$$
$$= 1 - \binom{10}{0}\left(\frac{1}{4}\right)^0\left(\frac{3}{4}\right)^{10} - \binom{10}{1}\left(\frac{1}{4}\right)^1\left(\frac{3}{4}\right)^9 = 0.756.$$

In this book, we will be careful about such vagaries. Be careful to clarify such issues if they arise in a real-world problem. ■

Example 3.11 In a field of 100 trees, each tree has a 10% chance of being infected by a root disease independently of other trees. What is the probability that more than five trees are infected?

Let X be the number of infected trees. Then X has a binomial distribution with $n = 100$ and $p = 0.10$. This gives

$$P(X > 5) = 1 - P(X \leq 5) = 1 - \sum_{k=0}^{5} P(X = k)$$

$$= 1 - \sum_{k=0}^{5} \binom{100}{k} (0.1)^k (0.9)^{100-k} = 0.942.$$

Note that independence here is a strong assumption, as you might expect clusters of infection among the trees. However, this assumption is needed for the binomial distribution to be applicable to the problem. ■

R: WORKING WITH PROBABILITY DISTRIBUTIONS

R has several commands for working with probability distributions like the binomial distribution. These commands are prefixed with `d`, `p`, and `r`. They take a suffix that describes the distribution. For future reference, this information is compiled in Appendix A. For the binomial distribution, these commands are the following:

R command	What it does
`dbinom(k,n,p)`	Computes $P(X = k)$
`pbinom(k,n,p)`	Computes $P(X \leq k)$
`rbinom(k,n,p)`	Simulates k random variables

To find the exact probability $P(X > 5)$ in the last Example 3.11, type

```
>  1-pbinom(5,100,0.10)
[1] 0.94242
> #or pbinom(5, 100, 0.10, lower.tail = FALSE)
```

To simulate the probability $P(X > 5)$ based on 10,000 repetitions, type

```
> runs <- 10000
> simlist <- rbinom(runs, 100, 0.10)
> sum(simlist > 5)/n
[1] 0.9456
```

■ **Example 3.12** According to Leder et al. [2002], many airlines consistently report that about 12% of all booked passengers do not show up to the gate due to cancellations and no-shows. If an airline sells 110 tickets for a flight that seats 100 passengers, what is the probability that the airline overbooked (sold more tickets than seats) in terms of the number of ticket holders who show up?

Let X be the number of ticket holders who arrive at the gate. If we assume that passengers' gate arrivals are independent, then X has a binomial distribution with $n = 110$ and $p = 1 - 0.12 = 0.88$. Overbooking occurs when more passengers show up than the number of available seats. Thus, the desired probability is

$$\begin{aligned} P(X > 100) &= \sum_{k=101}^{110} P(X = k) \\ &= \sum_{k=101}^{110} \binom{110}{k} (0.88)^k (0.12)^{110-k} \\ &= 0.137. \end{aligned}$$

In **R**, as $P(X > 100) = 1 - P(X \leq 100)$, type

```
> 1-pbinom(100, 110, 0.88)
[1] 0.1366599
```

For a 100-seat flight, suppose the airline would like to sell the maximum number of tickets such that the chance of overbooking is less than 5%. The airlines call this a "5% bump threshold" overbooking strategy. How many tickets should the airline sell?

Let n be the number of tickets sold. Find n so that $P(X > 100) \leq 0.05$, where $X \sim \text{Binom}(n, 0.88)$. We check in **R**.

```
> 1-pbinom(100, 108, 0.88)
[1] 0.04492587
> 1-pbinom(100, 109, 0.88)
[1] 0.08231748
```

Trial and error using **R** shows that for $n = 108$, $P(X > 100) = 0.0449$. And for $n = 109$, $P(X > 100) = 0.0823$. Therefore, sell $n = 108$ tickets. ■

R: SIMULATING THE OVERBOOKING PROBABILITY

To simulate the probability of overbooking when 108 tickets are sold, we could approach the problem in one of two ways: (i) generate 108 Bernoulli random variables with $p = 0.88$ or (ii) generate a single Binomial random variable with $n = 108$ and $p = 0.88$. Now that we have the Binomial framework, it is more convenient to use. Type

```
> rbinom(1, 108, 0.88)
```

To check if more than 100 tickets are sold and the airline overbooks, type

```
> if(rbinom(1, 108, 0.88) > 100) 1 else 0
```

which returns a 1 if too many tickets are sold, and 0 otherwise.

Here we simulate the probability of overbooking based on 10,000 trials. (Remember that a seed is being set for these!)

```
> simlist <- replicate(10000,
    if (rbinom(1, 108, 0.88) > 100) 1 else 0)
> mean(simlist)
[1] 0.0421
```

Example 3.13 The approximate frequencies of the four nucleotides in human DNA are given in Table 3.3 based on work by Piovesan et al. [2019].

In two DNA strands of length 10, what is the probability that nucleotides will match in exactly seven positions?

We make the assumption that the two strands are independent of each other. The chance that at any site, there will be a match is the probability that both sites are A, C, G, or T. The probability of a match is thus $(0.296)^2 + (0.204)^2 + (0.204)^2 + (0.296)^2 = 0.2585$. The number of matches is a binomial random variable with $n = 10$ and $p = 0.2585$. The desired probability is $P(X = 7) = 0.00377345$, where $X \sim \text{Binom}(10, 0.2585)$.

```
> dbinom(7, 10, 0.2585)
[1] 0.00377345
```

■

Example 3.14 Thea and Darius each toss four fair coins, independently. What is the probability that they get the same number of heads?

Let X be the number of heads Thea gets, and let Y be the number of heads Darius gets. The event that they get the same number of heads is

$$\{X = Y\} = \bigcup_{k=0}^{4} \{X = k, Y = k\}.$$

TABLE 3.3. Nucleotide frequencies in human DNA.

A	C	G	T
0.296	0.204	0.204	0.296

This gives

$$
\begin{aligned}
P(X=Y) &= P\left(\bigcup_{k=0}^{4}\{X=k, Y=k\}\right) \\
&= \sum_{k=0}^{4} P(X=k, Y=k) = \sum_{k=0}^{4} P(X=k)P(Y=k) \\
&= \sum_{k=0}^{4}\left[\binom{4}{k}\left(\frac{1}{2}\right)^4\right]^2 = \left(\frac{1}{2}\right)^8 \sum_{k=0}^{4}\binom{4}{k}^2 \\
&= \frac{1}{256}(1^2+4^2+6^2+4^2+1^2) = \frac{70}{256} = 0.273,
\end{aligned}
$$

where the third equality is due to independence. ■

Example 3.15 Random graphs. A *graph* is a set of vertices (or nodes), with edges joining them. An edge can be written as a pair of vertices. In mathematics, the study of graphs is called graph theory. Random graphs, meaning graphs generated using a probability model, have been used as models for Internet traffic, social networks, and the spread of infectious diseases. To get a sense of how this can work, we describe a model for a very simple *random graph*, the Erdős-Rényi graph.

Start with a graph on n vertices and no edges. For each pair of vertices, flip a coin with heads probability p. If the coin lands heads, place an edge between that pair of vertices. If tails, do not place an edge. Note the two extreme cases: if $p = 0$, there are no edges in the graph; if $p = 1$, every pair of vertices gets an edge in what is called the complete graph. The parameter p is called the *edge probability*. Properties of these graphs are studied for large n as the edge probability varies from 0 to 1 (see Fig. 3.2). Interested readers can learn more from Newman [2018].

Let X be the number of edges in a random graph of n vertices. There are $\binom{n}{2}$ ways to pick two vertices. Thus, there are $\binom{n}{2}$ possible edges in a graph on n vertices. The number of edges X thus has a binomial distribution with parameters $\binom{n}{2}$ and p.

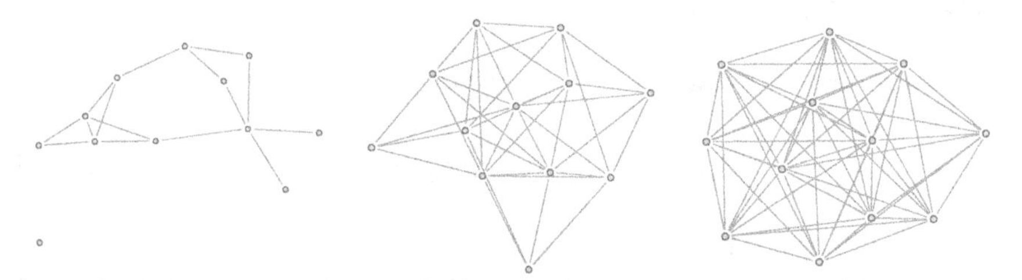

FIGURE 3.2: Three random graphs on $n = 12$ vertices generated, respectively, with $p = 0.2, 0.5$, and 0.9.

In a graph, the *degree* of a vertex is the number of edges that contain that vertex, also described as the number of edges incident to that vertex. Let $\deg(v)$ be the degree of vertex v in a random graph. There are $n-1$ possible edges incident to v, as there are $n-1$ vertices left in the graph, other than v. Each of those edges occurs with probability p. Thus, for any vertex v in our random graph, the degree $\deg(v)$ is a random variable that has a binomial distribution with parameters $n-1$ and p. That is,

$$P(\deg(v) = k) = \binom{n-1}{k} p^k (1-p)^{n-1-k}, \quad \text{for } k = 0, \ldots, n-1. \qquad \blacksquare$$

3.5 POISSON DISTRIBUTION

The binomial setting requires a fixed number n of independent trials. However, in many applications, we model counts of independent outcomes, where there is no prior constraint on the number of trials. Examples include

- The number of wrong numbers you receive on your cell phone over a month's time.
- The number of babies born on a maternity ward in one day.
- The number of blood cells recorded on a hemocytometer (a device used to count cells).
- The number of chocolate chips in a cookie.
- The number of accidents on a mile long stretch of highway.
- The number of soldiers killed by horse kick each year in each corps in the Prussian cavalry.

The last example may seem far-fetched, but it was actually one of the first uses of the distribution that we introduce in this section, called the Poisson distribution.

POISSON DISTRIBUTION

A random variable X has a *Poisson distribution with parameter* $\lambda > 0$ if

$$P(X = k) = \frac{e^{-\lambda}\lambda^k}{k!}, \quad \text{for } k = 0, 1, \ldots. \tag{3.4}$$

We write $X \sim \text{Pois}(\lambda)$.

See graphs of the Poisson distribution for four choices of λ in Figure 3.3.

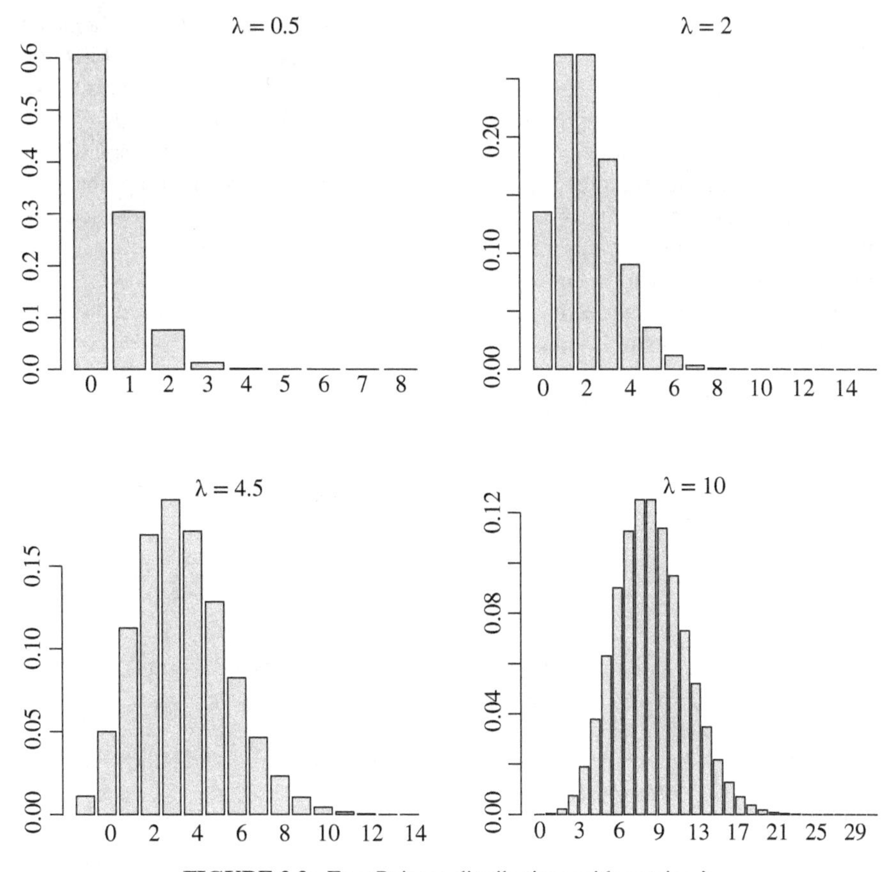

FIGURE 3.3: Four Poisson distributions with varying λ.

The probability function is nonnegative and sums to 1, as

$$\sum_{k=0}^{\infty} P(X = k) = \sum_{k=0}^{\infty} \frac{e^{-\lambda}\lambda^k}{k!} = e^{-\lambda} \sum_{k=0}^{\infty} \frac{\lambda^k}{k!} = e^{-\lambda}e^{\lambda} = 1.$$

(Not sure about the math above? Appendix C has useful results from calculus.)

The Poisson distribution is intimately connected to the binomial distribution introduced in the last section. To see how it arises in a real-world context, we elaborate on a specific example.

Consider developing a probability model for the number of babies born on a busy maternity ward in one day. Let X be the number of births. Note that this is not a binomial setting because there is neither a fixed number of trials nor at this stage even a probability to speak of. We give a heuristic argument to motivate the Poisson distribution.

Suppose that babies are born at some average rate of λ births per day. We break up the day into n subintervals of length $1/n$. Babies are born at the rate of λ/n births on each subinterval. For instance, if $n = 24$, babies are born at the rate of $\lambda/24$ births per hour. Let n get large. The lengths of the subintervals get small and *eventually* on each subinterval, it is very unlikely that two babies will be born during that small time period. In each small subinterval, either a baby is born or not born with probability λ/n. If we assume that births on different subintervals are independent of each other, then we can regard the occurrence of births on the subintervals as a sequence of n i.i.d. Bernoulli trials with success probability λ/n.

The total number of births in one day is the sum of the births on the n subintervals and has a binomial distribution with parameters n and λ/n. That is,

$$P(X = k) = \binom{n}{k}\left(\frac{\lambda}{n}\right)^k\left(1 - \frac{\lambda}{n}\right)^{n-k}, \quad \text{for } k = 0, \ldots, n.$$

The word "eventually" in the previous paragraph suggests a limiting process. Let n tend to infinity. We show in Section 3.5.2 that

$$\lim_{n\to\infty}\binom{n}{k}\left(\frac{\lambda}{n}\right)^k\left(1 - \frac{\lambda}{n}\right)^{n-k} = \frac{e^{-\lambda}\lambda^k}{k!}. \tag{3.5}$$

In other words, the number of babies born in one day has a Poisson distribution with parameter λ, where λ is the average rate of births per day.

Example 3.16 Data from a hospital maternity ward suggest that about 4.5 babies are born every day. What is the probability that there will be six births on the ward tomorrow?

Let X be the number of births on the ward tomorrow. Model X with a Poisson distribution with $\lambda = 4.5$. Then

$$P(X = 6) = \frac{e^{-4.5}(4.5)^6}{6!} = 0.128. \quad \blacksquare$$

The Poisson distribution is sometimes called the law of rare events. The "rarity" of the events does not refer to the number of events that occur, but rather to whether or not the event occurs in some small interval of time or space. What the examples at the beginning of this section have in common is that some event (e.g., births, traffic accidents, wrong numbers, and deaths) occur in some fixed region of time or space at a constant rate such that occurrences in disjoint subregions are independent. The parameter λ has the interpretation of the average number of occurrences per unit of time or space.

Example 3.17 According to the US Geological Survey, between 1970 and the start of 2016, there have been 33 *major earthquakes* (7.0 or greater on the Richter scale) in the United States, including the Aleutian Islands. Assuming successive major earthquakes are independent what is the probability there will be at least three major earthquakes in the United States next year?

Let X be the number of major earthquakes next year. The data are based on 47 years. The rate of earthquake occurrence is 33 per 47 years, or 33/47 per year. We model X with a Poisson distribution with $\lambda = 33/47$. Then

$$\begin{aligned} P(X \geq 3) &= 1 - P(X < 3) = 1 - P(X \leq 2) \\ &= 1 - P(X = 0) - P(X = 1) - P(X = 2) \\ &= 1 - e^{-33/47} - e^{-33/47}(33/47) - \frac{e^{-33/47}(33/47)^2}{2} \\ &= 0.0344. \end{aligned}$$

Geologists might argue whether successive earthquakes are in fact independent. After a big earthquake, it is very common for there to be aftershocks. The dependence of these residual quakes on the initial earthquake would violate the assumption of independence. More sophisticated models that incorporate some dependency structure are typically used to model seismic events. ■

Example 3.18 Death by horse kicks. In 1898, Ladislaus Bortkiewicz, a Polish statistician, authored a book about the Poisson distribution titled *The Law of Small Numbers*. In it, he studied the distribution of soldier deaths by horse kicks in the Prussian cavalry. Over 20 years, there were 122 deaths in 10 Prussian army corps. He divided the data into $20 \times 10 = 200$ corps-years. The average number of deaths per corps-year was $122/200 = 0.61$. Bortkiewicz modeled the number of deaths with a Poisson distribution with parameter $\lambda = 0.61$. Table 3.4 contains the data Borkiewicz worked with, including observed number of deaths and expected numbers predicted by the Poisson model. The expected number is the Poisson probability times 200. Observe how closely the model fits the data. ■

TABLE 3.4. Deaths by horse kicks in the Prussian cavalry.

Number of deaths	Observed	Poisson probability	Expected
0	109	0.543	108.7
1	65	0.331	66.3
2	22	0.101	20.2
3	3	0.021	4.1
4	1	0.003	0.6
5+	0	0.000	0.0

R: POISSON DISTRIBUTION

Commands for working with the Poisson distribution are

```
> dpois(x, lambda)     # Computes P(X = x)
> ppois(x, lambda)     # Computes P(X <= x)
> rpois(n, lambda)     # Generates n random numbers
```

To obtain the expected counts for the number of deaths by horse kicks in the last column of Table 3.4, type

```
> probs <- dpois(0:4, 0.61)
> probs <- c(probs,1 - ppois(4, 0.61))
> expected <- 200*probs
> expected
[1] 108.67  66.29  20.22   4.11   0.63   0.00
```

Example 3.19 The number of accidents per month at a busy intersection has a Poisson distribution with parameter $\lambda = 7.5$. Conditions at the intersection have not changed much over time. Suppose each accident costs local government about \$25,000 for clean-up. How much do accidents cost, on average, over a year's time?

With the interpretation that λ represents the average number of accidents per month, it is not hard to see that the average cost is about $7.5 \times 25{,}000 \times 12 =$ \$2,250,000. We approach the problem using simulation below. The following commands simulate 12 months of accidents and the associated annual cost.

R: SIMULATING ANNUAL ACCIDENT COST

```
> accidents <- rpois(12,7.5)
> cost <- sum(25000*accidents)
> cost
[1] 2200000
> simlist <- replicate(1000,sum(25000*rpois(12,7.5)))
> mean(simlist)
[1] 2258725
```

Local government can expect to pay about \$2.25 million in costs.

■

Example 3.20 The Poisson distribution is a common model in genetics. The distribution is used to describe occurrences of mutations and chromosome crossovers. *Crossovers* occur when two chromosomes break and then reconnect at different end pieces resulting in an exchange of genes. This process is known as genetic recombination.

Suppose a genetics lab has a means to count the number of crossovers between two genes on a chromosome. In 100 samples, 50 cells have no crossovers, 25 cells have one crossover, 20 cells have two crossovers, and 5 cells have three crossovers. Find the probability that a new sample will show at least one crossover.

The average number of crossovers from the sample is

$$50(0) + 25(1) + 20(2) + 5(3) = 80 \text{ crossovers per 100 cells,}$$

or 0.80 crossovers per cell. Model the number of crossovers on the chromosome with a Poisson distribution with $\lambda = 0.80$. This gives

$$P(\text{At least one crossover}) = 1 - P(\text{No crossovers}) = 1 - e^{-0.80} = 0.55. \qquad \blacksquare$$

3.5.1 Poisson Approximation of Binomial Distribution

We have seen the close connection between the binomial and Poisson distributions. The Poisson distribution arises as a limiting expression for the binomial distribution letting $n \to \infty$ and $p = \lambda/n \to 0$. For binomial problems with large n and small p, a Poisson approximation works well with $\lambda = np$. Different rules of thumb exist for what "large" means for n and what "small" means for p. As you can see in the following examples, we are examining very rare events to be sure the approximation is appropriate.

Example 3.21 We have not determined the accuracy of the following story, which has appeared in several blogs and websites, including Krantz [2005]:

> On October 9, 1972, the mathematician Dr. Jeffrey Hamilton from Warwick University wanted to show his students the effect of chance by tossing a coin. Taking a two pence coin out of his pocket, he tossed it. The class joined him in watching the coin flip over and over and then land on the floor—on its edge! Dozens of students witnessed the amazing event, and after a stunned silence, they all broke into wild applause. Hamilton later calculated that the chances of this happening are one in one billion.

Assume that the probability that a coin lands on its edge is one in one billion. Suppose everyone in the world flips a coin. With a population of over seven billion (still shy of eight billion) you would expect about seven coins to land on their edge. What is the probability that between six and eight coins land on their edge?

Let X be the number of coins that land on their edge. Then X has a binomial distribution with $n = 7{,}000{,}000{,}000$ and $p = 10^{-9} = 0.000000001$. The desired probability is

$$P(6 \le X \le 8) = \sum_{k=6}^{8} \binom{7 \times 10^9}{k} (10^{-9})^k \, (1 - 10^{-9})^{7\times 10^9 - k}.$$

Before calculators or computers, it would have been extremely difficult to compute such a probability. In fact, the Poisson approximation was discovered almost 200 years ago in order to solve problems like this.

Let $\lambda = np = (7 \times 10^9)10^{-9} = 7$. Then X has an approximate Poisson distribution with parameter $\lambda = 7$, and

$$P(6 \leq X \leq 8) \approx \frac{e^{-7}7^6}{6!} + \frac{e^{-7}7^7}{7!} + \frac{e^{-7}7^8}{8!} = 0.428.$$

In this example, the random variable X has an *exact* binomial distribution and an *approximate* Poisson distribution. ■

ON THE EDGE

Computing the exact probability of a coin landing on its edge is extremely hard, especially for thin coins. In Murray and Teare [1993], a model is presented and supported by numerical simulations. Extrapolations based on the model suggest that the probability of a US nickel landing on its edge is approximately 1 in 6,000.

Example 3.22 Mutations in DNA sequences occur from environmental factors, such as ultraviolet light and radiation, and mistakes that can happen when a cell copies its DNA in preparation for cell division. Nachman and Crowell [2000] estimate the mutation rate per nucleotide of human DNA as about 2.5×10^{-8}. There are about 3.3×10^9 nucleotide bases in the human DNA genome. Assume that whether or not a mutation occurs at a nucleotide site is independent of what occurs at other sites. We expect about $(3.3 \times 10^9)(2.5 \times 10^{-8}) = 82.5$ mutations. What is the probability that exactly 80 nucleotides will mutate in a person's DNA?

Let X be the number of mutations. Then X has an exact binomial distribution with $n = 3.3 \times 10^9$ and $p = 2.5 \times 10^{-8}$. Approximate X with a Poisson distribution with $\lambda = np = 82.5$. The approximate probability $P(X = 80)$ is

```
> dpois(80,82.5)
[1] 0.04288381408
```

The exact probability can be obtained in **R**.

```
> dbinom(80,3.3*10^9,2.5*10^(-8))
[1] 0.04288381456
```

The approximation is good to eight significant digits. The approximation works well because n is large and p is small.

But what if p is not small? Consider modeling "good" nucleotides, instead of mutations. The probability of a nucleotide not mutating is $1 - p$. What is the probability that $n - 80$ nucleotides do not mutate?

The number of "good" nucleotides has an exact binomial distribution with parameters n and $1 - p$. The exact probability is

```
> n <- 3.3*10^9
> p <- 2.5* 10^(-8)
> dbinom(n-80,n,1-p)
[1] 0.04288381
```

This is the same number obtained earlier, which makes sense as $n - 80$ "good" nucleotides is equivalent to 80 mutations. However, the Poisson approximation gives

```
> dpois(n-80,n*(1-p))
[1] 6.944694e-06
```

which is way off. The approximation is not good because $1 - p$ is large. ■

Example 3.23 Balls, bowls, and bombs. The following setting is very general. Suppose n balls are thrown into n/λ bowls so that each ball has an equal chance of landing in any bowl. If a ball lands in a bowl, call it a "hit." The chance that a ball hits a particular bowl is $1/(n/\lambda) = \lambda/n$. Keeping track of whether or not each ball hits that bowl, the successive hits form a Bernoulli sequence, and the number of hits has a binomial distribution with parameters n and λ/n. If n is large, the number of balls in each bowl is approximated by a Poisson distribution with parameter $n(\lambda/n) = \lambda$.

Many diverse applications can be fit into this general ball and bowl setting. In his classic analysis of Nazi bombing raids on London during World War II, Feller [1968] modeled bomb hits (balls) using a Poisson distribution. The city was divided into 576 small areas (bowls) of $1/4$ km^2. The number of areas hit exactly k times was counted. There were a total of 537 hits, so the average number of hits per area was $537/576 = 0.9323$. Feller gives the data in Table 3.5.

See the script **Balls.R** to simulate the London bombing example. Change the numbers in the script file to simulate your own balls and bowls experiment. ■

3.5.2 Poisson as Limit of Binomial Probabilities*

The Poisson probability function arises as the limit of binomial probabilities. Here we show the limit result

$$\lim_{n\to\infty} \binom{n}{k} \left(\frac{\lambda}{n}\right)^k \left(1 - \frac{\lambda}{n}\right)^{n-k} = \frac{e^{-\lambda}\lambda^k}{k!}$$

TABLE 3.5. Bomb hits over London during World War II.

Hits	0	1	2	3	4	≥ 5
Data	229	211	93	35	7	1
Expected	226.7	211.4	98.6	30.67	7.1	1.6

presented at the beginning of this section. Consider the product

$$\binom{n}{k}\left(\frac{\lambda}{n}\right)^k\left(1-\frac{\lambda}{n}\right)^{n-k}.$$

We can reexpress this as

$$\begin{aligned}
&= \frac{n(n-1)\cdots(n-k+1)}{k!}\left(\frac{\lambda}{n}\right)^k\left(1-\frac{\lambda}{n}\right)^{n-k} \\
&= \frac{n^k\left(1-\frac{1}{n}\right)\cdots\left(1-\frac{k-1}{n}\right)}{k!}\left(\frac{\lambda^k}{n^k}\right)\left(1-\frac{\lambda}{n}\right)^{-k}\left(1-\frac{\lambda}{n}\right)^{n} \\
&= \frac{\lambda^k}{k!}\left[\left(1-\frac{1}{n}\right)\cdots\left(1-\frac{k-1}{n}\right)\right]\left(1-\frac{\lambda}{n}\right)^{-k}\left(1-\frac{\lambda}{n}\right)^{n}. \qquad (3.6)
\end{aligned}$$

Taking limits as $n \to \infty$, consider the four factors in this expression.

(i) As λ and k are constants, $\lambda^k/k!$ stays unchanged in the limit.

(ii) For fixed k,

$$\left(1-\frac{1}{n}\right)\cdots\left(1-\frac{k-1}{n}\right) \to 1^{k-1} = 1, \quad \text{and}$$

(iii) $\left(1-\frac{\lambda}{n}\right)^{-k} \to 1^{-k} = 1.$

(iv) For the last factor $(1-\lambda/n)^n$, recall from calculus that the constant $e \approx 2.71828\ldots$ is defined as the limit

$$\lim_{x\to\infty}\left(1+\frac{1}{x}\right)^x = e.$$

Make the substitution $1/x = -\lambda/n$ so that $n = -\lambda x$. This gives

$$\lim_{n\to\infty}\left(1-\frac{\lambda}{n}\right)^n = \lim_{x\to\infty}\left(1+\frac{1}{x}\right)^{-\lambda x} = \left[\lim_{x\to\infty}\left(1+\frac{1}{x}\right)^x\right]^{-\lambda} = e^{-\lambda}.$$

Substituting in the four limits in Equation 3.6 shows us that the original product

$$= \frac{\lambda^k}{k!}\left[\left(1-\frac{1}{n}\right)\cdots\left(1-\frac{k-1}{n}\right)\right]\left(1-\frac{\lambda}{n}\right)^{-k}\left(1-\frac{\lambda}{n}\right)^{n} \to \frac{e^{-\lambda}\lambda^k}{k!},$$

as $n \to \infty$.

3.6 SUMMARY

Random variables are introduced. i.i.d. sequences are introduced in the context of Bernoulli trials and coin-flipping. The binomial distribution arises naturally as the distribution of the sum of n i.i.d. Bernoulli trials. Along with the binomial distribution, the Poisson distribution is one of the most important discrete probability distributions. It arises from the binomial setting when n, the number of fixed trials, is large and p, the success probability, is small.

- **Random variable X:** Assigns numbers to the outcomes of a random experiment. A real-valued function defined on the sample space.
- **Uniform distribution (discrete):** Let $S = \{s_1, \dots, s_k\}$. Then X is uniformly distributed on S if $P(X = s_i) = 1/k$, for $i = 1 \dots, k$.
- **Independent random variables:** Random variables X and Y are independent if $P(X = i, Y = j) = P(X = i)P(Y = j)$ for all i, j. Equivalently, $P(X \in A, Y \in B) = P(X \in A)P(Y \in B)$, for all $A, B \subseteq \mathfrak{R}$.
- **i.i.d. sequences:** A sequence of random variables is an i.i.d. sequence if the random variables are independent and all have the same distribution (including all distribution parameters).
- **Bernoulli distribution:** A random variable X has a Bernoulli distribution with parameter $0 < p < 1$, if $P(X = 1) = p = 1 - P(X = 0)$.
- **Binomial theorem:** For all x and y and nonnegative integer n,

$$(x + y)^n = \sum_{k=0}^{n} \binom{n}{k} x^k y^{n-k}.$$

- **Binomial distribution:** A random variable X has a binomial distribution with parameters n and p if

$$P(X = k) = \binom{n}{k} p^k (1 - p)^{n-k}, \quad \text{for } k = 0, 1, \dots, n.$$

- **Binomial setting:** The binomial distribution arises as the number of successes in n i.i.d. Bernoulli trials. The binomial setting requires: (i) a fixed number n of independent trials; (ii) trials take one of two possible values; and (iii) each trial has a constant probability p of success.
- **Poisson distribution:** A random variable X has a Poisson distribution with parameter $\lambda > 0$, if

$$P(X = k) = \frac{e^{-\lambda}\lambda^k}{k!}, \quad \text{for } k = 0, 1, \dots.$$

- **Poisson setting:** The Poisson setting arises in the context of discrete counts of "events" that occur over space or time with small probability and where successive events are independent.
- **Poisson approximation of binomial distribution:** Suppose $X \sim \text{Binom}(n, p)$ and $Y \sim \text{Pois}(\lambda)$. If $n \to \infty$ and $p \to 0$ in such a way so that $np \to \lambda > 0$, then for all k, $P(X = k) \to P(Y = k)$. The Poisson distribution with parameter

$\lambda = np$ serves as a good approximation for the binomial distribution when n is large and p is small.

EXERCISES

Random Variables and Independence

3.1 Your friend missed probability class today. Explain to your friend, in simple language, what a *random variable* is.

For the following problems 3.2–3.5, identify and define the random variable of interest. Then, express the probability in question in terms of the defined random variable, but do not compute the probability.

3.2 Roll four dice. Consider the probability of getting all fives.

3.3 A pizza shop offers three toppings: pineapple, peppers, and pepperoni. A pizza can have $0, 1, 2$, or 3 toppings. Consider the probability that a random customer asks for two toppings.

3.4 Bored one day, you decide to play the video game Angry Birds until you win. Every time you lose, you start over. Consider the probability that you win in less than 1000 tries.

3.5 In Julia's garden, there is a 3% chance that a tomato will be bad. Julia harvests 100 tomatoes and wants to know the probability that at most five tomatoes are bad.

3.6 Suppose X is a random variable that takes values on $\{0, 0.01, 0.02, \ldots, 0.99, 1\}$. If each outcome is equally likely, find

(a) $P(X \leq 0.33)$.

(b) $P(0.55 \leq X \leq 0.66)$.

3.7 Suppose X is a random variable that takes values on all positive integers. Let $A = \{2 \leq X \leq 4\}$ and $B = \{X \geq 4\}$. Describe the events (i) A^c; (ii) B^c; (iii) AB; and (iv) $A \cup B$.

3.8 Define two random variables X and Y that are not independent. Demonstrate the dependence between the random variables.

3.9 The original slot machine had 3 reels with 10 symbols on each reel. On each play of the slot machine, the reels spin and stop at a random position. Suppose each reel has one cherry on it. Let X be the number of cherries that show up from one play of the slot machine. Find $P(X = k)$, for $k = 0, 1, 2, 3$.

3.10 Suppose X_1, X_2 are i.i.d. random variables, each uniformly distributed on $\{-1, 0, 1\}$. Find the probability function for $X_1 + X_2$. That is, find $P(X_1 + X_2 = k)$, for $k = -2, \ldots, 2$.

3.11 Let X be a random variable such that $P(X = k) = k/10$, for $k = 1, 2, 3, 4$. Let Y be a random variable with the same distribution as X. Suppose X and Y are independent. Find $P(X + Y = k)$, for $k = 2, \ldots, 8$.

3.12 In role-playing games, different dice are used for different character actions. For dealing damage with a particular weapon in the game, a player rolls an eight-sided die (d8). On a critical hit, the number of damage dice rolled is doubled. Suppose X_1 and X_2 represent the die results of a critical hit damage roll with the weapon. Find the probability function for $X_1 + X_2$, the total amount of critical hit damage dealt by the player using the weapon.

3.13 Suppose X_1, X_2, X_3 are i.i.d. random variables, each uniformly distributed on $\{1, 2, 3\}$. Find the probability function for $X_1 + X_2 + X_3$. That is, find $P(X_1 + X_2 + X_3 = k)$, for $k = 3, \ldots, 9$.

Binomial Distribution

3.14 Pop quiz! Cora is surprised by a pop quiz 15 questions long in a class. Each question has four possible answers, and no question depends on the answer to another. Cora decides to guess her way through the quiz. Using random variables, write expressions for the following probabilities and solve them with **R**.

(a) The probability that Cora gets exactly 4 questions right.

(b) That probability that Cora gets at most 8 questions right.

(c) The probability that Cora gets between 3 and 10 questions right.

3.15 Every person in a group of 1000 people has a 1% chance of being infected by a virus. Assume that the process of being infected is independent from person to person. Using random variables, write expressions for the following probabilities and solve them with **R**.

(a) The probability that exactly 10 people are infected.

(b) That probability that at least 16 people are infected.

(c) The probability that between 12 and 14 people are infected.

(d) The probability that someone is infected.

3.16 **Newton–Pepys problem.** In 1693, Samuel Pepys wrote a letter to Isaac Newton posing the following question (Stigler 2006).
Which of the following three occurrences has the greatest chance of success?

1. Six fair dice are tossed and at least one 6 appears.
2. Twelve fair dice are tossed and at least two 6s appear.
3. Eighteen fair dice are tossed and at least three 6s appear.

Answer Mr. Pepys' question.

3.17 For the following situations, identify whether or not X has a binomial distribution. If it does, give n and p; if not, explain why not.

(a) Every day Alex goes out for lunch there is a 25% chance he will choose pizza. Let X be the number of times he chose pizza last week.

(b) Brandi plays basketball, and there is a 60% chance she makes each free throw. Let X be the number of successful free throw baskets she makes in a game.

(c) A bowl contains 100 red candies and 150 blue candies. Carlos reaches and takes out a sample of 10 candies. Let X be the number of red candies in his sample.

(d) Dwayne is reading a 600-page book. On even-numbered pages, there is a 1% chance of a typo. On odd-numbered pages, there is a 2% chance of a typo. Let X be the number of typos in the book.

(e) Evan is reading a 600-page book. The number of typos on each page has a Bernoulli distribution with $p = 0.01$. Let X be the number of typos in the book.

3.18 See Example 3.15. Consider a random graph on $n = 8$ vertices with edge probability $p = 0.25$.

(a) Find the probability that the graph has at least six edges.

(b) A vertex of a graph is said to be *isolated* if its degree is 0. Find the probability that a particular vertex is isolated.

3.19 A bag of 16 balls contains 1 red, 3 yellow, 5 green, and 7 blue balls. Suppose four balls are picked, sampling with replacement.

(a) Find the probability that the sample contains at least two green balls.

(b) Find the probability that each of the balls in the sample is a different color.

(c) Repeat parts (a) and (b) for the case when the sampling is without replacement.

3.20 Ecologists use *occupancy models* to study animal populations. Ecologists at the Department of Natural Resources use helicopter surveying methods to look for otter tracks in the snow along the Mississippi River to study which parts of the river are occupied by otter. The *occupancy rate* is the probability that an animal species is present in a region. The *detection rate* is the probability that the animal will be detected. (In this case, whether tracks will be seen from a helicopter.) If the animal is not detected, this might be due to the site not being occupied or because the site is occupied and the tracks were not detected.

A common model used by ecologists is a *zero-inflated binomial model*. If the region is occupied, then the number of detections is binomial with n the number of sites and p the detection rate. If the region is unoccupied, the number of detections is 0.

Let α be the occupancy rate, p the detection rate, and n the number of sites.

(a) Find the probability of zero detections. (Hint: Condition on occupancy.)

(b) DNR ecologists searched five sites along the Mississippi River for the presence of otter. Suppose $\alpha = 0.75$ and $p = 0.50$. Let Z be the number of observed detections. Give the probability function for Z.

3.21 **Sign test:** A new experimental drug is given to patients suffering from severe migraine headaches. Patients report their pain experience on a scale of 1–10 before and after the drug. The difference of their pain measurement is recorded. If their pain decreases, the difference will be positive (+); if the pain increases, the difference will be negative (−). The data are ignored if the difference is 0.

Under the *null hypothesis* that the drug is ineffective, and there is no difference in pain experience before and after the drug, the number of +'s will have a binomial distribution with n equal to the number of +'s and −'s; and $p = 1/2$. This is the basis of a statistical test called the sign test.

Suppose a random sample of 20 patients are given the new drug. Of 16 nonzero differences, 12 report an improvement (+). If one assumes that the drug is ineffective, what is the probability of obtaining 12 or more +'s, as observed in these data? Based on these data do you think the drug is ineffective?

3.22 I have two dice, one is a standard die. The other has three ones and three fours. I flip a coin. If heads, I will roll the standard die five times. If tails, I will roll the other die five times. Let X be the number of fours that appear. Find $P(X = 3)$. Does X have a binomial distribution?

Poisson Distribution

3.23 Suppose X has a Poisson distribution and $P(X = 2) = 2P(X = 1)$. Find $P(X = 3)$.

3.24 Computer scientists have modeled the length of search queries on the web using a Poisson distribution. Suppose the average search query contains about three words. Let X be the number of words in a search query. Because one cannot have a query consisting of zero words, we model X as a "restricted" Poisson distribution that does not take values of 0. That is, let $P(X = k) = P(Y = k|Y \neq 0)$, where $Y \sim \text{Pois}(3)$.

(a) Find the probability function of X.

(b) What is the probability of obtaining search queries longer than 10 words?

(c) Arampatzis and Kamps [2008] observe that many data sets contain very large queries that are not predicted by a Poisson model, such as queries of 10 words or more. They propose a restricted Poisson model for short queries, for instance, queries of six words or less. Find $P(Y = k|1 \leq Y \leq 6)$ for $k = 1, \ldots, 6$.

TABLE 3.6. No-hitter baseball games

Number of games	0	1	2	3	4	5	6	7
Seasons	18	30	21	21	6	3	3	2

3.25 The number of eggs a chicken hatches is a Poisson random variable. The probability that the chicken hatches no eggs is 0.10. What is the probability that she hatches at least two eggs?

3.26 Hemocytometer slides are used to count cells and other microscopic particles. They consist of glass engraved with a laser-etched grid. The size of the grid is known making it possible to count the number of particles in a specific volume of fluid. A hemocytometer slide used to count red blood cells has 160 squares. The average number of blood cells per square is 4.375. What is the probability that one square contains between 3 and 6 cells?

3.27 Cars pass a busy intersection at a rate of approximately 16 cars per minute. What is the probability that at least 1000 cars will cross the intersection in the next hour? (Hint: What is the rate per hour?)

3.28 Table 1 from Huber and Glen [2007] shows the number of no hitter baseball games that were pitched in the 104 ball seasons between 1901 and 2004. The data are reproduced in Table 3.6.

For instance, 18 seasons saw no no-hit games pitched; 30 seasons saw one no-hit game, etc. Use these data to model the number of no-hit games for a baseball season. Create a table that compares the observed counts with the expected number of no-hit games under your model.

3.29 Suppose $X \sim \text{Pois}(\lambda)$. Find the probability that X is odd. (Hint: Consider Taylor expansions of e^{λ} and $e^{-\lambda}$.)

3.30 If you take the red pill, the number of colds you get next winter will have a Poisson distribution with $\lambda = 1$. If you take the blue pill, the number of colds will have a Poisson distribution with $\lambda = 4$. Each pill is equally likely. Suppose you get three colds next winter. What is the probability you took the blue pill?

3.31 A physicist estimated that the probability of a US nickel landing on its edge is one in 6000. Suppose a nickel is flipped 10,000 times. Let X be the number of times it lands on its edge. Find the probability that X is between one and three using

(a) The exact distribution of X.

(b) An approximate distribution of X.

3.32 A chessboard is put on the wall and used as a dart board. Suppose 100 darts are thrown at the board and each of the 64 squares is equally likely to be hit.

(a) Find the exact probability that the left-top corner of the chessboard is hit by exactly two darts.

(b) Find an approximation of this probability using an appropriate distribution.

3.33 Give a probabilistic interpretation of the series

$$\frac{1}{e} + \frac{1}{2!e} + \frac{1}{4!e} + \frac{1}{6!e} + \cdots .$$

That is, pose a probability question for which the sum of the series is the answer.

3.34 Suppose that the number of eggs that an insect lays is a Poisson random variable with parameter λ. Further, the probability that an egg hatches and develops is p. Egg hatchings are independent of each other. Show that the total number of eggs that develop has a Poisson distribution with parameter λp. Hint: Condition on the number of eggs that are laid, and find expressions for the probability function of the number of eggs that develop using the conditioning.

Simulation and R

3.35 Let X be a random variable taking values 1, 4, 8, and 16 with respective probabilities $0.1, 0.2, 0.3, 0.4$. Show how to simulate X.

3.36 Modify the **Balls.R** script to simulate the distribution of 700 red blood cells on a hemocytometer slide as described in Exercise 3.26. Use your simulation to estimate the probability that a square contains between 3 and 6 cells, and compare to the exact solution.

3.37 Which is more likely: 5 heads in 10 coin flips, 50 heads in 100 coin flips, or 500 heads in 1000 coin flips? Use **R**'s `dbinom` command to find out.

3.38 Simulate the probability computed in part a of Exercise 3.19 when sampling is with replacement. Repeat for the case when sampling is without replacement.

3.39 Choose your favorite value of λ and let $X \sim \text{Pois}(\lambda)$. Simulate the probability that X is odd. See Exercise 3.29. Compare with the exact solution.

3.40 Write an **R** function `before(a,b)` to simulate the probability, in repeated independent throws of a pair of dice, that a appears before b, for $a, b = 2, \ldots, 12$.

3.41 Simulate a Prussian soldier's death by horse kick as in Example 3.18. Create a histogram based on 10,000 repetitions. Compare to a histogram of the Poisson distribution using the command `rpois(10000,0.61)`.

3.42 Poisson approximation of the binomial: Suppose $X \sim \text{Binom}(n, p)$. Write an **R** function `compare(n,p,k)` that computes (exactly) $P(X = k) - P(Y = k)$, where $Y \sim \text{Pois}(np)$. Try your function on numbers where you expect the Poisson probability to be a good approximation of the binomial. Also try it on numbers where you expect the approximation to be poor. Is your intuition correct?

Chapter Review

Chapter review exercises are available through the text website. The URL is `www.wiley.com/go/wagaman/probability2e`.

4

EXPECTATION AND MORE WITH DISCRETE RANDOM VARIABLES

Iacta alea est. (*The die is cast.*)

—From Julius Caesar, upon crossing the Rubicon.

Learning Outcomes

1. Apply definitions of probability mass function and expectation.
2. Solve problems involving joint distributions and marginal distributions.
3. Explain linearity of expectation and implications of independence in the discrete setting.
4. Compute expectations of functions of random variables, variance, standard deviation, covariance, and correlation.
5. Give examples of conditional distributions in the discrete setting.
6. (C) Verify analytical results via simulation for chapter concepts.

Introduction. Having introduced the binomial and Poisson distributions, two of the most important probability distributions for discrete random variables, we now look at more concepts related to random variables in general, still working in the discrete setting. We start with some terminology.

The probability function $P(X = x)$ of a discrete random variable X is called the *probability mass function (pmf)* of X. Pmfs are the central objects in discrete probability that allow one to compute probabilities. If we know the pmf of a random variable, in a sense we have "complete knowledge"—in a probabilistic sense—of the behavior of that random variable.

Probability: With Applications and R, Second Edition. Amy S. Wagaman and Robert P. Dobrow.

Companion Website: www.wiley.com/go/wagaman/probability2e

PROBABILITY MASS FUNCTION

For a random variable X that takes values in a set S, the *probability mass function of X* is the probability function

$$m(x) = P(X = x), \quad \text{for } x \in S,$$

and implicitly, 0, otherwise.

Often, we leave the implicitly 0, otherwise statement out when describing a pmf, but you should remember it is implied.

A Note on Notation

Although probability was studied for hundreds of years, many of the familiar symbols that we use today have their origin in the mid-twentieth century. In particular, their use was made popular in William Feller's remarkable probability textbook *An Introduction to Probability Theory and Its Applications* [1968], first published in 1950 and considered by many to be the greatest mathematics book of the twentieth century.

You have already seen many pmfs! Table 4.1 summarizes the common distributions that you have seen so far, together with their pmfs. See Appendix B for a more complete list of distributions covered in the text. You have also seen distributions that do not have a special name.

TABLE 4.1. Discrete probability distributions.

Distribution	Parameters	Probability mass function
Bernoulli	p	$P(X = k) = \begin{cases} p, & \text{if } k = 1 \\ 1 - p, & \text{if } k = 0 \end{cases}$
Binomial	n, p	$P(X = k) = \binom{n}{k} p^k (1 - p)^{n-k},\ k = 0, \ldots, n$
Poisson	λ	$P(X = k) = \frac{e^{-\lambda}\lambda^k}{k!},\ k = 0, 1, \ldots$
Uniform on $\{x_1, \ldots, x_n\}$		$P(X = x_k) = \frac{1}{n},\ k = 1, \ldots, n$

4.1 EXPECTATION

The expectation is a numerical measure that summarizes the typical, or average, behavior of a random variable. Expectation is a *weighted average* of the values of X, where the weights are the corresponding probabilities of those values. The expectation places more weight on values that have greater probability.

EXPECTATION

If X is a discrete random variable that takes values in a set S, its *expectation*, $E[X]$, is defined as

$$E[X] = \sum_{x \in S} xP(X = x).$$

The sum in the definition is over all values of X. The expectation will be finite provided that the series is absolutely convergent. If the series is not absolutely convergent, then we say that X has no finite expectation. Instead, it may be described as infinite or simply said to not exist.

In the case when X is uniformly distributed on a finite set $\{x_1, \ldots, x_n\}$, that is, all outcomes are equally likely,

$$E[X] = \sum_{i=1}^{n} x_i P(X = x_i) = \sum_{i=1}^{n} x_i \left(\frac{1}{n}\right) = \frac{x_1 + \cdots + x_n}{n}.$$

With equally likely outcomes, the expectation is just the regular average of the values.

Other names for expectation are *mean* and *expected value*. In the context of games and random experiments involving money, expected value is often used.

Example 4.1 Scrabble. In the game of Scrabble, there are 100 letter tiles with the distribution of point values given in Table 4.2. Let X be the point value of a random Scrabble tile. What is the expectation of X?

To compute the expectation, we convert the entries of Table 4.2 to probabilities by dividing the number of tiles by 100. The expected point value of a Scrabble tile is

$$E[X] = 0(0.02) + 1(0.68) + 2(0.07) + 3(0.08) + 4(0.10) + 5(0.01) + 8(0.02) + 10(0.02) = 1.87.$$

■

TABLE 4.2. Tile values in Scrabble.

Point value	0	1	2	3	4	5	8	10
Number of tiles	2	68	7	8	10	1	2	2

Note that in the last example, a random variable X was defined for us. If X had not been defined, and the question had asked for the expectation of the point value of a random Scrabble tile, it is useful to define a random variable to work with. We suggest defining the relevant random variables as a first step in most problems and exercises.

Example 4.2 Roulette. In the game of roulette, a ball rolls around a roulette wheel landing on one of 38 numbers. Eighteen numbers are red; 18 are black; and 2—0 and 00—are green. A bet of "red" costs \$1 to play and pays out \$2, for a net gain of \$1, if the ball lands on red.

Let X be a player's winnings at roulette after one bet of red. What is the distribution of X? What is the expected value, $E[X]$?

The player either wins or loses \$1. So $X = 1$ or -1, with $P(X = 1) = 18/38$ and $P(X = -1) = 20/38$. Now that the distribution of X is known, the expected value can be computed:

$$\begin{aligned} E[X] &= (1)P(X = 1) + (-1)P(X = -1) \\ &= (1)\frac{18}{38} + (-1)\frac{20}{38} = -\frac{2}{38} = -0.0526. \end{aligned}$$

The expected value of the game is about -5 cents. That is, the expected loss is about a nickel. ■

What does $E[X] = -0.0526$ or, in the previous example, $E[X] = 1.87$, really mean? You cannot lose 5.26 cents at roulette or pick a Scrabble tile with a value of 1.87.

We interpret $E[X]$ as a long-run average as it may not be a possible value of X. That is, if you pick Scrabble tiles repeatedly for a long time (with replacement), then the average of those tile values will be about 1.87. If you play roulette for a long time making many red bets, then the average of all your \$1 wins and losses will be about -5 cents. What that also means is that if you play, say, 10,000 times then your *total* loss will be about $5\times 10{,}000 = 50{,}000$ cents, or about \$500.

More formally, let $X_1, X_2, \ldots$ be an i.i.d. sequence of outcomes of roulette bets, where X_k is the outcome of the kth bet. Then the interpretation of expectation is that

$$E[X] \approx \frac{X_1 + \cdots + X_n}{n}, \tag{4.1}$$

when n is large. This gives a prescription for simulating the expectation of a random variable X: choose a large value of n, simulate n copies of X, and take the average as an approximation for $E[X]$.

R: PLAYING ROULETTE

The command

```
> sample(1:38,1)
```

simulates a uniform random integer between 1 and 38. Let the numbers 1 – 18 represent red. The command

```
> if (sample(1:38,1) <= 18) 1 else -1
```

returns a 1 if you roll red, and -1, otherwise.

Repeat 10,000 times and take the average to simulate the expected value.

```
> simlist <- replicate(10000,
    if (sample(1:38,1) <= 18) 1 else -1)
> mean(simlist)
[1] -0.0612
```

Example 4.3 Expectation of discrete uniform distribution. Let $X \sim \text{Unif}\{1, \ldots, n\}$. The expectation of X is

$$E[X] = \sum_{x=1}^{n} xP(X = x) = \sum_{x=1}^{n} \frac{x}{n} = \frac{1}{n}\left(\frac{(n+1)n}{2}\right) = \frac{n+1}{2}.$$

■

Example 4.4 Expectation of the Poisson distribution. Let $X \sim \text{Pois}(\lambda)$. What is $E[X]$? Before doing the calculation, think back to when the Poisson distribution was introduced, and make an educated guess at the expectation of X.

Working out the computation, we find

$$\begin{aligned} E[X] &= \sum_{k=0}^{\infty} kP(X = k) = \sum_{k=0}^{\infty} k\frac{e^{-\lambda}\lambda^k}{k!} \\ &= e^{-\lambda}\sum_{k=1}^{\infty} \frac{\lambda^k}{(k-1)!} = \lambda e^{-\lambda}\sum_{k=1}^{\infty} \frac{\lambda^{k-1}}{(k-1)!} \\ &= \lambda e^{-\lambda}\sum_{k=0}^{\infty} \frac{\lambda^k}{k!} = \lambda e^{-\lambda}e^{\lambda} = \lambda. \end{aligned}$$

The λ parameter of a Poisson distribution is the mean of the distribution. ■

4.2 FUNCTIONS OF RANDOM VARIABLES

Suppose X is a random variable and g is some function. Then $Y = g(X)$ is a random variable that is a function of X. The values of this new random variable are found as follows. If $X = x$, then $Y = g(x)$.

Functions of random variables, like X^2, e^X, and $1/X$, show up all the time in probability. Often, we apply some function to the outcomes of a random experiment, as we will see in many examples later. In statistics, it is common to transform data using an elementary function such as the log or exponential function.

Suppose X is uniformly distributed on $\{-2, -1, 0, 1, 2\}$. That is,

$$P(X = k) = \frac{1}{5}, \quad \text{for } k = -2, -1, 0, 1, 2.$$

Take $g(x) = x^2$, and let $Y = g(X) = X^2$. The possible outcomes of Y are $(-2)^2, (-1)^2, 0^2, 1^2, 2^2$; that is, 0, 1, and 4. As $Y = 0$ if and only if $X = 0$, $Y = 1$ if and only if $X = \pm 1$, and $Y = 4$ if and only if $X = \pm 2$, the pmf of Y is found as

$$P(Y = 0) = P(X^2 = 0) = P(X = 0) = \frac{1}{5},$$

$$P(Y = 1) = P(X^2 = 1) = P(X = \pm 1) = P(X = -1) + P(X = 1) = \frac{2}{5},$$

$$P(Y = 4) = P(X^2 = 4) = P(X = \pm 2) = P(X = -2) + P(X = 2) = \frac{2}{5}.$$

Here is one way to think of functions of random variables. Suppose there are two rooms labeled X and Y. In the first X room, a random experiment is performed. The outcome of the experiment is x. A messenger takes x and heads to the Y room. But before he gets there, he applies the g function to x and delivers $g(x)$ to the Y room. As the random experiment is repeated, an observer looking into the X room sees the x outcomes. An observer looking into the Y room sees the $g(x)$ outcomes. Several x's may map to the same y value.

Example 4.5 Tenzin spends \$2 in supplies to set up his lemonade stand. He charges 25 cents a cup. Suppose the number of cups he sells in a day has a Poisson distribution with $\lambda = 10$. Describe his profit as a function of a random variable and find the probability that the lemonade stand makes a positive profit.

Let X be the number of cups Tenzin sells in a day. Then $X \sim \text{Pois}(10)$. If he sells x cups, then his profit is $25x - 200$ cents. The random variable $Y = 25X - 200$ defines his profit as a function of X, the number of cups sold. To solve probability questions about Y, we reexpress them in terms of X, because we know its distribution.

The probability that Tenzin makes a positive profit is

$$P(Y > 0) = P(25X - 200 > 0) = P(X > 8)$$
$$= 1 - P(X \leq 8) = 1 - \sum_{k=0}^{8} \frac{e^{-10}10^k}{k!} = 0.667.$$

The probability is easily evaluated using **R**:

```
> 1-ppois(8, 10)
[1] 0.6671803
```

Observe carefully the use of algebraic operations with random variables. In the last example, we have $\{25X - 200 > 0\}$ if and only if $\{25X > 200\}$ if and only if $\{X > 8\}$. We can add, subtract, multiply, and do any allowable algebraic operation on both sides of the expression. You are beginning to see the power of working with random variables. ■

R: LEMONADE PROFITS

To simulate one day's profit in cents for the lemonade stand, type

```
> rpois(1,10)*25 - 200
```

To simulate the probability of making a positive profit, type

```
> reps <- 10000
> simlist <- rpois(reps, 10)*25 - 200
> sum(simlist > 0)/reps
[1] 0.6706
```

Next, we learn how to take the expectation of a function of a random variable.

EXPECTATION OF FUNCTION OF A RANDOM VARIABLE

Let X be a random variable that takes values in a set S. Let g be a function. Then,

$$E[g(X)] = \sum_{x \in S} g(x)P(X = x). \tag{4.2}$$

The result seems straightforward and perhaps even "obvious." But its proof is somewhat technical, and we leave the details for the interested reader at the end of the chapter. It is sometimes called "the law of the unconscious statistician."

Example 4.6 A number X is picked uniformly at random from 1 to 100. What is the expected value of X^2?

The random variable X^2 is equal to $g(X)$, where $g(x) = x^2$. This gives

$$\begin{aligned} E[X^2] &= \sum_{x=1}^{100} x^2 P(X = x) = \sum_{x=1}^{100} x^2 \left(\frac{1}{100}\right) \\ &= \left(\frac{1}{100}\right) \frac{100(101)(201)}{6} \\ &= \frac{(101)(201)}{6} = 3383.5, \end{aligned}$$

using the fact that the sum of the first n squares is $n(n+1)(2n+1)/6$.

You might have first thought that because $E[X] = 101/2 = 50.5$, then $E[X^2] = (101/2)^2 = 2550.25$. We see that this is not correct. It is not true that $E[X^2] = (E[X])^2$. Consider a simulation. We take 100,000 replications to get a good estimate.

```
> mean(sample(1:100, 100000, replace = TRUE)^2)
[1] 3389.155
```

The simulation should reinforce the fact that $E[X^2] \neq E[X]^2$. More generally, it is not true that $E[g(X)] = g(E[X])$. The operations of expectation and function evaluation cannot be interchanged. It is very easy, and common, to make this kind of a mistake. To be forewarned is to be forearmed. ■

Example 4.7 Create a "random sphere" whose radius R is determined by the roll of a six-sided die. Let V be the volume of the sphere. Find $E[V]$.

R is uniform on $\{1, 2, 3, 4, 5, 6\}$. The formula for the volume of a sphere as a function of radius is $v(r) = (4\pi/3)r^3$. The expected volume of the sphere is

$$\begin{aligned} E[V] &= E\left[\frac{4\pi}{3}R^3\right] = \sum_{r=1}^{6} \left(\frac{4\pi}{3}r^3\right) P(R = r) \\ &= \frac{4\pi}{3}\left(\frac{1}{6}\right)(1^3 + 2^3 + 3^3 + 4^3 + 5^3 + 6^3) \\ &= 98\pi. \end{aligned}$$

■

Example 4.8 Suppose X has a Poisson distribution with parameter λ. Find $E[1/(X+1)]$.

We proceed using the "law of the unconscious statistician," which gives

$$
\begin{aligned}
E\left[\frac{1}{X+1}\right] &= \sum_{k=0}^{\infty} \frac{1}{k+1} P(X=k) \\
&= \sum_{k=0}^{\infty} \frac{1}{k+1} \left(\frac{e^{-\lambda}\lambda^k}{k!}\right) \\
&= \frac{e^{-\lambda}}{\lambda} \sum_{k=0}^{\infty} \frac{\lambda^{k+1}}{(k+1)!} \\
&= \frac{e^{-\lambda}}{\lambda} \sum_{k=1}^{\infty} \frac{\lambda^k}{k!} = \frac{e^{-\lambda}}{\lambda}\left(\sum_{k=0}^{\infty} \frac{\lambda^k}{k!} - 1\right) \\
&= \frac{e^{-\lambda}}{\lambda}(e^{\lambda} - 1) = \frac{1-e^{-\lambda}}{\lambda}.
\end{aligned}
$$

Note how the series was reexpressed to make use of results from calculus. ■

We gave a dire warning a few paragraphs back that in general it is not true that $E[g(X)] = g(E[X])$. However, there is one notable exception. That is the case when g is a linear function.

EXPECTATION OF A LINEAR FUNCTION OF X

For constants a and b, and a random variable X with expectation $E[X]$,

$$E[aX+b] = aE[X] + b.$$

Let $g(x) = ax + b$. By the law of the unconscious statistician,

$$
\begin{aligned}
E[aX+b] &= \sum_x (ax+b)P(X=x) \\
&= a\sum_x xP(X=x) + b\sum_x P(X=x) \\
&= aE[X] + b.
\end{aligned}
$$

That is, the expectation of a linear function of X is that function evaluated at the expectation of X. This special case can be very useful, just remember that in general $E[g(X)] \neq g(E[X])$.

Example 4.9 A lab in France takes temperature measurements of data that are randomly generated. The mean temperature for their data is 5°C. The data are transferred to another lab in the United States, where temperatures are recorded in Fahrenheit. When the data are sent, they are first transformed using the Celsius to Fahrenheit conversion formula $f = 32 + (9/5)c$. Find the mean temperature of the Fahrenheit data.

Let F and C denote the random temperature measurements in Fahrenheit and Celsius, respectively. Then, because the conversion formula is a linear function,

$$E[F] = E\left[32 + \left(\frac{9}{5}\right)C\right] = 32 + \left(\frac{9}{5}\right)E[C]$$
$$= 32 + \left(\frac{9}{5}\right)5 = 41°\text{F}.$$

■

4.3 JOINT DISTRIBUTIONS

In the case of two random variables X and Y, a *joint distribution* specifies the values and probabilities for all pairs of outcomes. For two discrete variables X and Y, the *joint pmf of X and Y* is the function of two variables $P(X = x, Y = y)$.

The joint pmf is a probability function, and thus, it sums to 1. If X takes values in a set S and Y takes values in a set T, then

$$\sum_{x\in S}\sum_{y\in T} P(X = x, Y = y) = 1.$$

As in the one variable case, probabilities of events are obtained by summing over the individual outcomes contained in the event. For instance, for constants $a < b$ and $c < d$, possible values of X and Y, respectively,

$$P(a \le X \le b, c \le Y \le d) = \sum_{x=a}^{b}\sum_{y=c}^{d} P(X = x, Y = y).$$

A joint pmf can be defined for any finite collection of discrete random variables $X_1, \ldots, X_n$ defined on a common sample space. The joint pmf is the function of n variables $P(X_1 = x_1, \ldots, X_n = x_n)$.

Example 4.10 Suppose the joint pmf of X and Y is

$$P(X = x, Y = y) = cxy, \quad \text{for } x, y = 1, 2,$$

and 0, otherwise. (i) Find the constant c. (ii) Find $P(X \le 1, Y \ge 1)$.

(i) To find c, we use the fact that the sum of the probabilities is 1. Note that

$$1 = \sum_{x=1}^{2} \sum_{y=1}^{2} cxy = c(1 + 2 + 2 + 4) = 9c,$$

and thus $c = 1/9$.

(ii) The desired probability is

$$P(X \leq 1, Y \geq 1) = \sum_{x=1}^{1} \sum_{y=1}^{2} \frac{xy}{9} = \frac{1}{9}(1 + 2) = \frac{1}{3}.$$

■

■ **Example 4.11 Red ball, blue ball.** A bag contains four red, three white, and two blue balls. A sample of two balls is picked without replacement. Let R and B be the number of red and blue balls, respectively, in the sample. (i) Find the joint pmf of R and B. (ii) Use the joint pmf to find the probability that the sample contains at most one red and one blue ball.

(i) Consider the event $\{R = r, B = b\}$. The number of red and blue balls in the sample must be between 0 and 2. For $0 \leq r + b \leq 2$, if r red balls and b blue balls are picked, then $2 - r - b$ white balls must also be picked. Selecting r red, b blue, and $2 - r - b$ white balls can be done in $\binom{4}{r}\binom{2}{b}\binom{3}{2-r-b}$ ways. There are $\binom{9}{2} = 36$ ways to select two balls from the bag. Thus, the joint pmf of (R, B) is

$$P(R = r, B = b) = \binom{4}{r}\binom{3}{2-r-b}\binom{2}{b} \Big/ 36, \quad \text{for } 0 \leq r + b \leq 2.$$

The joint pmf of R and B is also described by the joint probability table:

		B	
R	0	1	2
0	3/36	6/36	1/36
1	12/36	8/36	0
2	6/36	0	0

where the pmf has been evaluated at all possible combinations of values for R and B.

(ii) The desired probability is

$$P(R \leq 1, B \leq 1) = \sum_{r=0}^{1} \sum_{b=0}^{1} P(R = r, B = b)$$
$$= \frac{3}{36} + \frac{12}{36} + \frac{6}{36} + \frac{8}{36} = \frac{29}{36}.$$

■

Marginal distributions. From the joint distribution of X and Y, one can obtain the univariate, or *marginal distribution* of each variable. As

$$\{X = x\} = \bigcup_{y \in T} \{X = x, Y = y\},$$

the pmf of X is

$$P(X = x) = P\left(\bigcup_{y \in T} \{X = x, Y = y\}\right) = \sum_{y \in T} P(X = x, Y = y).$$

The marginal distribution or marginal pmf of X is obtained from the joint distribution of X and Y by summing over the values of y. Similarly, the marginal pmf of Y is obtained by summing the joint pmf over the values of x.

MARGINAL DISTRIBUTIONS

If X takes values in a set S, and Y takes values in a set T, then the marginal distribution of X is

$$P(X = x) = \sum_{y \in T} P(X = x, Y = y) \quad \text{and}$$

the marginal distribution of Y is

$$P(Y = y) = \sum_{x \in S} P(X = x, Y = y).$$

Example 4.12 Red ball, blue ball, continued. For the last Example 4.11, (i) find the marginal distributions of the number of red and blue balls, respectively. (ii) Use these distributions to find the expected number of red balls and the expected number of blue balls in the sample.

(i) Given a joint probability table, the marginal distributions are obtained by summing over the rows and columns of the table, as shown below.

		B 0	1	2	
	0	3/36	6/36	1/36	10/36
R	1	12/36	8/36	0	20/36
	2	6/36	0	0	6/36
		21/36	14/36	1/36.	

That is,

$$P(R = r) = \begin{cases} 10/36, & \text{if } r = 0, \\ 20/36, & \text{if } r = 1, \\ 6/36, & \text{if } r = 2, \end{cases}$$

and

$$P(B = b) = \begin{cases} 21/36, & \text{if } b = 0, \\ 14/36, & \text{if } b = 1, \\ 1/36, & \text{if } b = 2. \end{cases}$$

(ii) For the expectations,

$$E[R] = \sum_{r=0}^{2} rP(R = r) = 0\left(\frac{10}{36}\right) + 1\left(\frac{20}{36}\right) + 2\left(\frac{6}{36}\right) = \frac{8}{9} \quad \text{and}$$

$$E[B] = 0\left(\frac{21}{36}\right) + 1\left(\frac{14}{36}\right) + 2\left(\frac{1}{36}\right) = \frac{4}{9}.$$

■

Example 4.13 A computer store has modeled the number of computers C it sells per day, together with the number of extended warranties W. The joint pmf is

$$P(C = c, W = w) = \left(\frac{5}{2}\right)^c \frac{e^{-5}}{w!(c-w)!},$$

for $c = 0, 1, \ldots,$ and $w = 0, \ldots, c$. We explore the marginal distributions of C and W.

To find the marginal distribution of C sum over all w. As $0 \leq w \leq c$,

$$\begin{aligned} P(C = c) &= \sum_w P(C = c, W = w) \\ &= \sum_{w=0}^{c} \left(\frac{5}{2}\right)^c \frac{e^{-5}}{w!(c-w)!} \\ &= \left(\frac{5}{2}\right)^c e^{-5} \sum_{w=0}^{c} \frac{1}{w!(c-w)!} \\ &= \left(\frac{5}{2}\right)^c \frac{e^{-5}}{c!} \sum_{w=0}^{c} \binom{c}{w} \\ &= \left(\frac{5}{2}\right)^c \frac{e^{-5}}{c!} 2^c = \frac{e^{-5}5^c}{c!}, \end{aligned}$$

for $c = 0, 1, \ldots$. The final expression is the pmf of a Poisson random variable with $\lambda = 5$. Thus, the marginal distribution of C, the number of computers sold, is a Poisson distribution with $\lambda = 5$.

For the marginal distribution of W, sum over c values. As $0 \leq w \leq c$, the sum is over all $c \geq w$, which gives

$$\begin{aligned}
P(W = w) &= \sum_c P(W = w, C = c) \\
&= \sum_{c=w}^{\infty} \left(\frac{5}{2}\right)^c \frac{e^{-5}}{w!(c-w)!} \\
&= \frac{e^{-5}}{w!} \sum_{c=w}^{\infty} \left(\frac{5}{2}\right)^c \frac{1}{(c-w)!} \\
&= \frac{e^{-5}}{w!} \left(\frac{5}{2}\right)^w \sum_{c=w}^{\infty} \left(\frac{5}{2}\right)^{c-w} \frac{1}{(c-w)!} \\
&= \frac{e^{-5}}{w!} \left(\frac{5}{2}\right)^w \sum_{c=0}^{\infty} \left(\frac{5}{2}\right)^c \frac{1}{c!} \\
&= \frac{e^{-5}}{w!} \left(\frac{5}{2}\right)^w e^{5/2} = \frac{e^{-5/2}(5/2)^w}{w!},
\end{aligned}$$

for $w = 0, 1, \ldots$. This gives the pmf of a Poisson random variable with $\lambda = 5/2$. That is, the marginal distribution of W, the number of extended warranties sold, is a Poisson distribution with $\lambda = 5/2$.

In summary, computers, according to the model, sell roughly at the rate of five per day, and warranties sell at half that rate.

Suppose the store sells c computers on a particular day, what is the probability they will sell w extended warranties, where $w \leq c$?

We have posed the conditional probability

$$\begin{aligned}
P(W = w | C = c) &= \frac{P(W = w, C = c)}{P(C = c)} \\
&= \frac{(5/2)^c e^{-5}/(w!(c-w)!)}{e^{-5}5^c/c!} \\
&= \binom{c}{w} \left(\frac{1}{2}\right)^c.
\end{aligned}$$

This is a binomial probability. We discovered that conditional on selling c computers, the number of warranties sold has a binomial distribution with parameters c and $p = 1/2$. This is the same distribution as the number of heads in c fair coin flips.

What we have learned is a probabilistic description of computer and warranty sales at the computer store. About five computers, on average, are sold per day according to a Poisson distribution. For each computer sold, there is a 50–50 chance that the extended warranty will be purchased, and thus about 2.5 warranties are sold, per day, on average, according to a Poisson distribution. ■

As demonstrated in the last example, learning to recognize common pmfs can be very useful. You can then harness other information you know about the distribution, such as its expectation. The same will be true when we turn our attention to continuous probability in Chapter 6.

Random variables can arise as functions of two or more random variables. Suppose $g(x, y)$ is a real-valued function of two variables. Then $Z = g(X, Y)$ is a random variable, which is a function of two random variables. For expectations of such random variables, there is a multivariate version of Equation 4.2.

EXPECTATION OF FUNCTION OF TWO RANDOM VARIABLES

$$E[g(X, Y)] = \sum_{x \in S} \sum_{y \in T} g(x, y) P(X = x, Y = y). \tag{4.3}$$

Example 4.14 In Example 4.10, we found a probability after finding $c = 1/9$. Now suppose we want to evaluate $E[XY]$. We know that $P(X = x, Y = y) = xy/9$. Apply the last result with $g(x, y) = xy$. Thus,

$$\begin{aligned} E[XY] &= \sum_{x=1}^{2} \sum_{y=1}^{2} xyP(X = x, Y = y) \\ &= \sum_{x=1}^{2} \sum_{y=1}^{2} \frac{x^2 y^2}{9} = \frac{1}{9}(1 + 4 + 4 + 16) = \frac{25}{9} = 2.78. \end{aligned}$$

In evaluating expectations of functions of random variables, note that $g(x, y)$ could even just be x or y. This shows you can obtain $E[X]$ and $E[Y]$ from joint distributions, not just marginals. ■

4.4 INDEPENDENT RANDOM VARIABLES

If X and Y are independent discrete random variables, then the joint pmf of X and Y has a particularly simple form. In the case of independence,

$$P(X = x, Y = y) = P(X = x)P(Y = y), \quad \text{for all } x \text{ and } y.$$

That is, the joint distribution is the product of the marginal distributions if $X \perp Y$.

Example 4.15 Kamile rolls a die four times and flips a coin twice. Let X be the number of ones she gets on the die. Let Y be the number of heads she gets on the coin. It seems reasonable to assume die rolls are independent of coin flips. (i) Find the joint pmf. (ii) Find the probability that Kamile gets the same number of ones on the die as heads on the coin.

(i) The joint pmf of X and Y is

$$P(X = x, Y = y) = P(X = x)P(Y = y)$$
$$= \binom{4}{x}\left(\frac{1}{6}\right)^x\left(\frac{5}{6}\right)^{4-x}\binom{2}{y}\left(\frac{1}{2}\right)^2,$$

for $x = 0, 1, 2, 3, 4$ and $y = 0, 1, 2$. In particular, $X \sim$ Binom(4, 1/6) and $Y \sim$ Binom(2, 1/2). The joint probability is the product of these two binomial probabilities.

(ii) The desired probability is

$$P(X = Y) = P(X = 0, Y = 0) + P(X = 1, Y = 1) + P(X = 2, Y = 2)$$
$$= \sum_{k=0}^{2}\binom{4}{k}\left(\frac{1}{6}\right)^k\left(\frac{5}{6}\right)^{4-k}\binom{2}{k}\left(\frac{1}{2}\right)^2$$
$$= 0.1206 + 0.1929 + 0.0289 = 0.3424.$$

■

R: DICE AND COINS

Here is a quick simulation of the probability $P(X = Y)$ in the die and coin example.

```
> n <- 100000
> sum(rbinom(n, 4, 1/6) == rbinom(n, 2, 1/2))/n
[1] 0.33958
```

If X and Y are independent random variables, then knowledge of whether or not X occurs gives no information about whether or not Y occurs. It follows that if g and h are functions, then $g(X)$ gives no information about whether or not $h(Y)$ occurs and, hence, $g(X)$ and $h(Y)$ are independent random variables.

FUNCTIONS OF INDEPENDENT RANDOM VARIABLES ARE INDEPENDENT

Suppose X and Y are independent random variables, and g and h are functions, then the random variables $g(X)$ and $h(Y)$ are independent.

The following result for independent random variables has wide application.

EXPECTATION OF A PRODUCT OF INDEPENDENT RANDOM VARIABLES

Let X and Y be independent random variables. Then for any functions g and h,

$$E[g(X)h(Y)] = E[g(X)]E[h(Y)]. \tag{4.4}$$

Letting g and h be the identity function gives the useful result that

$$E[XY] = E[X]E[Y]. \tag{4.5}$$

The expectation of a product of independent random variables is the product of their expectations. A proof of the special case of identify functions follows. If X and Y are independent,

$$\begin{aligned} E[XY] &= \sum_{x\in S}\sum_{y\in T} xyP(X = x, Y = y) \\ &= \sum_{x\in S}\sum_{y\in T} xyP(X = x)P(Y = y) \\ &= \sum_{x\in S} xP(X = x)\sum_{y\in T} yP(Y = y) \\ &= E[X]E[Y]. \end{aligned}$$

The general result $E[g(X)h(Y)] = E[g(X)]E[h(Y)]$ follows similarly.

Example 4.16 Random cone. Suppose the radius R and height H of a cone are independent and each uniformly distributed on $\{1, \ldots, 10\}$. Find the expected volume of the cone.

The volume of a cone is given by the formula $v(r, h) = \pi r^2 h/3$. Let V be the volume of the cone. Then $V = \pi R^2 H/3$ and

$$\begin{aligned} E[V] = E\left[\frac{\pi}{3}R^2H\right] &= \frac{\pi}{3}E[R^2H] = \frac{\pi}{3}E[R^2]E[H] \\ &= \frac{\pi}{3}\left(\sum_{r=1}^{10}\frac{r^2}{10}\right)\left(\sum_{h=1}^{10}\frac{h}{10}\right) \\ &= \frac{\pi}{3}\left(\frac{77}{2}\right)\left(\frac{11}{2}\right) = \frac{847\pi}{12} \approx 221.744, \end{aligned}$$

where the third equality uses the independence of R^2 and H, which follows from the independence of R and H. ■

R: EXPECTED VOLUME

The expectation is simulated with the commands

```
> simlist <- replicate(100000,
     (pi/3)*(sample(1:10, 1)^2)*sample(1:10, 1))
> mean(simlist)
[1] 221.955
```

4.4.1 Sums of Independent Random Variables

Sums of random variables figure prominently in probability and statistics. To find probabilities of the form $P(X+Y=k)$, observe that $X+Y=k$ if and only if $X=i$ and $Y=k-i$ for some i. This gives

$$P(X+Y=k)=P\left(\bigcup_i \{X=i, Y=k-i\}\right)=\sum_i P(X=i, Y=k-i). \quad (4.6)$$

If X and Y are independent, then

$$P(X+Y=k)=\sum_i P(X=i, Y=k-i)=\sum_i P(X=i)P(Y=k-i). \quad (4.7)$$

The limits of the sum $\sum_i$ will depend on the possible values of X and Y. For instance, if X and Y are nonnegative integers, then

$$P(X+Y=k)=\sum_{i=0}^{k} P(X=i, Y=k-i), \quad \text{for } k \geq 0.$$

Example 4.17 A nationwide survey collected data on TV usage in the United States. The distribution of US households by number of TVs per household is given in Table 4.3. If two households are selected at random, find the probability that there are a total of exactly two TVs in both households combined.

Let T_1 and T_2 be the number of TVs in the two households, respectively. Then,

$$\begin{aligned}
&P(T_1+T_2=2)\\
&\quad = P(T_1=0, T_2=2)+P(T_1=1, T_2=1)+P(T_1=2, T_2=0)\\
&\quad = P(T_1=0)P(T_2=2)+P(T_1=1)P(T_2=1)+P(T_1=2)P(T_2=0)\\
&\quad = (0.01)(0.33)+(0.21)(0.21)+(0.33)(0.01)=0.051.
\end{aligned}$$

■

TABLE 4.3. Distribution of US households by number of TVs.

TVs	0	1	2	3	4	5
Proportion of households	0.01	0.21	0.33	0.23	0.13	0.09

Example 4.18 During rush hour, the number of minivans M on a fixed stretch of highway has a Poisson distribution with parameter λ_M. The number of sports cars S on the same stretch has a Poisson distribution with parameter λ_S. If the number of minivans and sports cars is independent, find the pmf of the total number of these vehicles, $M + S$.

For $k \geq 0$,

$$\begin{aligned}
P(M + S = k) &= \sum_{i=0}^{k} P(M = i, S = k - i) \\
&= \sum_{i=0}^{k} P(M = i)P(S = k - i) \\
&= \sum_{i=0}^{k} \left(\frac{e^{-\lambda_M} \lambda_M^i}{i!} \right) \left(\frac{e^{-\lambda_S} \lambda_S^{k-i}}{(k-i)!} \right) \\
&= e^{-(\lambda_M + \lambda_S)} \sum_{i=0}^{k} \frac{\lambda_M^i \lambda_S^{k-i}}{i!(k-i)!} \\
&= \frac{e^{-(\lambda_M + \lambda_S)}}{k!} \sum_{i=0}^{k} \binom{k}{i} \lambda_M^i \lambda_S^{k-i} \\
&= \frac{e^{-(\lambda_M + \lambda_S)}}{k!} (\lambda_M + \lambda_S)^k.
\end{aligned}$$

The last equality follows from the binomial theorem. We see from the final form of the pmf that $M + S$ has a Poisson distribution with parameter $\lambda_M + \lambda_S$. ■

The last example illustrates a general result, which we will prove in Chapter 5.

THE SUM OF INDEPENDENT POISSON RANDOM VARIABLES IS POISSON

Let $X_1, \ldots, X_k$ be a sequence of independent Poisson random variables with respective parameters $\lambda_1, \ldots, \lambda_k$. Then

$$Y = X_1 + \cdots + X_k \sim \text{Pois}(\lambda_1 + \cdots + \lambda_k).$$

Example 4.19 Sum of uniforms. Let X and Y be independent random variables both uniformly distributed on $\{1, \ldots, n\}$. Find the pmf of $X + Y$.

The sum $X + Y$ takes values between 2 and $2n$. We have that

$$\begin{aligned} P(X + Y = k) &= \sum_i P(X = i, Y = k - i) \\ &= \sum_i P(X = i)P(Y = k - i), \quad \text{for } k = 2, \ldots, 2n. \end{aligned}$$

We need to take care with the limits of the sum $\sum_i$. The limits of i in the sum will depend on k, with two cases, the first when $2 \le k \le n$, and the second when $n + 1 \le k \le 2n$. We suggest making a table to examine the setting with $n = 3$ to reinforce this. For the first case, when $2 \le k \le n$,

$$P(X + Y = k) = \sum_{i=1}^{k-1} P(X = i)P(Y = k - i) = \sum_{i=1}^{k-1} \left(\frac{1}{n}\right)\left(\frac{1}{n}\right) = \frac{k-1}{n^2}.$$

For the second case, when $n + 1 \le k \le 2n$,

$$\begin{aligned} P(X + Y = k) &= \sum_{i=k-n}^{n} P(X = i)P(Y = k - i) \\ &= \sum_{i=k-n}^{n} \left(\frac{1}{n}\right)\left(\frac{1}{n}\right) = \frac{2n - k + 1}{n^2}. \end{aligned}$$

Summarizing,

$$P(X + Y = k) = \begin{cases} (k-1)/n^2, & \text{for } k = 2, \ldots, n, \\ (2n - k + 1)/n^2, & \text{for } k = n + 1, \ldots, 2n. \end{cases}$$

Observe the case $n = 6$ gives the pmf of the sum of two independent dice rolls. ■

4.5 LINEARITY OF EXPECTATION

A very important property for computing expectations is the *linearity* property of expectation.

LINEARITY OF EXPECTATION

For random variables X and Y, $E[X + Y] = E[X] + E[Y]$.

To prove this result in the discrete setting, we apply formula Equation 4.3 for the expectation of a function of two random variables with the function $g(x, y) = x + y$. This gives

$$\begin{aligned}
E[X+Y] &= \sum_{x \in S} \sum_{y \in T} (x+y) P(X=x, Y=y) \\
&= \sum_{x \in S} \sum_{y \in T} x P(X=x, Y=y) + \sum_{x \in S} \sum_{y \in T} y P(X=x, Y=y) \\
&= \sum_{x \in S} x \left(\sum_{y \in T} P(X=x, Y=y) \right) + \sum_{y \in T} y \left(\sum_{x \in S} P(X=x, Y=y) \right) \\
&= \sum_{x \in S} x P(X=x) + \sum_{y \in T} y P(Y=y) \\
&= E[X] + E[Y].
\end{aligned}$$

Thus, the expectation of a sum is equal to the sum of the expectations.

Linearity of expectation is an enormously useful result. Note carefully that it makes no assumptions about the distribution of X and Y. In particular, it does not assume independence. It applies to *all* random variables regardless of their joint distribution.

Linearity of expectation extends to finite sums. That is,

$$E[X_1 + \cdots + X_n] = E[X_1] + \cdots + E[X_n]. \tag{4.8}$$

It does not, however, extend to infinite sums in general. That is, it is not always true that $E[\sum_{x=1}^{\infty} X_i] = \sum_{x=1}^{\infty} E[X_i]$. However, if all the X_i's are nonnegative and if the infinite sum $\sum_{i=1}^{\infty} E[X_i]$ converges, we leave it to the reader to show that

$$E\left[\sum_{i=1}^{\infty} X_i\right] = \sum_{i=1}^{\infty} E[X_i].$$

Indicator variables. Next, we turn our attention to a discrete random variable that is useful for solving a wide range of problems. Given an event A, define a random variable I_A such that

$$I_A = \begin{cases} 1, & \text{if } A \text{ occurs,} \\ 0, & \text{if } A \text{ does not occur.} \end{cases}$$

Therefore, I_A equals 1, with probability $P(A)$, and 0, with probability $P(A^c)$. Such a random variable is called an *indicator variable.* An indicator is a Bernoulli random variable with $p = P(A)$.

The expectation of an indicator variable is important enough to highlight. Working from the definition of expectation, we see that

$$E[I_A] = (1)P(A) + (0)P(A^c) = P(A).$$

This is fairly simple, but nevertheless extremely useful and interesting, because it means that probabilities of events can be thought of as expectations of indicator random variables.

Often random variables involving *counts* can be analyzed by expressing the count as a sum of indicator variables. We illustrate this powerful technique in the next examples.

Example 4.20 Expectation of the binomial distribution. Let $I_1, \ldots, I_n$ be a sequence of i.i.d. Bernoulli (indicator) random variables with success probability p. Let $X = I_1 + \cdots + I_n$. Then X has a binomial distribution with parameters n and p. We want to find $E[X]$. By linearity of expectation,

$$E[X] = E\left[\sum_{k=1}^{n} I_k\right] = \sum_{k=1}^{n} E[I_k] = \sum_{k=1}^{n} p = np.$$

This result should be intuitive. For instance, if you roll 600 dice, you would expect 100 ones. The number of ones has a binomial distribution with $n = 600$, $p = 1/6$ and $np = 100$.

We emphasize the simplicity and elegance of the last derivation, a result of thinking probabilistically about the problem. Contrast this with the algebraic approach below. If X has a binomial distribution with parameters n and p, then by the definition of expectation,

$$\begin{aligned}
E[X] &= \sum_{k=0}^{n} kP(X = k) = \sum_{k=0}^{n} k\binom{n}{k} p^k(1-p)^{n-k} \\
&= \sum_{k=1}^{n} \frac{n!}{(k-1)!(n-k)!} p^k(1-p)^{n-k} \\
&= np\sum_{k=1}^{n} \frac{(n-1)!}{(k-1)!(n-k)!} p^{k-1}(1-p)^{n-k} \\
&= np\sum_{k=1}^{n} \binom{n-1}{k-1} p^{k-1}(1-p)^{n-k} \\
&= np\sum_{k=0}^{n-1} \binom{n-1}{k} p^{k}(1-p)^{n-k-1} \\
&= np(p + (1-p))^{n-1} = np,
\end{aligned}$$

where the next-to-last equality follows from the binomial theorem. ■

For the binomial example, the sequence of Bernoulli indicator variables is independent. But they need not be to use linearity of expectation, as illustrated next.

Example 4.21 Problem of coincidences. The problem of coincidences, also called the matching problem, was introduced by Pierre Rémond de Montmort in 1703. In the French card game Recontres (*Coincidences*), two persons, each having a full standard deck of cards, draw from their deck at the same time one card after the other, until they both draw the same card. Montmort asked for the probability that a match occurs. We consider the expected number of matches, with a modern twist.

At their graduation ceremony, a class of n seniors, upon hearing that they have graduated, throw their caps up into the air in celebration. Their caps fall back to the ground uniformly at random and each student picks up a cap. What is the expected number of students who get their original cap back (a "match")?

Let X be the number of matches. Define

$$I_k = \begin{cases} 1, & \text{if the } k\text{th student gets their cap back },\\ 0, & \text{if the } k\text{th student does not get their cap back,} \end{cases}$$

for $k = 1, \ldots, n$. Then $X = I_1 + \cdots + I_n$. The expected number of matches is

$$\begin{aligned} E[X] = E\left[\sum_{k=1}^{n} I_k\right] &= \sum_{k=1}^{n} E[I_k] \\ &= \sum_{k=1}^{n} P(k\text{th student gets their cap back}) \\ &= \sum_{k=1}^{n} \frac{1}{n} = n\left(\frac{1}{n}\right) = 1. \end{aligned}$$

The probability that the kth student gets their cap back is $1/n$ as there are n caps to choose from and only one belongs to the kth student.

Remarkably, the expected number of matches is one, independent of the number of people n. If everyone in the world throws their hat up in the air, on average about one person will get their hat back. ■

The indicator random variables $I_1, \ldots, I_n$ in the matching problem are not independent. In particular, if $I_1 = \cdots = I_{n-1} = 1$, that is, if the first $n-1$ people get their hats back, then necessarily $I_n = 1$, the last person must also get their hat back, as unlikely as that is to occur.

One can cast the matching problem in terms of permutations, which is useful when simulating. Given a permutation of $\{1, \ldots, n\}$, a *fixed point* is a number k such that the number k is in position k in the permutation. For instance, the permutation $(2, 4, 3, 1, 5)$ has two fixed points—3 and 5. The permutation $(3, 4, 5, 1, 2)$ has no

TABLE 4.4. Fixed points of permutations for $n = 3$.

Permutation	Number of fixed points
(1,2,3)	3
(1,3,2)	1
(2,1,3)	1
(2,3,1)	0
(3,1,2)	0
(3,2,1)	1

fixed points. Table 4.4 gives the permutations and number of fixed points for the case $n = 3$.

The number of fixed points in a permutation is equal to the number of matches in the problem of coincidences. If we pick a permutation above uniformly at random, then the pmf of F, the number of fixed points, is

$$P(F = k) = \begin{cases} 2/6 = 1/3, & \text{if } k = 0, \\ 3/6 = 1/2, & \text{if } k = 1, \\ 1/6, & \text{if } k = 3, \end{cases}$$

with expectation $E[F] = 0(2/6) + 1(3/6) + 3(1/6) = 1$.

R: SIMULATING THE MATCHING PROBLEM

In order to simulate the matching problem, we simulate random permutations with the command

```
> sample(n, n)
```

It is remarkable how fast this is done for even large n. Even though there are $n! \approx n^n$ permutations of an n-element set, the algorithm for generating a random permutation takes on the order of n steps, not n^n steps, and is thus extremely efficient. (You can learn more about computational efficiency in computer science courses.)

To count the number of fixed points in a uniformly random permutation, type

```
> sum(sample(n, n) == 1:n)
```

Here we simulate the expected number of matchings in the original game of Recontres with a standard deck of 52 cards. The result matches the analytical solution.

```
> n <- 52
> mean(replicate(10000, sum(sample(n, n) == (1:n))))
[1] 0.9992
```

Now pause to reflect upon what you have actually done in the simulation. You have simulated a random element from a sample space that contains $52! \approx 8 \times 10^{67}$ elements, checked the number of fixed points, and then repeated the operation 10,000 times, finally computing the average number of fixed points—all in a second or two on your computer. It would not be physically possible to write down a table of all 52! permutations, their number of fixed points, or their corresponding probabilities. And yet by generating random elements and taking simulations, the problem becomes computationally feasible. Of course, in this case, we already know the exact answer and do not need simulation to find the expectation. However, that is not the case with many complex, real-life problems. In many cases, simulation may be easier or it may even be the only way to go.

To learn more about the problem of coincidences, see Takacs [1980].

Example 4.22 St. Petersburg paradox. I offer you the following game. Flip a fair coin until heads appears. If it takes n tosses, I will pay you $\$2^n$. Thus, if heads comes up on the first toss, I pay you \$2. If it first comes up on the 10th toss, I pay you \$1024.

How much would you pay to play this game? Would you pay \$5, \$50, or \$500?

Let X be the payout. Your expected payout is

$$E[X] = \sum_{n=1}^{\infty} 2^n \frac{1}{2^n} = \sum_{n=1}^{\infty} 1 = +\infty.$$

The expectation is not finite. Indeed, it appears that the expected value is infinite.

This problem, discovered by the eighteenth-century Swiss mathematician Daniel Bernoulli, is the St. Petersburg paradox. The "paradox" is that most people would not pay very much to play this game, and yet the expected payout appears infinite. ■

4.6 VARIANCE AND STANDARD DEVIATION

Expectation is a measure of the *average* behavior of a random variable, often termed a measure of *center*. Variance and standard deviation are measures of *variability* or *spread*. They describe how near or far typical outcomes are to the expected value (the mean).

VARIANCE AND STANDARD DEVIATION

Let X be a random variable with mean $E[X] = \mu < \infty$. The *variance of* X is

$$V[X] = E[(X - \mu)^2] = \sum_x (x - \mu)^2 P(X = x). \tag{4.9}$$

The *standard deviation of* X is

$$\text{SD}[X] = \sqrt{V[X]}.$$

In the variance formula, $(x - \mu)$ is the difference or "deviation" of an outcome from the mean. Thus, the variance is a weighted average of the squared deviations from the mean. The standard deviation of X is the square root of the variance.

Variance and standard deviation are always nonnegative because the deviations from the mean are squared. The greater the variability of outcomes, the larger the deviations from the mean, and the greater the two measures. If X is a constant, and hence has *no* variability, then $X = E[X] = \mu$, and we see from the variance formula that $V[X] = 0$. The converse is also true. That is, if $V[X] = 0$, then X is almost certainly a constant. We leave the proof to Exercise 4.40.

The graphs in Figure 4.1 show four probability distributions all with expectations equal to 4, but with different variances.

Example 4.23 We find the variances for each of the distributions in Figure 4.1. Let W, X, Y, and Z be the corresponding random variables, respectively. We have that $E[W] = E[X] = E[Y] = E[Z] = 4$.

(a) W is a constant equal to 4. The variance is 0 because $(4 - 4)^2(1) = 0$.

(b) The pmf for X is

$$P(X = k) = \begin{cases} 1/25, & \text{if } k = 1, 7, \\ 3/25, & \text{if } k = 2, 6, \\ 5/25, & \text{if } k = 3, 5, \\ 7/25, & \text{if } k = 4, \end{cases}$$

with variance

$$V[X] = 2(1 - 4)^2 \frac{1}{25} + 2(2 - 4)^2 \frac{3}{25} + 2(3 - 4)^2 \frac{5}{25} = 2.08.$$

(c) Outcomes are equally likely. This gives

$$V[Y] = \sum_{k=1}^{7} (k - 4)^2 \frac{1}{7} = 4.$$

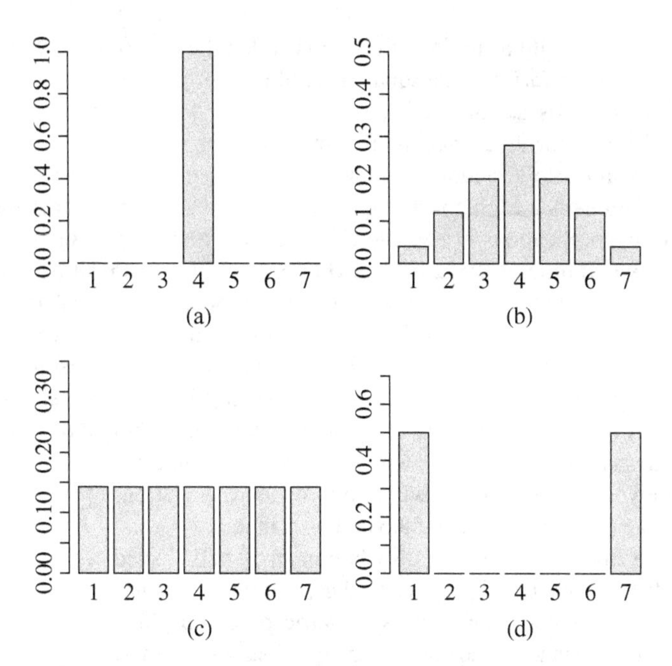

FIGURE 4.1: Four distributions with $\mu = 4$. Variances are (a) 0, (b) 2.08, (c) 4, and (d) 9.

(d) We have $P(Z = 1) = P(Z = 7) = 1/2$. This gives

$$V[Z] = (1 - 4)^2\frac{1}{2} + (7 - 4)^2\frac{1}{2} = 9. \qquad \blacksquare$$

The variance is a "cleaner" mathematical formula than the standard deviation because it does not include the square root. For simplicity and for connections with other areas of mathematics, mathematicians and probabilists often prefer variance when working with random variables.

However, the standard deviation can be easier to interpret particularly when working with data. In statistics, random variables are often used to model data, which have some associated units attached to their measurements. For instance, we might take a random person's height and assign it to a random variable H. The units are inches. The expected height $E[H]$ is also expressed in inches. Because of the square in the variance formula, the units of the variance $V[H]$ are square inches. The square root in the standard deviation brings the units back to the units of the variable.

In statistics, there are analogous definitions of mean, variance, and standard deviation for a collection of data. For a list of measurements $x_1, \ldots, x_n$, the *sample mean* is the average $\overline{x} = (x_1 + \cdots + x_n)/n$. The *sample variance* is defined as $\sum_{i=1}^{n} (x_i - \overline{x})^2/(n - 1)$. (The reason the denominator is $n - 1$ rather than n is a topic for a statistics class, it has to do with what is called the bias of the estimator.) The *sample standard deviation* is the square root of the sample variance.

The **R** commands `mean(vec)`, `var(vec)`, and `sd(vec)` compute these quantities for a vector `vec`. See the **R** supplement for an example.

When we simulate a random variable X with repeated trials, the sample mean of the replications is a Monte Carlo approximation of $E[X]$. Similarly, the sample variance of the replications is a Monte Carlo approximation of $V[X]$.

Within two standard deviations of the mean. Many probability distributions are fairly concentrated near their expectation in the sense that the probability that an outcome is within a few standard deviations from the mean is high. This is particularly true for symmetric, "bell-shaped" distributions such as the Poisson distribution when λ is large and the binomial distribution when p is close to 1/2. The same is true for many real world variables. For instance, the average height of women in the United States is about 63.6 inches with standard deviation 2.5 inches. Roughly 95% of all adult women's heights are within two standard deviations of the mean, which is within $63.6 \pm 2(2.5)$, between 58.6 and 68.6 inches.

When we discuss the normal distribution (the "bell-shaped curve") in Chapter 7, this will be made more precise. In the meantime, a rough "rule of thumb" is that for many symmetric and near-symmetric probability distributions, the mean and variance (or standard deviation) are good summary measures for describing the behavior of "typical" outcomes of the underlying random experiment. Typically, most outcomes from such distributions fall within two standard deviations of the mean.

Computational formula for variance. From the definition of the variance, a little manipulation goes a long way. A useful computational formula for the variance is

$$V[X] = E[X^2] - E[X]^2. \tag{4.10}$$

This is derived using properties of expectation. Remember that $\mu = E[X]$ is a constant. Then

$$\begin{aligned} V[X] &= E[(X-\mu)^2] = E[X^2 - 2\mu X + \mu^2] \\ &= E[X^2] - 2\mu E[X] + \mu^2 \\ &= E[X^2] - 2\mu^2 + \mu^2 = E[X^2] - \mu^2 \\ &= E[X^2] - E[X]^2. \end{aligned}$$

Example 4.24 Variance of uniform distribution. Suppose X is uniformly distributed on $\{1, \dots, n\}$. Find the variance of X.

Using the computational formula, rather than the definition of variance, we need to find $E[X^2]$, because we already know $E[X] = (n+1)/2$.

We find that

$$E[X^2] = \sum_{k=1}^{n} k^2 P(X = k) = \frac{1}{n} \sum_{k=1}^{n} k^2$$
$$= \left(\frac{1}{n}\right) \frac{n(n+1)(2n+1)}{6} = \frac{(n+1)(2n+1)}{6}.$$

Plugging in, this gives

$$V[X] = E[X^2] - E[X]^2 = \frac{(n+1)(2n+1)}{6} - \left(\frac{n+1}{2}\right)^2 = \frac{n^2-1}{12}.$$

For large n, the mean of the uniform distribution on $\{1, \ldots, n\}$ is $(n+1)/2 \approx n/2$, and the standard deviation is $\sqrt{(n^2-1)/12} \approx n/\sqrt{12} \approx n/3.5$. ■

Example 4.25 Variance of an indicator. For an event A, let I_A be the corresponding indicator random variable. Find $V[I_A]$.

As I_A only takes values 0 and 1, it follows that $(I_A)^2 = I_A$, which gives

$$V[I_A] = E[I_A^2] - E[I_A]^2$$
$$= E[I_A] - E[I_A]^2$$
$$= P(A) - P(A)^2$$
$$= P(A)(1 - P(A)) = P(A)P(A^c).$$

We summarize our results on indicator variables for later use. ■

EXPECTATION AND VARIANCE OF INDICATOR VARIABLE

$$E[I_A] = P(A) \qquad \text{and} \qquad V[I_A] = P(A)P(A^c).$$

Properties of variance. The linearity properties of expectation do *not* extend to the variance. What do we find instead?

If X is a random variable with expectation μ, and a and b are constants, then $aX + b$ has expectation $a\mu + b$ and

$$V[aX+b] = E[(aX + b - (a\mu + b))^2] = E[(aX - a\mu)^2]$$
$$= E[a^2(X-\mu)^2] = a^2 E[(X-\mu)^2] = a^2 V[X].$$

We summarize these important properties for reference.

PROPERTIES OF EXPECTATION, VARIANCE, AND STANDARD DEVIATION

Let X be a random variable, where $E[X]$ and $V[X]$ exist. For constants a and b,

$$E[aX + b] = aE[X] + b,$$
$$V[aX + b] = a^2 V[X], \text{ and}$$
$$\mathrm{SD}[aX + b] = |a|\mathrm{SD}[X].$$

Example 4.26 See Example 4.9. Suppose the French lab's temperature measurements have a variance of 2°C. Then upon conversion to Fahrenheit, the variance of the US lab's temperature measurements is

$$V[F] = V\left[32 + \left(\frac{9}{5}\right)C\right] = \left(\frac{9}{5}\right)^2 V[C] = 2\left(\frac{81}{25}\right) = \frac{162}{25},$$

with standard deviation $\mathrm{SD}[F] = \sqrt{162/25} = 2.55$°F. ■

Variance of the Poisson distribution. In Exercise 4.30, we invite you to show that the variance of a Poisson random variable X with parameter λ is equal to λ. Hence,

$$E[X] = V[X] = \lambda.$$

This is a special property of the Poisson distribution. Using the heuristic that *most* observations are within two standard deviations of the mean, it would follow that most outcomes of a Poisson random variable are contained in the interval $\lambda \pm 2\sqrt{\lambda}$. This is generally true, at least for large λ.

R: SIMULATION OF POISSON DISTRIBUTION

You can observe the phenomenon that most observations are within two standard deviations of the mean for a specific choice of λ for the Poisson distribution by typing

```
> lambda <- 25
> hist(rpois(100000, lambda), prob = TRUE)
> abline(v = lambda-2*sqrt(lambda))
> abline(v = lambda+2*sqrt(lambda))
```

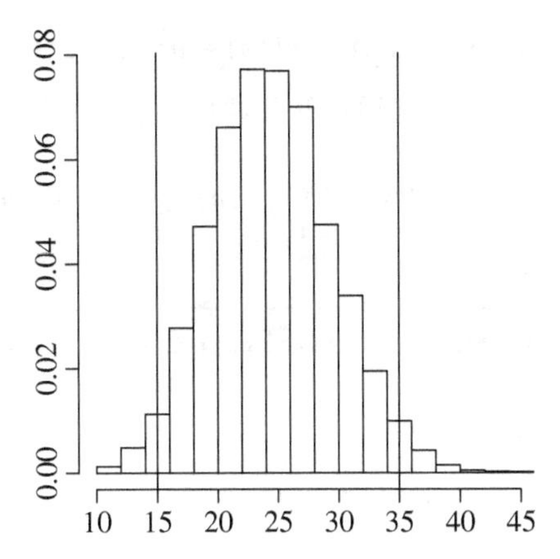

FIGURE 4.2: Simulation of Poisson(25) distribution. Vertical lines are drawn at $x = 15$ and $x = 35$, two standard deviations from the mean.

The output of the commands, with $\lambda = 25$, is shown in Figure 4.2. The graph is a histogram. Think of the rectangles as bins. For each x-value, the height of the rectangle corresponds to the number of outcomes of x. The vertical axis is scaled so that the areas of the rectangles sum to 1. The histogram is generated from 100,000 simulations of a Poisson(25) random variable. It thus simulates the probability distribution of X. Vertical lines are drawn at two standard deviations from the mean, that is, at $x = 15$ and $x = 35$.

Variances of sums and sums of variances. For random variables X and Y, consider the variance of the sum $V(X + Y)$. By definition, we have

$$V(X + Y) = E[(X + Y)^2] - (E[X + Y])^2.$$

Let $\mu_X = E[X]$ and $\mu_Y = E[Y]$. Then

$$E[(X + Y)^2] = E[X^2 + 2XY + Y^2] = E[X^2] + 2E[XY] + E[Y^2]$$

and

$$(E[X + Y])^2 = (\mu_X + \mu_Y)^2 = \mu_X^2 + 2\mu_X\mu_Y + \mu_Y^2.$$

Hence, by combining and rearranging terms,

$$\begin{aligned} V[X + Y] &= E[(X + Y)^2] - (E[X + Y])^2 \\ &= E[X^2] + 2E[XY] + E[Y^2] - (\mu_X^2 + 2\mu_X\mu_Y + \mu_Y^2) \end{aligned}$$

$$= (E[X^2] - \mu_X^2) + (E[Y^2] - \mu_Y^2) + 2(E[XY] - \mu_X\mu_Y)$$
$$= V[X] + V[Y] + 2(E[XY] - E[X]E[Y]). \quad (4.11)$$

If X and Y are independent, then $E[XY] = E[X]E[Y]$, making the last term zero, and the variance of $X + Y$ has a simple form.

VARIANCE OF THE SUM OF INDEPENDENT VARIABLES

If X and Y are independent, then

$$V[X + Y] = V[X] + V[Y]. \quad (4.12)$$

For the difference of independent random variables, it might be tempting to conclude that $V[X - Y] = V[X] - V[Y]$. But this is not true. Rather, if $X \perp Y$,

$$V[X - Y] = V[X + (-1)Y] = V[X] + V[(-1)Y]$$
$$= V[X] + (-1)^2 V[Y]$$
$$= V[X] + V[Y].$$

Variances always add. (If you are not completely convinced and have some nagging doubts about $V[X - Y]$, suppose that X and Y are independent with $V[X] = 1$ and $V[Y] = 2$. If $V[X - Y] = V[X] - V[Y] = 1 - 2 = -1$, we have a problem, because variances can never be negative.) These results extend to working with more than two random variables.

Example 4.27 Variance of binomial distribution. Recall how indicator variables are used to find the expectation of a binomial distribution. Use them again to find the variance.

Suppose $X = I_1 + \cdots + I_n$ is the sum of n independent indicator variables with success probability p. Then X has a binomial distribution with parameters n and p. As the I_k's are independent, the variance of the sum of indicators is equal to the sum of the variances and thus

$$V[X] = V\left[\sum_{k=1}^{n} I_k\right] = \sum_{k=1}^{n} V[I_k] = \sum_{k=1}^{n} p(1 - p) = np(1 - p).$$ ■

Example 4.28 Preston said he flipped 100 pennies and got 70 heads. Is this believable?

The number of heads, X, has a binomial distribution with parameters $n = 100$ and $p = 1/2$. The mean number of heads is $E(X) = np = 50$. The standard deviation is $\sqrt{np(1-p)} = \sqrt{25} = 5$. The distribution is symmetric and bell-shaped. Thus, we expect most outcomes from tossing 100 coins to fall with two standard deviations of the mean, that is, between 40 and 60 heads. As 70 heads represents an outcome that is four standard deviations from the mean, we are a little suspicious of Preston's claim. ■

The last example illustrated how knowledge of probability and probability models can be used to help assess claims. If you continue to study statistics, you will no doubt encounter the topics of hypothesis testing and confidence intervals. Within the realm of hypothesis testing, p-values are probabilities computed based on a probability model. This is one reason a solid foundation in probability is necessary for any statistician.

Example 4.29 Roulette continued—how the casino makes money. In Example 4.2, we found the expected value of a bet of "red" in roulette. We will shift gears and look at things from the casino's perspective. Your loss is their gain.

Let G be the casino's gain after a player makes one red bet. Then

$$P(G = 1) = \frac{20}{38} \text{ and } P(G = -1) = \frac{18}{38}$$

with $E[G] = 2/38$. The casino's expected gain from one red bet is about five cents.

For the variance, $E[G^2] = (1)(20/38) + (1)(18/38) = 1$, and thus,

$$V[G] = E[G^2] - E[G]^2 = 1 - (2/38)^2 = 0.99723,$$

with standard deviation $\sqrt{0.99723} = 0.998614$, almost \$1.

Suppose in one month, the casino expects customers to make n red bets. What is the expected value and standard deviation of the casino's total gain?

Let G_k be the casino's gain from the kth red bet of the month, for $k = 1, \ldots, n$. Let T be the casino's total gain. Write $T = G_1 + \cdots + G_n$. Unless someone is cheating, we can assume that the G_k's are independent.

By linearity of expectation,

$$E[T] = E[G_1 + \cdots + G_n] = E[G_1] + \cdots + E[G_n] = \frac{2n}{38},$$

or about n nickels. If one million bets are placed, that is an expected gain of \$52,631.58.

Of interest to the casino's accountants is the variability of total gain. If the variance is large, the casino might see big swings from month to month, where some months they make money and some months they do not.

By independence of the G_k's,

$$V[T] = V[G_1 + \cdots + G_n] = V[G_1] + \cdots + V[G_n] = (0.99723)n,$$

with standard deviation

$$\mathrm{SD}[T] = (0.998614)\sqrt{n}.$$

For one million bets, the standard deviation is \$998.61. This is about one-fiftieth the size of the mean.

By the heuristic that says that most observations are within about two standard deviations from the mean, it is virtually certain that every month, the casino will see a hefty positive gain of about \$52,000 give or take about \$2000 if one million bets are placed. This is why for the customers, roulette is risky entertainment, but for the casino, it is a business. ■

R: A MILLION RED BETS

Here we simulate the casino's gain from three different months of play, assuming one million red bets are placed each month.

```
> n <- 1000000
> probs <- c(18/38, 20/38)
> sum(sample(c(-1, 1), n, probs, replace = T))
[1] 53738
> sum(sample(c(-1, 1), n, probs, replace = T))
[1] 52140
> sum(sample(c(-1, 1), n, probs, replace = T))
[1] 52686
```

Note this example continues use of the `sample` command to simulate from the finite distribution of interest. It is a very useful command.

4.7 COVARIANCE AND CORRELATION

Having looked at measures of variability for individual and independent random variables, we now consider measures of variability between dependent random variables. The *covariance* is a measure of the association between two random variables.

COVARIANCE

For random variables X and Y, with respective means μ_X and μ_Y, the *covariance between X and Y* is

$$\text{Cov}(X, Y) = E[(X - \mu_X)(Y - \mu_Y)]. \tag{4.13}$$

Equivalently, an often more usable computational formula is

$$\text{Cov}(X, Y) = E[XY] - \mu_X\mu_Y = E[XY] - E[X]E[Y]. \tag{4.14}$$

We leave it as an exercise to show that Equation 4.14 follows from the definition Equation 4.13.

Covariance will be positive when large values of X are associated with large values of Y and small values of X are associated with small values of Y. In particular, for outcomes x and y, covariance is positive when products of the form $(x - \mu_X)(y - \mu_Y)$ in the covariance formula tend to be positive, meaning that both terms in the product are either positive or negative.

On the other hand, if X and Y are inversely related, most product terms $(x - \mu_X)(y - \mu_Y)$ will be negative, as when X takes values above the mean, Y will tend to fall below the mean, and vice versa. In this case, the covariance between X and Y will be negative.

To see examples of potential relationships, see Figure 4.3. Points on the graphs are simulations of the points (X, Y) from four joint distributions. Vertical and horizontal lines are drawn at the mean of the marginal distributions. In the graph in (a), the main contribution to the covariance is from points in the first and third quadrants, where products are positive. In panel (b), the main contribution is in the second and fourth quadrants, where products are negative. In panel (c), the random variables are independent, and each of the four quadrants are near equally represented so positive terms cancel negative terms and the covariance is 0.

Covariance is a measure of *linear association* between two variables. In a sense, the "less linear" the relationship, the closer the covariance is to 0. In the fourth graph (d), the covariance will be close to 0 as positive products tend to cancel out negative products, but here the random variables are not independent. There is a strong relationship between them, although it is not linear. Both large and small values of X are associated with small values of Y.

The sign of the covariance indicates whether two random variables are positively or negatively associated. But the magnitude of the covariance can be difficult to

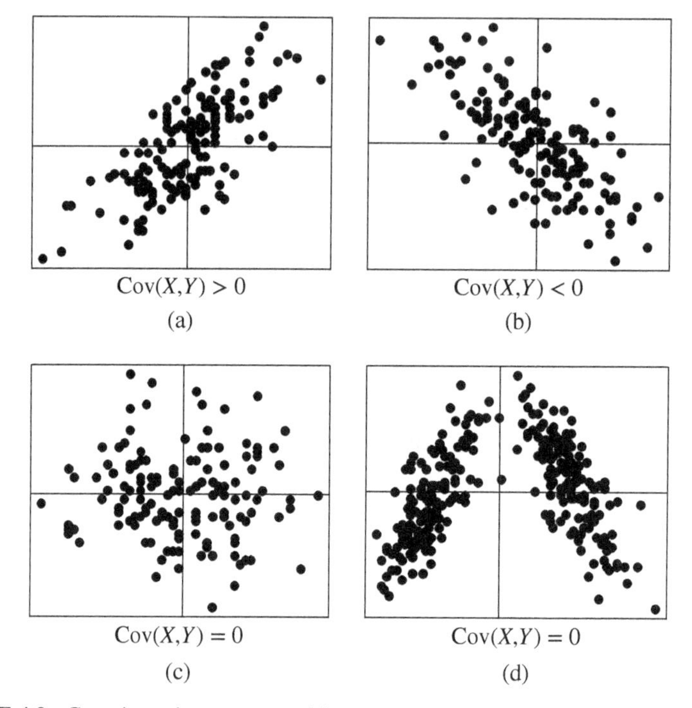

FIGURE 4.3: Covariance is a measure of linear association between two random variables. Vertical and horizontal lines are drawn at the mean of the marginal distributions.

interpret due to the scales of the original variables. The *correlation* is an alternative measure which is easier to interpret.

CORRELATION

The correlation between X and Y is

$$\text{Corr}(X, Y) = \frac{\text{Cov}(X, Y)}{\text{SD}[X]\text{SD}[Y]}.$$

Properties of correlation:

1. $-1 \leq \text{Corr}(X, Y) \leq 1$.
2. If $Y = aX + b$ is a linear function of X for constants a and b, then $\text{Corr}(X, Y) = \pm 1$, depending on the sign of a.

Correlation is a common summary measure in statistics. Dividing the covariance by the standard deviations creates a "standardized" covariance, which is a unitless measure that takes values between -1 and 1. The correlation is exactly equal to ± 1 if Y is a linear function of X. We prove these results in Section 4.9.

Random variables that have correlation, and covariance, equal to 0 are called *uncorrelated*.

UNCORRELATED RANDOM VARIABLES

We say random variables X and Y are uncorrelated if

$$E[XY] = E[X]E[Y],$$

that is, if $\text{Cov}(X, Y) = 0$.

For independent random variables, $E[XY] = E[X]E[Y]$ and thus $\text{Cov}(X, Y) = 0$. Thus, if random variables X and Y are independent, then they are also uncorrelated. However, the converse is not necessarily true, as shown in the next example.

Example 4.30 Let X be uniformly distributed on $\{-1, 0, 1\}$. Let $Y = X^2$. The two random variables are not independent as Y is a function of X. However,

$$\text{Cov}(X, Y) = \text{Cov}(X, X^2) = E[X^3] - E[X]E[X^2] = 0 - 0 = 0.$$

The random variables are uncorrelated. ■

Example 4.31 The number of defective parts in a manufacturing process is modeled as a binomial random variable with parameters n and p. Let X be the number of defective parts, and let Y be the number of nondefective parts. Find the covariance between X and Y.

Observe that $Y = n - X$ is a linear function of X. Thus, $\text{Corr}(X, Y) = -1$. Because "failures" for X are "successes" for Y, Y has a binomial distribution with parameters n and $1 - p$. To find the covariance, rearrange the correlation formula

$$\begin{aligned}\text{Cov}(X, Y) &= \text{Corr}(X, Y)\text{SD}[X]\text{SD}[Y] \\ &= (-1)\sqrt{np(1-p)}\sqrt{n(1-p)p} \\ &= -np(1-p).\end{aligned}$$

■

Example 4.32 Red ball, blue ball continued. For Example 4.11, find $\text{Cov}(R, B)$, the covariance between the number of red and blue balls.

Recall that the joint distribution table is

		B		
		0	1	2
R	0	3/36	6/36	1/36
	1	12/36	8/36	0
	2	6/36	0	0

and from the joint table, we find

$$E[RB] = \sum_r \sum_b rbP(R = r, B = b) = \frac{8}{36} = \frac{2}{9}.$$

Note that eight of the nine terms in the sum are equal to 0. Also, we previously found that $E[R] = 8/9$ and $E[B] = 4/9$. This gives

$$\begin{aligned}\text{Cov}(R, B) &= E[RB] - E[R]E[B] \\ &= \frac{2}{9} - \left(\frac{8}{9}\right)\left(\frac{4}{9}\right) \\ &= -\frac{14}{81} = -0.17284.\end{aligned}$$

The negative result should not be surprising. There is an inverse association between R and B as the more balls there are of one color in the sample the fewer balls there will be of another color. ■

One place where the covariance regularly appears is in the variance formula for sums of random variables. We summarize results for the variance of a sum.

GENERAL FORMULA FOR VARIANCE OF A SUM

For random variables X and Y with finite variance,

$$V[X + Y] = V[X] + V[Y] + 2\text{Cov}(X, Y) \text{ and}$$

$$V[X - Y] = V[X] + V[Y] - 2\text{Cov}(X, Y).$$

If X and Y are uncorrelated,

$$V[X \pm Y] = V[X] + V[Y].$$

Example 4.33 After a severe storm, the number of claims received by an insurance company for hail H and tornado T damage are each modeled with a Poisson distribution with respective parameters 400 and 100. The correlation between H and T is 0.75. Let Z be the total number of claims from hail and tornado damage. Find the variance and standard deviation of Z. Hint: Use what you know about the Poisson distribution.

We have

$$E[Z] = E[H + T] = E[H] + E[T] = 400 + 100 = 500,$$

and

$$\begin{aligned} V[Z] &= V[H + T] = V[H] + V[T] + 2\text{Cov}(H, T) \\ &= 400 + 100 + 2\text{Corr}(H, T)\text{SD}[H]\text{SD}[T] \\ &= 500 + 2(0.75)\sqrt{400}\sqrt{100} = 500 + 300 = 800, \end{aligned}$$

with standard deviation $\text{SD}[Z] = \sqrt{800} = 28.2843$. ■

The covariance of a random variable X with itself, $\text{Cov}(X, X)$, is just the variance $V[X]$, as $E[(X - \mu_X)(X - \mu_X)] = E[(X - \mu_X)^2]$. Also, the covariance is symmetric in its terms. That is, $\text{Cov}(X, Y) = \text{Cov}(Y, X)$. So another way of writing the variance of a sum of two random variables is

$$V[X + Y] = \text{Cov}(X, X) + \text{Cov}(Y, Y) + \text{Cov}(X, Y) + \text{Cov}(Y, X),$$

that is as a sum over all possible pairings of X and Y.

For the variance of a sum of more than two random variables, take covariances of all possible pairs.

VARIANCE OF SUM OF N RANDOM VARIABLES

$$\begin{aligned} V[X_1 + \cdots + X_n] &= \sum_{i=1}^{n}\sum_{j=1}^{n} \text{Cov}(X_i, X_j) \\ &= V[X_1] + \cdots + V[X_n] + \sum_{i \neq j} \text{Cov}(X_i, X_j) \\ &= V[X_1] + \cdots + V[X_n] + 2\sum_{i<j} \text{Cov}(X_i, X_j). \end{aligned}$$

The penultimate sum is over all pairs of indices $i \neq j$. For the last equality, because $\text{Cov}(X_i, X_j) = \text{Cov}(X_j, X_i)$, we can just consider indices such that $i < j$.

Example 4.34 Let X, Y, Z be random variables with variances equal to 1, 2, and 3, respectively. Also, $\text{Cov}(X, Y) = -1$, $\text{Cov}(X, Z) = 0$, and $\text{Cov}(Y, Z) = 3$. The variance of $X + Y + Z$ is

$$\begin{aligned} &V[X + Y + Z] \\ &= V[X] + V[Y] + V[Z] + 2(\text{Cov}(X, Y) + \text{Cov}(X, Z) + \text{Cov}(Y, Z)) \\ &= 1 + 2 + 3 + 2(-1 + 0 + 3) = 10. \end{aligned}$$

■

Example 4.35 Coincidences continued. In the matching problem, Example 4.21, the expectation of the number of students who get their cap back was shown to be one using the method of indicators. What is the variance?

Let X be the number of students who get their own cap back. Write $X = I_1 + \cdots + I_n$, where I_k is equal to 1, if the kth student gets their cap, and 0, otherwise, for $k = 1, \ldots, n$. As the I_k's are not independent, use the general sum formula

$$V[X] = V\left[\sum_{k=1}^{n} I_k\right] = \sum_{k=1}^{n} V[I_k] + \sum_{i \neq j} \text{Cov}(I_i, I_j).$$

We have

$$E[I_k] = P(k\text{th student gets their cap}) = \frac{1}{n} \text{ and}$$

$$\begin{aligned} V[I_k] &= P(k\text{th student gets their cap})P(k\text{th student does not get their cap}) \\ &= \left(\frac{1}{n}\right)\left(\frac{n-1}{n}\right) = \frac{n-1}{n^2}. \end{aligned}$$

For the covariance terms, consider $E[I_i I_j]$, where $i \neq j$. As I_i and I_j are 0–1 variables, the product $I_i I_j$ is equal to 1 if and only if both I_i and I_j are equal to 1, which happens if the ith and jth students both get their caps back. That is, $I_i I_j$ is the indicator of the event that the ith and jth students get their caps back. Thus,

$$E[I_i I_j] = P(i\text{th and } j\text{th students get their caps}) = P(I_i = 1, I_j = 1).$$

Conditional on the jth student getting their cap back, the probability that the ith student gets their cap is $1/(n-1)$ because there are $n - 1$ caps to choose from. Thus,

$$P(I_1 = 1, I_j = 1) = P(I_i = 1 | I_j = 1)P(I_j = 1) = \left(\frac{1}{n-1}\right)\left(\frac{1}{n}\right).$$

For $i \neq j$, this gives

$$\text{Cov}(I_i, I_j) = E[I_i I_j] - E[I_i]E[I_j] = \frac{1}{n(n-1)} - \frac{1}{n^2} = \frac{1}{n^2(n-1)}.$$

Putting all the pieces together, we have

$$V[X] = \sum_{i=1}^{n} \frac{n-1}{n^2} + \sum_{i \neq j} \frac{1}{n^2(n-1)} = \frac{n-1}{n} + \sum_{i \neq j} \frac{1}{n^2(n-1)}.$$

Finally, we need to count the number of terms in the last sum. There are n^2 terms in the full double sum $\sum_{i=1}^{n} \sum_{j=1}^{n}$, and there are n terms for which $i = j$. This leaves $n^2 - n$ terms for which $i \neq j$. Hence,

$$\begin{aligned} V[X] &= \frac{n-1}{n} + \sum_{i \neq j} \frac{1}{n^2(n-1)} \\ &= \frac{n-1}{n} + (n^2 - n)\left(\frac{1}{n^2(n-1)}\right) \\ &= \frac{n-1}{n} + \frac{1}{n} = 1. \end{aligned}$$

The variance, and standard deviation, of the number of matchings is also equal to 1. Thus, if everyone in the world throws their hat up in the air, we expect about one person to get their hat back give or take one or two. It would be extremely unlikely if four or more people get their hat back. ■

4.8 CONDITIONAL DISTRIBUTION

Chapter 2 introduced conditional probability. For jointly distributed random variables, we have the more general notion of a *conditional distribution*, where we can consider the distribution of one variable given a value of another. In the discrete setting, we define the conditional pmf.

CONDITIONAL PROBABILITY MASS FUNCTION

If X and Y are jointly distributed discrete random variables, then the *conditional probability mass function* of Y given $X = x$ is

$$P(Y = y|X = x) = \frac{P(X = x, Y = y)}{P(X = x)},$$

when $P(X = x) > 0$.

The conditional pmf *is* a pmf. It is nonnegative and sums to 1.

If $X \perp Y$, then the conditional pmf of Y given $X = x$ reduces to the regular pmf of Y, as $P(Y = y|X = x) = P(Y = y)$.

Example 4.36 In a study of geriatric health and the risk of hip fractures, Schechner et al. [2010] model the occurrence of falls and hip fractures for elderly individuals. They assume that the number of times a person falls during one year has a Poisson distribution with parameter λ. Each fall may independently result in a hip fracture with probability p.

Let X be the number of falls. If $X = x$, then conditional on the number of falls x, the number of hip fractures Y has a binomial distribution with parameters x and p. That is,

$$P(Y = y|X = x) = \binom{x}{y} p^y(1-p)^{x-y}, \quad \text{for } y = 0, \ldots, x.$$

We write $Y|X = x \sim \text{Binom}(x,p)$ or $\text{Bin}(x,p)$.

Now, we want to (i) find the joint pmf of X and Y, and (ii) find the marginal distribution of the number of hip fractures Y.

(i) Rearranging the conditional probability formula gives

$$\begin{aligned} P(X = x, Y = y) &= P(Y = y|X = x)P(X = x) \\ &= \binom{x}{y} p^y(1-p)^{x-y} \left(\frac{e^{-\lambda}\lambda^x}{x!}\right) \\ &= \frac{p^y(1-p)^{x-y}e^{-\lambda}\lambda^x}{y!(x-y)!}, \end{aligned}$$

for $x = 0, 1, \ldots,$ and $y = 0, \ldots, x$, and 0, implicitly otherwise.

(ii) The marginal distribution of Y is found by summing the joint pmf over values of x. As $0 \le y \le x$, we sum over $x \ge y$. This gives

$$\begin{aligned} P(Y = y) &= \sum_{x=y}^{\infty} P(X = x, Y = y) = \sum_{x=y}^{\infty} \frac{p^y(1-p)^{x-y}e^{-\lambda}\lambda^x}{y!(x-y)!} \\ &= \frac{e^{-\lambda}p^y}{y!} \sum_{x=y}^{\infty} \frac{(1-p)^{x-y}\lambda^x}{(x-y)!} = \frac{e^{-\lambda}(\lambda p)^y}{y!} \sum_{x=y}^{\infty} \frac{(\lambda(1-p))^{x-y}}{(x-y)!} \\ &= \frac{e^{-\lambda}(\lambda p)^y}{y!} \sum_{x=0}^{\infty} \frac{(\lambda(1-p))^x}{x!} \\ &= \frac{e^{-\lambda}(\lambda p)^y}{y!} e^{\lambda(1-p)} = \frac{e^{-\lambda p}(\lambda p)^y}{y!}, \end{aligned}$$

for $y = 0, 1, \ldots$. Thus, the marginal distribution of the number of hip fractures is Poisson with parameter λp. We can interpret this distribution as a "thinned" version of the distribution of the number of falls X. Both are Poisson distributions. The mean parameter of Y is p times the mean parameter of X. ■

Example 4.37 Red ball, blue ball continued. For Example 4.11, find the conditional pmf of the number of red balls given that there are no blue balls in the sample.

Recall that the joint distribution table is

		B: 0	1	2	
	0	3/36	6/36	1/36	10/36
R	1	12/36	8/36	0	20/36
	2	6/36	0	0	6/36
		21/36	14/36	1/36.	

We have

$$P(R = r|B = 0) = \frac{P(R = r, B = 0)}{P(B = 0)} = \frac{P(R = r, B = 0)}{21/36}.$$

This gives

$$P(R = r|B = 0) = \begin{cases} (3/36)/(21/36) = 1/7, & \text{if } r = 0, \\ (12/36)/(21/36) = 4/7, & \text{if } r = 1, \\ (6/36)/(21/36) = 2/7, & \text{if } r = 2. \end{cases}$$ ■

Example 4.38 During the day, Sam receives text message and phone calls. The numbers of each are independent Poisson random variables with parameters λ and μ, respectively. If Sam receives n texts and phone calls during the day, find the conditional distribution of the number of texts he receives.

Let T be the number of texts Sam receives. Let C be the number of phone calls. We showed in Section 4.4.1 that the sum of independent Poisson variables has a Poisson distribution, and thus $T + C \sim \text{Pois}(\lambda + \mu)$. For $0 \le t \le n$, the conditional pmf of T given $T + C = n$ is

$$\begin{aligned} P(T = t|T + C = n) &= \frac{P(T = t, T + C = n)}{P(T + C = n)} \\ &= \frac{P(T = t, C = n - t)}{P(T + C = n)} = \frac{P(T = t)P(C = n - t)}{P(T + C = n)} \\ &= \left(\frac{e^{-\lambda}\lambda^t}{t!}\right)\left(\frac{e^{-\mu}\mu^{n-t}}{(n-t)!}\right)\left(\frac{e^{-(\lambda+\mu)}(\lambda + \mu)^n}{n!}\right)^{-1} \end{aligned}$$

$$= \binom{n}{t} \frac{\lambda^t \mu^{n-t}}{(\lambda + \mu)^n} = \binom{n}{t} \left(\frac{\lambda}{\lambda + \mu}\right)^t \left(\frac{\mu}{\lambda + \mu}\right)^{n-t}.$$

The conditional distribution of T, given $T + C = n$, is binomial with parameters n and $p = \lambda/(\lambda + \mu)$. ■

Example 4.39 At the beginning of this section, we said that conditional distribution generalizes the idea of conditional probability. Let A and B be events. Define corresponding indicator random variables $X = I_A$ and $Y = I_B$. The conditional distribution of Y, given the outcomes of X is straightforward:

$$P(Y = y|X = 1) = \begin{cases} P(A|B), & \text{if } y = 1, \\ P(A^c|B), & \text{if } y = 0, \end{cases}$$

and

$$P(Y = y|X = 0) = \begin{cases} P(A|B^c), & \text{if } y = 1, \\ P(A^c|B^c), & \text{if } y = 0. \end{cases}$$

■

4.8.1 Introduction to Conditional Expectation

A *conditional expectation* is an expectation with respect to a conditional distribution.

CONDITIONAL EXPECTATION OF Y GIVEN $X = x$

For discrete random variables X and Y, the *conditional expectation* of Y given $X = x$ is

$$E[Y|X = x] = \sum_y yP(Y = y|X = x).$$

Similarly, a conditional variance is a variance with respect to a conditional distribution. Conditional expectation and conditional variance will be treated in depth in Chapter 9. We take a first look at this important concept in the discrete setting with several examples.

Example 4.40 In Example 4.38, we find that the conditional distribution of the number of Sam's texts T given $T + C = n$ is binomial with parameters n and $p = \lambda/(\lambda + \mu)$. From properties of the binomial distribution, it follows immediately that

$$E[T|T + C = n] = np = \frac{n\lambda}{\lambda + \mu} = n\left(\frac{\lambda}{\lambda + \mu}\right).$$

Similarly, the conditional variance is

$$V[T|T+C=n] = np(1-p) = n\left(\frac{\lambda}{\lambda+\mu}\right)\left(1-\frac{\lambda}{\lambda+\mu}\right) = \frac{n\lambda\mu}{(\lambda+\mu)^2}.$$

■

Example 4.41 The joint pmf of X and Y is given in the following joint table, along with the marginal probabilities. Find $E[Y|X=x]$.

		Y			
		1	2	6	
X	−1	0.0	0.1	0.1	0.2
	1	0.1	0.3	0.2	0.6
	4	0.1	0.1	0.0	0.2
		0.2	0.5	0.3.	

For $x = -1$,

$$\begin{aligned} E[Y|X=-1] &= \sum_{y\in\{1,2,6\}} yP(Y=y|X=-1) \\ &= \sum_{y\in\{1,2,6\}} y\frac{P(Y=y,X=-1)}{P(X=-1)} \\ &= 1(0.0/0.2) + 2(0.1/0.2) + 6(0.1/0.2) = 4. \end{aligned}$$

Similarly, for $x = 1$,

$$\begin{aligned} E[Y|X=1] &= 1(0.1/0.6) + 2(0.3/0.6) + 6(0.2/0.6) \\ &= 1.9/0.6 = 3.167. \end{aligned}$$

And for $x = 4$,

$$\begin{aligned} E[Y|X=4] &= 1(0.1/0.2) + 2(0.1/0.2) + 6(0.0/0.2) \\ &= 0.3/0.2 = 1.5 \end{aligned}$$

This gives

$$E[Y|X=x] = \begin{cases} 4, & \text{if } x=-1, \\ 3.167, & \text{if } x=1, \\ 1.5, & \text{if } x=4. \end{cases}$$

■

Example 4.42 Xavier picks a number X uniformly at random from $\{1, 2, 3, 4\}$. Having picked $X = x$, he shows the number to Yasmin, who picks a number Y uniformly at random from $\{1, \dots, x\}$. (i) Find the expectation of Yasmin's number if Xavier picks x. (ii) Yasmin picked the number 1. What do you think Xavier's number was?

(i) The problem asks for $E[Y|X = x]$. The conditional distribution of Y, given $X = x$, is uniform on $\{1, \dots, x\}$. The expectation of the uniform distribution gives

$$E[Y|X = x] = \frac{x+1}{2}.$$

(ii) If Yasmin picked one, we will infer Xavier's number based on the conditional expectation of X given $Y = 1$. We have

$$\begin{aligned} E[X|Y = 1] &= \sum_{x=1}^{4} xP(X = x|Y = 1) \\ &= \sum_{x=1}^{4} x\frac{P(X = x, Y = 1)}{P(Y = 1)}. \end{aligned}$$

In the numerator,

$$P(X = x, Y = 1) = P(Y = 1|X = x)P(X = x) = \frac{1}{x}\left(\frac{1}{4}\right) = \frac{1}{4x}.$$

For the denominator, condition on X and apply the law of total probability.

$$\begin{aligned} P(Y = 1) &= \sum_{x=1}^{4} P(Y = 1|X = x)P(X = x) \\ &= \sum_{x=1}^{4} \frac{1}{4x} = \frac{1}{4} + \frac{1}{8} + \frac{1}{12} + \frac{1}{16} = \frac{25}{48}. \end{aligned}$$

We thus have

$$E[X|Y = 1] = \sum_{x=1}^{4} x\left(\frac{1}{4x}\right)\frac{48}{25} = \sum_{x=1}^{4} \frac{12}{25} = \frac{48}{25} = 1.92.$$

A good guess at Xavier's number is 2! ■

R: SIMULATING A CONDITIONAL EXPECTATION

We simulate the conditional expectation $E[X|Y = 1]$ from the previous example.

```
> trials <- 10000
> ctr <- 0
> simlist <- numeric(trials)
> while (ctr < trials) {
   xav <- sample(1:4, 1)
   yas <- sample(1:xav, 1)
   if (yas == 1) {
    ctr <- ctr + 1
    simlist[ctr] <- xav}
   }
> mean(simlist)
[1] 1.9372
```

4.9 PROPERTIES OF COVARIANCE AND CORRELATION*

The covariance $\text{Cov}(X, Y)$ takes two arguments. In each argument, the function is linear in the following sense.

COVARIANCE PROPERTY: LINEARITY

For random variables X, Y, and Z, and constants a, b, c,

$$\text{Cov}(aX + bY + c, Z) = a\text{Cov}(X, Z) + b\text{Cov}(Y, Z) \tag{4.15}$$

and

$$\text{Cov}(X, aY + bZ + c) = a\text{Cov}(X, Y) + b\text{Cov}(X, Z). \tag{4.16}$$

Showing these properties is a straightforward application of the definition of covariance, which we leave to the exercises.

Given a random variable X with mean μ and variance σ^2, the *standardized* variable X^* is defined as

$$X^* = \frac{X - \mu}{\sigma}.$$

Observe that

$$E[X^*] = E\left[\frac{X-\mu}{\sigma}\right] = \frac{1}{\sigma}(E[X]-\mu) = \frac{1}{\sigma}(\mu-\mu) = 0$$

and

$$V[X^*] = V\left[\frac{X-\mu}{\sigma}\right] = \frac{1}{\sigma^2}(V[X-\mu]) = \frac{\sigma^2}{\sigma^2} = 1.$$

"Standardizing" a random variable in this way gives a new random variable with mean 0 and variance 1, often referred to in statistics as creating a "z-score" version of the variable. The standardization can be "undone" if desired as $X = \mu + \sigma X^*$.

The following theorem shows that for any random variables X and Y, the correlation $\text{Corr}(X, Y)$ is always between -1 and 1. Further, the correlation is equal to ± 1 if and only if one variable is a linear function of the other. Most proofs of this result use linear algebra. The following probabilistic treatment is based on Feller [1968].

CORRELATION RESULTS

Theorem. For random variables X and Y,

$$-1 \leq \text{Corr}(X, Y) \leq 1.$$

If $\text{Corr}(X, Y) = \pm 1$, then there exists constants $a \neq 0$ and b such that $Y = aX + b$.

Proof: Given X and Y, let X^* and Y^* be the standardized variables. Observe that

$$\begin{aligned}
\text{Cov}(X^*, Y^*) &= \text{Cov}\left(\frac{X-\mu_X}{\sigma_X}, \frac{Y-\mu_Y}{\sigma_Y}\right) \\
&= \frac{1}{\sigma_X\sigma_Y}\text{Cov}(X, Y) \\
&= \text{Corr}(X, Y).
\end{aligned}$$

Consider the variance of $X^* \pm Y^*$:

$$\begin{aligned}
V(X^* + Y^*) &= V(X^*) + V(Y^*) + 2\text{Cov}(X^*, Y^*) \\
&= 2 + 2\text{Corr}(X, Y)
\end{aligned}$$

and

$$\begin{aligned}
V(X^* - Y^*) &= V(X^*) + V(Y^*) - 2\text{Cov}(X^*, Y^*) \\
&= 2 - 2\text{Corr}(X, Y).
\end{aligned}$$

This gives

$$\text{Corr}(X, Y) = \frac{V(X^* + Y^*)}{2} - 1 \geq -1$$

and

$$\text{Corr}(X, Y) = -\frac{V(X^* - Y^*)}{2} + 1 \leq 1$$

because the variance is nonnegative. That is, $-1 \leq \text{Corr}(X, Y) \leq 1$.

Suppose $Y = aX + b$ is a linear function of X. In Exercise 4.45, you will show that $\text{Corr}(X, Y) = \pm 1$ depending upon the sign of a. Conversely, if $\text{Corr}(X, Y) = 1$, then $V(X^* - Y^*) = 0$ and thus $X^* - Y^*$ is a constant. Recall that

$$X^* - Y^* = \frac{X - \mu_X}{\sigma_X} - \frac{Y - \mu_Y}{\sigma_Y}.$$

Solving for X gives

$$X = \left(\frac{\sigma_X}{\sigma_Y}\right) Y + \text{constant}.$$

A similar argument holds when $\text{Corr}(X, Y) = -1$. □

4.10 EXPECTATION OF A FUNCTION OF A RANDOM VARIABLE*

Let X be a random variable. Let g be a function. Following is the proof that

$$E[g(X)] = \sum_x g(x)P(X = x),$$

the so-called "law of the unconscious statistician."

Proof: Let $Y = g(X)$. Then

$$\begin{aligned}
E[g(X)] = E[Y] &= \sum_y yP(Y = y) \\
&= \sum_y yP(g(X) = y) \\
&= \sum_y yP(X \in g^{-1}(y)) \\
&= \sum_y y \sum_{x : x \in g^{-1}(y)} P(X = x) \\
&= \sum_y y \sum_{x : g(x) = y} P(X = x)
\end{aligned}$$

$$= \sum_x \left(\sum_{y:y=g(x)} y \right) P(X = x)$$

$$= \sum_x g(x)P(X = x). \qquad \square$$

Note that if g is a one-to-one function and has an inverse g^{-1}, then $g^{-1}(y)$ is a single number. However, in general, $g^{-1}(y)$ is a set of numbers such that $x \in g^{-1}(y)$ if and only if $g(x) = y$.

4.11 SUMMARY

Random variables are the central objects of probability and this chapter introduces them in the discrete setting. Many important concepts are first introduced in this chapter. The pmf of a discrete random variable is its probability function $P(X = k)$. The expectation of a random variable is a measure of its "average" or typical value. The variance is a measure of spread and variability. We often work with functions of random variables $g(X)$, and these are presented along with methods for finding their expectation.

Numerous properties of expectation and variance are given. Most important is linearity of expectation: the expectation of a sum is equal to the sum of the expectations. The property holds without any condition on the distribution of the underlying random variables. If the random variables are also independent, then the variance of a sum is equal to the sum of the variances.

For two or more random variables defined on the same sample space, we have a joint distribution (generally speaking) and joint pmf (in the discrete case). The univariate marginal distributions are obtained from the joint distribution by summing over the other variable(s). Indicator random variables I_A are introduced and shown to be a useful method for representing counts. The covariance and correlation of two jointly distributed random variables is a measure of the linear association between them. The general formula for the variance of a sum requires a covariance term.

At the end of the chapter, conditional distributions are introduced. The conditional pmf is defined, along with conditional expectation. Several examples demonstrate the concept of a pmf for one variable conditional on another.

- **Probability mass function:** For a random variable X, the pmf is $m(x) = P(X = x)$.
- **Expectation:** $E[X] = \sum_x xP(X = x)$.
  1. For $X \sim \text{Unif}(\{1, \ldots, n\})$, $E[X] = (n + 1)/2$.
  2. For $X \sim \text{Ber}(p)$, $E[X] = p$
  3. For $X \sim \text{Binom}(n, p)$, $E[X] = np$.
  4. For $X \sim \text{Pois}(\lambda)$, $E[X] = \lambda$.

- **Functions of random variables:** If g is a function, then $g(X)$ is a random variable that takes values $g(x)$ whenever $X = x$.
- **Expectation of a function of a random variable and the "law of the unconscious statistician":** $E[g(X)] = \sum_x g(x)P(X = x)$. Warning: It is not true in general that $E[g(X)] = g(E[X])$. In particular, it is not true that $E[X^2] = E[X]^2$.
- **Expectation of a linear function:** For constants a, b, $E[aX + b] = aE[X] + b$.
- **Joint probability mass function:** For jointly distributed discrete random variables X and Y, the joint pmf is $P(X = x, Y = y)$.
- **Marginal distributions:** If X and Y are jointly distributed, the marginal distribution of X is found by summing over the Y variable in the discrete case. That is $P(X = x) = \sum_y P(X = x, Y = y)$. Similarly, $P(Y = y) = \sum_x P(X = x, Y = y)$.
- **Joint pmf for independent random variables:** If $X \perp Y$, $P(X = x, Y = y) = P(X = x)P(Y = y)$.
- **Expectation of a function of two random variables:**

$$E[g(X, Y)] = \sum_x \sum_y g(x, y)P(X = x, Y = y).$$

- **Independent random variables:** Functions of independent random variables are independent. That is, if g and h are functions of independent random variables X and Y, then $g(X)$ and $h(Y)$ are independent.
- **Expectation of a product of independent random variables:** If X and Y are independent, then $E[XY] = E[X]E[Y]$. Similarly,

$$E[g(X)h(Y)] = E[g(X)]E[h(Y)], \quad \text{if } X \perp Y.$$

- **Sum of independent random variables:** If X and Y are independent, then

$$P(X + Y = k) = \sum_i P(X = i)P(Y = k - i).$$

- **Linearity of expectation:** $E[X + Y] = E[X] + E[Y]$.
- **Variance:** $V[X] = E[(X - E[X])^2] = E[X^2] - (E[X])^2$.
  1. For $X \sim \text{Unif}(\{1, \ldots, n\})$, $V[X] = (n^2 - 1)/12$.
  2. For $X \sim \text{Ber}(p)$, $V[X] = p(1 - p)$.
  3. For $X \sim \text{Binom}(n, p)$, $V[X] = np(1 - p)$.
  4. For $X \sim \text{Pois}(\lambda)$, $V[X] = \lambda$.
- **Standard deviation:** $\text{SD}[X] = \sqrt{V[X]}$.
- **Properties of variance and standard deviation:**
  1. $V[X] = 0$ if and only if X is a constant.
  2. $V[aX + b] = a^2 V[X]$.
  3. $\text{SD}[aX + b] = |a|\text{SD}[X]$.

- **Within two standard deviation heuristic:** For many near-symmetric probability distributions, "most" outcomes are (roughly) within two standard deviations of the mean.
- **Indicator random variables:** For an event A, the indicator random variable is defined such that $I_A = 1$, if A occurs, and 0, if A does not occur.
- **Properties of indicators:** $E[I_A] = P(A)$ and $V[I_A] = P(A)P(A^c)$.
- **Covariance:**

$$\text{Cov}(X, Y) = E[(X - E[X])(Y - E[Y])] = E[XY] - E[X]E[Y].$$

  Covariance is a measure of linear association. Independent random variables have covariance equal to 0. If X and Y are inversely associated, the covariance between them is negative.
- **Correlation:** $\text{Corr}(X, Y) = \text{Cov}(X, Y)/(\text{SD}[X]\text{SD}[Y])$. Correlation always takes values between -1 and 1. If $Y = aX + b$ is a linear function of X, then $\text{Corr}(X, Y) = \pm 1$, with the sign determined by a.
- **Variance of sums:**
  1. $V[X + Y] = V[X] + V[Y] + 2\text{Cov}(X, Y)$.
  2. If X and Y are independent, $V[X \pm Y] = V[X] + V[Y]$.
  3. $V[X_1 + \cdots + X_n] = \sum_{k=1}^{n} V[X_k] + 2\sum_{i<j}\text{Cov}(X_i, X_j)$.
- **Uncorrelated:** Random variables X and Y are uncorrelated if $E[XY] = E[X]E[Y]$. Independent random variables are uncorrelated, but the converse is not necessarily true.
- **Conditional probability mass function:** For jointly distributed discrete random variables X and Y, the conditional pmf of Y, given $X = x$ is $P(Y = y|X = x) = P(X = x, Y = y)/P(X = x)$.
- **Conditional expectation:** $E[Y|X = x] = \sum_y yP(Y = y|X = x)$.
- **Problem-solving strategies—Using indicator random variables for counts:** Indicators are often used to model counts. For instance, if X is the number of successes in n trials, then write $X = I_1 + \cdots I_n$, where I_k is the indicator that success occurs on the kth trial. The representation of a count as a sum of indicators allows us to use linearity of expectation in finding the expectation of X, as

$$E[X] = E[I_1 + \cdots I_n] = E[I_1] + \cdots + E[I_n].$$

EXERCISES

Expectation

4.1 What is the average number of vehicles per household in the United States? Table 4.5 gives data from the 2010 US Census on the distribution of available vehicles per household. "Available vehicles" refers to the number of cars, vans, and pickup trucks kept at home and available for use by household members.

TABLE 4.5. Household size by vehicles available.

Available vehicles	0	1	2	3	4
Proportion of households	0.092	0.341	0. 376	0.135	0.056

Source: Data from U.S. Census.

Let X be the number of available vehicles for a randomly chosen household. Find the expectation of X.

4.2 Find the expectation of a random variable uniformly distributed on $\{a, \dots, b\}$.

4.3 The following dice game costs $10 to play. If you roll 1, 2, or 3, you lose your money. If you roll 4 or 5, you get your money back. If you roll a 6, you win $24.

(a) Find the distribution of your winnings W.

(b) Find the expected value of the game.

4.4 Let $X \sim \text{Unif}\{-2, -1, 0, 1, 2\}$.

(a) Find $E[X]$.

(b) Find $E[e^X]$.

(c) Find $E[1/(X+3)]$.

4.5 See Example 4.22 on the St. Petersburg paradox. Modify the game so that you only receive $\$2^{10} = \$1024$ if any number greater than or equal to 10 tosses are required to obtain the first tail. Find the expected value of this game.

4.6 Suppose $E[X^2] = 1$, $E[Y^2] = 2$, and $E[XY] = 3$. Find $E[(X+Y)^2]$.

4.7 Suppose $P(X = 1) = p$ and $P(X = 2) = 1 - p$. Show that there is no value of $0 < p < 1$ such that $E[1/X] = 1/E[X]$.

4.8 You have dealt five cards from a standard deck. Let X be the number of aces in your hand. Find $E[X]$.

4.9 Let $X \sim \text{Pois}(\lambda)$. Find $E[X!]$. For what values of λ does the expectation not exist?

4.10 On January 13, 2016, the jackpot of the powerball lottery was $1.586 billion. The website `https://www.powerball.com/games/home` has a link to the payouts and corresponding odds for the *Powerball* lottery. Recall that odds is the ratio of the probability that an event occurs to the probability that it will not occur. Find the expected value of the January 13, 2016 game.

Joint Distribution

4.11 The number of tornadoes T and earthquakes E over a month's time in a particular region is independent and has a Poisson distribution with parameters 4 and 2, respectively.

(a) Find the joint pmf of T and E.

(b) What is the probability of no tornadoes and no earthquakes in that region next month?

(c) What is the probability of at least two tornadoes and at least two earthquakes?

(d) What is the probability of at least two tornadoes or at most two earthquakes?

4.12 The joint pmf of X and Y is

$$P(X = x, Y = y) = \frac{x+1}{12},$$

for $x = 0, 1$ and $y = 0, 1, 2, 3$. Find the marginal distributions of X and Y. Describe their distributions *qualitatively.* That is, identify their distributions as one of the known distributions you have worked with (e.g., Bernoulli, binomial, Poisson, or uniform), including all distribution parameters.

4.13 Suppose X, Y, and Z have joint pmf

$$P(X = x, Y = y, Z = z) = c, \quad \text{for } x = 1, \ldots, n, \; y = 1, \ldots, x, \; z = 1, \ldots, y.$$

(a) Find the constant c. Of use will be the formula for the sum of the first n squares

$$\sum_{k=1}^{n} k^2 = \frac{n(n+1)(2n+1)}{6}.$$

(b) For $n = 4$, find $P(X \leq 3, Y \leq 2, Z = 1)$.

4.14 Suppose X, Y, and Z are independent random variables that take values 1 and 2 with probability 1/2 each. Find the pmf of (X, Y, Z).

4.15 Suppose X, Y, and Z are independent Bernoulli random variables with respective parameters $1/2$, $1/3$, and $1/4$.

(a) Find $E[XYZ]$.

(b) Find $E[e^{X+Y+Z}]$.

4.16 Suppose X and Y are independent random variables. Does $E[X/Y] = E[X]/E[Y]$? Either prove it true or exhibit a counterexample.

4.17 The joint pmf of (X, Y) is

$$P(X = x, Y = y) = \frac{1}{nx}, \quad \text{for } x = 1, \ldots, n, \; y = 1, \ldots, x.$$

(a) Find the marginal distribution of X. Describe the distribution qualitatively.

(b) Find $E[Y/(X+1)]$.

(c) For the case $n = 3$, write out explicitly the joint probability table and confirm your result in (b).

4.18 Let X and Y be the first and second numbers obtained in two draws from the set $\{1, 2, 3, 4\}$ sampling with replacement.

(a) Give the joint distribution table for X and Y.

(b) Find $P(X \leq Y)$.

(c) Repeat the previous two parts assuming the sampling is done without replacement.

4.19 A joint probability mass function is given by

$$P(X = 1, Y = 1) = \frac{1}{8} \quad P(X = 1, Y = 2) = \frac{1}{4}$$

$$P(X = 2, Y = 1) = \frac{1}{8} \quad P(X = 2, Y = 2) = \frac{1}{2}.$$

(a) Find the marginal distributions of X and Y.

(b) Are X and Y independent?

(c) Compute $P(XY \leq 3)$.

(d) Compute $P(X + Y > 2)$.

4.20 **Elevator problem.** An elevator containing p passengers is at the ground floor of a building with n floors. On its way to the top of the building, the elevator will stop if a passenger needs to get off. Passengers get off at a particular floor with probability $1/n$. Find the expected number of stops the elevator makes. (Hint: Use indicators, letting $I_k = 1$ if the kth floor is a stop. Be careful: more than one passenger can get off at a floor.)

4.21 A bag contains r red and g green candies. We draw n candies from the bag without replacement. Find the expected number of red candies drawn by using indicator variables.

4.22 Take an n-by-n board divided into one-by-one squares and color each square white or black uniformly at random. That is, for each square flip a coin and color it white, if heads, and black, if tails. Let X be the number of two-by-two square "subboards" of the chessboard that are all black (see Fig. 4.4). Use indicator variables to find the expected number of black two-by-two subboards.

4.23 In a class of 25 students, what is the expected number of months in which at least two students are born? Assume birth months are equally likely. Hint: Use indicators.

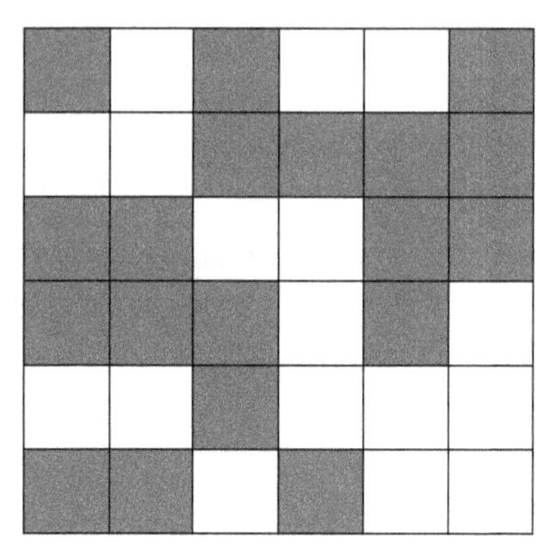

FIGURE 4.4: Randomly colored six-by-six board. There are 19 one-by-one black squares, 2 two-by-two black boards, and no three-by-three or larger black boards.

Variance, Standard Deviation, Covariance, Correlation

4.24 Find the variance for the outcome of a fair die roll two ways: (i) using the definition of variance and (ii) using the computational formula for variance, Equation 4.10. Which method do you prefer?

4.25 Suppose X takes values -1, 0, and 3, with respective probabilities 0.1, 0.3, and 0.6. Find $V[X]$.

4.26 Suppose $E[X] = a$, $E[Y] = b$, $V[X] = c$, and $V[Y] = d$. If X and Y are independent, find:

(a) $V[2X - 3Y + 4]$.

(b) $V[3X + Y - 2]$.

4.27 Find the variance of the sum of n independent tetrahedral dice rolls.

4.28 In a random experiment, let A and B be two independent events with $P(A) = P(B) = p$. In an outcome of the experiment, let X be the number of these events that occur (0, 1, or 2). Find $E[X]$ and $V[X]$.

4.29 Suppose $E[X] = 2$ and $V[X] = 3$. Find

(a) $E[(3 + 2X)^2]$.

(b) $V[4 - 5X]$.

4.30 Let X be a Poisson random variable with parameter λ. Find the variance of X. Hint: Write

$$X^2 = (X^2 - X) + X = X(X - 1) + X.$$

4.31 Suppose A and B are events with $P(A) = a$, $P(B) = b$, and $P(AB) = c$. Define indicator random variables I_A and I_B. Find $V[I_A + I_B]$.

4.32 Define a sequence $X_1, \ldots, X_n$ of independent random variables such that for each $k = 1, \ldots, n$, $X_k = \pm 1$, with probability 1/2 each. Let $S = X_1 + \cdots + X_n$. This model describes a *simple symmetric random walk* that starts at the origin and moves left or right at each unit of time. Find $E[S]$ and $V[S]$.

4.33 In the random walk model, suppose the distribution of the X_k's is given by $P(X_k = +1) = p$ and $P(X_k = -1) = 1 - p$. If $p > 1/2$, this describes a *random walk with positive drift*. Find $E[S]$ and $V[S]$.

4.34 A bag contains one red, two blue, three green, and four yellow balls. A sample of three balls is taken without replacement. Let B be the number of blue balls and Y the number of yellow balls in the sample.

(a) Find the joint probability table.

(b) Find $\text{Cov}(B, Y)$.

4.35 Random variables X and Y have joint distribution

$$P(X = i, Y = j) = c(i + j), \quad \text{for } i = 0, 1;\ j = 1, 2, 3$$

for some constant c.

(a) Find c.

(b) Find the marginal distributions of X and Y.

(c) Find $\text{Cov}(X, Y)$.

4.36 Suppose X and Y have the same distribution and $\text{Corr}(X, Y) = -0.5$. Find $V[X + Y]$.

4.37 Let D_1 and D_2 be the outcomes of two dice rolls. Let $X = D_1 + D_2$ be the sum of the two numbers rolled. Let $Y = D_1 - D_2$ be their difference. Show that X and Y are uncorrelated, but not independent.

4.38 Suppose X and Y are independent random variables with $V[X] = \sigma_X^2$, and $V[Y] = \sigma_Y^2$. Let $Z = wX + (1 - w)Y$, for $0 < w < 1$. Thus, Z is a weighted average of X and Y. Find the variance of Z. What value of w minimizes this variance?

4.39 Let $E[X] = 1$, $E[X^2] = 2$, $E[X^3] = 5$, and $E[X^4] = 15$. Also $E[Y] = 2$, $E[Y^2] = 6$, $E[Y^3] = 22$, and $E[Y^4] = 94$. Suppose X and Y are independent.

(a) Find $V[3X^2 - Y]$.

(b) Find $E[X^4 Y^4]$.

(c) Find $\text{Cov}(X, X^2)$.

(d) Find $V[X^2Y^2]$. (See also Exercise 4.58.)

4.40 It is apparent from the definition of variance that if X is a constant, then $V[X] = 0$. Here we show the converse.

(a) Suppose X is a nonnegative discrete random variable with $E[X] = 0$. Show that $P(X = 0) = 1$.

(b) Let X be a discrete random variable with $E[X] = \mu$ and $V[X] = 0$. Show that $P(X = \mu) = 1$.

4.41 Refer to Exercise 4.19. Find $V(X)$, $V(Y)$, and $\text{Cov}(X, Y)$.

4.42 Verify the linearity properties for covariance given in Equations 4.15 and 4.16.

4.43 Show that $\text{Cov}(X, Y) = E[XY] - E[X]E[Y]$, as a consequence of Equation 4.13.

4.44 Suppose $E[X] = 1$, $V[X] = 2$, $E[Y] = 3$, $V[Y] = 4$, and $\text{Cov}(X, Y) = -1$. Find $\text{Cov}(3X + 1, 2Y - 8)$.

4.45 Suppose $Y = aX + b$ for constants a and b. Find $\text{Cov}(X, Y)$.

Conditional Distribution, Expectation, Functions of Random Variables

4.46 The joint probability mass function of X and Y is

$$P(X = x, Y = y) = \frac{1}{e^2 y!(x - y)!}, \quad x = 0, 1, \ldots, \quad y = 0, 1, \ldots, x.$$

(a) Find the conditional distribution of Y given $X = x$.

(b) Describe the distribution in terms of distributions that you know.

(c) Without doing any calculations, find $E[Y|X = x]$ and $V[Y|X = x]$.

4.47 Leiter and Hamdan [1973] model traffic accidents and fatalities at a specific location in a given time interval. They suppose that the number of accidents X has a Poisson distribution with parameter λ. If $X = x$, then the number of fatalities Y has a binomial distribution with parameter p.

(a) Find the marginal distribution of the number of fatalities.

(b) Show that the conditional distribution of X given $Y = y$ is a Poisson distribution with parameter $\lambda(1 - p)$, which has been "shifted" y units to the right.

4.48 Let X be the first of two fair die rolls. Let M be the maximum of the two rolls.

(a) Find the conditional probability mass function of M given $X = x$.

(b) Find $E[M|X = x]$.

(c) Find the joint probability mass function of X and M.

4.49 Given events A and B, let I_A and I_B be the corresponding indicator variables. Find a simple expression for the conditional expectation of I_A given $I_B = 1$.

4.50 Suppose X and Y are independent with the following binomial distributions: $X \sim \text{Binom}(n, p)$ and $Y \sim \text{Binom}(m, p)$. Show that $X + Y$ has a binomial distribution and give the parameters. (Hint: Find $P(X + Y) = k$ and use Equation 1.8.)

4.51 Let X and Y be independent and identically distributed random variables. For each of the following questions, either show that it is true or exhibit a counter-example.

(a) Does $X + Y$ have the same distribution as $2X$?

(b) Does $X + Y$ have the same expectation as $2X$?

(c) Does $X + Y$ have the same variance as $2X$?

4.52 Let $X_1, \ldots, X_n$ be an i.i.d. Bernoulli sequence with parameter p.

(a) Find the conditional distribution of X_1 given $X_1 + \cdots + X_n = k$.

(b) Find $E[X_1|X_1 + \cdots + X_n = k]$ and $V[X_1|X_1 + \cdots + X_n = k]$.

4.53 Suppose X and Y are independent random variables both uniformly distributed on $\{1, \ldots, n\}$. Find the probability mass function of $X + Y$.

4.54 In the original matching problem, Montmort asked for the probability of at least one match. Find this probability and show that for large n, the probability of a match is about $1 - e^{-1} = 0.632$. Hint: Use inclusion–exclusion.

Simulation and R

4.55 Simulate the probability of obtaining at least one match in the problem of coincidences (see Exercise 4.54).

4.56 Simulate the dice game in Exercise 4.3. Estimate the expectation and variance of your winnings.

4.57 See Exercise 4.34. Simulate the mean and variance of the total number of blue and yellow balls in the sample.

4.58 In Exercise 4.39, the random variables X and Y have Poisson distributions with respective parameters 1 and 2. Simulate the results in that exercise.

4.59 Simulate the variance of the matching problem for a large value of n.

4.60 Let X be a random variable that takes values $x = (x_1, \ldots, x_n)$ with respective probabilities $p = (p_1, \ldots, p_n)$. Write two **R** functions `mymean(x,p)`, and `myvariance(x,p)`, which find the mean and variance of X, respectively. Use your function to find the mean and variance of the point value of a random Scrabble tile, as in Example 4.1.

4.61 In Texas hold 'em poker, players are initially dealt two cards each. (i) In a game of six players, simulate the probability that at least one of the players will have a pair. (ii) A hand is said to be *suited* if both cards are the same suit. Simulate the probability that at least one of the six players' hands is suited.

4.62 See Exercise 4.35. Write an **R** function `joint(i,j)` for computing $P(X = i, Y = j)$, for $i = 0, 1; j = 1, 2, 3$. Use this function to compute the covariance and correlation of X and Y.

Chapter Review

Chapter review exercises are available through the text website. The URL is `www.wiley.com/go/wagaman/probability2e`.

5

MORE DISCRETE DISTRIBUTIONS AND THEIR RELATIONSHIPS

To us, probability is the very guide of life.

—Marcus Tullius Cicero

Learning Outcomes

1. Distinguish common discrete distributions and their applications; Solve related problems.
2. Understand moment-generating functions and apply them to prove results.
3. Acquire insights into interesting problems such as coupon collecting and distributions of digits.*
4. (C) Use R to work with the distributions covered in the chapter.

Introduction. Previous chapters laid the foundations of probability and provided an introduction to discrete random variables including some common distributions. In this chapter, we present additional common discrete distributions, some of their applications, and use the new concept of moment-generating functions to highlight some relationships between them.

5.1 GEOMETRIC DISTRIBUTION

There are many questions one can ask about an underlying sequence of Bernoulli trials. The binomial distribution describes the number of successes in n trials. The geometric distribution describes the distribution of the number of trials until the first success occurs.

Probability: With Applications and R, Second Edition. Amy S. Wagaman and Robert P. Dobrow.

Companion Website: www.wiley.com/go/wagaman/probability2e

Juan likes to play the "pick-3" lottery game. In the "pick-3" game, you choose three single-digit numbers, each from 0 to 9, in order to match the winning number. There are 1000 possible choices so the probability of winning is $1/1000$.

If Juan buys a "pick-3" ticket every day, what is the probability he will win during the next year? And how many days can he expect to wait until he wins the lottery?

To model the number of days until Juan wins the lottery, consider successive outcomes as an independent and identically distributed (i.i.d.). Bernoulli sequence with success probability $p = 0.001$. Let X be the number of trials until the first success occurs. That is, the number of days until Juan wins the lottery. To find the probability mass function (pmf) of X observe that $X = k$ if success occurs on the kth trial and the first $k - 1$ trials are failures. This occurs with probability $(1 - p)^{k-1}p$.

GEOMETRIC DISTRIBUTION

The random variable X has a *geometric distribution with parameter p* if

$$P(X = k) = (1 - p)^{k-1}p, \quad \text{for } k = 1, 2, \ldots . \tag{5.1}$$

We write $X \sim \text{Geom}(p)$.

The parameter of the geometric distribution, p, is sometimes called the *success parameter*. The geometric pmf is nonnegative and sums to 1, as

$$\sum_{k=1}^{\infty} (1 - p)^{k-1}p = p\left(\frac{1}{1 - (1 - p)}\right) = 1.$$

Let X be the number of days until Juan wins the lottery. Then $X \sim \text{Geom}(0.001)$. The probability that Juan will win the lottery during the following year is

$$\begin{aligned} P(X \leq 365) &= \sum_{k=1}^{365} P(X = k) = \sum_{k=1}^{365} 0.999^{k-1}(0.001) \\ &= (0.001)\frac{1 - 0.999^{365}}{1 - 0.999} \\ &= 1 - 0.999^{365} = 0.3059, \end{aligned}$$

where we have used the partial sum of the geometric series formula

$$\sum_{k=1}^{n} r^{k-1} = \sum_{k=0}^{n-1} r^k = \frac{1 - r^n}{1 - r}, \quad \text{for } r \neq 1.$$

Another way to find the probability $P(X \leq 365)$ is to take complements and think probabilistically. Consider $P(X \leq 365) = 1 - P(X > 365)$. The event $\{X > 365\}$ is equal to the event that no lottery tickets are winners in the first 365 days. This occurs with probability $(1-p)^{365} = 0.999^{365}$. The desired probability is thus $1 - 0.999^{365}$.

This derivation gives the general closed form expression for the "tail probability" of a geometric random variable with success parameter p.

TAIL PROBABILITY OF GEOMETRIC DISTRIBUTION

If $X \sim \text{Geom}(p)$, then for $k > 0$,

$$P(X > k) = (1-p)^k.$$

Example 5.1 Expectation of the geometric distribution. Let $X \sim \text{Geom}(p)$. Then

$$E[X] = \sum_{k=1}^{\infty} kP(X=k) = \sum_{k=1}^{\infty} k(1-p)^{k-1}p = p\sum_{k=1}^{\infty} k(1-p)^{k-1}. \tag{5.2}$$

To make progress on the last sum, observe that the expression looks like a derivative. Consider the geometric series

$$\sum_{k=0}^{\infty} r^k = \frac{1}{1-r}, \quad \text{for } |r| < 1.$$

The series is absolutely convergent and can be differentiated termwise. That is, the derivative of the sum is equal to the sum of the derivatives. Differentiating with respect to r gives

$$\frac{d}{dr}\sum_{k=0}^{\infty} r^k = \sum_{k=1}^{\infty} kr^{k-1} = \frac{d}{dr}\left(\frac{1}{1-r}\right) = \frac{1}{(1-r)^2}.$$

Applying this result to Equation 5.2 with $r = 1-p$, we get

$$E[X] = p\sum_{k=1}^{\infty} k(1-p)^{k-1} = p\frac{1}{(1-(1-p))^2} = \frac{p}{p^2} = \frac{1}{p}.$$

■

Intuitively, this should make sense. The expected number of trials until the first success is the inverse of the success probability. You can expect to flip two coins, on average, until you see heads, and roll 36 pairs of dice until you see snake-eyes, and Juan can expect to wait $1/(0.001) = 1000$ days until he wins the lottery.

We leave for the reader the pleasure of finding the variance of a geometric distribution to be $\frac{1-p}{p^2}$. You will need to take two derivatives in a similar process (see Exercise 5.3).

5.1.1 Memorylessness

The geometric distribution has a unique property among discrete distributions. It is what is called *memoryless*. We illustrate the concept with an example.

Example 5.2 A traffic inspector is monitoring a busy intersection for moving violations. The inspector wants to know how many cars will pass by before the first moving violation occurs. In particular, she would like to find the probability that the first moving violation occurs after the 30th car from her arrival. Let X be the number of cars which pass the intersection until the first moving violation. The desired probability is $P(X > 30)$.

After 20 cars go by without a moving violation, a second inspector arrives on the scene. He would also like to know the probability that the first moving violation occurs after the 30th car. But given that 20 cars have gone by without any violations, this probability is equal to the conditional probability that 50 cars in total go by without moving violations, given that there were no violations among the first 20 cars, that is, $P(X > 50|X > 20)$.

If the distribution of X is memoryless, it means that these two probabilities are the same, that is, $P(X > 50|X > 20) = P(X > 30)$. The cars that go by once the second inspector arrives on the scene do not "remember" the 20 cars that went by before.

Assume that whether or not cars have moving violations is independent of each other, and that the probability that any particular car has a violation is p. Then the number of cars which pass the intersection until the first moving violation X has a geometric distribution with parameter p. In that case,

$$
\begin{aligned}
P(X > 50|X > 20) &= \frac{P(X > 50, X > 20)}{P(X > 20)} \\
&= \frac{P(X > 50)}{P(X > 20)} = \frac{(1-p)^{50}}{(1-p)^{20}} \\
&= (1-p)^{30} = P(X > 30),
\end{aligned}
$$

where we use the fact that if there are no violations among the first 50 cars, then there are no violations among the first 20 cars. That is, $\{X > 50\}$ implies $\{X > 20\}$ and thus $\{X > 50, X > 20\} = \{X > 50\}$. ■

MEMORYLESS PROPERTY

A random variable X has the *memoryless property* if for all $0 < s < t$,

$$P(X > t|X > s) = P(X > t - s).$$

We see that a geometric random variable is memoryless via the following proof. If $X \sim \text{Geom}(p)$, then for $0 < s < t$,

$$\begin{aligned} P(X > t | X > s) &= \frac{P(X > t, X > s)}{P(X > s)} = \frac{P(X > t)}{P(X > s)} \\ &= \frac{(1-p)^t}{(1-p)^s} = (1-p)^{t-s} \\ &= P(X > t - s). \end{aligned}$$

Example 5.3 Suppose a baseball player has a 30% chance of getting a hit during any at bat. After three times at bat, the player has not had a hit. What is the probability the player will not get a hit after seven times at bat?

The number of times at bat until getting a hit is a geometric random variable, X, with $p = 0.3$. By the memoryless property, the desired probability is $P(X > 7|X > 3) = P(X > 7 - 3) = (1 - p)^{7-3} = (0.70)^4 = 0.24$. ■

Example 5.4 **A hard day's night at the casino.** Brayden has been watching the roulette table all night, and suddenly sees the roulette ball land on black 10 times in a row. It is time for red, he reasons, by the "law of averages." But his friend Elijah, a student of probability, recognizes that the number of times until red comes up has a geometric distribution and is thus memoryless. It does not "remember" the last 10 outcomes. The probability that red will come up after the 11th bet given past history is the same as the probability of it coming up after the first. ■

Remarkably, the geometric distribution is the *only* discrete distribution that is memoryless. Here is a short derivation assuming X takes positive integer values.

Let $g(x) = P(X > x)$. The defining property of memorylessness gives that $g(t - s) = g(t)/g(s)$ or $g(t) = g(s)g(t - s)$ for all integers $0 < s < t$.

With $s = 1$ and $t = 2$, $g(2) = g(1)g(1) = g(1)^2$. With $t = 3$, $g(3) = g(1)g(2) = g(1)g(1)^2 = g(1)^3$. In general, we see that $g(n) = g(1)^n$, for all integer n. Now, we consider $P(X = n)$, which we can express in terms of $g(\ )$ as follows:

$$\begin{aligned} P(X = n) &= P(X > n - 1) - P(X > n) \\ &= g(n - 1) - g(n) = g(1)^{n-1} - g(1)^n = g(1)^{n-1}[1 - g(1)], \end{aligned}$$

for $n = 1, 2, \ldots$, which is the pmf of a geometric distribution with parameter $p = 1 - g(1) = P(X = 1)$.

5.1.2 Coupon Collecting and Tiger Counting

In a national park in India, an automatic camera photographs the number of tigers t passing by over a 12-month period. From the photographs, it was determined that the number of *different* tigers observed was d. The total number of tigers in the park is to be estimated from t and d (cited in Finkelstein et al. [1998]). What should the estimate be?

The problem of estimating the number of tigers in a park will introduce the *coupon collector's problem,* with many applications from wildlife management to electrical engineering. The coupon collector's problem asks: If one repeatedly samples with replacement from a collection of n distinct items ("coupons"), what is the expected number of draws required so that each item is selected at least once?

We can remember a certain fast-food chain which sold takeout meals with a Star Wars character in each meal for a limited time. There were 10 characters in all. The coupon collector's problem asks: How many meals need be bought to get a complete set of 10 characters?

Given n coupons, let X be the number of draws required to obtain a complete set sampling with replacement. We find $E[X]$.

As is often the case with random variables which represent counts, we will decompose the count into simpler units. The first coupon we get will be our first piece of the set. Let $G_1 = 1$ for that coupon. Now, we need to get the remainder of the set. For $k = 2, \ldots, n$, let G_k be the number of draws required to increase the set of distinct coupons from $k - 1$ to k. Then $X = G_1 + \cdots + G_n$.

For instance, if the set of coupons is $\{a, b, c, d\}$ and the sequence of successive draws is

$$a, a, d, b, d, a, d, b, c,$$

then $G_1 = 1, G_2 = 2, G_3 = 1, G_4 = 5$, and $X = G_1 + G_2 + G_3 + G_4 = 9$.

We find the distribution of the G_k's. Consider the process of collecting coupons. Once the first coupon is picked, successive draws might result in it being picked again, but eventually a second coupon will get picked. There are $n - 1$ possibilities for the second coupon. Successive draws are Bernoulli trials where "success" means picking a second coupon. The number of draws until the second new coupon is picked is a geometric random variable with parameter $p = (n - 1)/n$. That is, $G_2 \sim \text{Geom}((n - 1)/n)$.

Similarly, once the second coupon is picked, successive draws until the third coupon is picked form a Bernoulli sequence with $p = (n - 2)/n$, and thus $G_3 \sim \text{Geom}((n - 2)/n)$. In general, once the $(k - 1)$st coupon is picked, successive draws until the kth new coupon is picked form a Bernoulli sequence with parameter $p = (n - (k - 1))/n$, and $G_k \sim \text{Geom}((n - (k - 1))/n)$.

As the expectation of a geometric distribution is the inverse of the success probability,

$$\begin{aligned}
E[X] &= E[G_1 + G_2 + \cdots + G_n] \\
&= E[G_1] + E[G_2] + \cdots + E[G_n] \\
&= \frac{n}{n} + \frac{n}{n-1} + \cdots + \frac{n}{1} = n\left(\frac{1}{n} + \frac{1}{n-1} + \cdots + \frac{1}{1}\right)
\end{aligned}$$

$$= n\left(1 + \frac{1}{2} + \cdots + \frac{1}{n}\right). \tag{5.3}$$

For example, the expected number of meals needed to get the full set of 10 Star Wars characters is

$$E[X] = 10\left(1 + \frac{1}{2} + \cdots + \frac{1}{10}\right) = 29.3 \text{ meals}.$$

For a simulation to verify this result, see **CouponCollect.R**.

Observe that $G_1, \ldots, G_n$ are independent random variables. Consider G_k. Once $k - 1$ coupons have been picked, the number of additional trials needed to select the next coupon does not depend on the past history of selections. For each k, G_k is independent of the previous $G_1, \ldots, G_{k-1}$. The last expression in Equation 5.3 is the partial harmonic series. For large n, approximations for this series give $E(X) = n \ln n$.

For the tiger estimation problem introduced at the beginning of this section, we consider a slightly different version of the coupon collector's problem, as given in Ross [2012]. Suppose there are n coupons to choose from and t items have been selected after repeated selections with replacement. What is the expected number of distinct coupons in the group of t items?

For instance, for the Star Wars characters, if 12 meals are obtained what is the expected number of different characters collected? Here $n = 10$ and $t = 12$.

Let X be the number of different items collected after drawing t items. By the method of indicators write $X = I_1 + \cdots + I_n$, where

$$I_k = \begin{cases} 1, & \text{if coupon } k \text{ is contained in the set of } t \text{ items}, \\ 0, & \text{if coupon } k \text{ is not contained in the set of } t \text{ items}, \end{cases}$$

for $k = 1, \ldots, n$. Then

$$\begin{aligned} E[I_k] &= P(\text{Coupon } k \text{ is contained in the set of } t \text{ items}) \\ &= 1 - P(\text{Coupon } k \text{ is not contained in the set of } t \text{ items}) \\ &= 1 - \left(1 - \frac{1}{n}\right)^t, \end{aligned}$$

which gives

$$E[X] = E[I_1] + \cdots + E[I_n] = n\left[1 - \left(1 - \frac{1}{n}\right)^t\right]. \tag{5.4}$$

Continuing with the Star Wars story, if 12 meals are obtained, the expected number of different characters is

$$10\left[1 - \left(1 - \frac{1}{10}\right)^{12}\right] = 7.18.$$

So we expect about seven characters to have been collected.

What does all this have to do with the tigers we mentioned at the start of this section? The unknown quantity of interest is n the number of tigers in the park. A group of t tigers has been "collected" by the park's camera. The expectation in Equation 5.4 gives the expected number of distinct tigers in a group of t tigers seen on camera. Suppose the photographic analysis shows d distinct tigers. Then

$$d \approx E[X] = n\left[1 - \left(1 - \frac{1}{n}\right)^t\right].$$

The data consist of t and d. Solving for n gives an estimate of the total number of tigers in the park, which is a value of interest that is not observable itself.

Suppose the park's camera observes 100 tigers throughout the year, and identifies 50 distinct tigers from the photographs. To estimate the number of tigers in the park, solve

$$50 = n\left[1 - \left(1 - \frac{1}{n}\right)^{100}\right]. \tag{5.5}$$

R: NUMERICAL SOLUTION TO TIGER PROBLEM

There is no closed form algebraic solution to the nonlinear equation shown as Equation 5.5. But there are numerical ways to solve it. In **R**, the `uniroot` command finds the root of a general function. The script **tiger.R** finds the numerical solution.

```
# tiger.R
# Define function to solve
> func <- function(n) { n*(1-(1-1/n)^ 100) - 50}

# Find root numerically inside the interval (50, 200)
> uniroot(func, c(50, 200))[1] #root is first entry
[1] 62.40844
```

We estimate about 62–63 tigers in the park.

There are many generalizations and extensions of the coupon collector's problem. To learn more, see Dawkins [1991].

How R codes the geometric distribution. An alternate formulation of the geometric distribution counts the number of *failures* $\tilde{X}$ until the first success, rather than the number of trials required for the first success. As the number of trials required for the first success is the number of such failures plus one, $X = \tilde{X} + 1$. The pmf of $\tilde{X}$ is

$$P(\tilde{X} = k) = P(X - 1 = k) = P(X = k + 1) = (1 - p)^{(k+1)-1}p = (1 - p)^k p,$$

for $k = 0, 1, \ldots$. The expected number of failures until the first success is

$$E[\tilde{X}] = E[X - 1] = E[X] - 1 = \frac{1}{p} - 1 = \frac{1-p}{p}.$$

And the variance is

$$V[\tilde{X}] = V[X - 1] = V[X].$$

This variation of the geometric distribution is the one which **R** uses in its related commands. Bear this in mind for simulations involving geometric random variables.

R: GEOMETRIC DISTRIBUTION

The respective **R** commands for the geometric distribution are

```
> dgeom(k, p)
> pgeom(k, p)
> rgeom(n, p)
```

where k represents the number of failures before the first success.

If you are working with $k =$ the number of trials required for the first success, as we do in this section, then use the following modifications:

```
> dgeom(k-1, p)     # Computes P(X=x)
> pgeom(k-1, p)     # Computes P(X<=x)
> rgeom(n, p)+1     # Generates n random numbers
```

For example, recall Juan's issues with the lottery. To find the probability $P(X \leq 365)$ that Juan will win the lottery next year, type

```
> pgeom(364,0.001)
[1] 0.3059301
```

5.2 MOMENT-GENERATING FUNCTIONS

Some expectations have special names. For $k = 1, 2, \ldots$, the *kth moment* of a random variable X is $E[X^k]$. For instance, the first moment of X is the expectation $E[X]$. The moment-generating function (mgf), as the name suggests, can be used to generate the moments of a random variable. Mgfs are also useful for demonstrating some relationships between random variables.

MOMENT-GENERATING FUNCTION

Let X be a random variable. The *mgf* of X is the real-valued function

$$m(t) = E[e^{tX}],$$

defined for all real t when this expectation exists. Also written as $m_X(t)$.

Example 5.5 Geometric distribution. Let $X \sim \text{Geom}(p)$. The mgf of X is

$$\begin{aligned} m(t) = E[e^{tX}] &= \sum_{k=1}^{\infty} e^{tk}(1-p)^{k-1}p \\ &= pe^t \sum_{k=1}^{\infty} (e^t(1-p))^{k-1} = \frac{pe^t}{1-e^t(1-p)}. \end{aligned}$$

■

How do we get moments from the mgf? Moments of X are obtained from the mgf by successively differentiating $m(t)$ and evaluating at $t = 0$. We have

$$m'(t) = \frac{d}{dt}E[e^{tX}] = E\left[\frac{d}{dt}e^{tX}\right] = E[Xe^{tX}],$$

and $m'(0) = E[X]$.

Taking the second derivative gives

$$m''(t) = \frac{d}{dt}m'(t) = \frac{d}{dt}E[Xe^{tX}] = E\left[\frac{d}{dt}Xe^{tX}\right] = E[X^2e^{tX}],$$

and $m''(0) = E[X^2]$.

In general, the kth derivative of the mgf evaluated at $t = 0$ gives the kth moment as

$$m^{(k)}(0) = E[X^k], \quad \text{for } k = 1, 2, \ldots.$$

Remarks:

1. To define the mgf, it suffices that the expectation $E[e^{tX}]$ exists for values of t in an *interval* that contains zero.
2. For some distributions, the expectation $E[e^{tX}]$ is not finite. This is true for the Cauchy distribution and the t-distribution, used in statistics. However, another type of generating function called the *characteristic function* can be used. The characteristic function $E[e^{itX}]$ is defined for *all* distributions and requires the use of complex (imaginary) numbers.

3. In physics, the concept of moments is used to describe physical properties, with the first moment roughly equal to the center of mass. For a probability distribution, the first moment is the "center" of probability mass. Similar analogies can be made with higher moments.

Example 5.6 Let X be a random variable taking values -1, 0, and 1, with

$$P(X = -1) = p, \quad P(X = 0) = q, \quad \text{and} \quad P(X = 1) = r,$$

with $p + q + r = 1$. Find the mgf of X and use it to find the mean and variance of X.

The mgf of X is

$$m(t) = E[e^{tX}] = pe^{-t} + q + re^{t},$$

with

$$m'(t) = -pe^{-t} + re^{t} \quad \text{and} \quad m''(t) = pe^{-t} + re^{t}.$$

Thus,

$$E[X] = m'(0) = r - p,$$

and

$$V[X] = E[X^2] - E[X]^2 = m''(0) - [m'(0)]^2 = (r + p) - (r - p)^2.$$

■

Example 5.7 Poisson distribution. Let $X \sim \text{Pois}(\lambda)$. The mgf of X is

$$m(t) = E[e^{tX}] = \sum_{k=0}^{\infty} e^{tk} \frac{e^{-\lambda}\lambda^k}{k!} = e^{-\lambda} \sum_{k=0}^{\infty} \frac{(\lambda e^t)^k}{k!} = e^{-\lambda} e^{\lambda e^t} = e^{\lambda(e^t - 1)}.$$

Differentiating gives

$$m'(t) = \lambda e^t e^{\lambda(e^t - 1)}$$

with $m'(0) = \lambda = E[X]$; and

$$m''(t) = (\lambda e^t)(\lambda e^t + 1)(e^{\lambda(e^t - 1)})$$

with $m''(0) = \lambda^2 + \lambda = E[X^2]$. This gives $V[X] = \lambda^2 + \lambda - \lambda^2 = \lambda$, as expected. ■

Here are three important properties of the mgf.

PROPERTIES OF MOMENT GENERATING FUNCTIONS

1. If X and Y are independent random variables, then the mgf of their sum is the product of their mgfs.

That is,

$$\begin{aligned} m_{X+Y}(t) &= E[e^{t(X+Y)}] \\ &= E[e^{tX}e^{tY}] = E[e^{tX}]E[e^{tY}] \\ &= m_X(t)m_Y(t). \end{aligned}$$

2. Let X be a random variable with mgf $m_X(t)$ and constants $a \neq 0$ and b. Then

$$m_{aX+b}(t) = E[e^{t(aX+b)}] = e^{bt}E[e^{(ta)X}] = e^{bt}m_X(at).$$

3. Mgfs uniquely determine the underlying probability distribution. That is, if two random variables have the same mgf, then they have the same probability distribution.

Example 5.8 Sum of independent Poissons is Poisson. Let X and Y be independent Poisson random variables with respective parameters λ_1 and λ_2. Then

$$m_{X+Y}(t) = m_X(t)m_Y(t) = e^{\lambda_1(e^t-1)}e^{\lambda_2(e^t-1)} = e^{(\lambda_1+\lambda_2)(e^t-1)},$$

which is the mgf of a Poisson random variable with parameter $\lambda_1 + \lambda_2$.

This constitutes a proof that the sum of independent Poisson random variables is Poisson, a result derived by other means in Section 4.4.1. The extension to a set of k independent Poisson random variables follows similarly. ■

This last example gives a taste of the power of mgfs. Many results involving sums and limits of random variables can be proven using them. This allows us to explore relationships between distributions as well, as we will see in the next section.

5.3 NEGATIVE BINOMIAL—UP FROM THE GEOMETRIC

The geometric distribution counts the number of trials until the first success occurs in i.i.d. Bernoulli trials. The *negative binomial distribution* extends this, counting the number of trials until the rth success occurs.

NEGATIVE BINOMIAL DISTRIBUTION

A random variable X has the *negative binomial distribution with parameters r and p* if

$$P(X = k) = \binom{k-1}{r-1} p^r(1-p)^{k-r}, \quad r = 1, 2, \ldots, \; k = r, r+1, \ldots \tag{5.6}$$

We write $X \sim \text{NegBin}(r, p)$.

The pmf is derived by observing that if k trials are required for the rth success, then (i) the kth trial is a success and (ii) the previous $k-1$ trials have $r-1$ successes and $k-r$ failures. The first event (i) occurs with probability p. For (ii), there are $\binom{k-1}{r-1}$ outcomes of $k-1$ trials with $r-1$ successes. By independence, each of these outcomes occurs with probability $p^{r-1}(1-p)^{k-r}$.

Observe that the negative binomial distribution reduces to the geometric distribution when $r = 1$. Hence, if $X \sim \text{Geom}(p)$, then $X \sim \text{NegBin}(1, p)$ and vice versa.

Using **R** with the negative binomial distribution poses similar issues as with the geometric distribution. There are several alternate formulations of the negative binomial distribution, and the one that **R** uses is based on the number of *failures* $\tilde{X}$ until the rth success, where $\tilde{X} = X - r$.

R: NEGATIVE BINOMIAL DISTRIBUTION

The **R** commands are

```
> dnbinom(k, r, p)
> pnbinom(k, r, p)
> rnbinom(n, r, p)
```

where k represents the number of failures before r successes.

If you are working with k = the number of trials required for the first r successes, as we do in this section, then use the following modifications. When $X \sim \text{NegBin}(r, p)$,

```
> dnbinom(k-r, r, p)     # Computes P(X=k)
> pnbinom(k-r, r, p)     # Computes P(X<=k)
> rnbinom(n, r, p)+r     # Generates n random numbers
```

Example 5.9 Applicants for a new student internship are accepted with probability $p = 0.15$ independently from person to person, in the order their applications were received. Several hundred people are expected to apply. Find the probability that it will take no more than 100 applicants to find 10 students for the program.

Let X be the number of people who apply for the internship before the 10th student is accepted. Then X has a negative binomial distribution with parameters $r = 10$ and $p = 0.15$. The desired probability is

$$P(X \leq 100) = \sum_{k=10}^{100} P(X = k)$$

$$= \sum_{k=10}^{100} \binom{k-1}{r-1} p^r (1-p)^{k-r}$$

$$= \sum_{k=10}^{100} \binom{k-1}{9} (0.15)^{10}(0.85)^{k-10} = 0.945.$$

We used **R** to evaluate this summation by typing

```
> pnbinom(100-10, 10, 0.15)
[1] 0.9449054.
```

■

Example 5.10 World Series. Baseball's World Series is a best-of-seven playoff between the top teams in the American and National Leagues. The team that wins the series is the one which first wins four games. Thus, a series can go for 4, 5, 6, or 7 games. If both teams are evenly matched, what is the expected length of a World Series?

Let X be the number of games played in a World Series. Let Y be a random variable with a negative binomial distribution with parameters $r = 4$ and $p = 1/2$. For each $k = 4, 5, 6, 7$, the probability that team A wins the series in k games is equal to the probability that it takes k games for four "successes," which is $P(Y = k)$. As either team A or team B could win the series, for $k = 4, 5, 6, 7$,

$$\begin{aligned} P(X = k) &= 2P(Y = k) \\ &= 2\binom{k-1}{4-1}\left(\frac{1}{2}\right)^4\left(\frac{1}{2}\right)^{k-4} = 2\binom{k-1}{3}\left(\frac{1}{2}\right)^k, \end{aligned}$$

which gives

$$P(X = k) = \begin{cases} 0.125, & \text{if } k = 4, \\ 0.25, & \text{if } k = 5, \\ 0.3125, & \text{if } k = 6, 7. \end{cases}$$

The expected length of the World Series is

$$E[X] = 4(0.125) + 5(0.25) + 6(0.3125) + 7(0.3125) = 5.8125 \text{ games.}$$

The actual lengths of the 112 World Series held between 1903 and 2019, not counting the four series which went to eight games because ties were called is given in Table 5.1.

The average length is

$$\frac{1}{112}(4(19) + 5(28) + 6(25) + 7(40)) = 5.77 \text{ games.}$$

■

Expectation and variance. By expressing a negative binomial random variable as a sum of independent geometric random variables, we show how to find the expectation and variance of the negative binomial distribution.

TABLE 5.1. Lengths of 112 World Series, 1903–2019.

4	5	6	7
19	28	25	40

Consider an i.i.d. sequence of Bernoulli trials $X_1, X_2, \ldots$ with parameter p. Let G_1 be the number of trials required for the first success. Then $G_1 \sim \text{Geom}(p)$. Suppose it takes $G_1 = g$ trials for the first success. Let G_2 be the number of *additional* trials required for the second success. Then G_2 also has a geometric distribution with parameter p, and G_2 is independent of G_1. The reason is that after trial g, the sequence of random variables $X_{g+1}, X_{g+2}, \ldots$ is also an i.i.d. Bernoulli sequence. The number of trials required for the first success for this sequence is exactly the number of trials required for the second success for the original sequence. We say that the Bernoulli sequence "restarts" itself anew after trial g.

Continuing in this way, for each $k = 1, \ldots, r$, let G_k be the number of additional trials required after the $(k-1)$st success for the kth success to occur. Then $G_1, \ldots, G_r$ is an i.i.d. sequence of geometric random variables with parameter p.

Let X be the number of trials required for the rth success. Then X has a negative binomial distribution with parameters r and p, and $X = G_1 + \cdots + G_r$.

We can verify this relationship using mgfs. Previously, we found that if $G_i \sim \text{Geom}(p)$, then the mgf of G_i is

$$m(t) = E[e^{tG_i}] = \frac{pe^t}{1 - e^t(1-p)}.$$

Then because $X = G_1 + \cdots + G_r$, which are i.i.d. Geom(p), the mgf of X is

$$m_X(t) = m_{G_1}(t) \times \cdots \times m_{G_r}(t) = \left[\frac{pe^t}{1 - e^t(1-p)}\right]^r.$$

Now, we must verify that this is the mgf of a negative binomial random variable to establish the relationship. From the definition, the mgf of X is

$$\begin{aligned}
m(t) = E[e^{tX}] &= \sum_{k=r}^{\infty} e^{tk} P(X = k) = \sum_{k=r}^{\infty} e^{tk} \binom{k-1}{r-1} p^r (1-p)^{k-r} \\
&= (pe^t)^r \sum_{k=r}^{\infty} \binom{k-1}{r-1} (e^t)^{k-r} (1-p)^{k-r} = (pe^t)^r \sum_{k=r}^{\infty} \binom{k-1}{r-1} ((1-p)e^t)^{k-r} \\
&= (pe^t)^r \sum_{k=0}^{\infty} \binom{k+r-1}{r-1} ((1-p)e^t)^k = (pe^t)^r (1 - e^t(1-p))^{-r} \\
&= \left[\frac{pe^t}{1 - e^t(1-p)}\right]^r,
\end{aligned}$$

where the final summation is recognized as a negative binomial series. Now that we have established this relationship, we can use what we know about the geometric distribution to help us find the mean and variance for the negative binomial distribution.

The mean and variance of the geometric distribution are $1/p$ and $(1-p)/p^2$, respectively. Using the G_k's we constructed, it follows that for the negative binomial distribution,

$$E[X] = E[G_1 + \cdots + G_r] = E[G_1] + \cdots + E[G_r] = \frac{r}{p} \quad \text{and}$$
$$V[X] = V[G_1 + \cdots + G_r] = V[G_1] + \cdots + V[G_r] = \frac{r(1-p)}{p^2}.$$

Example 5.11 Sophia is making tomato sauce for dinner and needs 10 ripe tomatoes. In the produce department of the supermarket, there is a 70% chance that a tomato is ripe. (i) How many tomatoes can Sophia expect to sample until she gets what she needs? (ii) What is the standard deviation? (iii) What is the probability Sophia will need to sample more than 15 tomatoes before she gets 10 ripe tomatoes?

(i) and (ii) Let X be the number of tomatoes that Sophia needs to sample in order to get 10 ripe ones. Then $X \sim \text{NegBin}(10, 0.7)$, with expectation $E[X] = 10/(0.7) = 14.29$ and standard deviation $\text{SD}[X] = \sqrt{10(0.3)/(0.7)^2} = 2.47$.

(iii) The desired probability is

$$P(X > 15) = 1 - P(X \le 15)$$
$$= 1 - \sum_{k=10}^{15} \binom{k-1}{9} (0.7)^{10}(0.3)^{k-10} = 0.278.$$

To find this probability in **R**, type

```
> 1-pnbinom(15-10, 10, 0.7)
[1] 0.2783786
```

■

Why "negative binomial"? You might be wondering what "negative" and "binomial" have to do with the negative binomial distribution. The reason for the choice of words is that the distribution, in a sense, is inverse to the binomial distribution.

Consider an i.i.d. Bernoulli sequence with success probability p. The event (i) that there are r or fewer successes in the first n trials is equal to the event (ii) that the $(r+1)$st success occurs after the nth trial.

Let B be the number of successes in the first n trials. Then B has a binomial distribution with parameters n and p, and the probability of the first event (i) is

$P(B \leq r)$. Let Y be the number of trials required for the $(r+1)$st success. Then Y has a negative binomial distribution with parameters $r+1$ and p. The probability of the second event (ii) is $P(Y > n)$. Hence,

$$P(B \leq r) = P(Y > n), \tag{5.7}$$

giving

$$\sum_{k=0}^{r} \binom{n}{k} p^k (1-p)^{n-k} = \sum_{k=n+1}^{\infty} \binom{k-1}{r} p^{r+1} (1-p)^{k-r-1},$$

which is an interesting identity in its own right.

Example 5.12 Problem of points. This problem is said to have motivated the beginnings of modern probability through a series of letters between French mathematicians Blaise Pascal and Pierre de Fermat in the 1600s.

Two players are repeatedly playing a game of chance, say tossing a coin. The stakes are 100 francs. If the coin comes up heads, Player A gets a point. If it comes up tails, Player B gets a point. The players agree that the first person to get a set number of points will win the pot, but the game is interrupted. Player A needs a more points to win; Player B needs b more points. How should the pot be divided?

Pascal and Fermat not only solved the problem but in their correspondence also developed concepts that are fundamental to probability to this day. Their insight, remarkable for its time, was that the division of the stakes should not depend on the history of the game (i.e., not on what already took place), but what might have happened if the game were allowed to continue.

By today's standards, the solution is fairly simple, especially now that you know about the negative binomial distribution. We generalize the problem from its original formulation and allow the coin they are playing with to have probability p of coming up heads. Suppose the game were to continue. Let X be the number of coin tosses required for Player A to win a points. Then X has a negative binomial distribution with parameters a and p. As Player A needs a points and Player B needs b points, the game will be decided in $a+b-1$ plays. Hence, A will win the game if and only if $X \leq a+b-1$. The probability that A wins is

$$P(\text{A wins}) = P(X \leq a+b-1) = \sum_{k=a}^{a+b-1} \binom{k-1}{a-1} p^a (1-p)^{k-a}.$$

The pot would be divided according to this winning probability. That is, $100 \times P(\text{A wins})$ francs goes to Player A and the remainder $100 \times P(\text{B wins})$ goes to Player B.

R: PROBLEM OF POINTS

The exact probability that A wins is found by typing

```
> pnbinom(b-1, a, p)
```

See the script **Points.R** to play the "problem of points" for your choices of a, b, and p.

For an enjoyable book-length treatment of the history and mathematics of the problem of points, see Devlin [2008]. ■

5.4 HYPERGEOMETRIC—SAMPLING WITHOUT REPLACEMENT

Whereas the binomial distribution arises from sampling with replacement, the hypergeometric distribution often arises when sampling is without replacement from a finite population.

A bag of N balls contains r red balls and $N - r$ blue balls. A sample of n balls is picked. If the sampling is with replacement, then the number of red balls in the sample has a binomial distribution with parameters n and $p = r/N$. Here we consider when the sampling is without replacement. Let X be the number of red balls in the sample. Then X has a *hypergeometric distribution.*

The probability mass function of X is obtained by a straightforward counting argument. Consider the event $\{X = k\}$ that there are k red balls in the sample. The number of possible samples of size n with k red balls is $\binom{r}{k}\binom{N-r}{n-k}$. (Choose k reds from the r red balls in the bag, and choose the remaining $n - k$ blues from the $N - r$ blue balls in the bag.) There are $\binom{N}{n}$ possible samples of size n. Combining these, we obtain the pmf for the hypergeometric distribution.

HYPERGEOMETRIC DISTRIBUTION

A random variable X has the *hypergeometric distribution with parameters r, N, and n* if

$$P(X = k) = \frac{\binom{r}{k}\binom{N-r}{n-k}}{\binom{N}{n}},$$

for $\max(0, n - (N - r)) \leq k \leq \min(n, r)$. The values of k are restricted by the domain of the binomial coefficients as $0 \leq k \leq r$ and $0 \leq n - k \leq N - r$.

We write $X \sim \text{HyperGeo}(r, N, n)$.

The origins of this distribution go back to the 1700s when the word "hypergeometric numbers" was used for what we now call factorials.

R: HYPERGEOMETRIC DISTRIBUTION

The **R** commands are

```
> dhyper(k, r, N-r, n) # Computes P(X=k)
> phyper(k, r, N-r, n) # Computes P(X<=k)
> rhyper(k, r, N-r, n)
# Generates k random numbers from the distribution
```

Example 5.13 Independents. Suppose there are 100 political independents in the student body of 1000. A sample of 50 students is picked. What is the probability there will be six independents in the sample?

The number of independents in the sample I has a hypergeometric distribution with $n = 50$, $r = 100$, and $N = 1000$. The desired probability is

$$P(I = 6) = \frac{\binom{100}{6}\binom{900}{44}}{\binom{1000}{50}} = 0.158.$$

In **R**, type

```
> dhyper(6, 100, 900, 50)
[1] 0.1579155
```

■

Example 5.14 Inferring independents and maximum likelihood. Continuing from the last example, suppose the number of independents on campus, denoted c, is not known. A sample of 50 students yields six independents. How can we use these data to infer an estimate of c?

The probability of obtaining six independents in a sample of 50 students when c is unknown is

$$P(I = 6) = \frac{\binom{c}{6}\binom{1000-c}{44}}{\binom{1000}{100}}. \tag{5.8}$$

The *maximum likelihood method* in statistics says to estimate c with the value that will maximize this probability. That is, estimate c with the number which gives the highest probability (or "likelihood") of obtaining six independents in a sample of 50 students.

R: MAXIMIZING THE HYPERGEOMETRIC PROBABILITY

The **R** expression `0:1000` creates a vector of values from 0 to 1000. The command

```
> dhyper(6, 0:1000, 1000-(0:1000), 50)
```

creates a vector of probabilities for $P(I = 6)$ at values of c from 0 to 1000. To find the maximum value in that vector, use `max()`. To find the index of the vector for where that maximum is located, use `which.max()`.

The following commands find the value of c which maximizes the probability Equation 5.8.

```
> clist <- dhyper(6, 0:1000, 1000-(0:1000), 50)
> which.max(clist)-1
[1] 120
```

Note that one is subtracted in the second line because the vector `clist` starts at $c = 0$, rather than 1.

The maximum likelihood estimate is $c = 120$. We infer there are about 120 independents on campus. This estimate is intuitive because $6/50 = 12\%$ of the sample are independents, and 12% of the population size is 120. ■

As mentioned above, the hypergeometric distribution often arises when sampling is without replacement; the binomial distribution arises when sampling is with replacement. As discussed in Section 2.6, when the population size N is large, there is not much difference between sampling with and without replacement. One can show analytically that the hypergeometric probability mass function converges to the binomial pmf as $N \to \infty$, and the binomial distribution serves as a good approximation of the hypergeometric distribution when N is large. See Exercise 5.35, which includes a more precise statement of the limiting process.

Expectation and variance. The expectation of the hypergeometric distribution can be obtained by the method of indicator variables. Let N be the number of balls in the bag, r the number of red balls in the bag, and X the number of red balls in the sample. Define a sequence of 0–1 indicators $I_1, \ldots, I_n$, where

$$I_k = \begin{cases} 1, & \text{if the } k\text{th ball in the sample is red,} \\ 0, & \text{if the } k\text{th ball in the sample is not red,} \end{cases}$$

for $k = 1, \ldots, n$. Then $X = I_1 + \cdots + I_n$.

The probability that the kth ball in the sample is red is r/N. This can be shown most simply by a symmetry argument as the kth ball can be any of the N balls with equal probability. (If you need more convincing, use the law of total probability and find the probability that the second ball in the sample is red by conditioning on the first ball.) Now, we can compute $E[X]$ as

$$\begin{aligned}E[X] &= E[I_1 + \cdots + I_n] = E[I_1] + \cdots + E[I_n]\\ &= \sum_{k=1}^{n} P(k\text{th ball in the sample is red})\\ &= \sum_{k=1}^{n} \frac{r}{N} = \frac{nr}{N}.\end{aligned}$$

The variance of the hypergeometric distribution can also be found by the indicator method. However, the indicators are not independent. The derivation of the variance has similarities to the variance calculation in the matching problem (see Example 4.35). As $V[I_k] = (r/N)(1 - r/N)$, we have

$$\begin{aligned}V[X] &= V[I_1 + \cdots + I_n]\\ &= \sum_{i=1}^{n} V[I_i] + \sum_{i\neq j} \text{Cov}(I_i, I_j)\\ &= n\left(\frac{r}{N}\right)\left(1 - \frac{r}{N}\right) + (n^2 - n)(E[I_iI_j] - E[I_i]E[I_j])\\ &= \frac{nr(N-r)}{N^2} + (n^2 - n)\left(E[I_iI_j] - \frac{r^2}{N^2}\right).\end{aligned}$$

For $i \neq j$, the product I_iI_j is the indicator of the event that both the ith and jth balls in the sample are red. Thus,

$$E[I_iI_j] = P(I_i = 1, I_j = 1) = P(I_i = 1|I_j = 1)P(I_j = 1) = \left(\frac{r-1}{N-1}\right)\frac{r}{N},$$

because the sampling is without replacement. This gives

$$\begin{aligned}V[X] &= \frac{nr(N-r)}{N^2} + (n^2 - n)\left(E[I_iI_j] - \frac{r^2}{N^2}\right)\\ &= \frac{nr(N-r)}{N^2} + (n^2 - n)\left(\frac{r(r-1)}{N(N-1)} - \frac{r^2}{N^2}\right).\\ &= \frac{n(N-n)r(N-r)}{N^2(N-1)}.\end{aligned}$$

Example 5.15 In a bridge hand (13 cards from a standard 52-card deck), what is the mean and variance of the number of aces?

Let X be the number of aces in a bridge hand. Then X has a hypergeometric distribution. In our balls-in-the-bag imagery, the "bag" is the deck of cards with $N = 52$. The "red balls" are the aces with $r = 4$. The "sample" is the bridge hand with $n = 13$. This gives

$$E[X] = \frac{13(4)}{52} = 1$$

and

$$V[X] = \frac{13(52-13)(4)(52-4)}{52^2(51)} = \frac{12}{17} = 0.706.$$

■

R: SIMULATING ACES IN A BRIDGE HAND

To simulate, let 1, 2, 3, and 4 represent the aces in a 52-card deck. Type

```
> aces <-replicate(10000, sum(sample(1:52, 13)<=4))
> mean(aces)
[1] 0.985
> var(aces)
[1] 0.6968447
```

Example 5.16 **Counting the homeless.** So-called *capture–recapture* methods have been used for many years in ecology, public health, and social science to count rare and elusive populations. In order to estimate the size N of a population, like fish in a lake, researchers "capture" a sample of size r. Subjects are "tagged" and returned to the general population. Researchers then "recapture" a second sample of size n and count the number K which are found tagged.

If the first sample is sufficiently mixed up in the general population, then the proportion of those tagged in the second sample should be approximately equal to the proportion tagged in the population. That is,

$$\frac{K}{n} \approx \frac{r}{N}, \text{ so } N \approx \frac{nr}{K}$$

gives an estimate of the population size.

The number tagged in the sample K has a hypergeometric distribution and $E[K] = nr/N$. The approximation formula follows based on $K \approx E[K]$.

Williams [2010] describes such a methodology for estimating the homeless population in Plymouth, a city of a quarter million people in Britain. On a particular day, at various social service agency locations, homeless people are identified using several variables including sex and date of birth. At a later date, a sample of homeless people is taken and the number of people originally identified are counted.

Suppose on the first day, 600 homeless persons are counted and identified. At a later date, a sample of 800 homeless people is taken, which includes 100 of the original persons. Here $r = 600$, $K = 100$, and $n = 800$. This gives the estimate $N \approx (nr)/K = (800 \times 600)/100 = 4800$ for the size of the homeless population.

Williams [2010] points out many difficulties with trying to reliably count the homeless, but states that "yet for its shortcomings it remains possibly the most rigorous method of estimation. Some researchers argue that the homeless are a population we cannot reliably count but this is a doctrine of despair and though capture-recapture is far from a perfect method, like so many other research methods its use will undoubtedly lead to technical improvements" (page 55). ■

5.5 FROM BINOMIAL TO MULTINOMIAL

In a binomial setting, successive trials take one of two possible values (e.g., success or failure). The multinomial distribution is a generalization of the binomial distribution which arises when successive independent trials can take more than two values. The multinomial distribution is used to model such things as follows:

1. The number of ones, twos, threes, fours, fives, and sixes in 25 dice rolls.
2. The frequencies of r different alleles among n individuals.
3. The number of outcomes of an experiment that has m possible results when repeated n times.
4. The frequencies of six different colors in a sample of 10 candies.

Consider a random experiment repeated independently n times. At each trial, the experiment can assume one of r values. The probability of obtaining the kth value is p_k, for $k = 1, \ldots, r$, with $p_1 + \cdots + p_r = 1$. For each k, let X_k denote the number of times value k occurs. Then the collection of random variables, $(X_1, \ldots, X_r)$, has a multinomial distribution.

MULTINOMIAL DISTRIBUTION

Suppose $p_1, \ldots, p_r$ are nonnegative numbers such that $p_1 + \cdots + p_r = 1$. Random variables $X_1, \ldots, X_r$ have a *multinomial distribution with parameters* $n, p_1, \ldots, p_r$ if

$$P(X_1 = x_1, \ldots, X_r = x_r) = \frac{n!}{x_1! \cdots x_r!} p_1^{x_1} \cdots p_r^{x_r},$$

for nonnegative integers $x_1, \ldots, x_r$ such that $x_1 + \cdots + x_r = n$.

We write $(X_1, \ldots, X_r) \sim \text{Multi}(n, p_1, \ldots, p_r)$.

TABLE 5.2. Distribution of colors in a bag of candies.

Red	Orange	Yellow	Green	Blue	Purple
0.24	0.14	0.16	0.20	0.12	0.14

Example 5.17 In a bag of candies, colors are distributed according to the probability distribution in Table 5.2.

Denote colors with the letters *R*, *O*, *Y*, *G*, *B*, *P*. In a sample of 10 candies, let $X_R, X_O, X_Y, X_G, X_B, X_P$ denote the number of candies of the respective colors in the sample. Then $(X_R, X_O, X_Y, X_G, X_B, X_P)$ has a multinomial distribution with parameters $(10, 0.24, 0.14, 0.16, 0.20, 0.12, 0.14)$. ■

Deriving the joint pmf for the multinomial distribution is similar to the derivation of the binomial pmf. Consider the event

$$\{X_1 = x_1, X_2 = x_2, \ldots, X_r = x_r\}, \quad \text{for } x_1 + x_2 + \cdots + x_r = n. \tag{5.9}$$

Each outcome contained in this event can be represented as a sequence of length n with x_1 elements of type one, x_2 elements of type two, and so on. Count the number of such sequences. Of the n positions in the sequence, there are x_1 positions for the type one elements. Choose these positions in $\binom{n}{x_1}$ ways. Of the remaining $n - x_1$ positions, choose x_2 positions for the type two elements in $\binom{n-x_1}{x_2}$ ways. Continuing in this way gives the number of such sequences as

$$\begin{aligned}
&\binom{n}{x_1}\binom{n-x_1}{x_2}\cdots\binom{n-x_1-\cdots-x_{r-1}}{x_r} \\
&= \left(\frac{n!}{x_1!(n-x_1)!}\right)\left(\frac{(n-x_1)!}{x_2!(n-x_1-x_2)!}\right)\cdots\left(\frac{(n-x_1-\cdots-x_{r-1})!}{x_n!(n-x_1-\ldots-x_r)!}\right) \\
&= \frac{n!}{x_1!x_2!\cdots x_r!},
\end{aligned}$$

where the final simplification for the last equality happens because of the telescoping nature of the previous product.

Having counted the number of sequences in the event $\{X_1, X_2, \ldots, X_r\}$, by independence, the probability of each sequence is $p_1^{x_1} p_2^{x_2} \cdots p_r^{x_r}$. This gives the joint pmf for the multinomial distribution.

Observe how the multinomial distribution generalizes the binomial distribution. Consider a sequence of n i.i.d. Bernoulli trials with success parameter p. Let X_1 denote the number of successes in n trials. Let X_2 denote the number of failures.

Then $X_2 = n - X_1$ and

$$P(X_1 = k) = P(X_1 = k, X_2 = n-k) = \frac{n!}{k!(n-k)!}p^k(1-p)^{n-k}.$$

This shows that $X_1 \sim \text{Binom}(n,p)$ is equivalent to $(X_1, X_2) \sim \text{Multi}(n, p, 1-p)$, where $X_2 = n - X_1$.

Recall the binomial theorem. Here is the multinomial generalization.

MULTINOMIAL THEOREM

For any positive integer r and n, and real numbers $z_1, \ldots, z_r$,

$$(z_1 + \cdots + z_r)^n = \sum_{a_1+\cdots+a_r=n} \frac{n!}{a_1!a_2!\cdots a_r!} z_1^{a_1} z_2^{a_2} \cdots z_r^{a_r},$$

where the sum is over all lists of r nonnegative integers $(a_1, \ldots, a_r)$ which sum to n.

Proof: The proof is analogous to that given for the binomial theorem. □

It will be instructive for the reader to work through this identity for the case $r = 3$, $z_1 = z_2 = z_3 = 1$, and $n = 3$.

Verifying that the multinomial pmf sums to one is an application of the multinomial theorem as

$$\sum_{x_1+\cdots+x_r=n} P(X_1 = x_1, \ldots, X_r = x_r) = \sum_{x_1+\cdots+x_r=n} \frac{n!}{x_1!\cdots x_r!} p_1^{x_1}\cdots p_r^{x_r}$$

$$= (p_1 + \cdots + p_r)^n = 1^n = 1.$$

Multinomial coefficients. The quantity $n!/(x_1!\cdots x_r!)$ is known as a *multinomial coefficient* and generalizes the binomial coefficient, where $r = 2$. It is sometimes written as

$$\binom{n}{x_1, \ldots, x_r} = \frac{n!}{x_1!\cdots x_r!}.$$

Multinomial coefficients enumerate sequences of length n in which (i) each element can take one of r possible values and (ii) exactly x_k elements of the sequence take the kth value, for $k = 1, \ldots, r$. You can consider a set of n objects being put into r boxes, where you count the number of objects in each box.

Example 5.18

1. How many ways can 11 students fill 4 committees, of respective sizes 4, 2, 1, and 3?
2. How many ways can 11 balls be put into four boxes such that the first box gets four balls, the second box gets two balls, the third box gets one ball, and the fourth box gets four balls?
3. How many distinct ways can the letters M-I-S-S-I-S-S-I-P-P-I be permuted?

The answer to all three questions is

$$\binom{11}{4,2,1,4} = \frac{11!}{4!2!1!4!} = 34{,}650.$$

■

Example 5.19 In a parliamentary election, it is estimated that parties A, B, C, and D will receive 20, 25, 30, and 25% of the vote, respectively. In a sample of 10 voters, find the probability that there will be two supporters each of parties A and B, and three supporters each of parties C and D.

Let X_A, X_B, X_C, X_D denote the number of supporters in the sample of parties A, B, C, and D, respectively. Then

$$(X_A, X_B, X_C, X_D) \sim \text{Multi}(10, 0.20, 0.25, 0.30, 0.25).$$

The desired probability is

$$\begin{aligned} &P(X_A = 2, X_B = 2, X_C = 3, X_D = 3) \\ &= \frac{10!}{2!2!3!3!}(0.20)^2(0.25)^2(0.30)^3(0.25)^3 = 0.0266. \end{aligned}$$

■

R: MULTINOMIAL CALCULATION

The **R** commands for working with the multinomial distribution are `rmultinom` and `dmultinom`. The desired probability above is found by typing

```
> dmultinom(c(2,2,3,3),prob=c(0.20,0.25,0.30,0.25))
[1] 0.02657813
```

Example 5.20 **Genetics.** The Hardy–Weinberg principle in genetics states that the long-term gene frequencies in a population remain constant. An allele is a form of a gene. Suppose an allele takes one of two forms A and a. For a particular biological trait, you receive two alleles, one from each parent. The *genotypes* are the possible genetic makeups of a trait: *AA*, *Aa*, and *aa*.

For instance, fruit flies contain a gene for body color: *A* is the allele for black, and *a* for brown. Genotypes *AA* and *Aa* correspond to black body color, and *aa* corresponds to brown. We say black is dominant and brown is recessive.

Suppose an allele takes the form *A* with probability p, and *a* with probability $1-p$. The Hardy–Weinberg principle asserts that the proportion of individuals with genotype *AA*, *Aa*, and *aa* should occur in a population according to the frequencies p^2, $2p(1-p)$, and $(1-p)^2$, respectively. The frequency for *Aa*, for instance, is because you can get an *A* allele from your mother (with probability p) and an *a* allele from your father (with probability $1-p$), or vice versa.

In a sample of n fruit flies, let (X_1, X_2, X_3) be the number of flies with genotypes *AA*, *Aa*, and *aa*, respectively. Then

$$(X_1, X_2, X_3) \sim \text{Multi}(n, p^2, 2p(1-p), (1-p)^2).$$

Suppose that allele *A* occurs 60% of the time. What is the probability, in a sample of six fruit flies, that *AA* occurs twice, *Aa* occurs three times, and *aa* occurs once?

$$\begin{aligned} &P(X_1 = 2, X_2 = 3, X_3 = 1) \\ &\quad = \frac{6!}{2!3!1!}((0.60)^2)^2(2(0.60)(0.40))^3((0.40)^2)^1 = 0.1376. \end{aligned}$$

A common problem in statistical genetics is to estimate the allele probability p from a sample of n individuals, where the data consist of observed genotype frequencies. For instance, suppose in a sample of 60 fruit flies, we observe the gene distribution given in Table 5.3.

If p is unknown,

$$\begin{aligned} &P(X_1 = 35, X_2 = 17, X_3 = 8) \\ &\quad = \frac{60!}{35!17!8!}(p^2)^{35}(2p(1-p))^{17}((1-p)^2)^8 \\ &\quad = \text{constant} \times p^{87}(1-p)^{33}. \end{aligned} \tag{5.10}$$

The maximum likelihood principle, introduced in Example 5.14, says to estimate p with the value that maximizes the probability of obtaining the observed data, that is, the value of p that maximizes the probability given in Equation 5.10.

TABLE 5.3. Genotype frequencies for a sample of 60 fruit flies.

AA	*Aa*	*aa*
35	17	8

Differentiating this expression and setting it equal to 0 gives the equation

$$87p^{86}(1-p)^{33} = 33(1-p)^{32}p^{87}.$$

Solving for p gives the maximum likelihood estimate $\hat{p} = 87/120 = 0.725$. ■

Marginal distribution, expectation, variance. Let $(X_1, \ldots, X_r)$ have a multinomial distribution with parameters $n, p_1, \ldots, p_r$. The X_i's are not independent as they are constrained by $X_1 + \cdots + X_r = n$.

For each $k = 1, \ldots, r$, the marginal distribution of X_k is binomial with parameters n and p_k. We will derive this result two ways: easy and hard.

The easy way is a consequence of a simple probabilistic argument. Think of each of the underlying n independent trials as either resulting in outcome k or not. Then X_k counts the number of "successes" in n trials, which gives the binomial result $X_k \sim \text{Binom}(n, p_k)$.

The hard way is an exercise in working sums. To find a marginal distribution, we sum the joint probability mass function over all the other variables. Consider the marginal distribution of X_1. Summing over the outcomes of $X_2, \ldots, X_r$ gives

$$\begin{aligned}
P(X_1 = x_1) &= \sum_{x_2,\ldots,x_r} P(X_1 = x_1, X_2 = x_2, \ldots, X_r = x_r) \\
&= \sum_{x_2+\cdots+x_r=n-x_1} \frac{n!}{x_1!x_2!\cdots x_r!} p_1^{x_1} p_2^{x_2} \cdots p_r^{x_r} \\
&= \frac{n!}{x_1!} p_1^{x_1} \sum_{x_2+\cdots x_r=n-x_1} \frac{1}{x_2!\cdots x_r!} p_2^{x_2} \cdots p_r^{x_r} \\
&= \frac{n!}{x_1!(n-x_1)!} p_1^{x_1} \sum_{x_2+\cdots x_r=n-x_1} \frac{(n-x_1)!}{x_2!\cdots x_r!} p_2^{x_2} \cdots p_r^{x_r} \\
&= \frac{n!}{x_1!(n-x_1)!} p_1^{x_1} (p_2 + \cdots + p_r)^{n-x_1} \\
&= \frac{n!}{x_1!(n-x_1)!} p_1^{x_1} (1-p_1)^{n-x_1},
\end{aligned}$$

for $0 \le x_1 \le n$. The sum is over all nonnegative integers $x_2, \ldots, x_n$ which sum to $n - x_1$. The penultimate equality is because of the multinomial theorem.

We see that X_1 has a binomial distribution with parameters n and p_1. Similarly, for each $k = 1, \ldots, r$, $X_k \sim \text{Binom}(n, p_k)$.

The marginal result gives

$$E[X_k] = np_k \quad \text{and} \quad V[X_k] = np_k(1-p_k).$$

Covariance. Let $(X_1, \ldots, X_r)$ have a multinomial distribution with parameters $n, p_1, \ldots, p_r$. For $i \neq j$, consider $\text{Cov}(X_i, X_j)$. Use indicators and write $X_i = I_1 +$

$\cdots + I_n$, where

$$I_k = \begin{cases} 1, & \text{if the } k\text{th trial results in outcome } i, \\ 0, & \text{otherwise,} \end{cases}$$

for $k = 1, \ldots, n$. Similarly, write $X_j = J_1 + \cdots + J_n$, where

$$J_k = \begin{cases} 1, & \text{if the } k\text{th trial results in outcome } j, \\ 0, & \text{otherwise,} \end{cases}$$

for $k = 1, \ldots, n$.

Because of the independence of the Bernoulli trials, pairs of indicator variables involving different trials are independent. That is, I_g and J_h are independent if $g \neq h$, and hence, $\text{Cov}(I_g, J_h) = 0$. On the other hand, if $g = h$, then $I_g J_g = 0$, because the gth trial cannot result in both outcomes i and j. Thus,

$$\text{Cov}(I_g, J_g) = E[I_g J_g] - E[I_g]E[J_g] = -p_i p_j.$$

Using the linearity properties of covariance (see Section 4.9),

$$\begin{aligned} \text{Cov}(X_i, X_j) &= \text{Cov}\left(\sum_{g=1}^{n} I_g, \sum_{h=1}^{n} J_h\right) = \sum_{g=1}^{n}\sum_{h=1}^{n} \text{Cov}(I_g, J_h) \\ &= \sum_{g=1}^{n} \text{Cov}(I_g, J_g) = \sum_{g=1}^{n} (-p_i p_j) = -np_i p_j. \end{aligned}$$

Example 5.21 In Example 5.19 for the parliamentary election, the number of supporters for 4 parties out of 10 people was found to have a multinomial distribution. Find $\text{Cov}(X_A, X_C)$.

We know that

$$(X_A, X_B, X_C, X_D) \sim \text{Multi}(10, 0.20, 0.25, 0.30, 0.25).$$

Applying the formula just derived for covariance, we have that

$$\text{Cov}(X_A, X_C) = -np_A p_C = -10(0.20)(0.30) = -0.60.$$ ■

5.6 BENFORD'S LAW*

> It has been observed that the pages of a much used table of common logarithms show evidences of a selective use of the natural numbers. The pages containing the logarithms of the low numbers 1 and 2 are apt to be more stained and frayed by use than those of the higher numbers 8 and 9. Of course, no one could be expected to be greatly interested in the condition of a table of logarithms, but the matter may be considered more worthy of study when we recall that the table is used in the building

> up of our scientific, engineering, and general factual literature. There may be, in the relative cleanliness of the pages of a logarithm table, data on how we think and how we react when dealing with things that can be described by means of numbers.
>
> —Frank Benford [1938]

In this book's Introduction, we describe Benford's law and suggest the following classroom activity: Pick a random book in your backpack. Open up to a random page. Let your eyes fall on a random number in the middle of the page. Write down the number and circle the first digit, ignoring zeros.

We collect everybody's first digits, and before looking at the data, ask students to guess what the distribution of these numbers looks like. Many say that they will be roughly equally distributed between 1 and 9, following a discrete uniform distribution. Remarkably, they are not. Ones are most likely, then twos, which are more likely than threes, etc. The values tend to follow Benford's law, the distribution given in Table 5.4.

Benford's law is named after physicist Frank Benford who was curious about the wear and tear of the large books of logarithms which were widely used for scientific calculations before computers or calculators. Benford eventually looked at 20,229 data sets—everything from areas of rivers to death rates. In all cases, the first digit was one about 30% of the time, and the digits followed a remarkably similar pattern.

Benford eventually discovered the formula

$$P(d) = \log_{10}\left(\frac{d+1}{d}\right),$$

for the probability that the first digit is $d = 1, \ldots, 9$. Observe that P is in fact a probability distribution on $\{1, \ldots, 9\}$, as

$$\begin{aligned}\sum_{k=1}^{9} P(d) &= \sum_{k=1}^{9} \log_{10}\left(\frac{d+1}{d}\right) = \sum_{k=1}^{9} [\log_{10}(d+1) - \log_{10} d] \\ &= (\log_{10} 2 - \log_{10} 1) + (\log_{10} 3 - \log_{10} 2) + \cdots + (\log_{10} 10 - \log_{10} 9) \\ &= \log_{10} 10 - \log_{10} 1 = 1,\end{aligned}$$

as the terms are telescoping.

It is not easy to explain why Benford's law works for so many empirical datasets, and much of the mathematics is outside the scope of this book.

TABLE 5.4. Benford's law.

1	2	3	4	5	6	7	8	9
0.301	0.176	0.125	0.097	0.079	0.067	0.058	0.051	0.046

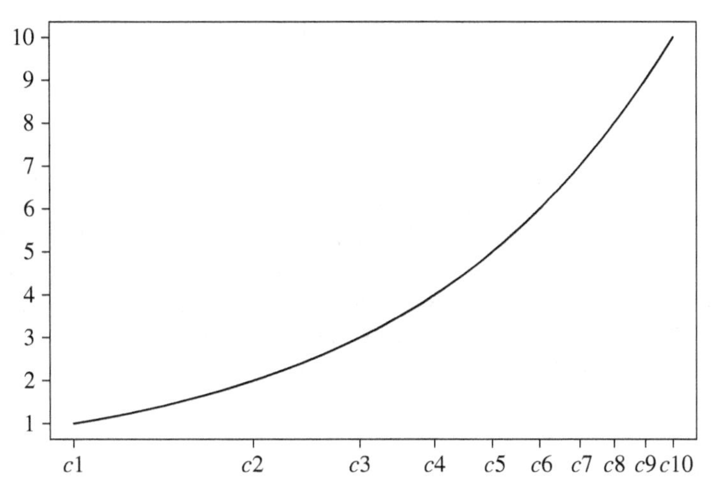

FIGURE 5.1: Graph of $P(n) = ar^n$.

Ross [2011] shows why phenomenon that exhibit exponential growth or decline—populations of cities, powers of 2, pollution levels, radioactive decay—exhibit Benford's law. As explained by Ross, suppose we have data of the form $P(n) = ar^n$ for constants a and r. For instance, $P(n)$ might be the population of a city at year n, or the half-life of a radioactive isotope after n minutes. Consider values of n such that $1 \leq P(n) < 10$. For each $d = 1, \ldots, 9$, let $[c_d, c_{d+1})$ be the interval of values for which the first digit of $P(n)$ is d (see Fig. 5.1).

For all values $c_1 \leq n < c_2$, the first digit of $P(n)$ is 1. For $c_2 \leq n < c_3$, the first digit of $P(n)$ is 2, and so on. For each $d = 1, \ldots, 9$, $P(c_d) = ar^{c_d} = d$. Taking logarithms (base 10) gives

$$\log a + c_d \log r = \log d.$$

Also $\log a + c_{d+1} \log r = \log(d + 1)$. Subtracting equations gives

$$(c_{d+1} - c_d)(\log r) = \log(d + 1) - \log d.$$

Thus, the length of each interval $[c_d, c_{d+1})$ is

$$c_{d+1} - c_d = \frac{\log(d + 1) - \log d}{\log r}.$$

The length of the entire interval $[c_1, c_{10})$ is

$$\sum_{d=1}^{9}(c_{d+1} - c_d) = \frac{\log 10 - \log 1}{\log r} = \frac{1}{\log r},$$

due to the telescoping terms. Thus, the "fraction of the time" that the first digit of ar^n is d is

$$\frac{c_{d+1} - c_d}{c_{10} - c_1} = \log(d+1) - \log d = \log\left(\frac{d+1}{d}\right).$$

The argument here for values of $P(n)$ between 1 and 10 actually applies to all intervals of values between 10^k and 10^{k+1}, for $k = 0, 1, \ldots$.

For a wonderful introduction to Benford's law, and its use in detecting fraud, hear the Radiolab podcast *From Benford to Erdős* at `www.radiolab.org`. For detailed treatment of Benford's law in literature, the reader is directed to Berger and Hill [2015] and Miller [2015]. The **R** supplements contain more examples and applications of **R** packages focused on Benford's law for the reader to explore.

5.7 SUMMARY

This chapter explores several new discrete probability distributions. Many of them are generalizations or extensions of the ones we have studied. Mgfs are introduced as a way of characterizing distributions and illustrating some of these relationships. The geometric distribution arises as the number of i.i.d. Bernoulli trials required for the first success to occur. The negative binomial generalizes the geometric distribution and counts the number of trials required for the rth first success. The hypergeometric distribution arises when sampling is done without replacement, in contrast to the binomial distribution which arises when sampling is done with replacement. The multinomial distribution is a multivariate distribution which generalizes the binomial distribution. Also introduced are multinomial coefficients and the multinomial theorem. Finally, Benford's law is a fascinating distribution which governs the distribution of first digits for many empirical datasets.

- **Geometric distribution:** A random variable X has a geometric distribution with parameter p, if $P(X = k) = (1-p)^{k-1}p$, for $k = 1, 2, \ldots$. The distribution arises as the number of trials required for the first success in repeated i.i.d. Bernoulli trials.
- **Memorylessness:** A random variable X has the memorylessness property if $P(X > s + t)|X > t) = P(X > s)$ for all $s, t > 0$. The geometric distribution is the only discrete distribution that is memoryless.
- **Properties of geometric distribution:**
  1. $E[X] = 1/p$.
  2. $V[X] = (1-p)/p^2$.
  3. $P(X > x) = (1-p)^x$, for $x > 0$.
- **Moments:** The kth moment of X is $E[X^k]$.
- **Moment-generating function:** $m(t) = E[e^{tX}]$ defined for real t when the expectation exists.

- **Properties of the moment-generating functions:**
  1. $E[X^k] = m^{(k)}(0)$.
  2. If X and Y are independent, then $m_{X+Y}(t) = m_X(t)m_Y(t)$.
  3. For constants $a \neq 0$ and b, $m_{aX+b}(t) = e^{tb}m_X(at)$.
  4. If two random variables have the same mgf, they have the same distribution.
- **Negative binomial distribution:** A random variable X has a negative binomial distribution with parameters r and p if

$$P(X = k) = \binom{k-1}{r-1} p^r(1-p)^{k-r}, \quad \text{for } r = 1, 2, \ldots, \; k = r, r+1, \ldots.$$

  The distribution arises as the number of trials required for the rth success in repeated i.i.d. Bernoulli trials. If $r = 1$, we get the geometric distribution.
- **Properties of negative binomial distribution:**
  1. $E[X] = r/p$.
  2. $V[X] = r(1-p)/p^2$.
- **Hypergeometric distribution:** Given a bag of N balls which contains r red balls and $N - r$ blue balls, let X be the number of red balls in a sample of size n taken without replacement. Then X has a hypergeometric distribution with pmf

$$P(X = k) = \frac{\binom{r}{k}\binom{N-r}{n-k}}{\binom{N}{n}}$$

  for $\max(0, n - (N - r)) \leq k \leq \min(n, r)$.
- **Properties of hypergeometric distribution:**
  1. $E[X] = nr/N$.
  2. $V[X] = n(N-n)r(N-r)/(N^2(N-1))$.
- **Multinomial distribution:** Random variables $X_1, \ldots, X_r$ have a multinomial distribution with parameters $n, p_1, \ldots, p_r$, where $p_1 + \cdots + p_r = 1$, if

$$P(X_1 = x_1, \ldots, X_r = x_r) = \frac{n!}{x_1! \cdots x_r!} p_1^{x_1} \cdots p_r^{x_r},$$

  for nonnegative integers $x_1, \ldots, x_r$ such that $x_1 + \cdots + x_r = n$. The distribution generalizes the binomial distribution (with $r = 2$) and arises when successive independent trials can take more than two values.
- **Multinomial counts:** The multinomial coefficient

$$\binom{n}{x_1, \ldots, x_r} = \frac{n!}{x_1! \cdots x_r!}$$

counts the number of n element sequences in which (i) each element can take one of r possible values and (ii) exactly x_k elements of the sequence take the kth value, for $k = 1, \ldots, r$.

- **Multinomial theorem:** For any positive integer r and n and real numbers $z_1, \ldots, z_r$,

$$(z_1 + \cdots + z_r)^n = \sum_{a_1+\cdots+a_r=n} \frac{n!}{a_1! \cdots a_r!} z_1^{a_1} \cdots z_r^{a_r},$$

where the sum is over all lists of r nonnegative integers $(a_1, \ldots, a_r)$ which sum to n.

- **Properties of multinomial distribution:**
  1. $X_k \sim \text{Binom}(n, p_k)$, for each $k = 1, \ldots, r$.
  2. $\text{Cov}(X_i, X_j) = -np_ip_j$.
- **Benford's law:** A random variable X has the Benford's law distribution if $P(X = d) = \log_{10}((d + 1)/d)$, for $d = 1, \ldots, 9$. The distribution arises as the distribution of the first digit for many datasets.

EXERCISES

Geometric Distribution

5.1 Rohit is playing five card poker with his friends. What is the expected number of hands it will take before he is dealt a full house? A full house is three cards of one face value and two of another face value.

5.2 What is the probability that it takes an even number of die rolls to get a four?

5.3 Find the variance of a geometric distribution. Hint: To find $E[X^2]$, write $k^2 = k^2 - k + k = k(k - 1) + k$. You will need to take two derivatives.

5.4 Suppose $X \sim \text{Geom}(0.3)$. Provide appropriate commands and compute the following probabilities in **R**. Pay attention to how **R** codes the Geometric distribution.

(a) $P(X = 3)$.

(b) $P(2 \leq X < 9)$.

(c) $P(X \geq 4)$.

5.5 Loki spends his weekends metal detecting along a beach for fun. He picks small areas to search, and from past experience, about 15% of these sites have resulted in "hits" (where he finds something). Assume he picks sites independently of each other.

(a) Find the probability that Loki gets his first hit on a weekend on the fifth site he visits.

(b) Find the probability that it takes at most five sites for Loki to get a hit.

(c) What is the expected number of sites it will take for Loki to get a hit?

5.6 A manufacturing process produces components which have a 1% chance of being defective. Successive components are independent.

(a) Find the probability that it takes exactly 110 components to be produced before a defective one occurs.

(b) Find the probability that it takes at least 110 components to be produced before a defective one occurs.

(c) What is the expected number of components that will be produced before a defective one occurs?

5.7 Dominic is applying to college and sending out many applications. He estimates there is a 25% chance that an application will be successful. How many applications should he send out so that the probability of at least one acceptance is at least 95%?

5.8 There are 15 professors in the math department. Every time Tina takes a math class each professor is equally likely to be the instructor. What is the expected number of math classes which Tina needs to take in order to be taught by every math professor?

5.9 In the coupon collector's problem, let X be the number of draws of n coupons required to obtain a complete set. Find the variance of X.

5.10 A bag has r red and b blue balls. Balls are picked at random without replacement. Let X be the number of selections required for the first red ball to be picked.

(a) Explain why X does not have a geometric distribution.

(b) Show that the probability mass function of X is

$$P(X = k) = \frac{\binom{r+b-k}{r-1}}{\binom{r+b}{r}} \quad \text{for } k = 1, 2, \ldots, b+1.$$

5.11 Let $X \sim \text{Geom}(p)$. Find $E[2^X]$ for those values of p for which the expectation exists.

5.12 Make up your own example to show that the Poisson distribution is *not* memoryless. That is, pick values for λ, s, and t and show that $P(X > t|X > s) \neq P(X > t - s)$.

MGFS

5.13 A random variable X has a mgf

$$m(t) = pe^{-t} + qe^{t} + 1 - p - q,$$

where $p + q = 1$. Find all the moments of X.

5.14 A random variable X takes values -1, 0, and 2, with respective probabilities 0.2, 0.3, and 0.5. Find the mgf of X. Use the mgf to find the first two moments of X.

5.15 Find the mean of a geometric distribution with parameter p using mgfs.

5.16 Find the mgf of a Bernoulli random variable with parameter p. Use this to find the mgf of a binomial random variable with parameters n and p.

5.17 Identify the distributions of the following random variables based on their provided mgfs:

(a) $m_X(t) = 0.8e^t + 0.2$.

(b) $m_Y(t) = \frac{0.1e^t}{1-0.9e^t}$.

(c) $m_W(t) = (0.3e^t + 0.7)^{14}$.

5.18 Suppose $X \sim \text{Geom}(0.2)$, and $Y = 4X + 2$. Find the mgf of Y.

5.19 Let X and Y be independent binomial random variables with parameters (m, p) and (n, p), respectively. Use mgfs to show that $X + Y$ is a binomial random variable with parameters $m + n$ and p.

Negative Binomial Distribution

5.20 Suppose $X \sim \text{NegBin}(3, 0.4)$. Provide appropriate commands and compute the following probabilities in **R**. Pay attention to how **R** codes the Negative Binomial distribution.

(a) $P(X = 6)$.

(b) $P(5 \leq X < 11)$.

(c) $P(X \geq 8)$.

5.21 A fair coin is tossed until heads appears four times.

(a) Find the probability that it took exactly 10 flips.

(b) Find the probability that it took at least 10 flips.

(c) Let Y be the number of tails that occur. Find the pmf of Y.

5.22 Baseball teams A and B face each other in the World Series. For each game, the probability that A wins the game is p, independent of other games. Find the expected length of the series.

5.23 Let X and Y be independent geometric random variables with parameter p. Find the pmf of $X + Y$. You should know what distribution will result.

5.24 Each student in a campus club is asked to sell 12 candy bars to support a cause. From previous drives, the students know the probability that they make a candy bar sale going dorm room to dorm room is about 0.2 per room.

- **(a)** What is the probability a student sells all 12 bars by visiting at most 50 dorm rooms?
- **(b)** How many dorm rooms should each student expect to visit to sell their 12 candy bars? (Assume they pick different dorms.)

5.25 People whose blood type is O-negative are universal donors—anyone can receive a blood transfusion of O-negative blood. In the United States, 7.2% of the people have O-negative blood. A blood donor clinic wants to find 10 O-negative individuals. In repeated screening, what is the chance of finding such individuals among the first 100 people screened?

5.26 Using the relationship in Equation 5.7 between the binomial and negative binomial distributions, recast the solution to the Problem of Points in terms of the binomial distribution. Give the exact solution in terms of a binomial probability.

5.27 Aidan and Bethany are playing a game worth \$100. They take turns flipping a penny. The first person to get 10 heads will win. But they just realized that they have to be in class right away and are forced to stop the game. Aidan had four heads and Bethany had seven heads. How should they divide the pot?

5.28 **Banach's matchbox problem.** This famous problem was posed by mathematician Hugo Steinhaus as an affectionate honor to fellow mathematician Stefan Banach, who was a heavy pipe smoker. A smoker has two matchboxes, one in each pocket. Each box has n matches in it. Whenever the smoker needs a match, he reaches into a pocket at random and takes a match from the box. Suppose he reaches into a pocket and finds that the matchbox is empty, find the probability that the box in the other pocket has exactly k matches left.

- **(a)** Let X be the number of matches in the right box when the left box is found empty. Show that X has a negative binomial distribution with parameters $n + 1$ and $1/2$.
- **(b)** Show that the desired probability is $2 \times P(X = 2n + 1 - k)$.
- **(c)** Work out the problem in detail for the case $n = 1$.

5.29 Let $R \sim \text{Geom}(p)$. Conditional on $R = r$, suppose X has a negative binomial distribution with parameters r and p. Show that the marginal distribution of X is geometric. What is the parameter?

Hypergeometric Distribution

5.30 There are 500 deer in a wildlife preserve. A sample of 50 deer are caught and tagged and returned to the population. Suppose that 20 deer are caught later and examined to see if they are tagged.

(a) Find the mean and standard deviation of the number of tagged deer in the sample.

(b) Find the probability that the sample contains at least three tagged deer.

5.31 **The Lady Tasting Tea.** This is one of the most famous experiments in the founding history of statistics. In his 1935 book *The Design of Experiments* [1935], Sir Ronald A. Fisher writes,

> A lady declares that by tasting a cup of tea made with milk she can discriminate whether the milk or the tea infusion was first added to the cup. We will consider the problem of designing an experiment by means of which this assertion can be tested ... Our experiment consists in mixing eight cups of tea, four in one way and four in the other, and presenting them to the subject for judgment in a random order. ... Her task is to divide the 8 cups into two sets of 4, agreeing, if possible, with the treatments received.

Consider such an experiment. Four cups are poured milk first and four cups are poured tea first and presented to a friend for tasting. Let X be the number of milk-first cups that your friend correctly identifies as milk-first.

(a) Identify the distribution of X.

(b) Find $P(X = k)$ for $k = 0, 1, 2, 3, 4$.

(c) If in reality, your friend had no ability to discriminate and actually guessed, what is the probability they would correctly identify all four cups correctly?

5.32 In a town of 20,000, there are 12,000 voters, of whom 5000 are registered democrats and 6000 are registered republicans. An exit poll is taken of 200 voters. Assume all registered voters actually voted. Use **R** to find:

(a) The mean and standard deviation of the number of democrats in the sample.

(b) The probability that more than half the sample are republicans.

5.33 In the card game bridge, four players are dealt 13 cards each from a standard deck. Find the following probabilities for a bridge hand of 13 cards.

(a) The probability of being dealt exactly two red cards.

(b) The probability of being dealt four spades, four clubs, three hearts, and two diamonds.

5.34 Candice has 8 marigolds and 15 impatiens plants available to plant in her garden. In one area of the garden, she has room for six plants. She has decided to plant randomly this year. Let X be the number of marigolds planted in the area with six plants. Find the following:

(a) $P(X = 3)$.

(b) $P(X \leq 2)$.

(c) $E(X)$.

5.35 Consider the hypergeometric distribution with parameters r, n, and N. Suppose r depends on N in such a way that $r/N \to p$ as $N \to \infty$, where $0 < p < 1$. Show that the mean and variance of the hypergeometric distribution converges to the mean and variance, respectively, of a binomial distribution with parameters n and p, as $N \to \infty$.

Multinomial Distribution

5.36 A Halloween bag contains three red, four green, and five blue candies. Tom reaches in the bag and takes out three candies. Let R, G, and B denote the number of red, green, and blue candies, respectively, that Tom got.

(a) Is the distribution of (R, G, B) multinomial? Explain.

(b) What is the probability that Tom gets one of each color?

5.37 In a city of 100,000 voters, 40% are Democrat, 30% Republican, 20% Green, and 10% Undecided. A sample of 1000 people is selected.

(a) What is the expectation and variance for the number of Greens in the sample?

(b) What is the expectation and variance for the number of Greens and Undecideds in the sample?

5.38 A random experiment takes r possible values, with respective probabilities $p_1, \ldots, p_r$. Suppose the experiment is repeated N times, where N has a Poisson distribution with parameter λ. For $k = 1, \ldots, r$, let N_k be the number of occurrences of outcome k. In other words, if $N = n$, then $(N_1, \ldots, N_r)$ has a multinomial distribution.

Show that $N_1, \ldots, N_r$ form a sequence of independent Poisson random variables and for each $k = 1, \ldots, r$, $N_k \sim \text{Pois}(\lambda p_k)$.

5.39 Suppose a gene allele takes two forms A and a, with $P(A) = 0.20 = 1 - P(a)$. Assume a population is in Hardy–Weinberg equilibrium.

(a) Find the probability that in a sample of eight individuals, there is one AA, two Aa's, and five aa's.

(b) Find the probability that there are at least seven aa's.

5.40 In 10 rolls of a fair die, let X be the number of fives rolled, let Y be the number of even numbers rolled, and let Z be the number of odd numbers rolled.

(a) Find $\text{Cov}(X, Y)$.

(b) Find $\text{Cov}(X, Z)$.

5.41 Prove the identity

$$4^n = \sum_{a_1+\cdots a_4=n} \frac{n!}{a_1!a_2!a_3!a_4!},$$

where the sum is over all nonnegative integers that sum to 4.

Benford's Law

5.42 Find the expectation and variance of the Benford's law distribution.

5.43 Find a real-life dataset to test whether Benford's law applies. Report your findings.

5.44 Find a real-life journal article that uses Benford's law. How was the law used? What were the findings? Cite your article.

5.45 Investigate whether there are similar laws for second digits. What do you find? Cite your sources.

Other

For the next four problems 5.46–5.49, identify a random variable and describe its distribution before doing any computations. (For instance, for Problem 5.46, start your solution with "Let X be the number of days when there is no homework. Then X has a binomial distribution with $n = 42$ and $p = \ldots$.")

5.46 A professor starts each class by picking a number from a hat that contains the numbers 1–30. If a prime number is chosen, there is no homework that day. There are 42 class periods in the semester. How many days can the students expect to have no homework?

5.47 Suppose eight cards are drawn from a standard deck *with replacement*. What is the probability of obtaining two cards from each suit?

5.48 Among 30 raffle tickets six are winners. Jenna buys 10 tickets. Find the probability that she got three winners.

5.49 A teacher writes an exam with 20 problems. There is a 5% chance that any problem has a mistake. The teacher tells the class that if the exam has three or more problems with mistakes he will give everyone an A. The teacher repeats this in 10 different classes. Find the probability that the teacher gave out all As at least once. Hint: How many random variables are involved here?

Simulation and R

5.50 Conduct a study to determine how well the binomial distribution approximates the hypergeometric distribution. Consider a bag with n balls, 25% of which are red. A sample of size $(0.10)n$ is taken. Let X be the number of red balls in the sample. Find $P(X \leq (0.02)n)$ for increasing values of n when sampling is (i) with replacement and (ii) without replacement. Use **R**.

5.51 Write a function `coupon(n)` for simulating the coupon collector's problem. That is, let X be the number of draws required to obtain all n items when sampling with replacement. Use your function to simulate the mean and standard deviation of X for $n = 10$ and $n = 52$.

5.52 Let $p = (p_1, \ldots, p_n)$ be a list of probabilities with $p_1 + \cdots + p_n = 1$. Write a function `coupon(n, p)` which generalizes the function above, and simulates the coupon collector's problem for unequal probabilities, where the probability of choosing item i is p_i. For $n = 10$, let p be a list of binomial probabilities with parameters 10 and 1/2. Use your function to simulate the mean and standard deviation of X, the number of draws required to obtain all n items when sampling with replacement.

5.53 Read about the World Series in Example 5.10. Suppose the World Series is played between two teams A and B such that for any matchup between A and B, the probability that A wins is $0 < p < 1$. For $p = 0.25$ and $p = 0.60$, simulate the expected length and standard deviation of the length of the series.

Chapter Review

Chapter review exercises are available through the text website. The URL is `www.wiley.com/go/wagaman/probability2e`.

6

CONTINUOUS PROBABILITY

The theory of probabilities is at bottom nothing but common sense reduced to calculus; it enables us to appreciate with exactness that which accurate minds feel with a sort of instinct for which ofttimes they are unable to account.

—Laplace

Learning Outcomes

1. Define the terms: continuous RV, probability density function, and cumulative density function.
2. Compute expectation and variance for continuous RVs.
3. Apply uniform and exponential distributions to appropriate problems.
4. Solve problems involving joint and marginal distributions in the continuous setting.
5. Extend independence, covariance, and correlation concepts to the continuous setting.
6. (C) Work with uniform and exponential distributions in **R** to find probabilities and simulate problems.

Introduction. Picking a real number between 0 and 1 is an example of a random experiment where the sample space is a continuum of values. Such sets have no gaps between elements; they are not discrete. The elements are uncountable and cannot be listed. We call such sample spaces *continuous*. The most common continuous sample spaces in one dimension are intervals such as (a, b), $(-\infty, c]$, and $(-\infty, \infty)$.

Probability: With Applications and R, Second Edition. Amy S. Wagaman and Robert P. Dobrow.

Companion Website: www.wiley.com/go/wagaman/probability2e

In two dimensions, regions in the Cartesian plane are often the sample spaces. Consider the following example.

An archer shoots an arrow at a target. The target C is a circle of radius 1. The bullseye B is a smaller circle in the center of the target of radius $1/4$. What is the probability $P(B)$ of hitting the bullseye?

Our probability model will be *uniform* on the target—all points are equally likely. But how to make sense of that?

We start with some intuitive discussion because this example can be solved without using multivariate calculus, if thought through. The technical details about using double integrals will come later for examples where this is not the case.

If this were a discrete finite problem, the solution would be to count the number of points in the bullseye and divide by the number of points in the target. But the sets are uncountable. The way to "count" points in continuous sets is by integration. As all points on the target are equally likely we will integrate in such a way so that no point gets more "weight" than any other—all points are treated equal. That is, we integrate a constant c over the bullseye region. This leads to

$$P(B) = \iint_B c\, dx\, dy = c[\text{Area } (B)] = \frac{c\pi}{16}. \tag{6.1}$$

What is c? As this is a probability model, the points in the sample space should "add up" or integrate to 1. The sample space is C, the target. This gives

$$1 = P(C) = \iint_C c\, dx\, dy = c[\text{Area } (C)] = c\pi$$

and thus $c = 1/\pi$. Plugging in c to Equation 6.1 gives

$$P(B) = \frac{1}{\text{Area } (C)} \iint_B dx\, dy = \frac{1}{16},$$

the proportion of the total area of the target taken up by the bullseye.

This gives the beginnings of a *continuous uniform probability model*. If Ω is a continuous set with all points equally likely, then for subsets $S \subseteq \Omega$,

$$P(S) = \frac{1}{\text{Area } (\Omega)} \iint_S dx\, dy = \frac{\text{Area } (S)}{\text{Area } (\Omega)}.$$

In one dimension, the double integral becomes a single integral and area becomes length. In three dimensions, we have a triple integral and volume.

The following examples are all meant to be approached intuitively (no calculus required!).

Example 6.1

- A real number is picked uniformly at random from the interval $(-5, 2)$. The probability the number is positive is

$$\frac{\text{Length }(0, 2)}{\text{Length }(-5, 2)} = \frac{2}{7}.$$

- A sphere of radius 1 is inscribed in a cube with side length 2. A point in the cube is picked uniformly at random. The probability that the point is contained in the sphere is

$$\frac{\text{Volume (sphere)}}{\text{Volume (cube)}} = \frac{4\pi/3}{8} = \frac{\pi}{6}.$$

- A student arrives to class at a uniformly random time between 8:45 and 9:05 a.m. Class starts at 9:00 a.m. The probability that the student arrives on time is

$$\frac{15 \text{ minutes}}{20 \text{ minutes}} = \frac{3}{4}.$$

■

All of these examples can be cast in terms of random variables. A variable X might be a point on a target or a time in a continuous interval. But to work with random variables in the continuous world will require some mathematical tools, including integrals and integration techniques such as integration by parts.

6.1 PROBABILITY DENSITY FUNCTION

A continuous random variable X is a random variable that takes values in a continuous set. If S is a subset of the real numbers, then $\{X \in S\}$ is the event that X takes values in S. For instance, $\{X \in (a, b)\} = \{a < X < b\}$, and $\{X \in (-\infty, c]\} = \{X \leq c\}$.

In the discrete setting, to compute $P(X \in S)$, we add up values of the probability mass function (pmf). That is, $P(X \in S) = \sum_{x \in S} P(X = x)$.

If X, however, is a continuous random variable, to compute $P(X \in S)$ we integrate the *probability density function* (pdf) over S. The pdf plays the role of the pmf. It is the function used to compute probabilities. Observe the similarities between the pdf for continuous random variables and the pmf.

PROBABILITY DENSITY FUNCTION

Let X be a continuous random variable. A function f is a *probability density function* of X if

1. $f(x) \geq 0$, for all $-\infty < x < \infty$.
2. $\int_{-\infty}^{\infty} f(x)\, dx = 1$.
3. For $S \subseteq \Re$,

$$P(X \in S) = \int_S f(x)\, dx. \tag{6.2}$$

Note: And 0, Otherwise.

When providing pdfs (and similarly with pmfs), we often describe the range (or region) where the pdf takes positive value. Recall that the pdf must be nonnegative. Thus, if a range of values is given where the pdf is positive, it is implied that the pdf takes the value of 0 for any other value of x. Fully specifying a pdf includes stating this explicitly, which often means a statement that the pdf is "0, otherwise." In the many examples to come, you will see a mix of the pdf being fully specified and not. Just remember that it is implied that the pdf is 0 outside the described range, called its *support*.

When writing density functions we sometimes use subscripts, e.g., $f_X(x) = f(x)$, to identify the associated random variable. The integral $\int_S f(x)\, dx$ is taken over the set of values in S. For instance, if $S = (a, b)$, then

$$P(X \in S) = P(a < X < b) \quad \text{and} \quad \int_S f(x)\, dx = \int_a^b f(x)\, dx.$$

If $S = (-\infty, c]$, then

$$P(X \in S) = P(X \leq c) = \int_{-\infty}^{c} f(x)\, dx.$$

Note that for any real number a,

$$P(X = a) = P(X \in \{a\}) = \int_a^a f(x)\, dx = 0.$$

That is, for a continuous random variable, the probability of any particular number occurring is 0. Nonzero probabilities are assigned to intervals, not to individual or discrete outcomes.

As $[a, b) = (a, b) \cup \{a\}$ and $P(X = a) = 0$, it follows that

$$P(X \in [a, b)) = P(a \leq X < b) = P(a < X < b).$$

Similarly,

$$P(a < X < b) = P(a \leq X < b) = P(a < X \leq b) = P(a \leq X \leq b).$$

Example 6.2 A random variable X has density function $f(x) = 3x^2/16$, for $-2 < x < 2$, and 0, otherwise. Find $P(X > 1)$.

We have

$$P(X > 1) = \int_1^2 f(x)\, dx = \int_1^2 \frac{3x^2}{16}\, dx = \frac{1}{16}\,(x^3)\Big|_1^2 = \frac{7}{16}.$$

The density function is shown in Figure 6.1, with the shaded area denoting $P(X > 1)$. ■

Example 6.3 A random variable X has density function of the form $f(x) = ce^{-|x|}$, for all x. (i) Find c. (ii) Find $P(0 < X < 1)$.

(i) To find c, we use the fact that the total area under the curve integrates to 1. Solve

$$1 = \int_{-\infty}^{\infty} ce^{-|x|}\, dx = \int_{-\infty}^{0} ce^{x}\, dx + \int_0^{\infty} ce^{-x}\, dx$$

$$= 2c \int_0^{\infty} e^{-x}\, dx = 2c(-e^{-x})\Big|_0^{\infty} = 2c,$$

giving $c = 1/2$.

(ii)

$$P(0 < X < 1) = \int_0^1 \frac{e^{-|x|}}{2}\, dx = \frac{1}{2}(-e^{-x})\Big|_0^1 = \frac{1 - e^{-1}}{2} = 0.316.$$ ■

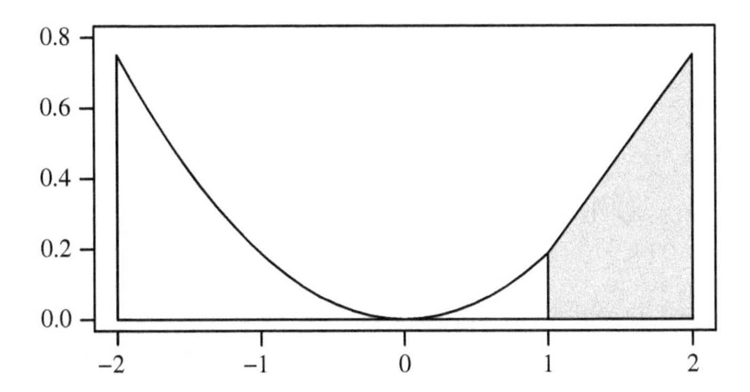

FIGURE 6.1: Density with shaded area indicating $P(X > 1)$.

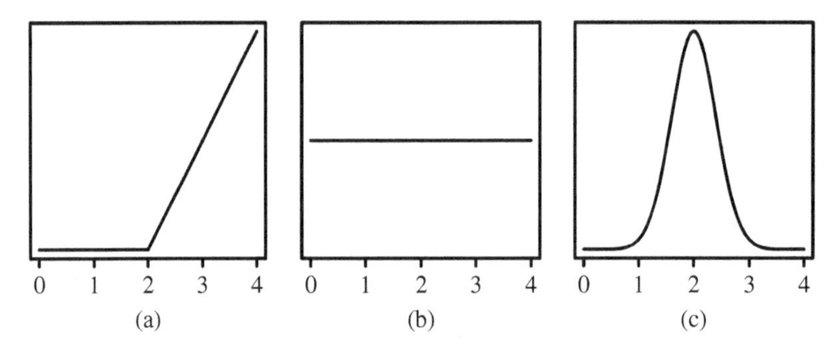

FIGURE 6.2: Three density shapes.

Pdf and pmf—similarities and differences. It is important to understand the similarities and differences between probability density and mass functions. Both give measures of how likely or "probable" a particular value is. The graph of a density function is comparable to a "smooth" probability histogram for a pmf.

See the graphs of the three density functions in Figure 6.2. The y-axis in each is the density value. In (a), the density models a random variable which takes values between 2 and 4, with values near 4 most likely and those near two least likely. In (b), values between 0 and 4 are equally likely. In (c), outcomes near 2 are most likely, with probability decreasing quickly for outcomes far from 2.

Unlike the pmf, however, the density $f(x)$ is *not* the probability that X is equal to x. That probability is always 0 for continuous variables. The value of $f(x)$ is a unitless measure of probability "mass" with the property that the total probability mass, that is, the area under the density curve, is equal to 1.

For more insight into what $f(x)$ "measures," consider an interval $(x - \epsilon/2, x + \epsilon/2)$ centered at x of length ϵ, where ϵ is small. If f is continuous at x then the probability that X falls in the interval is

$$P\left(x - \frac{\epsilon}{2} < X < x + \frac{\epsilon}{2}\right) = \int_{x-\epsilon/2}^{x+\epsilon/2} f(t)\, dt \approx f(x)\epsilon,$$

where the integral is approximated by the area of the rectangle of height $f(x)$ and width ϵ. This gives

$$f(x) \approx \frac{1}{\epsilon} P\left(x - \frac{\epsilon}{2} < X < x + \frac{\epsilon}{2}\right). \tag{6.3}$$

In physics, "density" is a measure of mass per unit volume, area, or length. In probability, Equation 6.3 shows that the density function is a measure of "probability mass" per unit length.

A function f is said to be *proportional* to a function g if the ratio of the two functions is constant, that is, $f(x)/g(x) = c$ for some constant c. The constant c is called the *proportionality constant*. We write $f(x) \propto g(x)$ to signify this relationship. Probability density functions are often specified up to a proportionality constant.

Example 6.4 Suppose the random time T of a radioactive emission is proportional to a decaying exponential function of the form $e^{-\lambda t}$ for $t > 0$, where $\lambda > 0$ is a constant. Find the pdf of T.

Write $f(t) = ce^{-\lambda t}$, for $t > 0$. Solving for the proportionality constant c gives

$$1 = \int_0^\infty ce^{-\lambda t}\, dt = \frac{c}{\lambda}\left(-e^{-\lambda t}\right)\Big|_0^\infty = \frac{c}{\lambda}.$$

Thus, $c = \lambda$ and $f(t) = \lambda e^{-\lambda t}$, for $t > 0$, and 0, otherwise. ■

Example 6.5 A random variable X has density function proportional to $x + 1$ on the interval $(2, 4)$. Find $P(2 < X < 3)$.

Write $f(x) = c(x + 1)$. To find c set

$$1 = \int_2^4 c(x+1)\, dx = c\left(\frac{x^2}{2} + x\right)\Bigg|_2^4 = 8c,$$

giving $c = 1/8$. Then,

$$P(2 < X < 3) = \int_2^3 \frac{1}{8}(x+1)\, dx = \frac{1}{8}\left(\frac{x^2}{2} + x\right)\Bigg|_2^3 = \frac{1}{8}\left(\frac{7}{2}\right) = \frac{7}{16}.$$ ■

Problem-solving strategy—reminder. In tackling this problem, note that there are two key steps. First, you must find the proportionality constant. Then you can find the specified probability. When approaching any problem, it can help to break it down into key steps. These can include identifying and defining a random variable, figuring out which equation or formula you need, and what components in that formula you need to compute. If you read over any example in the text, be sure you understand why each step is needed, as this can help you figure out strategies for tackling similar problems on your own.

6.2 CUMULATIVE DISTRIBUTION FUNCTION

One way to connect and unify the treatment of discrete and continuous random variables is through the *cumulative distribution function* (cdf) which is defined for *all* random variables.

CUMULATIVE DISTRIBUTION FUNCTION

Let X be a random variable. The *cdf* of X is the function

$$F(x) = P(X \leq x),$$

defined for all real numbers x.

The cdf plays an important role for continuous random variables in part because of its relationship to the density function. For a continuous random variable X

$$F(x) = P(X \leq x) = \int_{-\infty}^{x} f(t)\, dt.$$

If F is differentiable at x, then taking derivatives on both sides and invoking the fundamental theorem of calculus gives

$$F'(x) = f(x).$$

The pdf is the derivative of the cdf. Given a density for X we can, in principle, obtain the cdf, and vice versa. Thus either function can be used to specify the probability distribution of X.

If two random variables have the same density function, then they have the same probability distribution. Similarly, if they have the same cdf, they have the same distribution.

If we know the cdf of a random variable, we can compute probabilities on intervals. For $a < b$, observe that $(-\infty, b] = (\infty, a] \cup (a, b]$. This gives

$$\begin{aligned} P(a < X \leq b) &= P(X \in (a, b]) \\ &= P(X \in (-\infty, b]) - P(X \in (-\infty, a]) \\ &= P(X \leq b) - P(X \leq a) \\ &= F(b) - F(a). \end{aligned}$$

For continuous variables,

$$\begin{aligned} F(b) - F(a) &= P(a < X \leq b) = P(a < X < b) \\ &= P(a \leq X < b) = P(a \leq X \leq b). \end{aligned}$$

You have already seen cdfs in the context of discrete random variables when working in **R**. Functions such as `pbinom` and `ppois` evaluated the cdf for their respective distributions.

Example 6.6 The density for a random variable X is

$$f(x) = 2xe^{-x^2}, \quad \text{for } x > 0.$$

Find the cdf of X and use it to compute $P(1 < X < 2)$.

For $x > 0$,

$$F(x) = P(X \le x) = \int_0^x 2te^{-t^2}\, dt = \int_0^{x^2} e^{-u}\, du = 1 - e^{-x^2},$$

where we use the u-substitution $u = t^2$. For $x \le 0$, $F(x) = 0$.

We could find the probability $P(1 < X < 2)$ by integrating the density function over $(1, 2)$. But because we have already found the cdf there is no need. Note that

$$\begin{aligned} P(1 < X < 2) &= F(2) - F(1) \\ &= (1 - e^{-4}) - (1 - e^{-1}) = e^{-1} - e^{-4} = 0.350. \end{aligned}$$

■

Example 6.7 A random variable X has density function

$$f(x) = \begin{cases} 2/5, & \text{if } 0 < x \le 1 \\ 2x/5, & \text{if } 1 \le x < 2 \\ 0, & \text{otherwise.} \end{cases}$$

(i) Find the cdf of X. (ii) Find $P(0.5 < X < 1.5)$.

(i) We need to take care as this density is defined differently on two intervals. If $0 < x < 1$,

$$F(x) = P(X \le x) = \int_0^x \frac{2}{5}\, dt = \frac{2x}{5}.$$

If $1 < x < 2$,

$$\begin{aligned} F(x) = P(X \le x) &= \int_0^1 \frac{2}{5}\, dt + \int_1^x \frac{2t}{5}\, dt \\ &= \frac{2}{5} + \frac{x^2 - 1}{5} = \frac{x^2 + 1}{5}. \end{aligned}$$

This gives

$$F(x) = \begin{cases} 0, & \text{if } x \le 0 \\ 2x/5, & \text{if } 0 < x \le 1 \\ (x^2 + 1)/5, & \text{if } 1 < x \le 2 \\ 1, & \text{if } x > 2. \end{cases}$$

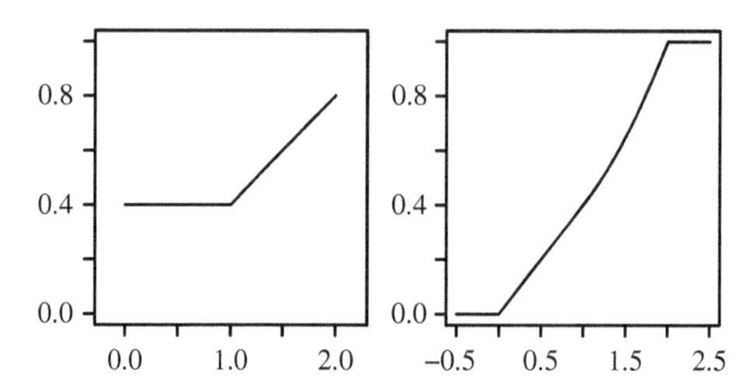

FIGURE 6.3: Density function and cdf.

Observe that $F(x)$ is continuous at all points x. See the graphs of f and F in Figure 6.3.

(ii) The desired probability is

$$P(0.5 < X < 1.5) = F(1.5) - F(0.5) = \frac{(1.5)^2 + 1}{5} - \frac{2(0.5)}{5} = 0.45 \quad \blacksquare$$

Cumulative distribution functions for discrete random variables. Continuous random variables have continuous cdfs. However, the cdf of a discrete random variable has points of discontinuity at the discrete values of the variable, yielding a step function.

For example, suppose $X \sim \text{Bin}(2, 1/2)$. Then the pmf of X is

$$P(X = x) = \begin{cases} 1/4, & \text{if } x = 0 \\ 1/2, & \text{if } x = 1 \\ 1/4, & \text{if } x = 2. \end{cases}$$

The cdf of X, graphed in Figure 6.4, is

$$F(x) = P(X \leq x) = \begin{cases} 0, & \text{if } x < 0 \\ 1/4, & \text{if } 0 \leq x < 1 \\ 3/4, & \text{if } 1 \leq x < 2 \\ 1, & \text{if } x \geq 2, \end{cases}$$

with points of discontinuity at $x = 0, 1,$ and 2. To fully specify the cdf (similar to the pdf), do not forget the regions where the cdf is 0 and 1. Again, this is implied if a range is given, but do not forget this implication.

In general, a cdf need not be continuous. However, it is always right-continuous. The cdf is also an increasing function. That is, if $a \leq b$, then $F(a) \leq F(b)$. This holds because if $a \leq b$, the event $\{X \leq a\}$ implies $\{X \leq b\}$ and thus $P(X \leq a) \leq P(X \leq b)$.

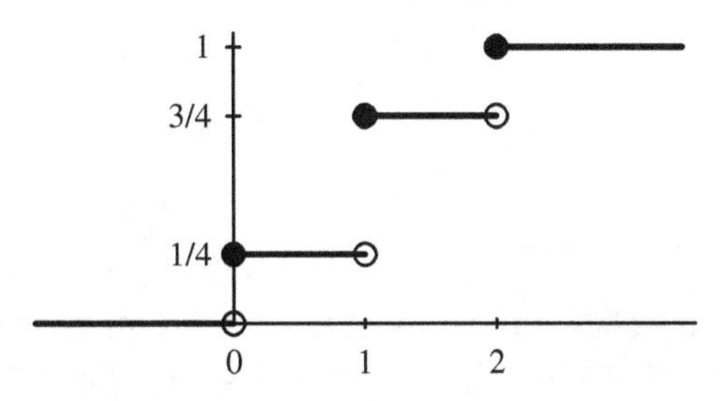

FIGURE 6.4: Cumulative distribution function $P(X \leq x)$ for $X \sim \text{Binom}(2, 1/2)$. The cdf has points of discontinuity at $x = 0, 1, 2$.

The cdf is a probability and takes values between 0 and 1. It has the properties that $F(x) \to 0$, as $x \to -\infty$, and $F(x) \to 1$, as $x \to +\infty$.

We summarize the four defining properties of a cdf.

CUMULATIVE DISTRIBUTION FUNCTION

A function F is a cdf, that is, there exists a random variable X whose cdf is F, if it satisfies the following properties.

1. $\lim_{x \to -\infty} F(x) = 0.$
2. $\lim_{x \to +\infty} F(x) = 1.$
3. If $a \leq b$, then $F(a) \leq F(b)$.
4. F is right-continuous. That is, for all real a,

$$\lim_{x \to a+} F(x) = F(a).$$

6.3 EXPECTATION AND VARIANCE

Formulas for expectation and variance for continuous random variables follow as expected from the discrete formulas: integrals replace sums and pdfs replace pmfs.

EXPECTATION AND VARIANCE FOR CONTINUOUS RANDOM VARIABLES

For random variable X with density function f,

$$E[X] = \int_{-\infty}^{\infty} x f(x)\, dx$$

and

$$V[X] = \int_{-\infty}^{\infty} (x - E[X])^2 f(x)\, dx.$$

Note that these integrals must be absolutely convergent for the expectation and variance to exist, respectively. If they are not absolutely convergent, these values do not exist, and you will see an example where this is the case.

Properties of expectation and variance introduced in Chapter 4 for discrete random variables transfer to continuous random variables. We remind the reader of the most important of these.

PROPERTIES OF EXPECTATION AND VARIANCE

For constants a and b, and random variables X and Y,

- $E[aX + b] = aE[X] + b$,
- $E[X + Y] = E[X] + E[Y]$,
- $V[X] = E[X^2] - E[X]^2$, and
- $V[aX + b] = a^2 V[X]$.

Example 6.8 Random variable X has density proportional to x^{-4}, for $x > 1$. Find $V[1 - 4X]$. As you solve this problem, think carefully about the steps needed to solve it.

First find the constant of proportionality. Solve

$$1 = \int_1^{\infty} \frac{c}{x^4}\, dx = c\left(\frac{1}{-3x^3}\right)\Bigg|_1^{\infty} = \frac{c}{3}$$

giving $c = 3$. To find the variance, first find $E[X]$ and $E[X^2]$. We find

$$E[X] = \int_1^{\infty} x\left(\frac{3}{x^4}\right) dx = \int_1^{\infty} \frac{3}{x^3}\, dx = 3\left(\frac{1}{-2x^2}\right)\Bigg|_1^{\infty} = \frac{3}{2}, \text{ and}$$

$$E[X^2] = \int_1^{\infty} x^2\left(\frac{3}{x^4}\right) dx = \int_1^{\infty} \frac{3}{x^2}\, dx = 3\left(\frac{-1}{x}\right)\Bigg|_1^{\infty} = 3.$$

Via the computational formula for variance, this gives

$$V[X] = E[X^2] - (E[X])^2 = 3 - \left(\frac{3}{2}\right)^2 = \frac{3}{4}.$$

Therefore,

$$V[1 - 4X] = (-4)^2 V[X] = 16V[X] = 16\left(\frac{3}{4}\right) = 12. \qquad ■$$

The expectation of a function of a random variable is given similarly as in the discrete case.

EXPECTATION OF FUNCTION OF CONTINUOUS RANDOM VARIABLE

If X has density function f, and g is a function, then

$$E[g(X)] = \int_{-\infty}^{\infty} g(x)f(x)\,dx.$$

Example 6.9 Let X be a random variable with pdf given by $f(x) = 3x^2, 0 < x < 1$, and 0, otherwise. Suppose we want to find the moment-generating function of X. Remember that the mgf is a special expectation, $E(e^{tX})$, which is a function of X. Thus, we apply the last result to find the expectation as

$$m_X(t) = E(e^{tX}) = \int_0^1 e^{tx} 3x^2\,dx.$$

The integration here may be solved by using integration by parts, and results in finding the mgf of X to be

$$m_X(t) = \frac{3}{t^3}(e^t(t^2 - 2t + 2) - 2). \qquad ■$$

A common mistake made when computing the expectation of a function of a random variable is that you forget to include $g(x)$ in the integral. This will make the integral evaluate to 1, as you are integrating the pdf over its range. Be sure you put both the function of the random variable, $g(x)$, and the pdf, $f(x)$, in the integral.

Next we look at two common continuous distributions.

6.4 UNIFORM DISTRIBUTION

A uniform model on the interval (a, b) is described by a density function that is flat. That is, $f(x)$ is constant for all $a < x < b$, i.e., $f(x) = c$. As the density integrates to 1, to find that constant solve

$$1 = \int_a^b c\,dx = c(b - a),$$

giving $c = 1/(b - a)$, the reciprocal of the length of the interval.

UNIFORM DISTRIBUTION

A random variable X is *uniformly distributed on* (a, b) if the density function of X is

$$f(x) = \frac{1}{b-a}, \quad \text{for } a < x < b,$$

and 0, otherwise. We write $X \sim \text{Unif}(a, b)$.

If X is uniformly distributed on (a, b), the cdf of X is

$$F(x) = P(X \leq x) = \begin{cases} 0, & \text{if } x \leq a \\ (x-a)/(b-a), & \text{if } a < x \leq b \\ 1, & \text{if } x > b. \end{cases}$$

For the case when X is uniformly distributed on $(0, 1)$, $F(x) = x$, for $0 < x < 1$.

Probabilities in the uniform model reduce to length. That is, if $X \sim \text{Unif}(a, b)$ and $a < c < d < b$, then

$$P(c < X < d) = F(d) - F(c) = \frac{d-a}{b-a} - \frac{c-a}{b-a} = \frac{d-c}{b-a} = \frac{\text{Length } (c, d)}{\text{Length } (a, b)}.$$

Example 6.10 Esme is taking a train from Minneapolis to Boston, a distance of roughly 1400 miles. Her position is uniformly distributed between the two cities. What is the probability that she is past Chicago, which is 400 miles from Minneapolis?

Let X be Esme's location. Then $X \sim \text{Unif}(0, 1400)$. The probability is

$$P(X > 400) = \frac{1400 - 400}{1400} = \frac{5}{7} = 0.7143.$$

■

Expectation and variance. Let $X \sim \text{Unif}(a, b)$. Then

$$E[X] = \int_a^b x\left(\frac{1}{b-a}\right) dx = \left(\frac{1}{b-a}\right) \frac{x^2}{2}\Big|_a^b = \frac{b+a}{2},$$

the midpoint of the interval (a, b). Also, using the computational formula for variance,

$$E[X^2] = \int_a^b x^2\left(\frac{1}{b-a}\right) dx = \left(\frac{1}{b-a}\right) \frac{x^3}{3}\Big|_a^b = \frac{b^2 + ab + a^2}{3},$$

which yields

$$V[X] = E[X^2] - E[X]^2 = \frac{b^2 + ab + a^2}{3} - \left(\frac{b+a}{2}\right)^2 = \frac{(b-a)^2}{12}.$$

It is worthwhile to remember these results as they occur frequently. The mean of a uniform distribution is the midpoint of the interval. The variance is one-twelfth the square of its length. These properties are summarized for common continuous distributions in Appendix B.

R: UNIFORM DISTRIBUTION

The **R** commands for the continuous uniform distribution on (a, b) are

```
> dunif(x, a, b)     #   f(x) ≠ P(X = x)
> punif(x, a, b)     #   P(X ≤ x)
> runif(n, a, b)     #   Simulates n random variables.
```

Default parameters are $a = 0$ and $b = 1$. To generate a uniform variable on $(0, 1)$, type

```
> runif(1)
[1] 0.973387
```

Example 6.11 Commercial washing machines use roughly 2.5–3.5 gallons of water per load of laundry. Suppose X, the amount of water for one load used, is uniform over that range. Carter finds a broken washer is using three times as much water as expected. What is the expectation and standard deviation of how much water the broken washer is using?

Let Y be the amount of water used by the broken washer. Then $Y = 3X$, and $X \sim \text{Unif}(2.5, 3.5)$. Based on the results above, $E(X) = 3$, and $V(X) = 1/12$. Thus, we find that the expectation of Y is $E(Y) = E(3X) = 3E(X) = 9$ gallons. For the standard deviation, we find the variance of Y first as $V(Y) = V(3X) = 9V(X) = 3/4$. The standard deviation of Y is therefore the square root of $3/4$, or roughly 0.866 gallons. ■

Example 6.12 A balloon has radius uniformly distributed on $(0, 2)$. Find the expectation and standard deviation of the volume of the balloon.

Let V be the volume. Then $V = (4\pi/3)R^3$, where $R \sim \text{Unif}(0, 2)$. This gives $f(r) = 1/2, 0 < r < 2$, and 0, otherwise. The expected volume is

$$E[V] = E\left[\frac{4\pi}{3}R^3\right] = \frac{4\pi}{3}E[R^3] = \frac{4\pi}{3}\int_0^2 r^3\left(\frac{1}{2}\right)\, dr$$

$$= \frac{2\pi}{3}\left(\frac{r^4}{4}\right)\Bigg|_0^2 = \frac{8\pi}{3} = 8.378.$$

We find $E[V^2]$ to use the computational formula for variance as

$$E[V^2] = E\left[\left(\frac{4\pi}{3}R^3\right)^2\right] = \frac{16\pi^2}{9}E[R^6]$$

$$= \frac{16\pi^2}{9}\int_0^2 \frac{r^6}{2}\,dr = \frac{16\pi^2}{9}\left(\frac{128}{14}\right) = \frac{1024\pi^2}{63}.$$

This gives

$$\text{Var}[V] = E[V^2] - (E[V])^2 = \frac{1024\pi^2}{63} - \left(\frac{8\pi}{3}\right)^2 = \frac{64\pi^2}{7},$$

with standard deviation $\text{SD}[V] = 8\pi/\sqrt{7} \approx 9.50$. ■

R: SIMULATING BALLOON VOLUME

Simulating the balloon volume is easy for **R**.

```
> volume <-(4/3)*pi*runif(100000, 0, 2)^3
> mean(volume)
[1] 8.368416
> sd(volume)
[1] 9.511729
```

6.5 EXPONENTIAL DISTRIBUTION

What is the size of a raindrop? When will your next text message arrive? How long does a bee spend gathering nectar at a flower? The applications of the exponential distribution are vast. The distribution is one of the most important in probability both for its practical and theoretical use. It is often used to model lifetimes or times between events.

EXPONENTIAL DISTRIBUTION

A random variable X has an *exponential distribution with parameter* $\lambda > 0$ if its density function has the form

$$f(x) = \lambda e^{-\lambda x}, \quad \text{for } x > 0,$$

and 0, otherwise. We write $X \sim \text{Exp}(\lambda)$.

The cdf of the exponential distribution is

$$F(x) = \int_0^x \lambda e^{-\lambda t}\, dt = \lambda \left(\frac{-1}{\lambda} e^{-\lambda t} \right) \Bigg|_0^x = 1 - e^{-\lambda x},$$

for $x > 0$, and 0, otherwise. The "tail probability" formula

$$\boxed{P(X > x) = 1 - F(x) = e^{-\lambda x}, \quad \text{for } x > 0,}$$

is commonly used. We leave it to the exercises to show that

$$E[X] = \frac{1}{\lambda} \quad \text{and} \quad V[X] = \frac{1}{\lambda^2}.$$

For the exponential distribution, the mean is equal to the standard deviation.

Example 6.13 A school's help desk receives calls throughout the day. The time T (in minutes) between calls is modeled with an exponential distribution with mean 4.5. A call just arrived. What is the probability no call will be received in the next 5 minutes?

The parameter of the exponential distribution is $\lambda = 1/4.5$. If no call is received in the next 5 minutes, then the time of the next call is greater than five. The desired probability is

$$P(T > 5) = e^{-5/4.5} = 0.329.$$

■

R: EXPONENTIAL DISTRIBUTION

The **R** commands for the exponential distribution are

```
> dexp(x,lambda)   # Computes density value
> pexp(x,lambda)   # Computes P(X <= x)
> rexp(n,lambda)   # Generates n random numbers
```

Alternative parameterization. When working with the exponential distribution, be sure to check the form of the pdf. An alternative parameterization exists with $\beta = 1/\lambda$. This parameterization is sometimes preferred for computational problems.

6.5.1 Memorylessness

An important property of the exponential distribution is memorylessness. You were introduced to this property in the discrete setting for the geometric distribution.

The exponential distribution is the only continuous distribution that is memoryless. Here is the general property.

MEMORYLESSNESS FOR EXPONENTIAL DISTRIBUTION

Let $X \sim \text{Exp}(\lambda)$. For $0 < s < t$,

$$P(X > s + t | X > s) = \frac{P(X > s + t)}{P(X > s)} = \frac{e^{-\lambda(s+t)}}{e^{-\lambda s}} = e^{-\lambda t} = P(X > t).$$

To illustrate in the continuous setting, suppose Adrian and Zoe are both waiting for a bus. Buses arrive about every 30 minutes according to an exponential distribution. Adrian gets to the bus stop at time $t = 0$. The time until the next bus arrives has an exponential distribution with $\lambda = 1/30$.

Zoe arrives at the bus stop 10 minutes later, at time $t = 10$. The memorylessness of the exponential distribution means that the time that Zoe waits for the bus will also have an exponential distribution with $\lambda = 1/30$. They will *both* wait about the same amount of time.

Memorylessness means that if the bus does not arrive in the first 10 minutes then the probability that it will arrive after time $t = 10$ (for Zoe) is the same as the probability that a bus arrives after time t (for Adrian). After time $t = 10$, the next bus "does not remember" what happened in the first 10 minutes.

This may seem amazing, even paradoxical. Run the script file **Memory.R** to convince yourself it is true. See Figure 6.5 for the simulated distributions of Adrian's and Zoe's waiting times.

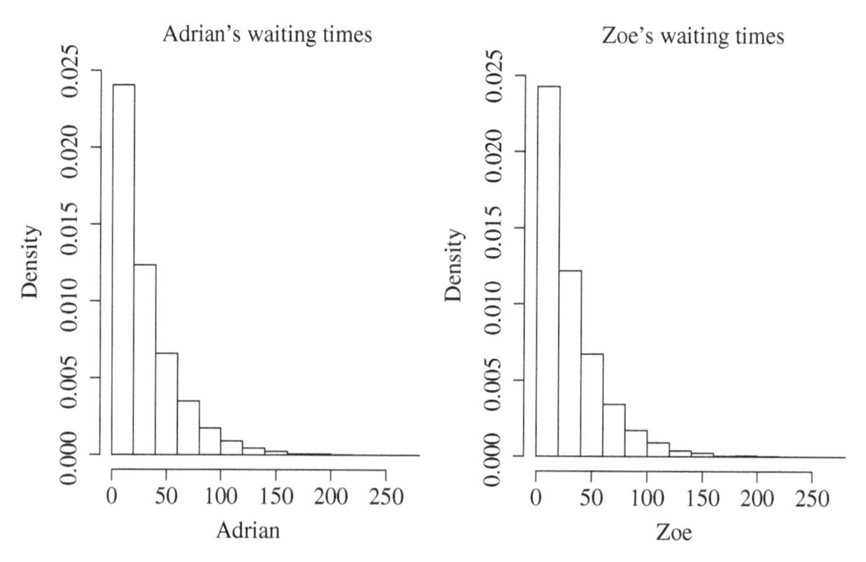

FIGURE 6.5: Zoe arrives at the bus station 10 minutes after Adrian. But the distribution of their waiting times is the same.

As long as the bus does not come in the first 10 minutes, Zoe waits more than t minutes for the bus if and only if Adrian waits more than $t + 10$ minutes. Letting A and Z denote Adrian and Zoe's waiting times, respectively, this gives,

$$P(Z > t) = P(A > t + 10|A > 10) = \frac{P(A > t + 10)}{P(A > 10)}$$

$$= \frac{e^{-(t+10)/30}}{e^{-10/30}} = e^{-t/30} = P(A > t).$$

R: BUS WAITING TIME

```
# Memory.R
   # Adrian arrives at time t=0
   # Zoe arrives at time t=10
> n <- 10000
> mu <- 30
> bus <- rexp(n, 1/mu)
> Adrian <- bus
> Zoe <- bus[bus > 10]-10
> mean(Adrian)
[1] 30.23546
> mean(Zoe)
[1] 29.77742
> par(mfrow=c(1,2))
> hist(Adrian, prob = T)
> hist(Zoe, prob = T); par(mfrow=c(1,1))
```

Many random processes that evolve in time and/or space exhibit a lack of memory in the sense described here, which is key to the central role of the exponential distribution in applications. Novel uses of the exponential distribution include:

- The diameter of a raindrop plays a fundamental role in meteorology, and can be an important variable in predicting rain intensity in extreme weather events. Marshall and Palmer [1948] proposed an exponential distribution for raindrop size. The parameter λ is a function of rainfall intensity (see Example 7.14).The empirically derived model has held up for many years and has found wide application in hydrology.
- Dorsch et al. [2008] use exponential distributions to model patterns in ocean storms off the southern coast of Australia as a way to study changes in ocean "storminess" as a result of global climate change. They fit storm duration and time between storms to exponential distributions with means 21.1 hours and 202.0 hours, respectively.
- The "bathtub curve" is used in reliability engineering to represent the lifetime of a product. The curve models the failure rate, which is the frequency with which a product or component fails, often expressed as failures per time unit.

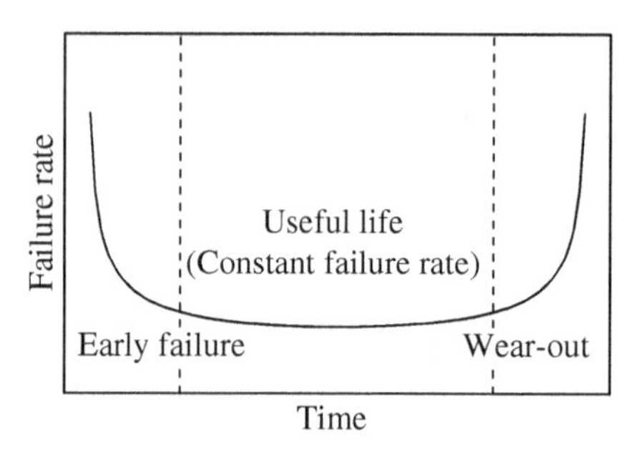

FIGURE 6.6: Bathtub curves are used in reliability engineering to model failure rates of a product or component.

See Figure 6.6. In the middle of the curve, the failure rate is constant, and the exponential distribution often serves as a good model for the *time until failure*.

Exponential random variables often arise in applications as a sequence of successive times between *arrivals*. These times might represent arrivals of phone calls, buses, accidents, or component failures.

Let $X_1, X_2, \ldots$ be an independent sequence of Exp(λ) random variables where X_k is the time between the $(k-1)$st and kth arrival. As the common expectation of the X_k's is $1/\lambda$ we expect arrival times to be about $1/\lambda$ time units apart, and there are about λ arrivals per one unit of time. Thus, λ represents the rate of arrivals (number of arrivals per unit time). For this reason, the parameter λ is often called the *rate* of the exponential distribution.

Example 6.14 The time it takes for each customer to be served at a local restaurant has an exponential distribution. Serving times are independent of each other. Typically customers are served at the rate of 20 customers per hour. Arye is waiting to be served. What is the mean and standard deviation of his serving time? Find the probability that he will be served within 5 minutes.

Model Arye's serving time S with an exponential distribution with $\lambda = 20$. Then in hour units, $E[S] = \text{SD}[S] = 1/20$, or 3 minutes. As the given units are hours, we must convert the 5 minutes into hours. This means the desired probability is

$$P\left(S \leq \frac{5}{60}\right) = F_S\left(\frac{1}{12}\right) = 1 - e^{-20/12} = 0.811.$$

In **R**, the solution is obtained by typing

```
> pexp(1/12,20)
[1] 0.8111244
```

■

Sequences of i.i.d. exponential random variables as described above form the basis of an important class of random processes called the Poisson process, introduced in Chapter 7.

6.6 JOINT DISTRIBUTIONS

For two or more random variables, the *joint density function* or joint *pdf* plays the role of the joint pmf for discrete variables. Single integrals become double integrals for two variables and multiple integrals when working with even more. If you are not familiar with double integrals, and want more practice after reading the examples below, see Appendix D.

JOINT DENSITY FUNCTION

For continuous random variables X and Y defined on a common sample space, the *joint density function* $f(x, y)$ of X and Y has the following properties.

1. $f(x, y) \geq 0$ for all real numbers x and y.
2. The total probability is still 1. That is,

$$\int_{-\infty}^{\infty} \int_{-\infty}^{\infty} f(x, y)\, dx\, dy = 1.$$

3. For $S \subseteq \Re^2$,

$$P((X, Y) \in S) = \iint_S f(x, y)\, dx\, dy. \tag{6.4}$$

Typically in computations, the integral in Equation 6.4 will be an iterated integral whose limits of integration are determined by S. As the joint density is a function of two variables, its graph is a *surface* over a two-dimensional domain.

For continuous random variables $X_1, \ldots, X_n$ defined on a common sample space, the joint density function $f(x_1, \ldots, x_n)$ is defined similarly.

Example 6.15 Suppose the joint density of X and Y is

$$f(x, y) = cxy, \text{ for } 1 < x < 4 \quad \text{and} \quad 0 < y < 1,$$

and 0, otherwise. Find c and compute $P(2 < X < 3, Y > 1/4)$.

First find c using property 2,

$$1 = \int_0^1 \int_1^4 cxy\, dx\, dy = c \int_0^1 y \left(\int_1^4 x\, dx \right) dy$$

$$= c \int_0^1 y \left(\frac{15}{2} \right) dy = c \left(\frac{15}{2} \right) \left(\frac{1}{2} \right) = \frac{15c}{4},$$

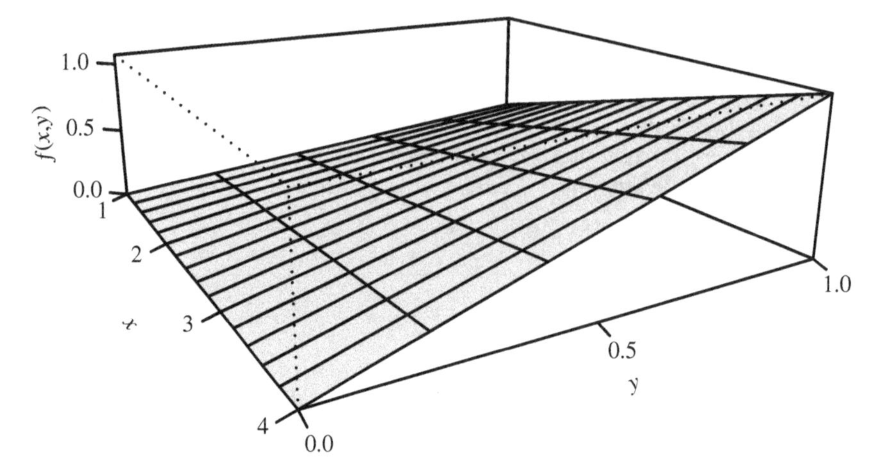

FIGURE 6.7: Joint density $f(x, y) = 4xy/15, 1 < x < 4, 0 < y < 1$.

giving $c = 4/15$. Then,

$$P(2 < X < 3, Y > 1/4) = \int_{1/4}^{1} \int_{2}^{3} \frac{4xy}{15}\, dx\, dy = \int_{1/4}^{1} \frac{4y}{15} \left(\frac{x^2}{2}\right)\Bigg|_2^3 dy$$

$$= \left(\frac{4}{15}\right)\left(\frac{5}{2}\right) \int_{1/4}^{1} y\, dy = \left(\frac{4}{15}\right)\left(\frac{5}{2}\right)\left(\frac{15}{32}\right) = \frac{5}{16}.$$

The joint density function is graphed in Figure 6.7. The density is a surface over the domain $1 < x < 4$ and $0 < y < 1$. ■

JOINT CUMULATIVE DISTRIBUTION FUNCTION

If X and Y have joint density function f, the *joint cumulative distribution function* of X and Y is

$$F(x, y) = P(X \leq x, Y \leq y) = \int_{-\infty}^{x} \int_{-\infty}^{y} f(s, t)\, dt\, ds$$

defined for all x and y. Differentiating with respect to both x and y gives

$$\frac{\partial^2}{\partial x\, \partial y} F(x, y) = f(x, y). \tag{6.5}$$

Example 6.16 Suppose X and Y have joint cdf for $2 < x < 4$ and $y > 0$ given by

$$F(x, y) = \frac{(x-2)}{2}(1 - e^{-y/2}).$$

Find the joint pdf of X and Y.

We need to take the partial derivatives of the joint cdf with respect to both x and y. We start with the partial derivative with respect to x, as it is the easier of the two.

$$f(x,y) = \frac{\partial^2}{\partial x\, \partial y} F(x,y) = \frac{\partial^2}{\partial x\, \partial y} \frac{(x-2)}{2}(1-e^{-y/2})$$
$$= \frac{\partial}{\partial y} \frac{1}{2}(1-e^{-y/2})$$
$$= \frac{1}{2}\left(\frac{1}{2}e^{-y/2}\right) = \frac{1}{4}e^{-y/2}.$$

This is correct for this example, because X was uniform on $(2,4)$, and Y was exponentially distributed with $\lambda = 2$, with X independent of Y. We leave it to the reader to check that the joint pdf is the product of the specified marginals.

For an additional example for practice with partial derivatives, see Example D.4 in Appendix D. ■

Uniform model. Let S be a bounded region in the plane. By analogy with the development of the one-dimensional uniform model, if (X, Y) is distributed uniformly on S, the joint density will be constant on S. That is, $f(x,y) = c$ for all $(x,y) \in S$. Solving

$$1 = \iint_S f(x,y)\, dx\, dy = \iint_S c\, dx\, dy = c[\text{Area } (S)]$$

gives $c = 1/\text{Area } (S)$ and the uniform model on S.

UNIFORM DISTRIBUTION IN TWO DIMENSIONS

Let S be a bounded region in the plane. Then random variables (X, Y) are uniformly distributed on S if the joint density function of X and Y is

$$f(x,y) = \frac{1}{\text{Area } (S)}, \quad \text{for } (x,y) \in S,$$

and 0, otherwise. We write $(X, Y) \sim \text{Unif}(S)$.

Example 6.17 Suppose X and Y have joint density $f(x,y)$. (i) Find a general expression for $P(X < Y)$. (ii) Solve for the case when (X, Y) is uniformly distributed on the circle centered at the origin of radius one.

(i) The region determined by the event $\{X < Y\}$ is the set of all points in the plane (x,y) such that $x < y$. Setting up the double integral gives

$$P(X < Y) = \iint_{\{(x,y):x<y\}} f(x,y)\, dx\, dy = \int_{-\infty}^{\infty} \int_{-\infty}^{y} f(x,y)\, dx\, dy.$$

(ii) For the special case when (X, Y) is uniformly distributed on the circle, first draw the picture of the domain. See Figure 6.8. The joint density is $f(x,y) = 1/\pi$, if

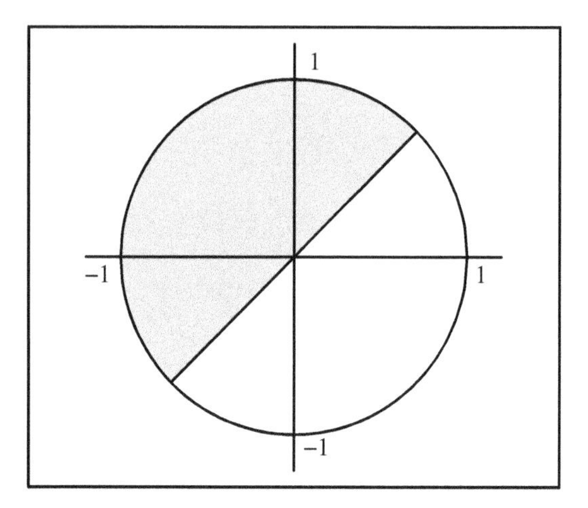

FIGURE 6.8: Shaded region defined by $P(X < Y)$.

(x, y) is in the circle, and 0, otherwise. This gives

$$P(X < Y) = \int_{-1}^{-1/\sqrt{2}} \int_{-\sqrt{1-s^2}}^{\sqrt{1-s^2}} \frac{1}{\pi} \, dt \, ds + \int_{-1/\sqrt{2}}^{1/\sqrt{2}} \int_{s}^{\sqrt{1-s^2}} \frac{1}{\pi} \, dt \, ds.$$

Solving the integral requires trigonometric substitution. But a much easier route is to recognize that because (X, Y) has a uniform distribution, the problem reduces to finding areas. And then it is clear from the picture that the desired probability is 1/2. ■

Problems involving two or more continuous random variables will require multiple integrals. Setting these up correctly can be challenging. As an example, consider the probability $P(X < Y)$ from the last Example 6.17, but where (X, Y) has joint density

$$f(x, y) = \frac{3x^2 y}{64}, \quad \text{for } 0 < x < 2, 0 < y < 4.$$

The distribution here is not uniform so the problem cannot be reduced to finding areas. The graph of the density function, as shown in Figure 6.9a, is a surface over the rectangle $[0, 2] \times [0, 4]$.

The event $\{X < Y\}$ determines the shaded region in the domain in Figure 6.9b. For solving problems such as this, the most important step is to draw the picture of the domain and identify the region corresponding to the desired probability. The desired probability is the integral of the density $f(x, y)$ over this shaded region.

In setting up the double integral be aware of *all* constraints on each variable. If, say, x is chosen as the variable for the outer integral, then the limits of integration for

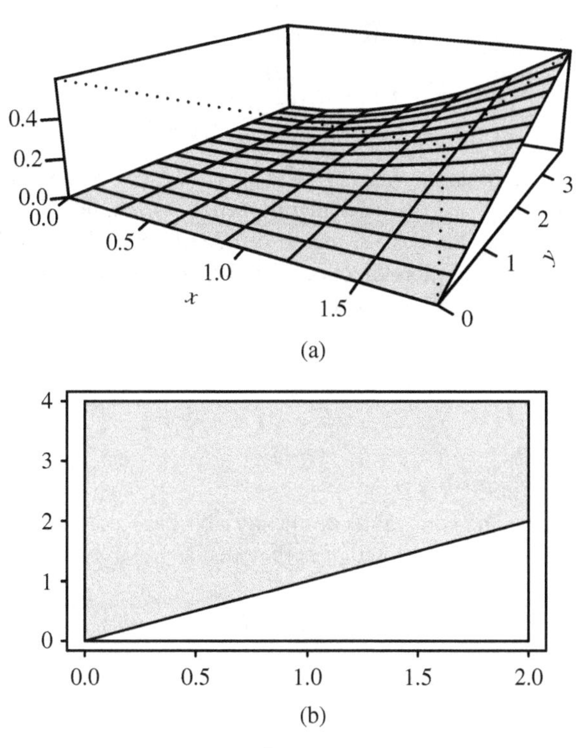

FIGURE 6.9: (a) Joint density $f(x, y) = 3x^2y/64$. (b) Domain of joint density. Shaded region shows event $\{X < Y\}$.

the outer integral will not depend on y. Otherwise, y will appear in the final answer, which does not make sense, as the final value should be a probability.

To set up the double integral for $P(X < Y)$, the limits for the outer integral are $0 < x < 2$. For the inner integral, the constraints on y are $0 < y < 4$ and $y > x$ which together give $x < y < 4$. Thus,

$$P(X < Y) = \int_{x=0}^{2} \int_{y=x}^{4} \frac{3x^2y}{64}\, dy\, dx.$$

Notice that we have explicitly written x and y in the limits of integration to help keep track of which variable corresponds to which integral.

The multiple integral can also be set up with y on the outside integral and x on the inside. If y is the first variable, then the limits for the outer integral are $0 < y < 4$. For the inner integral, the constraints on x are $x < y$ and $0 < x < 2$. This is equivalent to $0 < x < \min(y, 2)$. The limits of integration for x depend on y. You can see this in the figure as the region of interest being composed of a triangle and a rectangle. We need to break up the outer integral into two parts. For $0 < y < 2$, where we are in the triangular region, x ranges from 0 to y. For $2 < y < 4$, we are in the rectangle and x ranges from 0 to 2. This gives

$$P(X < Y) = \int_{y=0}^{2} \int_{x=0}^{y} \frac{3x^2y}{64}\, dx\, dy + \int_{y=2}^{4} \int_{x=0}^{2} \frac{3x^2y}{64}\, dx\, dy.$$

In both cases, the final answer is $P(X < Y) = 17/20$, but the computation was much easier with only one double integral needed. Often, drawing the picture of the domain and finding the desired region of integration will show that one setup is easier than another. For more practice with this concept, particularly if you have not had a multivariate calculus course, see Appendix D.

Example 6.18 The joint density of X and Y is

$$f(x, y) = \frac{1}{x^2y^2}, \quad x > 1, y > 1,$$

and 0, otherwise. Find $P(X \geq 2Y)$.

See Figure 6.10. The constraints on the variables are $x > 1$, $y > 1$, and $x \geq 2y$. Setting up the multiple integral with y as the outside variable gives

$$P(X \geq 2Y) = \int_{y=1}^{\infty} \int_{x=2y}^{\infty} \frac{1}{x^2y^2}\, dx\, dy = \int_{y=1}^{\infty} \frac{1}{y^2} \left(\frac{-1}{x}\right)\Big|_{2y}^{\infty} dy$$
$$= \int_{y=1}^{\infty} \frac{1}{y^2}\left(\frac{1}{2y}\right) dy = \left(\frac{-1}{4y^2}\right)\Big|_{1}^{\infty} = \frac{1}{4}.$$

To set up the integral with x on the outside consider the constraints $x > 1, x \geq 2y$, and $y > 1$. As the minimum value of y is 1 and $x > 2y$, we must have $x > 2$ for the outer integral. For the inside variable, we have $y > 1$ and $y \leq x/2$. That is, $1 < y \leq x/2$. This gives

$$P(X \geq 2Y) = \int_{x=2}^{\infty} \int_{y=1}^{x/2} \frac{1}{x^2y^2}\, dy\, dx = \int_{x=2}^{\infty} \frac{1}{x^2} \left(\frac{-1}{y}\right)\Big|_{1}^{x/2} dx$$
$$= \int_{x=2}^{\infty} \frac{1}{x^2}\left(1 - \frac{2}{x}\right) dx = \left(\frac{-1}{x} + \frac{1}{x^2}\right)\Big|_{2}^{\infty} = \frac{1}{2} - \frac{1}{4} = \frac{1}{4}. \quad \blacksquare$$

Example 6.19 There are many applications in reliability theory of two-unit systems in which the lifetimes of the two units are described by a joint probability distribution. See Harris [1968] for examples. Suppose such a system depends on components A and B whose respective lifetimes X and Y are jointly distributed with density function

$$f(x, y) = e^{-y}, \quad \text{for } 0 < x < y < \infty.$$

Find the probability (i) that component B lasts at least three time units longer than component A, and (ii) that both components last for at least two time units.

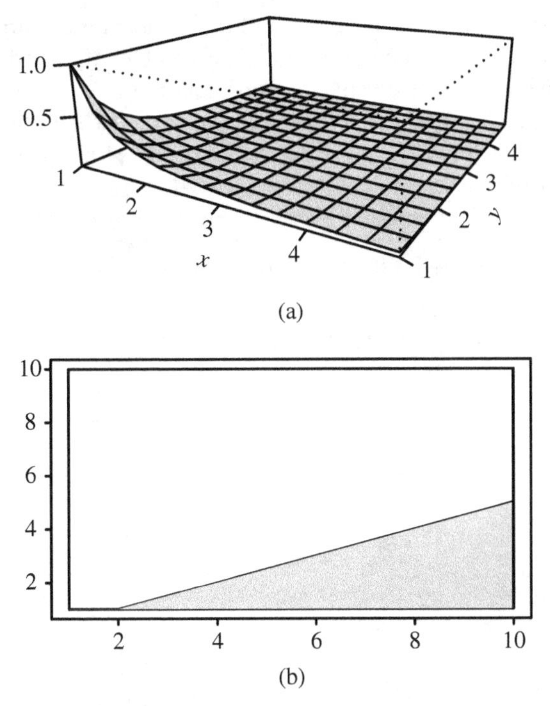

FIGURE 6.10: (a) Joint density. (b) Shaded region shows event $\{X \geq 2Y\}$ in the part of the domain displayed.

For the problem, before any constraints related to the probability desired are examined, sketch the domain. You will note that the domain is a triangle in the first quadrant of the Cartesian plane. Be sure you know which of the two possible triangles you are in, to help set up bounds for the integrals. Then, for each desired probability, consider what additional constraints are added.

(i) The desired probability is

$$P(Y > X + 3) = \int_{x=0}^{\infty} \int_{y=x+3}^{\infty} e^{-y}\, dy\, dx$$

$$= \int_{x=0}^{\infty} e^{-(x+3)}\, dx = e^{-3} = 0.050.$$

(ii) The desired probability is

$$P(X > 2, Y > 2) = \int_{y=2}^{\infty} \int_{x=2}^{y} e^{-y}\, dx\, dy$$

$$= \int_{y=2}^{\infty} (y-2)e^{-y}\, dy = e^{-2} = 0.135.$$

■

The joint density of X and Y captures all the "probabilistic information" about X and Y. In principle, it can be used to find any probability which involves these variables. From the joint density, the marginal densities are obtained by integrating out the extra variable. (In the discrete case, we sum over the other variable.) Be careful when bounds for one variable depend on the other, as this information must be incorporated appropriately.

MARGINAL DISTRIBUTIONS FROM JOINT DENSITIES

$$f_X(x) = \int_{-\infty}^{\infty} f(x, y)\, dy \quad \text{and} \quad f_Y(y) = \int_{-\infty}^{\infty} f(x, y)\, dx.$$

To see why these equations hold, consider

$$\{X \leq x\} = \{X \leq x, -\infty < Y < \infty\}.$$

Hence,

$$P(X \leq x) = P(X \leq x, -\infty < Y < \infty) = \int_{-\infty}^{x} \left[\int_{-\infty}^{\infty} f(s, y)\, dy\right] ds.$$

Differentiating with respect to x, applying the fundamental theorem of calculus, gives

$$f_X(x) = \int_{-\infty}^{\infty} f(x, y)\, dy,$$

and similarly for $f_Y(y)$.

Example 6.20 Consider the joint density function from the last Example 6.19

$$f(x, y) = e^{-y}, \quad \text{for } 0 < x < y < \infty,$$

and 0, otherwise. Find the marginal densities of X and Y.

The marginal density of X is

$$f_X(x) = \int_{-\infty}^{\infty} f(x, y)\, dy = \int_{x}^{\infty} e^{-y}\, dy = e^{-x}, \quad \text{for } x > 0,$$

and 0, otherwise. Note that once we integrate out the y variable, the domain of the x variable is all positive real numbers. The dependence between x and y shows in the limits of the integral. From the form of the density function we see that X has an exponential distribution with parameter $\lambda = 1$.

Similarly, the marginal density of Y is

$$f_Y(y) = \int_0^y e^{-y}\, dx = ye^{-y}, \text{ for } y > 0,$$

and 0, otherwise. ■

Computing expectations of functions of two or more continuous random variables should offer no surprises as we use the continuous form of the "law of the unconscious statistician."

EXPECTATION OF FUNCTION OF JOINTLY DISTRIBUTED RANDOM VARIABLES

If X and Y have joint density f, and $g(x, y)$ is a function of two variables, then

$$E[g(X, Y)] = \int_{-\infty}^{\infty} \int_{-\infty}^{\infty} g(x, y) f(x, y)\, dx\, dy.$$

The expected product of random variables X and Y is thus

$$E[XY] = \int_{-\infty}^{\infty} \int_{-\infty}^{\infty} xy f(x, y)\, dx\, dy.$$

Be sure that you have both the desired function $g(x, y)$ and the joint pdf $f(x, y)$ in the double integral before evaluating. A common mistake is to forget $g(x, y)$, which results in the integrals evaluating to 1.

Example 6.21 Suppose (X, Y) is uniformly distributed on the circle of radius 1 centered at the origin. Find the expected distance $D = \sqrt{X^2 + Y^2}$ to the origin.

Let C denote the circle. Because the area of C is π, the joint density function of X and Y is

$$f(x, y) = \frac{1}{\pi}, \quad \text{for } (x, y) \in C,$$

and 0, otherwise. This gives

$$\begin{aligned} E[D] &= E\left[\sqrt{X^2 + Y^2}\right] \\ &= \int_{-\infty}^{\infty} \int_{-\infty}^{\infty} \sqrt{x^2 + y^2} f(x, y)\, dx\, dy = \iint_C \sqrt{x^2 + y^2} \frac{1}{\pi}\, dx\, dy. \end{aligned}$$

Changing to polar coordinates (r, θ) gives

$$E[D] = \int_0^{2\pi} \int_0^1 \sqrt{r^2} \left(\frac{1}{\pi}\right) r\, dr\, d\theta = \int_0^{2\pi} \frac{1}{3\pi}\, d\theta = \frac{2}{3}.$$

■

6.7 INDEPENDENCE

As in the discrete case, if random variables X and Y are independent, then for all A and B,

$$P(X \in A, Y \in B) = P(X \in A)P(Y \in B).$$

In particular, for all x and y,

$$F(x, y) = P(X \leq x, Y \leq y) = P(X \leq x)P(Y \leq y) = F_X(x)F_Y(y).$$

Take derivatives with respect to x and y on both sides of this equation. This gives the following characterization of independence for continuous random variables in terms of pdfs.

INDEPENDENCE AND DENSITY FUNCTIONS

Continuous random variables X and Y are independent if and only if their joint density function is the product of their marginal densities. That is,

$$f(x, y) = f_X(x)f_Y(y), \quad \text{for all } x, y.$$

More generally, if $X_1, \ldots, X_n$ are jointly distributed with joint density function f, then the random variables are mutually independent if and only if

$$f(x_1, \ldots, x_n) = f_{X_1}(x_1) \cdots f_{X_n}(x_n), \quad \text{for all } x_1, \ldots, x_n.$$

Example 6.22 Caleb and Destiny are at the airport terminal buying tickets. Caleb is in the regular ticket line where the waiting time C has an exponential distribution with mean 10 minutes. Destiny is in the express line, where the waiting time D has an exponential distribution with mean 5 minutes. Waiting times for the two lines are independent. What is the probability that Caleb gets to the ticket counter before Destiny?

The desired probability is $P(C < D)$. By independence, the joint density of C and D is

$$f(c, d) = f_C(c)\, f_D(d) = \frac{1}{10}e^{-c/10}\frac{1}{5}e^{-d/5}, \quad \text{for } c > 0, d > 0.$$

Then,

$$P(C < D) = \iint_{\{(c,d):c<d\}} f(c, d)\, dc\, dd$$

$$= \int_0^\infty \int_0^d \frac{1}{50} e^{-c/10} e^{-d/5} \, dc \, dd$$

$$= \frac{1}{50} \int_0^\infty e^{-d/5} \left(-10e^{-c/10}\right)\Big|_0^d \, dd$$

$$= \frac{1}{5} \int_0^\infty e^{-d/5}(1 - e^{-d/10}) \, dd$$

$$= \frac{1}{5}\left(\frac{5}{3}\right) = \frac{1}{3}.$$

■

Example 6.23 Let X, Y, and Z be i.i.d. random variables with common marginal density $f(t) = 2t$, for $0 < t < 1$, and 0, otherwise. Find $P(X < Y < Z)$.

Here we have three jointly distributed random variables. Results for joint distributions of three or more random variables are natural extensions of the two variable case. You might envision a triple integral to find the desired probability. We also show an alternative solution that does not require any integration. To find the desired probability, we first need the joint pdf. By independence, the joint density of (X, Y, Z) is,

$$f(x, y, z) = f_X(x) f_Y(y) f_Z(z) = (2x)(2y)(2z) = 8xyz,$$

for $0 < x, y, z < 1$, and 0, otherwise.

To find $P(X < Y < Z)$ integrate the joint density function over the three-dimensional region $S = \{(x, y, z) : x < y < z\}$.

$$P(X < Y < Z) = \iiint_S f(x, y, z) \, dx \, dy \, dz = \int_0^1 \int_0^z \int_0^y 8xyz \, dx \, dy \, dz$$

$$= \int_0^1 \int_0^z 8yz \left(\frac{y^2}{2}\right) dy \, dz = \int_0^1 4z \left(\frac{z^4}{4}\right) dz = \frac{1}{6}.$$

An alternate solution appeals to symmetry. There are $3! = 6$ ways to order X, Y, and Z. As the marginal densities are all the same, the probabilities for each ordered relationship are the same. And thus

$$P(X < Y < Z) = P(X < Z < Y) = \cdots = P(Z < Y < X) = \frac{1}{6}.$$

Note that in this last derivation we have not used the specific form of the density function f. The result holds for *any* three continuous i.i.d. random variables.

By extension, if $X_1, \ldots, X_n$ are continuous i.i.d. random variables then for any permutation $(i_1, \ldots, i_n)$ of $\{1, \ldots, n\}$,

$$P(X_{i_1} < \cdots < X_{i_n}) = \frac{1}{n!}.$$

■

If X and Y are independent then their joint density function factors into a product of two functions—one that depends on x and one that depends on y. Conversely, suppose the joint density of X and Y has the form $f(x, y) = g(x)h(y)$ for some functions g and h. Then it follows that X and Y are independent with $f_X(x) \propto g(x)$ and $f_Y(y) \propto h(y)$. We leave the proof to the reader.

Example 6.24 Are X and Y independent or dependent?

(i) The joint density of X and Y is

$$f(x, y) = 6e^{-(2x+3y)}, \quad \text{for } x, y > 0.$$

The joint density can be written as $f(x, y) = g(x)h(y)$, where $g(x) = \text{constant} \times e^{-2x}$, for $x > 0$; and $h(y) = \text{constant} \times e^{-3y}$, for $y > 0$. It follows that X and Y are independent. Furthermore, $X \sim \text{Exp}(2)$ and $Y \sim \text{Exp}(3)$. This becomes clear if we write $g(x) = 2e^{-2x}$ for $x > 0$, and $h(y) = 3e^{-3y}$ for $y > 0$.

(ii) Consider the joint density function

$$f(x, y) = 15e^{-(2x+3y)}, \quad \text{for } 0 < x < y.$$

Superficially the joint density appears to have a similar form as the density in (i). However, the constraints on the domain $x < y$ show that X and Y are not independent. The domain constraints are part of the function definition. We cannot factor the joint density into a product of two functions which only depend on x and y, respectively. ■

6.7.1 Accept–Reject Method

Rectangles are "nice" sets to have as a domain because their areas are easy to compute. If (X, Y) is uniformly distributed on the rectangle $[a, b] \times [c, d]$, then the joint pdf of (X, Y) is

$$f(x, y) = \frac{1}{(b - a)(d - c)}, \quad \text{for } a < x < b \quad \text{and} \quad c < y < d.$$

The density factors as

$$f(x, y) = f_X(x)f_Y(y),$$

where f_X is the uniform density on (a, b), and f_Y is the uniform density on (c, d).

This observation suggests how to simulate a uniformly random point in the rectangle: Generate a uniform number X in (a, b). Independently generate a uniform number Y in (c, d). Then (X, Y) gives the desired uniform point on the rectangle.

In **R**, type

```
> c(runif(1,a,b),runif(1,c,d))
```

to generate $(X, Y) \sim \text{Unif}([a, b] \times [c, d])$.

How can we generate a point uniformly distributed on some "complicated" set S? The idea is intuitive. Assume S is bounded, and enclose the set in a rectangle.

1. Generate a uniformly random point in the rectangle.
2. If the point is contained in S *accept* it as the desired point.
3. If the point is not contained in S, *reject* it and generate another point and keep doing so until a point is contained in S. When it does, *accept* that point.

We show that the accepted point is uniformly distributed in S. This method is known as the *accept–reject method.*

Let R be a rectangle that encloses S. Let r be a point uniformly distributed in R. And let T be the point obtained by the accept–reject method.

Proposition. The random point T generated by the accept–reject method is uniformly distributed on S.

Proof: To show that $T \sim \text{Unif}(S)$, we need to show that for all $A \subseteq S$,

$$P(T \in A) = \frac{\text{Area }(A)}{\text{Area }(S)}.$$

Let $A \subseteq S$. If $T \in A$, then necessarily a point r was accepted. Furthermore, $r \in A$. That is,

$$\begin{aligned} P(T \in A) &= P(r \in A | r \text{ is accepted}) \\ &= \frac{P(r \in A, r \text{ is accepted})}{P(r \text{ is accepted})} \\ &= \frac{P(r \in A)}{P(r \text{ is accepted})} \\ &= \frac{\text{Area }(A)/\text{Area }(R)}{\text{Area }(S)/\text{Area }(R)} = \frac{\text{Area }(A)}{\text{Area }(S)}. \end{aligned}$$

□

Example 6.25 See Figure 6.11. The shape in the top left is the region S between the functions

$$f_1(x) = \frac{-20x^2}{9\pi^2} + 6 \quad \text{and} \quad f_2(x) = \cos x + \cos 2x + 2.$$

The region is contained in the rectangle $[-5, 5] \times [0, 6]$. Points are generated uniformly on the rectangle with the commands

```
> x <- runif(1,-5,5)
> y <- runif(1,0,6)
```

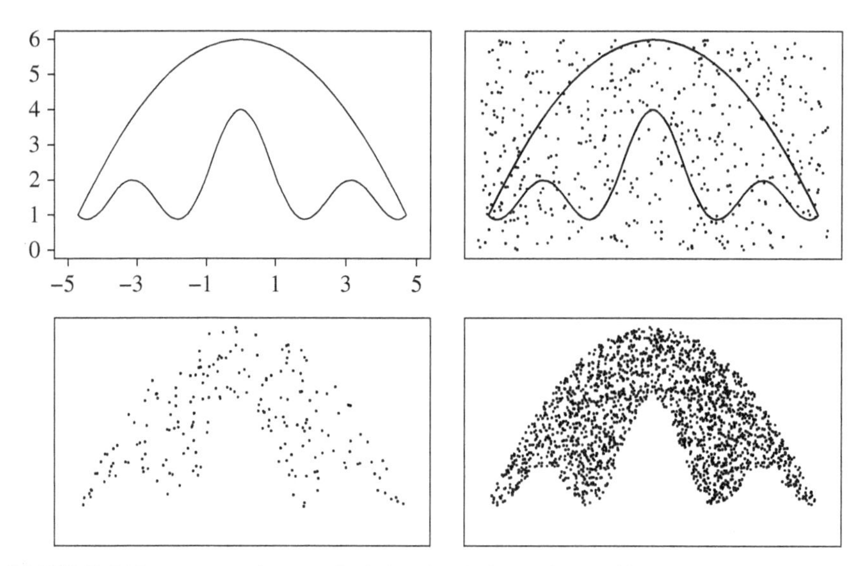

FIGURE 6.11: Accept–reject method for simulating points uniformly distributed in the top-left set S. In the top-right, 500 points are generated uniformly on the rectangle. The 205 points inside S are accepted in the bottom left. In bottom right, the accept–reject method is used with an initial 5000 points of which 2008 are accepted.

and accepted if they fall inside S. The accepted points are uniformly distributed on S. See the script **AcceptReject.R**.

The accept–reject method for generating uniform points in planar regions extends in a natural way to three (and higher) dimensions. ■

Example 6.26 Let (X, Y, Z) be a point uniformly distributed on the sphere of radius 1 centered at the origin. Estimate the mean and standard deviation for the distance from the point to the origin.

Let $D = \sqrt{X^2 + Y^2 + Z^2}$ be the distance from (X, Y, Z) to the origin. We will simulate D by evaluating the distance function for a uniformly random point in the sphere.

Enclose the sphere in a cube of side length 2 centered at the origin. The command

```
> pt <- runif(3,-1,1)
```

generates a point uniformly distributed in the cube. A point (x, y, z) is contained in the unit sphere if $x^2 + y^2 + z^2 < 1$. The command

```
> if ((pt[1]^2 + pt[2]^2 + pt[3]^2) < 1) 1 else 0
```

checks whether pt lies in the sphere or not.

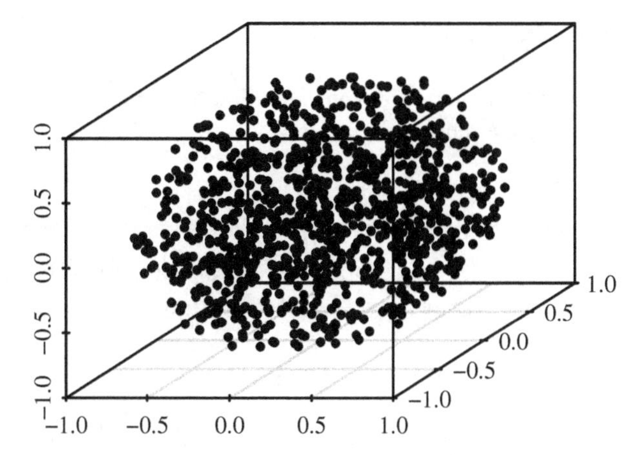

FIGURE 6.12: One thousand points generated in the unit sphere using the accept–reject method.

Here is the simulation (**Distance.R**). The points generated in the sphere are shown in Figure 6.12. To estimate the mean distance to the origin simply compute the distance to the origin for each of the simulated points and take their average. Similarly for the standard deviation.

R: DISTANCE IN UNIT SPHERE—DISTANCE.R

```
# Distance in unit sphere
> n <- 10000
> # initialize n by 3 matrix
> mat <- matrix(rep(0,3*n),nrow=n)
> i <- 1
> while (i <= n) {
     pt <- runif(3,-1,1)
     if ((pt[1]^2 + pt[2]^2 + pt[3]^2) < 1) {
         mat[i,] <- pt
   i <- i+1  { {
> d <-sqrt(mat[,1]^2 + mat[,2]^2 + mat[,3]^2)
> mean(d)
[1] 0.7522634
> sd(d)
[1] 0.1923916
```

The exact theoretical values for the mean and standard deviation of distance are $E[D] = 3/4$ and $\text{SD}[D] = \sqrt{3/80} \approx 0.194$. ■

6.8 COVARIANCE, CORRELATION

For jointly distributed continuous random variables, the covariance and correlation are defined as for discrete random variables.

COVARIANCE

Let X and Y be jointly distributed continuous random variables with joint density function f. Let $\mu_X = E[X]$ and $\mu_Y = E[Y]$. The *covariance of X and Y* is

$$\begin{aligned}\text{Cov}(X, Y) &= E[(X - \mu_X)(Y - \mu_Y)] \\ &= \int_{-\infty}^{\infty}\int_{-\infty}^{\infty}(x - \mu_X)(y - \mu_Y)f(x, y)\,dx\,dy.\end{aligned}$$

The *correlation of X and Y* is

$$\text{Corr}(X, Y) = \frac{\text{Cov}(X, Y)}{\text{SD}[X]\text{SD}[Y]}.$$

For all jointly distributed random variables—continuous and discrete—we have the computational formula for covariance: $\text{Cov}(X, Y) = E[XY] - E[X]E[Y]$. We remind the reader that correlation is a standardized version of covariance.

Example 6.27 A point (X, Y) is uniformly distributed on the triangle with vertices $(0, 0)$, $(1, 0)$, and $(1, 1)$. Find the covariance and correlation between X and Y.

The area of the triangle is 1/2 and is described by the constraints $0 < y < x < 1$. The joint density function is thus

$$f(x, y) = 2, \text{ for } 0 < y < x < 1,$$

and 0, otherwise. Recall that to use the computational formula for covariance we need to find $E[XY]$, $E[X]$, and $E[Y]$. We have

$$E[XY] = \int_0^1\int_0^x 2xy\,dy\,dx = \int_0^1 2x\left(\frac{x^2}{2}\right)dx = \int_0^1 x^3\,dx = \frac{1}{4}.$$

For the marginal density of X, integrate out the y variable, giving

$$f_X(x) = \int_y f(x, y)\,dy = \int_0^x 2\,dy = 2x, \text{ for } 0 < x < 1.$$

Thus,

$$E[X] = \int_0^1 x(2x)\,dx = \frac{2}{3}.$$

For the Y variable, the marginal is

$$f_Y(y) = \int_y^1 2\,dx = 2(1-y), \text{ for } 0 < y < 1,$$

and

$$E[Y] = \int_0^1 y\,2(1-y)\,dy = \frac{1}{3}.$$

This gives

$$\mathrm{Cov}(X,Y) = E[XY] - E[X]E[Y] = \frac{1}{4} - \left(\frac{2}{3}\right)\left(\frac{1}{3}\right) = \frac{1}{36}.$$

For the correlation, we need the marginal standard deviations, which we get from the variances. We find $E[X^2]$ first.

$$E[X^2] = \int_0^1 x^2(2x)\,dx = \frac{1}{2},$$

giving

$$V[X] = E[X^2] - E[X]^2 = \frac{1}{2} - \left(\frac{2}{3}\right)^2 = \frac{1}{18} = 0.0278.$$

We leave it to the reader to verify that $V[Y] = 1/18$, also. This gives

$$\mathrm{Corr}(X,Y) = \frac{\mathrm{Cov}(X,Y)}{\mathrm{SD}[X]\mathrm{SD}[Y]} = \frac{1/36}{1/18} = \frac{1}{2}.$$

■

R: SIMULATION OF COVARIANCE, CORRELATION

The accept–reject method is used to simulate uniform points in the triangle in the script **Triangle.R**. The sample covariance and correlation of the simulated points, computed with the **R** commands `cov(x,y)` and `cor(x,y)`, give Monte Carlo estimates of the theoretical covariance and correlation.

```
> xsim <- c()
> ysim <- c()
> x <- runif(10000)
> y <- runif(10000)
```

```
> xsim <- x[y < x]
> ysim <- y[y < x]
> cov(xsim,ysim)
[1] 0.02789259
> cor(xsim,ysim)
[1] 0.4893518
```

6.9 SUMMARY

Continuous probability is introduced in this chapter. In the continuous setting, the sample space is uncountable. Integrals replace sums and density functions replace pmfs. Many concepts first introduced for discrete probability, such as expectation, variance, joint and conditional distributions, extend naturally to the continuous framework. The density function plays a similar role as that of the discrete pmf: we integrate the density to compute probabilities. However, unlike the pmf, the density is not a probability.

For all random variables X, discrete and continuous, there is a cdf $F(x) = P(X \leq x)$. For continuous random variables, the density function is the derivative of the cdf.

The continuous uniform and exponential distributions are introduced in this chapter. The uniform distribution on an interval (a, b) has density function constant on that interval. The exponential distribution is the only continuous distribution that is memoryless.

For two or more jointly distributed random variables, there is a joint density function. Probabilities which involve multiple random variables will require multiple integrals. For independent random variables the joint density $f(x, y)$ is a product of the marginal densities $f_X(x)f_Y(y)$. Covariance and correlation extend naturally to the continuous case.

- **Continuous random variable:** A random variable which takes values in a continuous set.
- **Probability density function:** A function f is the density function of a continuous random variable if
    1. $f(x) \geq 0$ for all x.
    2. $\int_{-\infty}^{\infty} f(x)\,dx = 1$.
    3. For all $S \subseteq \Re$, $P(X \in S) = \int_S f(x)\,dx$.
- **Cumulative distribution function:** The cdf of X is $F(x) = P(X \leq x)$, defined for all real x.
- **Pdf and cdf:** $F'(x) = f(x)$.
- **Properties of cdf:**

1. $\lim_{x\to\infty} F(x) = 1$.
2. $\lim_{x\to-\infty} F(x) = 0$.
3. $F(x)$ is right-continuous at all x.
4. $F(x)$ is an increasing function of x.

- **Expectation:** $E[X] = \int_{-\infty}^{\infty} xf(x)\,dx$.
- **Variance:** $V[X] = \int_{-\infty}^{\infty} (x - E[X])^2 f(x)\,dx$.
- **Law of unconscious statistician:** If g is a function, then

$$E[g(X)] = \int_{-\infty}^{\infty} g(x)f(x)\,dx.$$

- **Uniform distribution:** A continuous random variable X is uniformly distributed on (a, b) if the density function of X is

$$f(x) = \frac{1}{b-a}, \quad \text{for } a < x < b,$$

and 0, otherwise.

- **Uniform setting:** The uniform distribution arises as a model for equally likely outcomes. Properties of the continuous uniform distribution include:
  1. $E[X] = (b + a)/2$.
  2. $V[X] = (b - a)^2/12$.
  3. $F(x) = P(X \le x) = (x - a)/(b - a)$, if $a < x < b$, 0, if $x \le a$, and 1, if $x \ge b$.
- **Exponential distribution:** The distribution of X is exponential with parameter $\lambda > 0$ if the density of X is $f(x) = \lambda e^{-\lambda x}$, for $x > 0$.
- **Exponential setting:** The exponential distribution is often used to model arrival times—the time until some event occurs, such as phone calls, traffic accidents, component failures, etc. Properties of the exponential distribution include:
  1. $E[X] = 1/\lambda$.
  2. $V[X] = 1/\lambda^2$.
  3. $F(x) = P(X \le x) = 1 - e^{-\lambda x}$.
  4. The exponential distribution is the only continuous distribution which is memoryless.
- **Joint probability density function:** For jointly continuous random variables the joint density $f(x, y)$ has similar properties as the univariate density function:
  1. $f(x, y) \ge 0$, for all x and y.
  2. $\int_{-\infty}^{\infty}\int_{-\infty}^{\infty} f(x, y) = 1$.
  3. For all $S \subseteq \Re^2$, $P((X, Y) \in S) = \iint_S f(x, y)\,dx\,dy$.

- **Joint cumulative distribution function:** $F(x, y) = P(X \leq x, Y \leq y)$, defined for all real x and y.
- **Joint cdf and joint pdf:** $\frac{\partial^2}{\partial x\, \partial y} F(x, y) = f(x, y)$.
- **Expectation of function of two random variables:** If $g(x, y)$ is a function of two variables, then $E[g(X, Y)] = \int_{-\infty}^{\infty} \int_{-\infty}^{\infty} g(x, y)f(x, y)\, dx\, dy$.
- **Independence:** If X and Y are jointly continuous and independent, with marginal densities f_X and f_Y, respectively, then the joint density of X and Y is $f(x, y) = f_X(x)f_Y(y)$.
- **Accept–reject method:** Suppose S is a bounded set in the plane. The method gives a way to simulate from the uniform distribution on S. Enclose S in a rectangle R. Generate a point uniformly distributed in R. If the point is in S, "accept"; if the point is not in S, "reject" and try again. The first accepted point will be uniformly distributed on S.
- **Covariance:**

$$\begin{aligned} \text{Cov}(X, Y) &= E[(X - E[X])(Y - E[Y])] \\ &= \int_{-\infty}^{\infty} \int_{-\infty}^{\infty} (x - E[X])(y - E[Y])f(x, y)\, dx\, dy. \end{aligned}$$

- **Setting up multiple integrals:** Many continuous problems involving two random variables or functions of random variables will involve multiple integrals. Make sure to define the limits of integration carefully. Several examples in the book and Appendix D, such as $P(X < Y)$ for different distributions of X and Y, are good problems to practice on.

EXERCISES

Density, cdf, Expectation, Variance

6.1 A random variable X has density function

$$f(x) = cx, \quad \text{for } 0 < x < 1.$$

(a) Find c.

(b) Find $P(X < 0.5)$.

(c) Find $E[X]$.

(d) Find the mgf of X, $E(e^{tX})$.

6.2 A random variable X has density function

$$f(x) = cxe^{-x/2}, \quad \text{for } 0 < x < \infty.$$

(a) Find c.

(b) Find $E[X]$.

6.3 A nonnegative continuous random variable X has cdf $F(x)$. Find the following, either by supplying a numeric answer or an expression in terms of $F(x)$.

(a) $F(-5)$.

(b) $P(3 < X < 11)$.

(c) limit of $F(x)$ as x goes to infinity.

(d) $P(X > 12|X > 6)$.

6.4 Suppose that the cdf of X, time in months from diagnosis age until death for a population of cancer patients, is $F(x) = 1 - \exp(-0.03x^{1.2})$, for $x > 0$.

(a) Find the probability of surviving at least 12 months.

(b) Find the pdf of X.

6.5 A random variable X has density function

$$f(x) = ce^x, \text{ for } -2 < x < 2.$$

(a) Find c.

(b) Find $P(X < -1)$.

(c) Find $E[X]$.

6.6 A random variable X has density function proportional to x^{-5} for $x > 1$.

(a) Find the constant of proportionality.

(b) Find and graph the cdf of X.

(c) Use the cdf to find $P(2 < X < 3)$.

(d) Find the mean and variance of X.

6.7 The cumulative distribution function for a random variable X is

$$F(x) = \begin{cases} 0, & \text{if } x \leq 0 \\ \sin x, & \text{if } 0 < x \leq \pi/2 \\ 1, & \text{if } x > \pi/2. \end{cases}$$

(a) Give the density of X.

(b) Find $P(0.1 < X < 0.2)$.

(c) Find $E[X]$.

6.8 The Laplace distribution, also known as the double exponential distribution, has density function proportional to $e^{-|x|}$ for all real x. Find the mean and variance of the distribution.

6.9 The random variable X has density f that satisfies

$$f(x) \propto \frac{1}{1+x^2}$$

on the real numbers.

(a) Find $P(X > 1)$.

(b) Show that the expectation of X does not exist.

6.10 Show that

$$f(x) = e^x e^{-e^x}, \quad \text{for all } x,$$

is a probability density function. If X has such a density function find the cdf of X.

6.11 Let $X \sim \text{Unif}(a, b)$. Find a general expression for the *kth moment* $E[X^k]$.

6.12 An isosceles right triangle has side length uniformly distributed on (0, 1). Find the expectation and variance of the length of the hypotenuse.

6.13 Suppose $f(x)$ and $g(x)$ are probability density functions. Under what conditions on the constants α and β will the function $\alpha f(x) + \beta g(x)$ be a probability density function?

6.14 Some authors take the following as the definition of continuous random variables: A random variable is continuous if the cdf $F(x)$ is continuous for all real x. Show that if X is a discrete random variable, then the cdf of X is not continuous.

6.15 For continuous random variable X and constants a and b, prove that $E[aX + b] = aE[X] + b$.

Exponential Distribution

6.16 It is 9:00 p.m. The time until Julian receives his next text message has an exponential distribution with mean 5 minutes.

(a) Find the probability that he will not receive a text in the next 10 minutes.

(b) Find the probability that the next text arrives between 9:07 and 9:10 p.m.

(c) Find the probability that a text arrives before 9:03 p.m.

(d) A text has not arrived for 5 minutes. Find the probability that none will arrive for 7 minutes.

6.17 Let $X \sim \text{Exp}(\lambda)$. Suppose $0 < s < t$. As X is memoryless, is it true that $\{X > s + t\}$ are $\{X > t\}$ are independent events?

6.18 Derive the mean of the exponential distribution with parameter λ.

6.19 Derive the variance of the exponential distribution with parameter λ.

6.20 Suppose $X \sim \text{Exp}(1/20)$. Provide appropriate commands and compute the following probabilities in **R**.

(a) $P(X < 18)$.

(b) $P(15 \leq X \leq 23)$.

(c) $P(X \geq 26)$.

6.21 The time each student takes to finish an exam has an exponential distribution with mean 45 minutes. In a class of 10 students, what is the probability that at least one student will finish in less than 20 minutes? Assume students' times are independent.

6.22 For a continuous random variable X, the number m such that

$$P(X \leq m) = \frac{1}{2}$$

is called the *median* of X.

(a) Find the median of an $\text{Exp}(\lambda)$ distribution.

(b) Give a simplified expression for the difference between the mean and the median. Is the difference positive or negative?

6.23 Find the probability that an exponential random variable is within two standard deviations of the mean. That is, compute

$$P(|X - \mu| \leq 2\sigma),$$

where $\mu = E[X]$ and $\sigma = \text{SD}[X]$.

6.24 Solve these integrals without calculus. (Hint: Think exponential distribution.)

(a) $\int_0^\infty e^{-3x/10}\, dx$.

(b) $\int_0^\infty te^{-4t}\, dt$.

(c) $\int_0^\infty z^2 e^{-2z}\, dz$.

Joint Distributions, Independence, Covariance

6.25 Suppose X and Y have joint pdf $f(x, y) = 8xy$, for $0 \leq x \leq y \leq 1$, and 0, otherwise.

(a) Find $P(Y < 1/2)$.

(b) Find $E(X)$.

(c) Find $E(XY)$.

6.26 Let X and Y be jointly distributed random variables with joint pdf given by

$$f(x, y) = 2(x + y), \quad \text{for } 0 < x < y < 1.$$

(a) Find the marginal of Y.
(b) Find $P(X < 3/4)$.
(c) Find $P(Y > 2X)$.

6.27 Refer to the joint density in Exercise 6.26.
(a) Find $E(X)$.
(b) Find $E(XY)$.
(c) Find $\text{Cov}(X, Y)$.

6.28 The joint density of X and Y is

$$f(x, y) = \frac{2x}{9y}, \quad \text{for } 0 < x < 3, 1 < y < e.$$

(a) Are X and Y independent?
(b) Find the joint cumulative distribution function.
(c) Find $P(1 < X < 2, Y > 2)$.

6.29 The joint density of X and Y is

$$f(x, y) = 2e^{-(x+2y)}, \quad \text{for } x > 0, y > 0.$$

(a) Find the joint cumulative distribution function.
(b) Find the cumulative distribution function of X.
(c) Find $P(X < Y)$.

6.30 The joint density of X and Y is

$$f(x, y) = ce^{-2y}, \quad \text{for } 0 < x < y < \infty.$$

(a) Find c.
(b) Find the marginal densities of X and Y. Do you recognize either of these distributions?
(c) Find $P(Y < 2X)$.

6.31 See the joint density in Exercise 6.30. Find the covariance of X and Y.

6.32 See the joint density in Exercise 6.25. Find the covariance of X and Y.

6.33 Suppose the joint density function of X and Y is

$$f(x, y) = 2e^{-6x}, \quad \text{for } x > 0, 1 < y < 4.$$

By noting the form of the joint density and without doing any calculations show that X and Y are independent. Describe their marginal distributions.

6.34 The time until the light in Savanna's office fails is exponentially distributed with mean 2 hours. The time until the computer crashes in Savanna's office is exponentially distributed with mean 3 hours. Failure and crash times are independent.

(a) Find the probability that neither the light nor computer fail in the next 2 hours.

(b) Find the probability that the computer crashes at least 1 hour after the light fails.

6.35 Let (X, Y, Z) be uniformly distributed on a three-dimensional box with side lengths 3, 4, and 5. Find $P(X < Y < Z)$.

6.36 A stick of unit length is broken into two pieces. The break occurs at a location uniformly at random on the stick. What is the expected length of the longer piece?

6.37 Suppose (X, Y) are distributed uniformly on the circle of radius 1 centered at the origin. Find the marginal densities of X and Y. Are X and Y independent?

6.38 Suppose X and Y have joint probability density $f(x, y) = g(x)h(y)$ for some functions g and h which depend only on x and y, respectively. Show that X and Y are independent with $f_X(x) \propto g(x)$ and $f_Y(y) \propto h(y)$.

Simulation and R

6.39 Simulate the expected length of the hypotenuse of the isosceles right triangle in Exercise 6.12.

6.40 Let X and Y be independent exponential random variables with parameter $\lambda = 1$. Simulate $P(X/Y < 1)$.

6.41 Use the accept–reject method to simulate points uniformly distributed on the circle of radius 1 centered at the origin. Use your simulation to approximate the expected distance of a point inside the circle to the origin (see Example 6.21).

6.42 Simulate the probabilities computed in Exercise 6.34.

6.43 See the **R** script **AcceptReject.R**. Make up your own interesting shape S and use the accept–reject method to generate uniformly distributed points in S.

Chapter Review

Chapter review exercises are available through the text website. The URL is `www.wiley.com/go/wagaman/probability2e`.

7

CONTINUOUS DISTRIBUTIONS

At this point an enigma presents itself which in all ages has agitated inquiring minds. How can it be that mathematics, being after all a product of human thought which is independent of experience, is so admirably appropriate to the objects of reality?

—Albert Einstein

Learning Outcomes

1. Distinguish common continuous distributions and their applications, including the normal, gamma, and beta distributions.
2. Solve problems with the different RVs involved in Poisson processes.
3. Demonstrate relationships between RVs using moment-generating functions (mgfs).
4. (C) Find probabilities involving normal, gamma, and beta distributions using **R**.

7.1 NORMAL DISTRIBUTION

I know of scarcely anything so apt to impress the imagination as the wonderful form of cosmic order expressed by "the law of error." The law would have been personified by the Greeks and deified, if they had known of it. It reigns with severity in complete self-effacement amidst the wildest confusion. The huger the mob and the greater the anarchy the more perfect is its sway. Let a large sample of chaotic elements be taken and marshalled in order of their magnitudes, and then, however wildly irregular they appeared, an unexpected and most beautiful form of regularity proves to have been present all along.

—Sir Francis Galton

Probability: With Applications and R, Second Edition. Amy S. Wagaman and Robert P. Dobrow.

Companion Website: www.wiley.com/go/wagaman/probability2e

The "law of error" in the quotation is now known as the normal distribution. It is perhaps the most important distribution in statistics, is ubiquitous as a model for natural phenomenon, and arises as the limit for many random processes and distributions throughout probability and statistics. It is sometimes called the Gaussian distribution, after Carl Friedrich Gauss, one of the greatest mathematicians in history, who discovered its utility as a model for astronomical measurement errors. Adolphe Quetelet, the father of quantitative social science, was the first to apply it to human measurements, including his detailed study of the chest circumferences of 5738 Scottish soldiers. Statistician Karl Pearson penned the name "normal distribution" in 1920, although he did admit that it "had the disadvantage of leading people to believe that all other distributions of frequency are in one sense or another *abnormal*."

NORMAL DISTRIBUTION

A random variable X has the *normal distribution with parameters* μ *and* σ^2, if the density function of X is

$$f(x) = \frac{1}{\sigma\sqrt{2\pi}} e^{-\frac{(x-\mu)^2}{2\sigma^2}}, -\infty < x < \infty.$$

We write $X \sim \text{Norm}(\mu, \sigma^2)$ or $\text{N}(\mu, \sigma^2)$.

The shape of the density curve is the famous "bell curve" (see Figure 7.1). The parameters μ and σ^2 are, respectively, the mean and variance of the distribution.

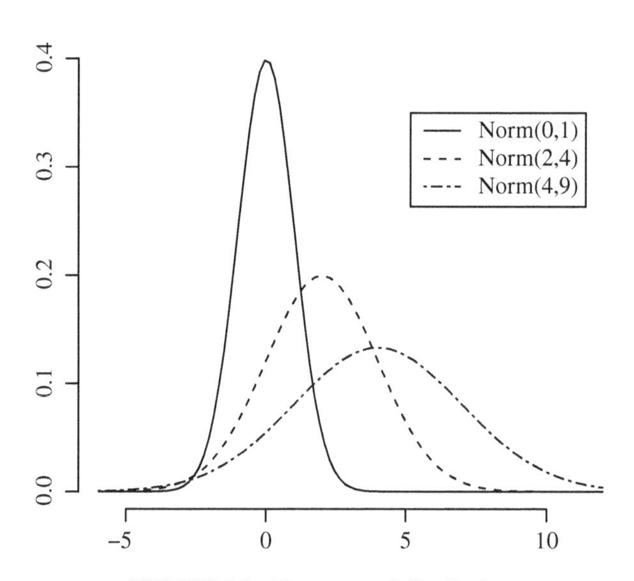

FIGURE 7.1: Three normal distributions.

The density curve is symmetric about the line $x = \mu$ and takes values in the set of all real numbers. The inflection points, where the curvature of the density function changes, occur one standard deviation unit from the mean, that is, at the points $x = \mu \pm \sigma$.

Surprisingly there is no closed form expression for the normal cumulative distribution function (cdf)

$$F(x) = \int_{-\infty}^{x} \frac{1}{\sigma\sqrt{2\pi}} e^{-\frac{(t-\mu)^2}{2\sigma^2}}\, dt.$$

The integral has no antiderivative that is expressible with elementary functions. So numerical methods must be used to find normal probabilities.

R: NORMAL DISTRIBUTION

The **R** commands for working with the normal distribution are

```
> dnorm(x, mu, sigma)   # f(x); density value
> pnorm(x, mu, sigma)   # Computes P(X <= x)
> rnorm(n, mu, sigma)   # Generates n random numbers
```

Default values for parameters are $\mu = 0$ and $\sigma = 1$.

Note that **R** uses the standard deviation σ in specifying the normal distribution, not the variance σ^2 as we do. This convention differs from source to source. Always check the syntax for your source!

Although in general it is not possible to get exact closed form expressions for normal probabilities of the form

$$P(a < X < b) = \int_{a}^{b} f(x)\, dx,$$

it is possible with basic calculus to show that the density integrates to 1 on $(-\infty, \infty)$. It requires working in two dimensions with polar coordinates. Write

$$I = \int_{-\infty}^{\infty} \frac{1}{\sigma\sqrt{2\pi}} e^{-\frac{(t-\mu)^2}{2\sigma^2}}\, dt.$$

Change variables by setting $x = (t - \mu)/\sigma$ to get

$$I = \frac{1}{\sqrt{2\pi}} \int_{-\infty}^{\infty} e^{-\frac{x^2}{2}}\, dx.$$

Now consider

$$I^2 = \frac{1}{2\pi} \int_{-\infty}^{\infty} \int_{-\infty}^{\infty} e^{-\frac{x^2+y^2}{2}}\, dx\, dy.$$

Work in polar coordinates, setting $x^2 + y^2 = r^2$ and $dx\,dy = r\,dr\,d\theta$. Then

$$I^2 = \frac{1}{2\pi}\int_0^\infty \int_0^{2\pi} re^{-\frac{r^2}{2}}\,dr\,d\theta = \int_0^\infty re^{-\frac{r^2}{2}}\,dr.$$

Solve this integral with the substitution $z = r^2/2$ and $dz = r\,dr$, giving

$$I^2 = \int_0^\infty e^{-z}\,dz = 1,$$

and thus $I = 1$.

Mgf of the normal distribution. The mgf of the normal distribution is useful to know to show several results. If $X \sim \text{Norm}(\mu, \sigma^2)$, then the mgf of X is

$$m(t) = E[e^{tX}] = e^{\mu t + \sigma^2 t^2/2}.$$

The derivation itself is left for the reader, as it is largely an exercise in algebra and involves completing the square within the exponent.

7.1.1 Standard Normal Distribution

For real μ and $\sigma > 0$, suppose $X \sim \text{Norm}(\mu, \sigma^2)$. Define the *standardized* random variable

$$Z = \frac{X - \mu}{\sigma} = \left(\frac{1}{\sigma}\right)X - \frac{\mu}{\sigma}.$$

We show that Z is normally distributed with mean 0 and variance 1, using mgfs in two ways: using the definition, and relying on properties of mgfs for linear functions of random variables.

Let $Z \sim \text{Norm}(0, 1)$. Using the definition of mgf, the mgf of Z is

$$\begin{aligned} m(t) = E[e^{tZ}] &= \int_{-\infty}^{\infty} (e^{tz})\frac{1}{\sqrt{2\pi}}e^{-z^2/2}\,dz \\ &= \int_{-\infty}^{\infty} \frac{1}{\sqrt{2\pi}}e^{-(z^2-2tz)/2}\,dz \\ &= e^{t^2/2}\int_{-\infty}^{\infty} \frac{1}{\sqrt{2\pi}}e^{-(z-t)^2/2}\,dz = e^{t^2/2}, \end{aligned}$$

achieved by completing the square so that the last integral gives the density of a normal distribution with mean t and variance one, and thus, integrates to 1.

As an alternative derivation, recall that for linear functions of random variables, the mgf can be found as a function of the mgf of the original variable. In this context, the constants a and b are respectively $1/\sigma$ and $-\mu/\sigma$. Thus, the mgf of Z is

$$m_Z(t) = e^{tb} m_X(at) = e^{-\mu t/\sigma} e^{\mu t/\sigma + \sigma^2 t^2/2\sigma^2} = e^{t^2/2} = e^{0t + 1^2 t^2/2},$$

which is the mgf of a normal distribution with mean 0 and variance 1, as emphasized in the final equality. We call this the *standard normal distribution* and reserve the letter Z for a standard normal random variable henceforth. Many other texts have common notations for the standard normal probability distribution function (pdf) and cdf of $\phi(z)$ and $\Phi(z)$, respectively.

Any normal random variable $X \sim \text{Norm}(\mu, \sigma^2)$ can be transformed to a standard normal variable Z by the linear function $Z = (1/\sigma)X - \mu/\sigma$. In fact, *any* linear function of a normal random variable is normally distributed. We leave the proof of that result to the exercises.

LINEAR FUNCTION OF NORMAL RANDOM VARIABLE

Let X be normally distributed with mean μ and variance σ^2. For constants $a \neq 0$ and b, the random variable $Y = aX + b$ is normally distributed with mean

$$E[Y] = E[aX + b] = aE[X] + b = a\mu + b$$

and variance

$$V[Y] = V[aX + b] = a^2 V[X] = a^2\sigma^2.$$

That is, $Y = aX + b \sim \text{Norm}(a\mu + b, a^2\sigma^2)$.

The ability to transform any normal distribution to a standard normal distribution by a change of variables makes it possible to simplify many computations. Many statistics textbooks include tables of standard normal probabilities, and problems involving normal probabilities can be solved by first standardizing the variables to work with the standard normal distribution. Today this is a largely outdated practice due to technology.

As a standard normal distribution has mean 0 and standard deviation 1,

$$P(|Z| \leq z) = P(-z \leq Z \leq z)$$

is the probability that Z is within z standard deviation units from the mean. It is also the probability that *any* normal random variable $X \sim \text{Norm}(\mu, \sigma^2)$ is within z standard deviation units from its mean since

$$P(|X - \mu| \leq \sigma z) = P\left(\left|\frac{X - \mu}{\sigma}\right| \leq z\right) = P(|Z| \leq z).$$

The probability that a normal random variable is within one, two, and three standard deviations from the mean, respectively, is found in **R**.

```
> pnorm(1)-pnorm(-1)
[1] 0.6826895
> pnorm(2)-pnorm(-2)
[1] 0.9544997
> pnorm(3)-pnorm(-3)
[1] 0.9973002
```

This gives the so-called "68-95-99.7 rule" or empirical rule: For the normal distribution, the probability of being within one, two, and three standard deviations from the mean is, respectively, about 68, 95, and 99.7%. This rule can be a valuable tool for doing "back of the envelope" calculations when technology is not available, and holds generally for bell-shaped distribution curves, not just the normal.

Example 7.1 Babies' birth weights are normally distributed with mean $\mu = 120$ and standard deviation $\sigma = 20$ ounces. What is the probability that a random baby's birth weight will be greater than 140 ounces?

Let X represent a baby's birth weight. Then $X \sim \text{Norm}(120, 20^2)$. The desired probability is

$$P(X > 140) = P\left(\frac{X - \mu}{\sigma} > \frac{140 - 120}{20}\right) = P(Z > 1).$$

The weight $140 = 120 + 20$ is one standard deviation above the mean. As 68% of the probability mass is within one standard deviation of the mean, the remaining 32% is evenly divided between the outer two halves of the distribution. The desired probability is about 0.16. ■

7.1.2 Normal Approximation of Binomial Distribution

One of the first uses of the normal distribution was to approximate binomial probabilities for large n. Before the use of computers or technology, calculations involving binomial coefficients with large factorials were extremely hard to compute. The approximation made it possible to numerically solve many otherwise intractable problems and also gave theoretical insight into the importance of the normal distribution.

If $X \sim \text{Binom}(n, p)$ and n is large, then the distribution of X is approximately normal with mean np and variance $np(1 - p)$. Equivalently, the standardized random variable

$$\frac{X - np}{\sqrt{np(1 - p)}}$$

has an approximate standard normal distribution.

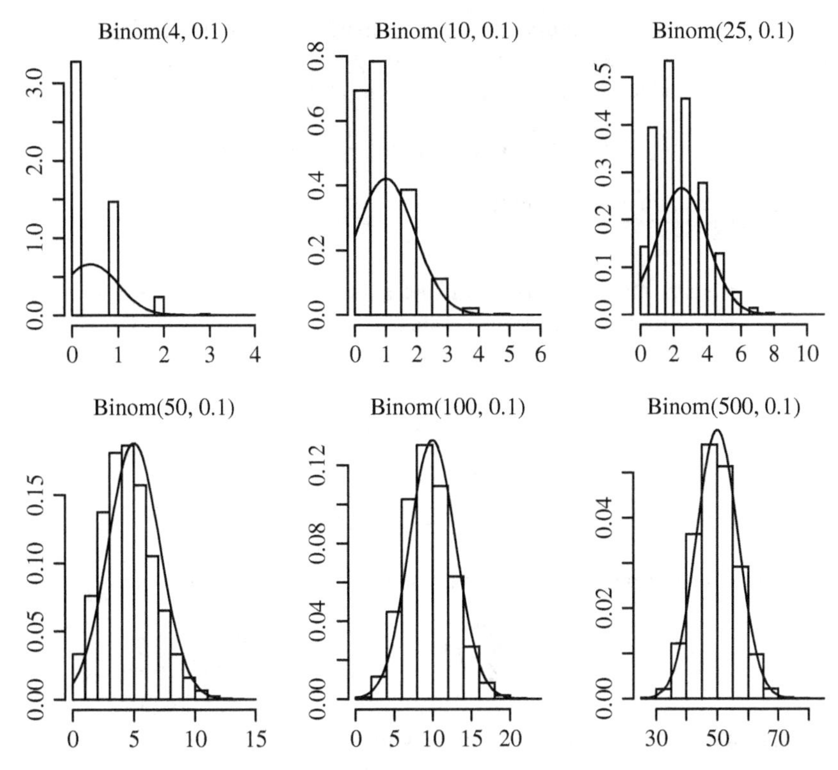

FIGURE 7.2: Normal approximation of the binomial distribution. For fixed p, as n gets large the binomial distribution tends to a normal distribution.

The result, discovered by Abraham de Moivre in early 1700s and generalized by Laplace 100 years later, says that if X is a binomial random variable with parameters n and p, then for $a < b$,

$$\lim_{n\to\infty} P\left(a \le \frac{X - np}{\sqrt{np(1-p)}} \le b\right) = \frac{1}{\sqrt{2\pi}} \int_a^b e^{-x^2/2}\, dx.$$

We see the normal approximation in Figure 7.2. The six binomial distributions all have parameter $p = 0.10$ with differing n's of $4, 10, 25, 50, 100$, and 500. Each super-imposed curve is a normal density with mean np and variance $np(1 - p)$.

What does n is "large" mean in this context? Broadly speaking, the approximation holds well when np and $n(1 - p)$ are both greater than or equal to 10. See Exercise 7.18 for more insight.

Example 7.2 In 600 rolls of a die, what is the probability of rolling between 90 and 110 fours?

Let X be the number of fours in 600 rolls. Then $X \sim \text{Binom}(600, 1/6)$. Let Y be a normal random variable with mean $np = 600(1/6) = 100$ and variance $np(1-p) = 500/6$. The normal approximation of the binomial gives

$$P(90 \leq X \leq 110) \approx P(90 \leq Y \leq 110) = 0.7267.$$

```
> pnorm(110,100,sqrt(500/6))-pnorm(90,100,sqrt(500/6))
[1] 0.7266783
```

An alternate derivation works with the standardized random variable (which results in minor differences due to rounding) to obtain

$$\begin{aligned} P(90 \leq X \leq 110) &= P\left(\frac{90-100}{\sqrt{500/6}} \leq \frac{X-np}{\sqrt{np(1-p)}} \leq \frac{110-100}{\sqrt{500/6}}\right) \\ &\approx P(-1.095 \leq Z \leq 1.095) \\ &= F_Z(1.095) - F_Z(-1.095) = 0.7265. \end{aligned}$$

Compare the normal approximation with the exact binomial probability

$$P(90 \leq X \leq 110) = \sum_{k=90}^{110} \binom{600}{k} \left(\frac{1}{6}\right)^{90} \left(\frac{5}{6}\right)^{110} = 0.7501.$$

This probability can be found using **R** with

```
> pbinom(110,600,1/6)-pbinom(89,600,1/6)
[1] 0.7501249
```

■

Continuity correction. Accuracy can often be improved in the normal approximation by accounting for the fact that we are using the area under a continuous density curve to approximate a discrete sum.

If X is a discrete random variable taking integer values, then for integers $a < b$ the probabilities

$$P(a \leq X \leq b), \quad P(a-1/2 \leq X \leq b+1/2), \quad \text{and} \quad P(a-1 < X < b+1)$$

are all equal to each other. However, if Y is a continuous random variable with density f, then the corresponding probabilities

$$P(a \leq Y \leq b), \quad P(a-1/2 \leq Y \leq b+1/2), \quad \text{and} \quad P(a-1 < Y < b+1)$$

are all different, as the integrals

$$\int_a^b f(y)\,dy, \quad \int_{a-1/2}^{b+1/2} f(y)\,dy, \quad \text{and} \quad \int_{a-1}^{b-1} f(y)\,dy$$

have different intervals of integration.

The *continuity correction* uses the middle integral for the approximation. That is,

$$P(a \le X \le b) = P(a - 1/2 \le Y \le b + 1/2) \approx \int_{a-1/2}^{b+1/2} f(y)\, dy,$$

taking the limits of integration out one-half unit to the left of a and to the right of b. The selection of $1/2$ as the value to shift by, rather than $1/3$ or $1/4$ is due to the integer scale of X.

For example, suppose the normal distribution is used to approximate $P(11 \le X \le 13)$, where X is a binomial random variable with parameters $n = 20$ and $p = 0.5$. Let Y be a normal random variable with the same mean and variance as X. The curve in the graphs in Figure 7.3 is part of the normal density. The rectangles are part of the probability mass function of X.

The light gray area represents the binomial probability $P(11 \le X \le 13)$. The dark area in the first panel represents the normal probability $P(11 \le Y \le 13)$. However, because the interval of integration is $[11, 13]$, the area under the normal density curve does not capture the area of the binomial probability to the left of 11 and to the right of 13. In the second panel, the dark area represents $P(10.5 \le Y \le 13.5)$, and visually we can see that this will give a better approximation of the desired binomial probability.

In fact, the exact binomial probability is $P(11 \le X \le 13) = 0.3542$. Compare to the normal approximations $P(11 \le Y \le 13) = 0.2375$ and $P(10.5 \le Y \le 13.5) = 0.3528$ to see the significant improvement using the continuity correction. In **R**,

```
> sum(dbinom(11:13, 20, 0.5)) #exact
[1] 0.3542423
> pnorm(13, 10, sqrt(5))-pnorm(11, 10, sqrt(5)) #no CC
```

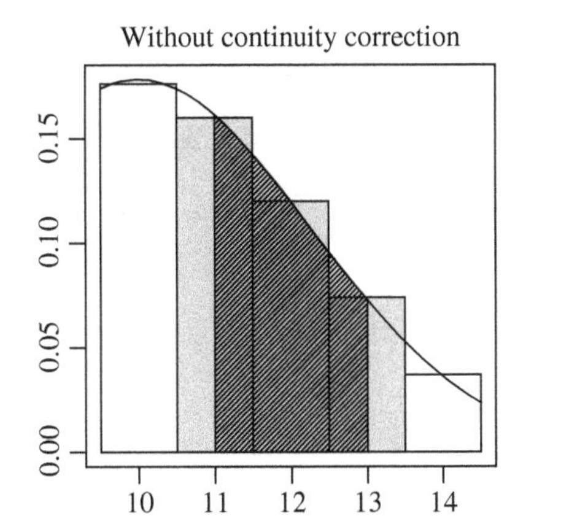

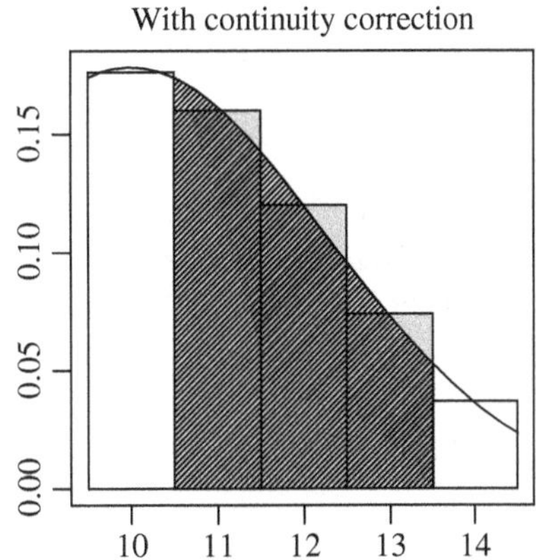

FIGURE 7.3: Approximating the binomial probability $P(11 \le X \le 13)$ using a normal distribution with and without the continuity correction.

```
[1] 0.2375042
> pnorm(13.5, 10, sqrt(5))-pnorm(10.5, 10, sqrt(5)) #CC
[1] 0.3527692
```

Example 7.3 Dice, continued. In Example 7.2, applying the continuity correction gives

$$\begin{aligned}P(90 \leq X \leq 110) &= P(89.5 \leq X \leq 110.5) \\ &= P\left(\frac{89.5 - 100}{9.13} \leq \frac{X - np}{\sqrt{np(1-p)}} \leq \frac{110.5 - 100}{9.13}\right) \\ &\approx P(-1.15 \leq Z \leq 1.15) \\ &= F_Z(1.15) - F_Z(-1.15) = 0.7499,\end{aligned}$$

which improves upon the previous estimate 0.727. The exact answer is 0.7501. ■

7.1.3 Quantiles

It is hard to escape taking standardized tests in the United States. Typically after taking such a test when your scores are reported, they usually include your percentile ranking. If you scored at the 80th percentile, also called the 80th quantile, this means that you scored as well, or better, than 80% of all test takers.

QUANTILE

Let $0 < p < 1$. If X is a continuous random variable, then the *pth quantile* is the number q that satisfies

$$P(X \leq q) = p/100.$$

That is, the pth quantile separates the bottom p percent of the probability mass from the top $(1 - p)$%.

The **R** command for finding quantiles of a probability distribution is obtained by prefacing the distribution name with the letter `q`. Normal quantiles are found with the command `qnorm`. To find the 25th quantile of the standard normal distribution, type

```
> qnorm(0.25)
[1] -0.6744898
```

As seen in the definition, the quantile function is inverse to the cdf for continuous random variables. The cdf evaluated at a quantile returns the original $p/100$.

```
> pnorm(-0.6744899)
[1] 0.25
```

When finding quantiles for discrete distributions, be aware that the definition changes slightly. We encourage the reader to investigate further if interested.

Example 7.4 The 68-95-99.7 rule says that the area under the standard normal curve between $z = -2$ and $z = 2$ is about 0.95. The area to the left of $z = -2$ is about 0.025. Thus, the total area to the left of $z = 2$ is $0.95 + 0.025 = 0.975$. Hence, $q = 2$ is the approximate 97.5th quantile of the standard normal distribution.

In **R**, we find the exact 97.5th quantile of a standard normal distribution.

```
> qnorm(0.975)
[1] 1.959964
```

■

Example 7.5 IQR. The *interquartile range (IQR)* of a distribution is the difference between the 75th and 25th quantiles, also called the third and first quartiles. It is the range of the middle 50% of values in the distribution. If $X \sim \text{Norm}(\mu, \sigma^2)$, find the IQR of X.

Let q_{75} be the 75th quantile of the given normal distribution. Since

$$0.75 = P(X \le q_{75}) = P\left(Z \le \frac{q_{75} - \mu}{\sigma}\right),$$

$(q_{75} - \mu)/\sigma$ is the 75th quantile of the standard normal distribution. In **R**, we find

```
> qnorm(0.75)
[1] 0.6744898.
```

Thus, $(q_{75} - \mu)/\sigma = 0.6745$, which gives $q_{75} = \mu + 0.6745\sigma$. The normal distribution is symmetric about μ. It follows that the 25th quantile is $q_{25} = \mu - 0.6745\sigma$. The IQR is

$$\text{IQR} = q_{75} - q_{25} = (\mu + 0.6745\sigma) - (\mu - 0.6745\sigma) = 1.35\sigma.$$

Many statistical software programs will flag an observation as an *outlier* by the so-called $1.5 \times \text{IQR}$ rule: an observation is labeled an outlier if it is more than $1.5 \times$ IQR units above the 75th quantile, or $1.5 \times$ IQR units below the 25th quantile. As

$$q_{75} + (1.5)\text{IQR} = (\mu + 0.6745\sigma) + (1.5)(1.35\sigma) = \mu + 2.7\sigma$$

and

$$q_{25} - (1.5)\text{IQR} = (\mu - 0.6745\sigma) - (1.5)(1.35\sigma) = \mu - 2.7\sigma,$$

an observation X from a normal distribution would be labeled an outlier if it is more than 2.7 standard deviations from the mean. The probability of this occurring is

$$P(|Z| > 2.7) = 2P(Z > 2.7) = 0.007.$$

While this probability is relatively small, it also means that in a dataset of 1000 observations taken from a normal distribution, the software would label about seven points as outliers. ■

Example 7.6 Let X be the number of heads in 10,000 coin tosses. We expect $X \approx 5000$. What range of values of X are typically observed with high probability, say 0.99? More precisely, what number t satisfies

$$P(5000 - t \leq X \leq 5000 + t) = 0.99?$$

As $X \sim \text{Bin}(10{,}000, 0.5)$, the distribution of X is approximated by a normal distribution with mean 5000 and standard deviation $\sqrt{10{,}000(0.5)(0.5)} = 50$. n is fairly large, so we will ease notation and omit the continuity correction, as it would have a negligible effect. We have

$$\begin{aligned}
P(5000 - t \leq X \leq 5000 + t) &= P\left(\frac{-t}{50} \leq \frac{X - 5000}{50} \leq \frac{t}{50}\right) \\
&\approx P\left(\frac{-t}{50} \leq Z \leq \frac{t}{50}\right) \\
&= F_Z\left(\frac{t}{50}\right) - F_Z\left(\frac{-t}{50}\right) \\
&= F_Z\left(\frac{t}{50}\right) - \left[1 - F_Z\left(\frac{t}{50}\right)\right] \\
&= 2F_Z\left(\frac{t}{50}\right) - 1,
\end{aligned}$$

where the next-to-last equality is from the symmetry of the standard normal density about 0. Setting the last expression equal to 0.99 gives

$$F_Z\left(\frac{t}{50}\right) = P\left(Z \leq \frac{t}{50}\right) = \frac{1 + 0.99}{2} = 0.995.$$

Thus, $t/50$ is the 99.5th quantile of the standard normal distribution. We find

```
> qnorm(0.995)
[1] 2.575829
```

This gives $t = 50(2.576) = 128.8 \approx 130$. In 10,000 coin tosses, we expect to observe 5000 ± 130 heads with high probability. A claim of tossing a fair coin 10,000 times and obtaining less than 4870 or more than 5130 heads would be highly suspect. ■

10,000 COIN FLIPS

Mathematician John Kerrich actually flipped 10,000 coins when he was interned in a Nazi prisoner of war camp during World War II. Kerrich was visiting Copenhagen at the start of the war, just when the German army was occupying Denmark. The data from Kerrich's coin tossing experiment are given in Freedman et al. [2007]. Kerrich's results are in line with expected numbers. He got 5067 heads. In Chapter 10, we use Kerrich's data to illustrate the law of large numbers.

Kerrich spent much of his time in the war camp conducting experiments to demonstrate laws of probability. He is reported to have used ping-pong balls to illustrate Bayes formula.

7.1.4 Sums of Independent Normals

The normal distribution has the property that the sum of independent normal random variables is normally distributed. We show this is true for the sum of two independent normal variables using mgfs.

Let X and Y be independent normal variables, with respective means μ_X and μ_Y and variances σ_X^2 and σ_Y^2.

Then using properties of mgfs, the mgf for $X + Y$ is

$$\begin{aligned} m_{X+Y}(t) &= m_X(t)m_Y(t) = e^{\mu_X t + \sigma_X^2 t^2/2} e^{\mu_Y t + \sigma_Y^2 t^2/2} \\ &= e^{(\mu_X + \mu_Y)t + (\sigma_X^2 + \sigma_Y^2)t^2/2}, \end{aligned}$$

which is the mgf for a normal random variable with mean $\mu_X + \mu_Y$ and variance $\sigma_X^2 + \sigma_Y^2$. Thus, $X + Y \sim \mathrm{N}(\mu_X + \mu_Y, \sigma_X^2 + \sigma_Y^2)$.

The general result stated next can be derived similarly with mathematical induction.

SUM OF INDEPENDENT NORMAL RANDOM VARIABLES IS NORMAL

Let $X_1, \ldots, X_n$ be a sequence of independent normal random variables with

$$X_k \sim \text{Norm}\,(\mu_k, \sigma_k^2), \text{for } k = 1, \ldots, n.$$

Then

$$X_1 + \cdots + X_n \sim \text{Norm}\,(\mu_1 + \cdots + \mu_n, \sigma_1^2 + \cdots + \sigma_n^2).$$

Example 7.7 The mass of a cereal box is normally distributed with mean 385g and standard deviation 5g. What is the probability that 10 boxes will contain less than 3800 g of cereal?

Let $X_1, \ldots, X_{10}$ denote the respective cereal weights in each of the 10 boxes. Then $T = X_1 + \cdots + X_{10}$ is the total weight. We have that T is normally distributed with mean $E[T] = 10(385) = 3850$ and variance $V[T] = 10(5^2) = 250$.

The desired probability $P(T < 3800)$ can be found directly with **R**.

```
> pnorm(3800,3850,sqrt(250))
[1] 0.0007827011
```

An alternate solution is to standardize T, giving

$$P(T < 3800) = P\left(\frac{T-\mu}{\sigma} < \frac{3800 - 3850}{\sqrt{250}} \right) = P(Z < -3.16).$$

We see that 3800 g is a little more than three standard deviations below the mean. By the 68-95-99.7 rule, the probability is less than $0.003/2 = 0.0015$, which was confirmed via the calculations above. ■

Example 7.8 According to the National Center for Health Statistics, the mean male adult height in the United States is $\mu_M = 69.2$ inches with standard deviation $\sigma_M = 2.8$ inches. The mean female adult height is $\mu_F = 63.6$ inches with standard deviation $\sigma_F = 2.5$ inches. Assume male and female heights are normally distributed and independent of each other. If a man and woman are chosen at random, what is the probability that the woman will be taller than the man?

Let M and F denote male and female height, respectively. The desired probability is $P(F > M) = P(F - M > 0)$. One approach to find this probability is to obtain the joint density of F and M and then integrate over the region $\{(f, m) : f > m\}$.

But a much easier approach is to recognize that because F and M are independent normal variables, $F - M$ is normally distributed with mean

$$E[F - M] = E[F] - E[M] = 63.6 - 69.2 = -5.6$$

and variance

$$V[F - M] = V[F] + V[M] = (2.5)^2 + (2.8)^2 = 14.09.$$

This gives

$$P(F > M) = P(F - M > 0) = P\left(Z > \frac{0 - (-5.6)}{\sqrt{14.09}} \right)$$
$$= P(Z > 1.492) = 0.068.$$

■

Averages. Averages of i.i.d. random variables figure prominently in probability and statistics. If $X_1, \ldots, X_n$ is an i.i.d. sequence (not necessarily normal), let $S_n = X_1 + \cdots + X_n$. The average of the X_i's is S_n/n. When working with observations from a sample, you might see notation of $\overline{X_n}$ for S_n/n. Results for averages are worthwhile to highlight and remember.

AVERAGES OF i.i.d. RANDOM VARIABLES

Let $X_1, \ldots, X_n$ be an i.i.d. sequence of random variables with common mean μ and variance σ^2. Let $S_n = X_1 + \cdots + X_n$. Then

$$E\left[\frac{S_n}{n}\right] = \mu \quad \text{and} \quad V\left[\frac{S_n}{n}\right] = \frac{\sigma^2}{n}.$$

If the X_i's are normally distributed, then $S_n/n \sim \text{Norm}(\mu, \sigma^2/n)$.

The mean and variance of the average are

$$E\left[\frac{S_n}{n}\right] = E\left[\frac{X_1 + \cdots + X_n}{n}\right] = \frac{1}{n}(\mu + \cdots + \mu) = \frac{n\mu}{n} = \mu$$

and

$$V\left[\frac{S_n}{n}\right] = \frac{1}{n^2}V[X_1 + \cdots + X_n] = \frac{1}{n^2}(\sigma^2 + \cdots + \sigma^2) = \frac{n\sigma^2}{n^2} = \frac{\sigma^2}{n}.$$

Example 7.9 Averages are better than single measurements. The big metal spring scale at the fruit stand is not very precise and has a significant measurement error. When measuring fruit, the measurement error is the difference between the true weight and what the scale says. Measurement error is often modeled with a normal distribution with mean 0.

Suppose the scale's measurement error M is normally distributed with $\mu = 0$ and $\sigma = 2$ ounces. If a piece of fruit's true weight is w, then the *observed weight* of the fruit is what the customer sees on the scale—the sum of the true weight and the measurement error. Let X be the observed weight. Then $X = w + M$. As w is a constant, X is normally distributed with mean $E[X] = E[w + M] = w + E[M] = w$, and variance $V[X] = V[w + M] = V[M] = 4$.

When a shopper weighs their fruit, the probability that the observed measurement is within 1 ounce of the true weight is

$$\begin{aligned} P(|X - w| \leq 1) &= P(-1 \leq X - w \leq 1) \\ &= P\left(\frac{-1}{2} \leq \frac{X - w}{\sigma} \leq \frac{1}{2}\right) \\ &= P\left(\frac{-1}{2} \leq Z \leq \frac{1}{2}\right) = 0.383. \end{aligned}$$

Equivalently, there is a greater than 60% chance that the scale will show a weight that is off by more than 1 ounce.

However, a savvy shopper decides to take n independent measurements $X_1, \ldots, X_n$ and rely on the average $S_n/n = (X_1 + \cdots + X_n)/n$. As $S_n/n \sim$ Norm $(w, 4/n)$, the probability that the average measurement is within 1 ounce of the true weight is

$$\begin{aligned} P(|S_n/n - w| \leq 1) &= P(-1 \leq S_n/n - w \leq 1) \\ &= P\left(\frac{-1}{\sqrt{4/n}} \leq \frac{S_n/n - w}{\sigma/\sqrt{n}} \leq \frac{1}{\sqrt{4/n}}\right) \\ &= P\left(\frac{-\sqrt{n}}{2} \leq Z \leq \frac{\sqrt{n}}{2}\right) \\ &= 2F\left(\frac{\sqrt{n}}{2}\right) - 1. \end{aligned}$$

If the shopper wants to be "95% confident" that the average is within 1 ounce of the true weight, that is, $P(|S_n/n - w| \leq 1) = 0.95$, how many measurements should they take? Solve $0.95 = 2F(\sqrt{n}/2) - 1$ for n. We have

$$F\left(\frac{\sqrt{n}}{2}\right) = \frac{1 + 0.95}{2} = 0.975.$$

The 97.5th quantile of the standard normal distribution is 1.96. Thus, $\sqrt{n}/2 = 1.96$ and $n = (2 \times 1.96)^2 = 7.68$. The shopper should take eight measurements. ■

7.2 GAMMA DISTRIBUTION

The gamma distribution is a family of positive, continuous distributions with two parameters. The density curve can take a wide variety of shapes, which allows the distribution to be used to model variables that exhibit skewed and nonsymmetric behavior. See Figure 7.4 for example gamma densities.

GAMMA DISTRIBUTION

A random variable X has a *gamma distribution with parameters* $a > 0$ *and* $\lambda > 0$ if the density function of X is

$$f(x) = \frac{\lambda^a x^{a-1} e^{-\lambda x}}{\Gamma(a)}, \quad \text{for } x > 0,$$

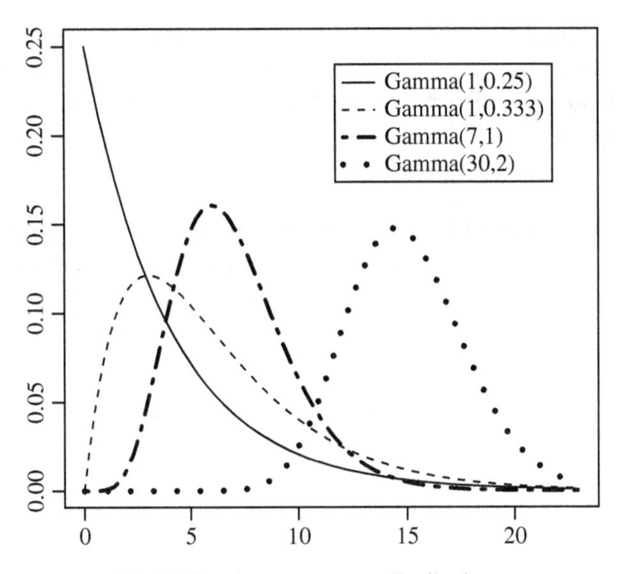

FIGURE 7.4: Four gamma distributions.

where

$$\Gamma(a) = \int_0^\infty t^{a-1}e^{-t}\, dt.$$

We write $X \sim \text{Gamma}(a, \lambda)$.

The function Γ is the *gamma function*. The function is continuous, defined by an integral, and arises in many applied settings. Observe that $\Gamma(1) = 1$. Integration by parts (see Exercise 7.22) gives

$$\Gamma(x) = (x-1)\Gamma(x-1), \quad \text{for all } x. \tag{7.1}$$

If x is a positive integer, unwinding Equation 7.1 shows that

$$\begin{aligned}\Gamma(x) &= (x-1)\Gamma(x-1)\\ &= (x-1)(x-2)\Gamma(x-2)\\ &= (x-1)(x-2)\cdots(1)\\ &= (x-1)!\end{aligned}$$

Another useful result is that $\Gamma(1/2) = \sqrt{\pi}$.

Note that if the first parameter a of the gamma distribution is equal to one, the gamma density function reduces to an exponential density. The gamma distribution is a generalization of the exponential distribution, with an additional parameter.

In many applications, exponential random variables are used to model *interarrival times* between events, such as the times between successive highway accidents, component failures, telephone calls, or bus arrivals. The nth occurrence, or time of the nth arrival, is the sum of n interarrival times. It turns out that the sum of n i.i.d. exponential random variables has a gamma distribution.

Example 7.10 Sum of i.i.d. exponentials is gamma. Here we show that the sum of i.i.d. exponential random variables has a gamma distribution using mgfs. First, find the mgf of the exponential distribution.

Let $X \sim Exp(\lambda)$. The mgf of X is

$$\begin{aligned} m_X(t) = E[e^{tX}] &= \int_0^\infty e^{tx} \lambda e^{-\lambda x}\, dx \\ &= \frac{\lambda}{\lambda - t} \int_0^\infty (\lambda - t) e^{-(\lambda - t)x}\, dx \\ &= \frac{\lambda}{\lambda - t}, \end{aligned}$$

defined for all $0 < t < \lambda$.

Now find the mgf of the gamma distribution. Let $Y \sim$ Gamma(a, λ). The mgf of Y is

$$\begin{aligned} m_Y(t) = E[e^{tY}] &= \int_0^\infty e^{ty} \frac{\lambda^a y^{a-1} e^{-\lambda y}}{\Gamma(a)}\, dy \\ &= \int_0^\infty \frac{\lambda^a y^{a-1} e^{-(\lambda - t)y}}{\Gamma(a)}\, dy \\ &= \left(\frac{\lambda}{\lambda - t}\right)^a \int_0^\infty \frac{(\lambda - t)^a y^{a-1} e^{-(\lambda - t)y}}{\Gamma(a)}\, dy \\ &= \left(\frac{\lambda}{\lambda - t}\right)^a. \end{aligned}$$

That is, $m_Y(t) = [m_X(t)]^a$. For a positive integer a, this is the mgf of the sum of a i.i.d. exponential random variables, $X_1, \ldots, X_a$, with parameter λ. ■

SUM OF i.i.d. EXPONENTIALS HAS GAMMA DISTRIBUTION

Let $E_1, \ldots, E_n$ be an i.i.d. sequence of exponential random variables with parameter λ. Let $S = E_1 + \cdots + E_n$. Then S has a gamma distribution with parameters n and λ.

We demonstrate the last result with a simulation.

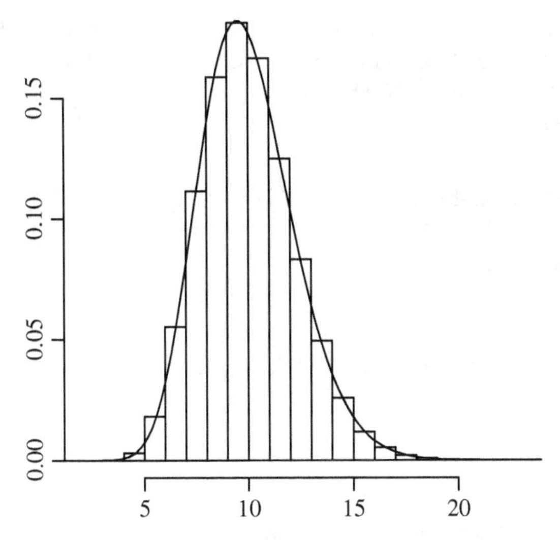

FIGURE 7.5: The histogram is from simulating the sum of 20 exponential variables with $\lambda = 2$. The curve is the density of a gamma distribution with parameters $a = 20$ and $\lambda = 2$.

R: SIMULATING THE GAMMA DISTRIBUTION FROM A SUM OF EXPONENTIALS

We simulate the sum of 20 independent exponential random variables with $\lambda = 2$ and compare to the gamma density with $a = 20$ and $\lambda = 2$ (see Figure 7.5).

```
> simlist <- replicate(10000, sum(rexp(20, 2)))
> hist(simlist, prob = T)
> curve(dgamma(x, 20, 2), 0, 20, add = T)
```

Application. Zwahlen et al. [2000] explores the process by which at-risk individuals take and retake HIV tests. The time between successive retaking of the HIV test is modeled with an exponential distribution. The time that an individual retakes the test for the nth time is fitted to a gamma distribution with parameters n and λ. Men and women retake the test at different rates, and statistical methods are used to estimate λ for men and women and for different populations. Understanding the process of test-taking can help public health officials treat at-risk populations more effectively.

Example 7.11 The times between insurance claims following a natural disaster are modeled as i.i.d. exponential random variables. Claims arrive during the first few days at a rate of about four claims per hour. (i) What is the probability that the

100th insurance claim will not arrive during the first 24 hours? (ii) The probability is at least 95% that the 100th claim will arrive before what time?

Let S_{100} be the time of the 100th claim (in hours). Then S_{100} has a gamma distribution with parameters $n = 100$ and $\lambda = 4$.

(i) The desired probability is $P(S_{100} > 24)$. In **R**, type

```
> 1-pgamma(24,100,4)
[1] 0.6450564
```

(ii) Solve $P(S_{100} \leq t) \geq 0.95$. Find the 95th quantile of S_{100}. Type

```
> qgamma(0.95,100,4)
[1] 29.24928
```

The probability is at least 95% that the 100th claim will arrive before time $t = 29.25$, which is 5 hours and 15 minutes into the second day. ■

Expectation and variance. For the gamma distribution with parameters a and λ, the expectation μ and variance σ^2 are

$$\mu = \frac{a}{\lambda} \quad \text{and} \quad \sigma^2 = \frac{a}{\lambda^2}.$$

These should not be surprising. If a is a positive integer, consider a sum of a i.i.d. exponential random variables with parameter λ. Each exponential variable has expectation $1/\lambda$, and the sum has a gamma distribution. Similarly, the variance of each exponential variable is $1/\lambda^2$, and the variance of the sum is a/λ^2. The expectation for the gamma distribution is derived in the next section.

7.2.1 Probability as a Technique of Integration

Many integrals can be solved by recognizing that the integrand is proportional to a known probability density, such as the uniform, exponential, or normal density. You have seen this "trick" before, such as in the last section with the normal distribution, and in the evaluation of the exponential and gamma mgfs.

Example 7.12 Solve

$$I = \int_0^\infty e^{-3x/5}\, dx.$$

This is an "easy" integral with a substitution. But before solving it with calculus, recognize that the integrand is "almost" an exponential density. It is proportional to an exponential density with $\lambda = 3/5$. Putting in the necessary constant to make it so gives

$$I = \frac{5}{3} \int_0^\infty \frac{3}{5} e^{-3x/5} \, dx = \frac{5}{3},$$

because the integrand is now a probability density, which integrates to one. ■

Example 7.13 Solve

$$I = \int_{-\infty}^{\infty} e^{-x^2} \, dx. \tag{7.2}$$

The integral looks like a normal density function, but is missing some components. As the exponential factor of the normal density has the form $e^{-(x-\mu)^2/2\sigma^2}$, in order to match up with the integrand take $\mu = 0$ and $2\sigma^2 = 1$, that is, $\sigma^2 = 1/2$. Reexpressing the integral as a normal density gives

$$I = \int_{-\infty}^{\infty} e^{-x^2} \, dx = \sqrt{\pi} \int_{-\infty}^{\infty} \frac{1}{\sqrt{2\pi}\sqrt{1/2}} e^{-\frac{x^2}{2(1/2)}} \, dx = \sqrt{\pi},$$

where the last equality follows as the final integrand is a normal density with $\mu = 0$ and $\sigma^2 = 1/2$ and thus integrates to one. ■

Example 7.14 Meteorologists and hydrologists use probability to model numerous quantities related to rainfall. *Rainfall intensity*, as discussed by Watkins (http://www.sjsu.edu/faculty/watkins/raindrop.htm) is equal to the raindrop volume times the intensity of droplet downfall. Let R denote rainfall intensity. (The units are usually millimeters per hour.) A standard model is to assume that the intensity of droplet downfall k (number of drops per unit area per hour) is constant and that the radius of a raindrop D has an exponential distribution with parameter λ. This gives $R = 4k\pi D^3/3$. The expected rainfall intensity is

$$E[R] = E\left[\frac{4k\pi D^3}{3}\right] = \frac{4k\pi}{3} E[D^3] = \frac{4k\pi}{3} \int_0^\infty x^3 \lambda e^{-x\lambda} \, dx.$$

The integral can be solved by three applications of integration by parts. But avoid the integration by recognizing that the integrand is proportional to a gamma density with parameters $a = 4$ and λ. The integral is equal to

$$\int_0^\infty \lambda x^3 e^{-\lambda x} \, dx = \frac{\Gamma(4)}{\lambda^3} \int_0^\infty \frac{\lambda^4 x^3 e^{-\lambda x}}{\Gamma(4)} \, dx = \frac{\Gamma(4)}{\lambda^3} = \frac{6}{\lambda^3}.$$

The expected rainfall intensity is

$$E[R] = \frac{4k\pi}{3}\left(\frac{6}{\lambda^3}\right) = \frac{8k\pi}{\lambda^3}.$$

■

Expectation of the gamma distribution. The expectation of the gamma distribution is obtained similarly. Let $X \sim \text{Gamma}(a, \lambda)$. Then

$$E[X] = \int_0^\infty x \frac{\lambda^a x^{a-1} e^{-\lambda x}}{\Gamma(a)}\, dx = \int_0^\infty \frac{\lambda^a x^a e^{-\lambda x}}{\Gamma(a)}\, dx.$$

Because of the $x^a = x^{(a+1)-1}$ term in the integrand, make the integrand look like the density of a gamma distribution with parameters $a + 1$ and λ. This gives

$$\int_0^\infty \frac{\lambda^a x^a e^{-\lambda x}}{\Gamma(a)}\, dx = \frac{\Gamma(a+1)}{\lambda\Gamma(a)} \int_0^\infty \frac{\lambda^{a+1} x^a e^{-\lambda x}}{\Gamma(a+1)}\, dx = \frac{\Gamma(a+1)}{\lambda\Gamma(a)},$$

as the last integral integrates to one. From the relationship in Equation 7.1 for the gamma function, we have

$$E[X] = \frac{\Gamma(a+1)}{\lambda\Gamma(a)} = \frac{a\Gamma(a)}{\lambda\Gamma(a)} = \frac{a}{\lambda}.$$

We invite the reader to use this technique to derive the variance of the gamma distribution $V[X] = a/\lambda^2$ in Exercise 7.23.

7.3 POISSON PROCESS

Consider a process whereby "events"—also called "points" or "arrivals"—occur randomly in time or space. Examples include phone calls throughout the day, car accidents along a stretch of highway, component failures, service times, and radioactive particle emissions.

For many applications, it is reasonable to model the times between successive events as memoryless (e.g., a phone call doesn't "remember" when the last phone call took place). Model these interarrival times as an independent sequence $E_1, E_2, \ldots$ of exponential random variables with parameter λ, where E_k is the time between the $(k-1)$st and kth arrival. Set $S_0 = 0$ and let

$$S_n = E_1 + \cdots + E_n,$$

for $n = 1, 2, \ldots$. Then, S_n is the time of the nth arrival and $S_n \sim \text{Gamma}(n, \lambda)$. The sequence $S_0, S_1, S_2, \ldots$ is the arrival sequence, i.e., the sequence of arrival times.

For each time $t \geq 0$, let N_t be the number of arrivals that occur up through time t. Then for each t, N_t is a discrete random variable. We now show that N_t has a Poisson distribution. The collection of N_t's forms a random process called a *Poisson process with parameter λ*. It is an example of what is called a *stochastic process*. Formally, a stochastic process is a collection of random variables defined on a common sample space.

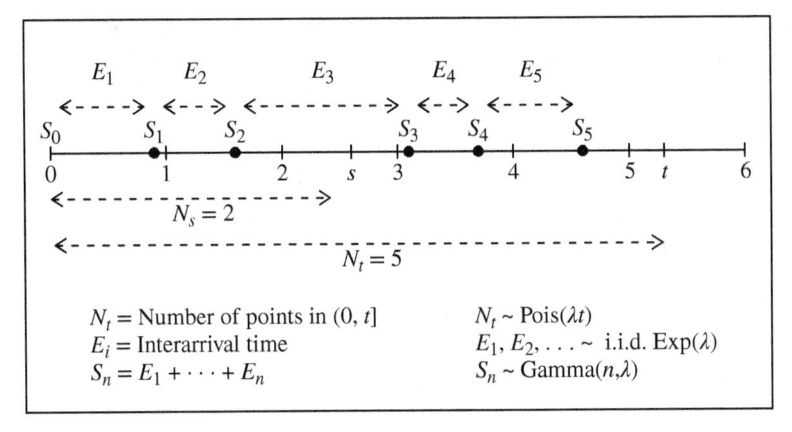

FIGURE 7.6: Poisson process: relationships of underlying random variables.

Throughout this book, we have considered sequences of random variables as models for random processes. However, the collection of N_t's in a Poisson process is not a sequence because N_t is defined for all real t. The N_t's form an uncountable collection of random variables. We write $(N_t)_{t\geq 0}$ to denote a Poisson process.

To understand the Poisson process, it is important to understand the relationship between the interarrival times (E_k's), the arrival times (S_n's), and the number of arrivals (N_t) (see Figure 7.6).

N_t **has a Poisson distribution.** Although the *time* of arrivals is continuous, the *number* of arrivals is discrete. We show that N_t, the number of arrivals up to time t, has a Poisson distribution with parameter λt.

DISTRIBUTION OF N_T FOR POISSON PROCESS WITH PARAMETER λ

Let $(N_t)_{t\geq 0}$ be a Poisson process with parameter λ. Then

$$P(N_t = k) = \frac{e^{-\lambda t}(\lambda t)^k}{k!}, \text{ for } k = 0, 1, \ldots.$$

Consider the event $\{N_t = k\}$ that there are k arrivals in $(0, t]$. This occurs if and only if (i) the kth arrival occurs by time t and (ii) the $(k + 1)$st arrival occurs after time t. That is, $\{N_t = k\} = \{S_k \leq t, S_{k+1} > t\}$. Since

$$S_{k+1} = E_1 + \cdots + E_k + E_{k+1} = S_k + E_{k+1},$$

we have

$$P(N_t = k) = P(S_k \leq t, S_k + E_{k+1} > t).$$

The sum S_k is a function of $E_1, \ldots, E_k$, which are independent of E_{k+1}. Thus, S_k and E_{k+1} are independent random variables, and the joint density of (S_k, E_{k+1}) is the product of their marginal densities. That is,

$$f_{S_k,E_{k+1}}(x,y) = f_{S_k}(x)f_{E_{k+1}}(y) = \frac{\lambda^k x^{k-1} e^{-\lambda x}}{(k-1)!}\lambda e^{-\lambda y}, \text{ for } x, y > 0.$$

The region defined by

$$\{S_k \le t, S_k + E_{k+1} > t\} = \{S_k \le t, E_{k+1} > t - S_k\}$$

is the set of all (x, y) such that $0 < x \le t$ and $y > t - x$. This gives

$$\begin{aligned}
P(N_t = k) &= P(S_k \le t, S_k + E_{k+1} > t)\\
&= \int_0^t \int_{t-x}^{\infty} \frac{\lambda^k x^{k-1} e^{-\lambda x}}{(k-1)!}\lambda e^{-\lambda y}\, dy\, dx\\
&= \int_0^t \frac{\lambda^k x^{k-1} e^{-\lambda x}}{(k-1)!}\left(\int_{t-x}^{\infty} \lambda e^{-\lambda y}\, dy\right) dx\\
&= \int_0^t \frac{\lambda^k x^{k-1} e^{-\lambda x}}{(k-1)!} e^{-(t-x)\lambda}\, dx\\
&= \frac{e^{-t\lambda}\lambda^k}{(k-1)!}\int_0^t x^{k-1}\, dx\\
&= \frac{e^{-t\lambda}(\lambda t)^k}{k!}, \text{ for } k = 0, 1, \ldots,
\end{aligned}$$

recognizable as the Poisson distribution, which establishes the result.

Example 7.15 Marketing. A commonly used model in marketing is the so-called NBD model, introduced in late 1950s by Ehrenberg [1959] and still popular today. (NBD stands for negative binomial distribution, but we will not discuss the role of that distribution in our example.) Individual customers' purchasing occasions are modeled as a Poisson process. Different customers purchase at different rates λ and often the goal is to estimate such λ. Many empirical studies show a close fit to a Poisson process.

Suppose the time scale for such a study is in days. We assume that Chase's purchases form a Poisson process with parameter $\lambda = 0.5$. Consider the following questions of interest.

1. What is the average rate of purchases?
2. What is the probability that Chase will make at least three purchases within the next 7 days?

3. What is the probability that his 10th purchase will take place within the next 20 days?
4. What is the expected number of purchases Chase will make next month? Assume the month has 30 days.
5. What is the probability that Chase will not buy anything for the next 5 days given that he will not buy anything in the next 2 days?

Solutions:

1. Chase purchases at the rate of $\lambda = 0.5$ items per day.
2. The desired probability is $P(N_7 \geq 3)$, where N_7 has a Poisson distribution with parameter $\lambda t = (0.5)7 = 3.5$. This gives

$$P(N_7 \geq 3) = 1 - P(N_7 \leq 2) = 1 - P(N_7 = 0) - P(N_7 = 1) - P(N_7 = 2)$$
$$= 1 - e^{-7/2} - \frac{7e^{-7/2}}{2} - \frac{49e^{-7/2}}{8} = 0.679.$$

3. The time of Chase's 10th purchase is S_{10}. The desired probability is $P(S_{10} \leq 20)$. Since $S_{10} \sim \text{Gamma}(10, 0.5)$, we have

$$P(S_{10} \leq 20) = \int_0^{20} \frac{(1/2)^{10} x^9 e^{-x/2}}{\Gamma(10)}\, dx = 0.542,$$

   where the probability is evaluated using **R**.
4. The expectation is $E[N_{30}]$. The variable N_{30} has a Poisson distribution with parameter $(0.5)30 = 15$. Thus, the expected number of purchases next month is $E[N_{30}] = 15$.
5. If we consider the present time as $t = 0$, then the time of Chase's next purchase is E_1. The desired probability is $P(E_1 > 5 | E_1 > 2)$. By the memoryless property of the exponential distribution, this is equal to $P(E_1 > 3) = e^{-3\lambda} = e^{-1.5} = 0.223$. ■

Properties of the Poisson process. Suppose calls come in to a call center starting at 8 a.m. according to a Poisson process with parameter λ. Jack and Jill work at the center. Jill gets in to work at 8 a.m. Jack does not start until 10 a.m.

If we consider the process of phone call arrivals that Jack sees starting at 10 a.m., that arrival process is also a Poisson process with parameter λ. Because of the memoryless property of the exponential distribution, the process started at 10 a.m. is a "translated" version of the original process, shifted over 2 hours.

The number of calls that Jack sees in the hour between 11 a.m. and noon has the same distribution as the number of calls that Jill sees between 9 and 10 a.m. (We did not say that the number of calls between 11 a.m. and noon is *equal* to the number of calls between 9 and 10 a.m. Rather that their *distributions* are the same.)

This is the *stationary increments property* of a Poisson process and is a consequence of memorylessness. The distribution of the number of arrivals in an interval only depends on the *length* of the interval, not on the location of the interval.

In addition, whether or not calls come in between 9 and 10 a.m. has no influence on the distribution of calls between 11 a.m. and noon. The number of calls in each disjoint interval is independent of each other. This is called the *independent increments property* of a Poisson process.

PROPERTIES OF POISSON PROCESS

Let $(N_t)_{t\geq 0}$ be a Poisson process with parameter λ.

Independent increments

If $0 < q < r < s < t$, then $N_r - N_q$ and $N_t - N_s$ are independent random variables. That is,

$$P(N_r - N_q = m, N_t - N_s = n) = P(N_r - N_q = m)P(N_t - N_s = n),$$

for all $m, n = 0, 1, 2, \ldots.$

Stationary increments

For all $0 < s < t$, $N_{t+s} - N_t$ and N_s have the same distribution. That is,

$$P(N_{t+s} - N_t = k) = P(N_s = k) = \frac{e^{-\lambda s}(\lambda s)^k}{k!}, \text{ for } k = 0, 1, 2, \ldots.$$

Example 7.16 Starting at 6:00 a.m., birds perch on a power line according to a Poisson process with parameter $\lambda = 8$ birds per hour.

1. Find the probability that at most two birds arrive between 7 and 7:15 a.m. We start at time $t = 0$ and keep track of hours after 6:00 a.m. The desired probability is $P(N_{1.25} - N_1 \leq 2)$. By the stationary increments property, this is equal to the probability that at most two birds perch during the first 15 minutes (one-fourth of an hour). The desired probability is

$$\begin{aligned} P(N_{1.25} - N_1 \leq 2) &= P(N_{0.25} \leq 2) \\ &= e^{-8/4}[1 + 2 + 2^2/2] \\ &= 5e^{-2} = 0.677. \end{aligned}$$

2. Between 7 and 7:30 a.m, five birds arrive on the power line. What is the expectation and standard deviation for the number of birds that arrive from 3:30 to 4:00 p.m?

By the independent increments property, the number of birds that perch in the afternoon is independent of the number that perch in the morning, so the fact that 5 birds arrived in that half-hour in the morning is irrelevant. By the stationary increments property, the number of birds within each half-hour period has a Poisson distribution with parameter $\lambda/2 = 4$. Hence, the desired expectation is four birds and the standard deviation is two birds. ■

Example 7.17 The number of goals scored during a soccer match is modeled as a Poisson process with parameter λ. Suppose five goals are scored in the first t minutes. Find the probability that two goals are scored in the first s minutes, $s < t$.

The event that two goals are scored in the time interval $[0, s]$ and five goals are scored in $[0, t]$ is equal to the event that two goals are scored in $[0, s]$ and three goals are scored in $(s, t]$. Thus,

$$\begin{aligned} P(N_s = 2|N_t = 5) &= \frac{P(N_s = 2, N_t = 5)}{P(N_t = 5)} \\ &= \frac{P(N_s = 2, N_t - N_s = 3)}{P(N_t = 5)} \\ &= \frac{P(N_s = 2)P(N_t - N_s = 3)}{P(N_t = 5)} \\ &= \frac{P(N_s = 2)P(N_{t-s} = 3)}{P(N_t = 5)} \\ &= \frac{(e^{-\lambda s}(\lambda s)^2/2!)(e^{-\lambda(t-s)}(t-s)^3/3!)}{e^{-\lambda t}(\lambda t)^5/5!} \\ &= \binom{5}{2}\frac{s^2(t-s)^3}{t^5} = \binom{5}{2}\left(\frac{s}{t}\right)^2\left(1-\frac{s}{t}\right)^3, \end{aligned}$$

where the third equality is from independent increments and the fourth equality is from stationary increments.

The final expression is a binomial probability. Extrapolating from this example we obtain a general result: Let $(N_t)_{t\geq 0}$ be a Poisson process with parameter λ. For $0 < s < t$, the conditional distribution of N_s given $N_t = n$ is binomial with parameters n and $p = s/t$. ■

R: SIMULATING A POISSON PROCESS

One way to simulate a Poisson process is to first simulate exponential interarrival times and then construct the arrival sequence. To generate n arrivals of a Poisson process with parameter λ, type

```
> n <- 30 # choose a reasonable n
```

```
> lambda <- 1/2
> inter <- rexp(n, lambda)
> arrive <- cumsum(inter)
```

The `cumsum` command returns a cumulative sum from the interarrival times. The vector `arrive` contains the times of successive arrivals. The last arrival time is

```
> last <- tail(arrive, 1)
```

For $t \leq$ `last`, the number of arrivals up to time t, N_t, is simulated by

```
> sum(arrive <= t)
```

For example, consider a Poisson process with parameter $\lambda = 1/2$. We simulate the probability $P(N_5 - N_2 = 1)$:

```
> reps <- 10000
> simlist <- numeric(reps)
> for (i in 1:n) {
inter <- rexp(n,lambda)
arrive <- cumsum(inter)
nt <- sum( 2 <= arrive & arrive <= 5)
simlist[i] <- if (nt == 1) 1 else 0 }
> mean(simlist)
[1] 0.3335
```

As $P(N_5 - N_2 = 1) = P(N_3 = 1)$, here is the exact answer:

```
> dpois(1,3/2)
[1] 0.3346952
```

Example 7.18 Waiting time paradox. The following classic is from Feller [1968]. Buses arrive at a bus stop according to a Poisson process with parameter λ. The expected time between buses is $1/\lambda$. Suppose you get to the bus stop at noon. How long can you expect to wait for a bus? Here are two possible answers.

1. The memoryless property of the Poisson process means that the distribution of your waiting time should not depend on when you arrive at the bus stop. Thus, your expected waiting time is $1/\lambda$.
2. You arrive at some time between two consecutive buses. By symmetry your expected waiting time should be half the expected time between two consecutive buses, that is, $1/(2\lambda)$.

We explore the issue with a simulation, assuming that buses arrive on average every 20 minutes and $\lambda = 1/20$. Assume you arrive at the bus stop at time $t = 200$. The variable `wait` is your waiting time. The simulation repeats the experiment 10,000 times. See **bus.R**.

R: WAITING TIME PARADOX

```
> mytime <- 200  # arbitrary time you get to bus stop
> n <- 50
> lambda <- 1/20
> reps <- 10000
> simlist <- numeric(reps)
> for (i in 1:reps) {
arrivals <- cumsum(rexp(n, lambda))
wait <- arrivals[arrivals > mytime][1] - mytime
simlist[i] <- wait  }
> mean(simlist)
[1] 19.86749
```

It appears the average wait is around 20 minutes.

The simulation suggests that the expected waiting time is $20 = 1/\lambda$ minutes. So what is the fallacy of the second approach? Buses arrive on average, say, every 20 minutes. But the time between buses is random, buses may arrive one right after the other, or there may be a long time between consecutive buses. When you get to the bus stop, it is more likely that you get there during a longer interval between buses rather than a shorter interval.

This phenomenon is known as *length-biased* or *size-biased* sampling. If you reach into a bag containing pieces of string of different lengths and pick one "at random," you tend to pick a longer rather than shorter piece. Bus interarrival time intervals are analogous to pieces of string.

Here is another example of size-biased sampling. Suppose you want to estimate how much time people spend working out at the gym. If you go to the gym to ask people at random how long they work out, you are likely to get a biased estimate that is too big. You are more likely to sample someone who works out a lot. The people who go to the gym for brief periods of time are likely not to be there when you go to sample.

In the bus waiting problem, it turns out that the expected length of an interarrival time interval *which contains a fixed time t* is actually about $2/\lambda$, which is twice as large as the expected interarrival time. To verify, we modify the simulation code, keeping track of bus arrival times before and after the time we get to the bus stop.

R: WAITING TIME SIMULATION CONTINUED

```
> mytime <- 200
> for (i in 1:reps) {
arrivals <- cumsum(rexp(n, lambda))
indx <- which(arrivals > mytime)[1]
lengthintrvl <- arrivals[indx] - arrivals[indx-1]
simlist[i] <- lengthintrvl   }
> mean(simlist)
[1] 39.99006
```

Repeated twice more, we find averages of 40.89851 and 39.85409.

So the second answer was not entirely wrong. Symmetry says that your expected waiting time should be half the expected time between the two buses that arrive before and after the time you get there. This gives the expected waiting time for a bus as $(1/2)(2/\lambda) = 1/\lambda$.

Finally, there is nothing special about the time $t = 200$. The result holds for *any* fixed time. You can try out other times in the **R** supplement, available online. ■

7.4 BETA DISTRIBUTION

The beta distribution generalizes the uniform distribution. A beta random variable takes values between zero and one. The distribution is parametrized by two positive numbers a and b (see Figure 7.7 for examples).

A random variable X has a *beta distribution with parameters* $a > 0$ *and* $b > 0$ if the density function of X is

$$f(x) = \frac{\Gamma(a+b)}{\Gamma(a)\Gamma(b)} x^{a-1}(1-x)^{b-1}, \; 0 < x < 1,$$

and 0, otherwise. We write $X \sim \text{Beta}(a, b)$.

When $a = b = 1$, the beta distribution reduces to the uniform distribution on $(0, 1)$. Be aware that other sources may use α and β for the distribution parameters.

Using integration by parts, one can show that

$$\int_0^1 x^{a-1}(1-x)^{b-1}\, dx = \frac{\Gamma(a)\Gamma(b)}{\Gamma(a+b)}, \tag{7.3}$$

for $a, b > 0$, and thus the beta density integrates to one.

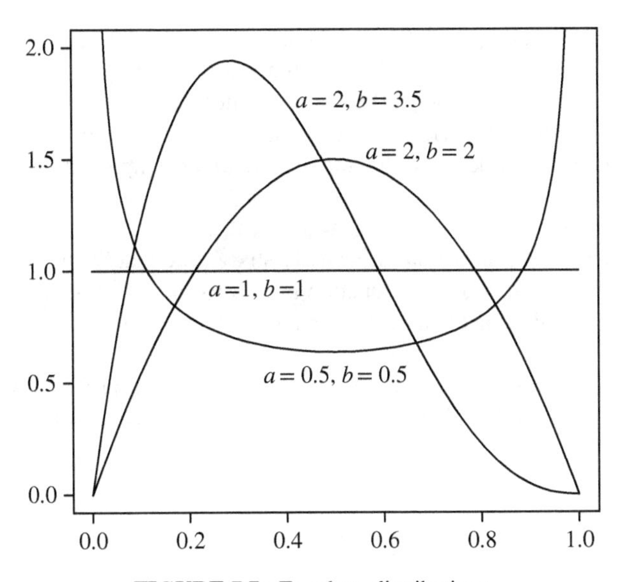

FIGURE 7.7: Four beta distributions.

To find the expectation of a beta distribution, use the methods of Section 7.2.1, making integrals look like probability distributions so that they integrate to one. Let $X \sim \text{Beta}(a, b)$. Then

$$E[X] = \int_0^1 (x)\frac{\Gamma(a+b)}{\Gamma(a)\Gamma(b)}x^{a-1}(1-x)^{b-1}\,dx = \frac{\Gamma(a+b)}{\Gamma(a)\Gamma(b)}\int_0^1 x^a(1-x)^{b-1}\,dx.$$

The integrand is proportional to a beta density with parameters $a + 1$ and b. This gives

$$\begin{aligned} E[X] &= \frac{\Gamma(a+b)}{\Gamma(a)\Gamma(b)}\int_0^1 x^a(1-x)^{b-1}\,dx \\ &= \frac{\Gamma(a+b)}{\Gamma(a)\Gamma(b)}\left(\frac{\Gamma(a+1)\Gamma(b)}{\Gamma(a+1+b)}\right)\int_0^1 \frac{\Gamma(a+1+b)}{\Gamma(a+1)\Gamma(b)}x^a(1-x)^{b-1}\,dx \\ &= \frac{\Gamma(a+b)}{\Gamma(a)\Gamma(b)}\left(\frac{\Gamma(a+1)\Gamma(b)}{\Gamma(a+1+b)}\right) = \frac{a}{a+b}, \end{aligned}$$

using the property $\Gamma(x+1) = x\Gamma(x)$ of the gamma function.

We leave it to the reader to derive the variance of the beta distribution which is

$$V[X] = \frac{ab}{(a+b)^2(a+b+1)},$$

using similar techniques in Exercise 7.38.

As probabilities and proportions are numbers between zero and one, the beta distribution is widely used to model an unknown probability or proportion p. For instance, Chia and Hutchinson [1991] model the fraction of daylight hours not receiving bright sunshine with a beta distribution. Gupta and Nadarajah [2004] is a book-length treatment of the many and diverse applications of the beta distribution.

Example 7.19 The proportion of his daily study time that Sean devotes to probability is modeled with a density function proportional to $x^3(1-x)$, for $0 < x < 1$. What is the probability that Sean will spend more than half his study time tomorrow on probability?

Let X be the proportion of Sean's study time spent on probability. The given density function is that of a beta distribution with parameters $a = 4$ and $b = 2$. The desired probability is

$$P(X > 1/2) = \int_{1/2}^{1} \frac{\Gamma(6)}{\Gamma(4)\Gamma(2)} x^3(1-x)\, dx = \int_{1/2}^{1} 20x^3(1-x)\, dx = \frac{13}{16}.$$

In **R**, type

```
> 1-pbeta(1/2,4,2)
[1] 0.8125
```

■

Example 7.20 Clarissa is responsible for monitoring a newly installed sensor network to find faulty sensors that need replaced. Suppose the proportion of faulty sensors is modeled by a Beta distribution with parameters $a = 1$ and $b = 200$. What is the expected proportion of faulty sensors?

We have already calculated the mean of the Beta distribution as $a/(a+b)$, so the expected proportion is $1/201$. However, if you did not remember this result, we want to emphasize the power of being able to recognize density functions for solving integrals. If the expectation was set up as defined, you would be solving

$$E(X) = \int_{0}^{1} \frac{\Gamma(201)}{\Gamma(1)\Gamma(200)} x^2(1-x)^{200}\, dx.$$

Note that while the constants do not match, the structure of $x^2(1-x)^{200}$ looks like another beta distribution. By multiplying by conveniently chosen constants, you can obtain an integrand that will integrate to one, with constants outside to evaluate, as shown.

$$E(X) = \frac{\Gamma(201)\Gamma(2)}{\Gamma(1)\Gamma(202)} \int_{0}^{1} \frac{\Gamma(202)}{\Gamma(2)\Gamma(200)} x^2(1-x)^{200}\, dx = \frac{\Gamma(201)\Gamma(2)}{\Gamma(1)\Gamma(202)} = \frac{1}{201}.$$

■

Beta distributions are often used to model continuous random variables that take values in some bounded interval. The beta distribution can be extended to the interval (s, t) by the scale change

$$Y = (t - s)X + s.$$

If X has a beta distribution on $(0, 1)$, then Y has an extended beta distribution on (s, t). This result is an exercise for the reader in Chapter 8.

7.5 PARETO DISTRIBUTION*

The normal, exponential, gamma, Poisson, and geometric distributions all have probability functions that contain an exponential factor of the form a^{-x}, where a is a positive constant. What this means is that the probability of a "large" outcome far from the mean in the "tail" of the distribution is essentially negligible. For instance, the probability that a normal distribution takes a value greater than six standard deviations from the mean is about two in a billion.

In many real-world settings, however, variables take values over a wide range covering several orders of magnitude. For instance, although the average size of US cities is about 6000, there are American cities whose populations are a thousand times that much.

Tens of thousands of earthquakes occur every day throughout the world, measuring less than 3.0 on the Richter magnitude scale. Yet each year there are about a hundred earthquakes of magnitude between 6.0 and 7.0, and one or two greater than 8.0.

The Pareto distribution is an example of a "power–law" distribution whose density function is proportional to the power function x^{-a}. Such distributions have "heavy tails," which give nonnegligible probability to extreme values. The Pareto distribution has been used to model population, magnitude of earthquakes, size of meteorites, maximum one-day rainfalls, price returns of stocks, Internet traffic, and wealth in America.

The distribution was discovered by the Italian economist Vilfredo Pareto who used it to model income distribution in Italy at the beginning of the twentieth century.

PARETO DISTRIBUTION

A random variable X has a *Pareto distribution with parameters $m > 0$ and $a > 0$* if the density function of X is

$$f(x) = a\frac{m^a}{x^{a+1}}, \text{ for } x > m.$$

The distribution takes values above a minimum positive number m. We write $X \sim \text{Pareto}(m, a)$.

TABLE 7.1. Comparison of tail probabilities for normal and Pareto distributions

	$P(\lvert X-\mu\rvert > k\sigma)$	
k	Normal(3/2, 3/4)	Pareto(1, 3)
1	3.173×10^{-1}	7.549×10^{-2}
2	4.550×10^{-2}	2.962×10^{-2}
3	2.700×10^{-3}	1.453×10^{-2}
4	6.334×10^{-5}	8.175×10^{-3}
5	5.733×10^{-7}	5.046×10^{-3}
6	1.973×10^{-9}	3.331×10^{-3}
7	2.560×10^{-12}	2.312×10^{-3}

Example 7.21 Consider a Pareto distribution with parameters $m = 1$ and $a = 3$. The expectation and variance of this distribution are $\mu = 3/2$ and $\sigma^2 = 3/4$. In Table 7.1, we compare the tail probability $P(|X - \mu| > k\sigma)$ of falling more than k standard deviations from the mean for this distribution and for a normal distribution with the same mean and variance.

The data highlight the heavy tail property of the Pareto distribution. In a billion observations of a normally distributed random variable, you would expect about $1.973 \times 10^{-9} \times 10^9 = 1.973 \approx 2$ observations greater than six standard deviations from the mean. But for a Pareto distribution with the same mean and variance, you would expect to see about $3.31 \times 10^{-3} \times 10^9 = 3{,}310{,}000$ such outcomes. ■

The Pareto distribution is characterized by a special *scale-invariance* property. Intuitively this means that the shape of the distribution does not change by changing the scale of measurements. For instance, it is reasonable to model income with a scale-invariant distribution because the distribution of income does not change if units are measured in dollars, or converted to euros or to yen. You can see this phenomenon in Figure 7.8. The plots show the density curve for a Pareto distribution with $m = 1$ and $a = 1.16$ over four different intervals of the form $c < x < 5c$. The shape of the curve does not change for any value of c.

SCALE-INVARIANCE

A probability distribution with density f is *scale-invariant* if for all positive constants c,

$$f(cx) = g(c)f(x),$$

where g is some function that does not depend on x.

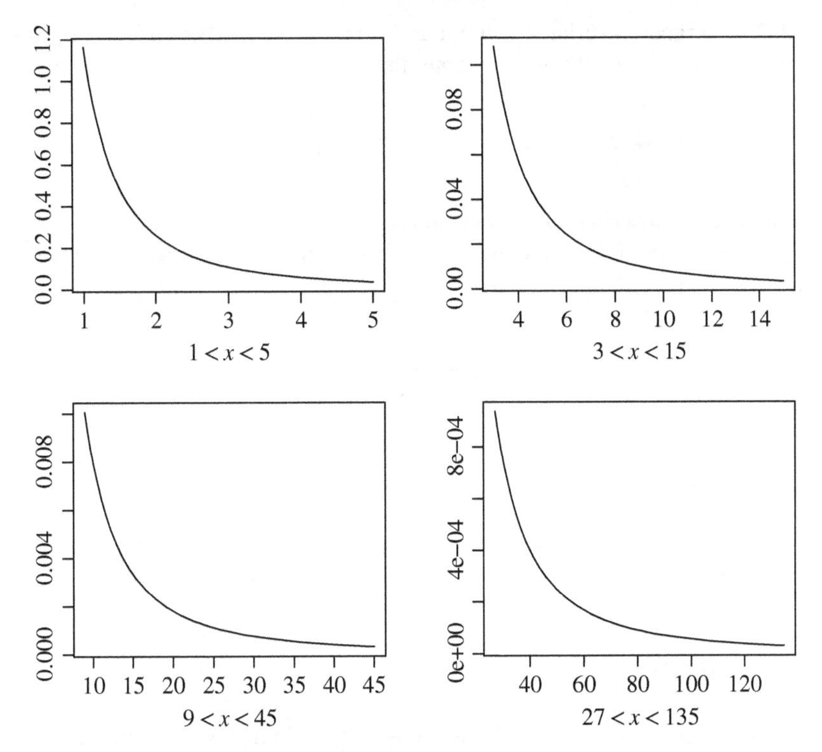

FIGURE 7.8: Scale-invariance of the Pareto distribution. The density curve is for a Pareto distribution with $m = 1$ and $a = 1.16$. The curve is shown on intervals of the form $c < x < 5c$, for $c = 1, 3, 9, 27$.

For the Pareto distribution,

$$f(cx) = a\frac{m^a}{(cx)^{a+1}} = \left(\frac{1}{c^{a+1}}\right) a\frac{m^a}{x^{a+1}} = \frac{1}{c^{a+1}} f(x).$$

Hence, the distribution is scale-invariant.

Newman [2005] is an excellent source for background on the Pareto distribution. He shows that a power–law distribution is the only probability distribution that is scale-invariant.

The 80-20 rule. Ever noticed that it seems to take 20% of your time to do 80% of your homework? Or you wear 20% of your clothes 80% of the time? The numbers 80 and 20 seem to be empirically verified by many real-life phenomena. Pareto observed that 80% of the land in Italy was owned by 20% of the people. The so-called 80-20 rule is characteristic of the Pareto distribution.

Suppose income is modeled with a Pareto distribution. Further suppose that income follows an 80-20 rule with 80% of the wealth owned by 20% of the people. How should the parameter a be determined?

Let X be a random variable with the given Pareto distribution. Then the proportion of the population with income larger than x is

$$P(X > x) = \int_x^\infty a\frac{m^a}{t^{a+1}}\,dt = \left(\frac{m}{x}\right)^a, \quad \text{for } x > m.$$

Solving $0.20 = (m/x)^a$ gives $x = m5^{1/a}$.

The proportion of total income in the hands of those people is estimated in economics by the *Lorenz function*

$$L(x) = \frac{\int_x^\infty tf(t)\,dt}{\int_m^\infty tf(t)\,dt} = \left(\frac{m}{x}\right)^{a-1}.$$

This gives

$$L(m5^{1/a}) = 5^{(1-a)/a}.$$

If 80% of the income is owned by 20% of the people, then $0.80 = 5^{(1-a)/a}$, which gives $\log 4/5 = (1-a)/a \log 5$, and thus $a = \log 5/\log 4 = 1.16$.

7.6 SUMMARY

Several families of continuous distributions and random processes are introduced in this chapter. The normal distribution is perhaps the most important distribution in statistics. Many key properties of the distribution are presented, such as (i) linear functions of normal random variables are normal; (ii) sums of independent normal variables are normally distributed; and (iii) the so-called "68-95-99.7 rule," which quantifies the probability that a normal variable is within one, two, and three standard deviations from the mean. The normal approximation of the binomial distribution is discussed, as well as the continuity correction used when a continuous density curve is used to approximate a discrete probability. Quantiles are introduced, as well as general results for sums and averages of i.i.d. random variables.

The gamma distribution arises as the sum of i.i.d. exponential random variables. This flexible two-parameter family of distributions generalizes the exponential distribution. The Poisson process, a continuous time stochastic (random) process, is presented. The process sees the interrelationship between gamma, exponential, and Poisson random variables. The process arises as a model for the "arrival" of "events" (e.g., phone calls at a call center, accidents on the highway, and goals at a soccer game) in continuous time. The key properties of a Poisson process—stationary and independent increments—are explained.

Also introduced in this chapter is the beta distribution, a two-parameter family of distributions on (0, 1) that generalizes the uniform distribution. Finally, the Pareto distribution, part of a larger class of scale-invariant, power–law distributions, is presented. Connections are drawn with the so-called 80-20 rule.

- **Normal distribution:** A random variable X is normally distributed with parameters μ and σ^2 if the density of X is

$$f(x) = \frac{1}{\sigma\sqrt{2\pi}} e^{-(x-\mu)^2/2\sigma^2}, \quad \text{for } -\infty < x < \infty.$$

  Parameters μ and σ^2 are, respectively, the mean and variance of the distribution.
- **Normal setting:** The bell-shaped, symmetric distribution arises as a simple model for many complex phenomena. It has been used to model measurement error, standardized test scores, and human characteristics like height and weight.
- **68-95-99.7 rule:** For any normal distribution, the probability that an outcome is within one, two, and three standard deviations from the mean is, respectively, about 68, 95, and 99.7%.
- **Standard normal:** A normal random variable with mean $\mu = 0$ and variance $\sigma^2 = 1$ is said to have a standard normal distribution.
- **Linear function:** If $X \sim \text{Norm}(\mu, \sigma^2)$, then for constants $a \neq 0$ and b, the random variable $aX + b$ is normally distributed with mean $a\mu + b$ and variance $a^2\sigma^2$. It follows that $(X - \mu)/\sigma \sim \text{Norm}(0, 1)$.
- **Normal approximation of the binomial distribution:** If $X \sim \text{Binom}(n, p)$, and n is large, then X has an approximate normal distribution with mean $E[X] = np$ and variance $V[X] = np(1 - p)$. In particular, for $a < b$,

$$\lim_{n\to\infty} P\left(a \leq \frac{X - np}{\sqrt{np(1-p)}} \leq b \right) = \frac{1}{\sqrt{2\pi}} \int_a^b e^{-x^2/2}\, dx.$$

- **Continuity correction:** Suppose X is a discrete random variable and we wish to approximate $P(a \leq X \leq b)$ using a continuous distribution whose density function is f. Then take

$$P(a \leq X \leq b) \approx \int_{a-1/2}^{b+1/2} f(x)\, dx.$$

- **Quantiles:** If X is a continuous random variable and $0 < p < 1$, then the pth quantile is the number q that satisfies $P(X \leq q) = p/100$.
- **Averages of i.i.d. random variables:** Suppose $X_1, \ldots, X_n$ are i.i.d. with mean μ and variance σ^2. Let $S_n = X_1 + \cdots + X_n$. Then S_n/n is the average and
  1. $E[S_n/n] = \mu$.
  2. $V[S_n/n] = \sigma^2/n$.
  3. If the X_i's are normally distributed, then so is S_n/n. That is, $S_n/n \sim \text{Norm}(\mu, \sigma^2/n)$.

- **Gamma distribution:** A random variable X has a gamma distribution with parameters a and λ if the density of X is

$$f(x) = \frac{\lambda^a x^{a-1} e^{-\lambda x}}{\Gamma(a)}, \quad \text{for } x > 0,$$

where $\Gamma(a) = \int_0^\infty t^{a-1} e^{-t}\, dt$ is the gamma function.

- **Gamma setting:** The distribution is a flexible, two-parameter distribution defined on the positive reals. A sum of n i.i.d. exponential random variables with parameter λ has a gamma distribution with parameters n and λ. The gamma distribution generalizes the exponential distribution (with $a = 1$).
- **Gamma distribution properties:** If $X \sim \text{Gamma}(a, \lambda)$, then
  1. $E[X] = a/\lambda$.
  2. $V[X] = a/\lambda^2$.
- **Poisson process:** This is a model for the distribution of "events" or "arrivals" in space and time. Times between arrivals are modeled as i.i.d. exponential random variables with parameter λ. Let N_t be the number of arrivals up to time t, defined for all nonnegative t. Then $N_t \sim \text{Pois}(\lambda t)$. The collection of random variables $(N_t)_{t\geq 0}$ is a Poisson process with a parameter λ.
- **Properties of Poisson process:**
  1. **Stationary increments:** For $0 < s < t$, the distribution of $N_{t+s} - N_t$, the number of arrivals between times t and $t + s$, only depends on the length of the interval $(t + s) - t = s$. That is, the distribution of $N_{t+s} - N_t$ is the same as the distribution of N_s. Hence,

$$P(N_{t+s} - N_t = k) = \frac{e^{-\lambda s}(\lambda s)^k}{k!}, \quad \text{for } k = 0, 1, \ldots$$

  2. **Independent increments:** For $0 < a < b < c < d$, $N_b - N_a$ and $N_d - N_c$ are independent random variables. The number of arrivals in disjoint intervals is an independent random variable.
- **Beta distribution:** A random variable X has a beta distribution with parameters $a > 0$ and $b > 0$ if the density of X is

$$f(x) = \frac{\Gamma(a+b)}{\Gamma(a)\Gamma(b)} x^{a-1}(1-x)^{b-1}, \quad \text{for } 0 < x < 1.$$

- **Beta setting:** The distribution is a flexible two-parameter family of distributions on $(0, 1)$. It is often used to model an unknown proportion or probability. The distribution generalizes the uniform distribution on $(0, 1)$ (with $a = b = 1$).

- **Beta distribution properties:** If $X \sim \text{Beta}(a, b)$, then
  1. $E[X] = a/(a+b)$.
  2. $V[X] = ab/((a+b)^2(a+b+1))$.
- **Pareto distribution:** A random variable X has a Pareto distribution with parameters $m > 0$ and $a > 0$ if the density of X is

$$f(x) = a\frac{m^a}{x^{a+1}}, \quad \text{for } x > m.$$

- **Pareto setting:** The distribution is used to model heavily skewed data such as population and income. It is an example of a "power–law" distribution with the so-called heavy tails.
- **Scale-invariance:** A distribution with density function f is scale-invariant if for all constants c, $f(cx) = g(c)f(x)$, where g is some function that does not depend on x. The Pareto distribution is scale-invariant.

EXERCISES

Normal Distribution

7.1 Let $X \sim \text{Norm}(-4, 25)$. Approximate the following probabilities without using software:

(a) $P(X > 6)$.

(b) $P(-9 < X < 1)$.

(c) $P(\sqrt{X} > 1)$.

7.2 Let $X \sim \text{Norm}(4, 4)$. Find the following probabilities using R:

(a) $P(|X| < 2)$.

(b) $P(e^X < 1)$.

(c) $P(X^2 > 3)$.

7.3 Babies' birth weights are normally distributed with mean 120 ounces and standard deviation 20 ounces. *Low birth weight* is an important indicator of a newborn baby's chances of survival. One definition of low birth weight is that it is the fifth percentile of the weight distribution.

(a) Babies who weigh less than what amount would be considered low birth weight?

(b) *Very low birth weight* is used to described babies who are born weighing less than 52 ounces. Find the probability that a baby is born with very low birth weight.

(c) Given that a baby is born with low birth weight, what is the probability that they have very low birth weight?

7.4 An elevator's weight capacity is 1000 pounds. Three men and three women are riding the elevator. Adult male weight is normally distributed with mean 172 pounds and standard deviation 29 pounds. Adult female weight is normally distributed with mean 143 pounds and standard deviation 29 pounds. Find the probability that the passengers' total weight exceeds the elevator's capacity.

7.5 The rates of return of ten stocks are normally distributed with mean $\mu = 2$ and standard deviation $\sigma = 4$ (units are percentages). Rates of return are independent from stock to stock. Each stock sells for \$10 a share. Gabrielle, Hasan, and Ian have \$100 each to spend on their stock portfolio. Gabrielle buys one share of each stock. Hasan buys two shares from five different companies. Ian buys ten shares from one company. For each person, find the probability that their average rate of return is positive.

7.6 The two main standardized tests in the United States for high school students are the ACT and SAT. ACT scores are normally distributed with mean 18 and standard deviation 6. SAT scores are normally distributed with mean 500 and standard deviation 100. Suppose Vanessa takes an SAT exam and scores 680. Wyatt plans to take the ACT. What score does Wyatt need to get so that his standardized score is the same as Vanessa's standardized score? What percentile do they each score at?

7.7 In 2011, the SAT exam was a composite of three exams—in reading, math, and writing. For 2011 college-bound high school seniors, Table 7.2 gives the mean and standard deviation of the three exams. The data are from the College Board [2011].

Let R, M, and W denote the reading, math, and writing scores. The College Board also estimates the following correlations between the three exams:

$$\text{Corr}(R, M) = \text{Corr}(W, M) = 0.72, \ \text{Corr}(R, W) = 0.84.$$

(a) Find the mean and standard deviation of the total composite SAT score $T = R + M + W$.

(b) Assume total composite score is normally distributed. Find the 80th and 90th percentiles.

(c) Chana took the SAT in 2011. Find the probability that the average of her three exam scores is greater than 600.

TABLE 7.2. SAT statistics for 2011 college-bound seniors

	Reading	Math	Writing
Mean	497	514	489
SD	114	117	113

7.8 Let $X \sim \text{Norm}(\mu, \sigma^2)$. Show that $E[X] = \mu$.

7.9 Find all the inflection points of a normal density curve. Show how this information can be used to draw a normal curve given values of μ and σ.

7.10 Suppose $X_i \sim \text{Norm}(i, i)$ for $i = 1, 2, 3, 4$. Further assume all the X_i's are independent. Find $P(X_1 + X_3 < X_2 + X_4)$.

7.11 Let $X_1, \ldots, X_n$ be an i.i.d. sequence of normal random variables with mean μ and variance σ^2. Let $S_n = X_1 + \cdots + X_n$.
(a) Suppose $\mu = 5$ and $\sigma^2 = 1$. Find $P(|X_1 - \mu| \leq \sigma)$.
(b) Suppose $\mu = 5$, $\sigma^2 = 1$, and $n = 9$. Find $P(|S_n/n - \mu| \leq \sigma)$.

7.12 Let $X \sim \text{Norm}(\mu, \sigma^2)$. Suppose $a \neq 0$ and b are constants. Show that $Y = aX + b$ is normally distributed using mgfs.

7.13 Let $X_1, \ldots, X_n$ be i.i.d. normal random variables with mean μ and standard deviation σ. Recall that $\overline{X} = (X_1 + \cdots + X_n)/n$ is the sample average. Let

$$S = \frac{1}{n-1} \sum_{i=1}^{n} (X_i - \overline{X})^2.$$

Show that $E[S] = \sigma^2$. In statistics, S is called the *sample variance*.

7.14 Let $Z \sim \text{Norm}(0, 1)$. Find $E[|Z|]$.

7.15 If Z has a standard normal distribution, find the distribution of Z^2 using mgfs.

7.16 If X and Y are independent standard normal random variables, show that $X^2 + Y^2$ has an exponential distribution using mgfs.

7.17 If X, Y, Z are independent standard normal random variables, find the distribution of $X^2 + Y^2 + Z^2$ using mgfs.

7.18 Validate the normal approximation to the binomial requirement that np and $n(1-p)$ must be greater than or equal to 10 for the approximation to work well. Hint: think about where the majority of values for both distributions are.

7.19 Let X be a random variable with mean μ and standard deviation σ. The *skewness* of X is defined as

$$\text{skew}(X) = \frac{E[(X-\mu)^3]}{\sigma^3}.$$

Skewness is a measure of the asymmetry of a distribution. Distributions that are symmetric about μ have skewness equal to 0. Distributions that are right skewed have positive skewness. Left-skewed distributions have negative skewness.

(a) Show that

$$\text{skew}(X) = \frac{E[X^3] - 3\mu\sigma^2 - \mu^3}{\sigma^3}.$$

(b) Use the mgfs to find the skewness of the exponential distribution with parameter λ.

(c) Use the mgfs to find the skewness of a normal distribution.

7.20 Let X be a random variable with mean μ and standard deviation σ. The *kurtosis* of X is defined as

$$\text{kurtosis}(X) = \frac{E[(X-\mu)^4]}{\sigma^4}.$$

The kurtosis is a measure of the "peakedness" of a probability distribution. Use the mgfs and find the kurtosis of a standard normal distribution.

Gamma Distribution, Poisson Process

7.21 Your friend missed probability class today. Explain to your friend, in simple language, what random variables are involved in a Poisson process.

7.22 Using integration by parts, show that the gamma function satisfies

$$\Gamma(a) = (a-1)\Gamma(a-1).$$

7.23 Let $X \sim \text{Gamma}(a, \lambda)$. Find $V[X]$ using the methods of Section 7.2.1.

7.24 May is tornado season in Oklahoma. According to the National Weather Service, the rate of tornadoes in Oklahoma in May is 21.7 per month, based on data from 1950 to the present. Assume tornadoes follow a Poisson process.

(a) What is the probability that next May there will be more than 25 tornadoes?

(b) What is the probability that the 10th tornado in May occurs before May 15?

(c) What is the expectation and standard deviation for the number of tornadoes during 7 days in May?

(d) What is the expected number of days between tornadoes in May? What is the standard deviation?

7.25 Starting at 9 a.m., students arrive to class according to a Poisson process with parameter $\lambda = 2$ (units are minutes). Class begins at 9:15 a.m. There are 30 students.

(a) What is the expectation and variance of the number of students in class by 9:15 a.m.?

(b) Find the probability there will be at least 10 students in class by 9:05 a.m.

(c) Find the probability that the last student who arrives is late.

(d) Suppose exactly six students are late. Find the probability that exactly 15 students arrived by 9:10 a.m.

(e) What is the expected time of arrival of the seventh student who gets to class?

7.26 Solve the following integrals without calculus by recognizing the integrand as related to a known probability distribution and making the necessary substitution(s).

(a)

$$\int_{-\infty}^{\infty} e^{-(x+1)^2/18}\, dx,$$

(b)

$$\int_{-\infty}^{\infty} xe^{-(x-1)^2/2}\, dx,$$

(c)

$$\int_{0}^{\infty} x^4 e^{-2x}\, dx,$$

(d)

$$\int_{0}^{\infty} x^s e^{-tx}\, dx,$$

for positive integer s and positive real t.

7.27 Suppose $(N_t)_{t\geq 0}$ is a Poisson process with parameter $\lambda = 2$.

(a) Find $P(N_4 = 7)$.

(b) Find $P(N_4 = 7, N_5 = 9)$.

(c) Find $P(N_4 = 7|N_5 = 9)$.

7.28 Suppose $(N_t)_{t\geq 0}$ is a Poisson process with parameter $\lambda = 1$.

(a) Find $P(N_3 = 4)$.

(b) Find $P(N_3 = 4, N_5 = 8)$.

(c) Find $P(N_3 = 4|N_5 = 8)$.

(d) Find $P(N_3 = 4, N_5 = 8, N_6 = 10)$.

7.29 Suppose $(N_t)_{t\geq 0}$ is a Poisson process with parameter λ. Find $P(N_s = k|N_t = n)$ when $s > t$.

7.30 The number of accidents on a highway is modeled as a Poisson process with parameter λ. Suppose exactly one accident has occurred by time t. If $0 < s < t$, find the probability that accident occurred by time s.

7.31 Suppose $(N_t)_{t\geq 0}$ is a Poisson process with parameter λ. For $s < t$, find $\text{Cov}(N_s, N_t)$.

7.32 If X has a gamma distribution and c is a positive constant, show that cX has a gamma distribution using mgfs. Find the parameters.

7.33 Let $X \sim \text{Gamma}(a, \lambda)$. Let $Y \sim \text{Gamma}(b, \lambda)$. If X and Y are independent show that $X + Y$ has a gamma distribution using mgfs.

Beta Distribution

7.34 A density function is proportional to $f(x) = x^5(1 - x)$, for $0 < x < 1$.

(a) Find the constant of proportionality.

(b) Find the mean and variance of the distribution.

(c) Use **R** to find $P(X < 0.4)$.

7.35 A density function is proportional to $f(x) = x^3(1 - x)^7$, for $0 < x < 1$.

(a) Find the constant of proportionality.

(b) Find the mean and variance of the distribution.

(c) Use **R** to find $P(X > 0.5)$.

7.36 A random variable X has density function proportional to

$$f(x) = \frac{1}{\sqrt{x(1-x)}}, \qquad \text{for } 0 < x < 1.$$

Use **R** to find $P(1/8 < X < 1/4)$.

7.37 Solve the following integrals without calculus by recognizing the integrand as related to a known probability distribution and making the necessary substitution(s).

(a)

$$\int_0^1 \frac{1}{16} x^6 (1-x)^3 \, dx.$$

(b)

$$\int_0^1 3(1-x)^7 \, dx.$$

(c)

$$\int_0^1 12x^4 (1-x)^5 \, dx.$$

7.38 Find the variance of a beta distribution with parameters a and b.

7.39 The proportion of time that an office machine is in use during a usual 40-hour work week, X, has a pdf which is proportional to $x^2(1 - x)$, for $0 < x < 1$.

(a) Find the constant of proportionality.

(b) How many hours during the week should the office manager expect the machine to be in use?

(c) A report from the manufacturer indicates that the machine may need early repairs if used more than 35 hours per week. Would you worry about needing early repairs as the office manager? Explain, using probability to support your response.

7.40 Let $X \sim \text{Beta}(a, b)$, for $a > 1$. Find $E[1/X]$.

Pareto, Scale-invariant Distribution

7.41 Let $X \sim \text{Pareto}(m, a)$.

(a) For what values of a does the mean exist? For such a, find $E[X]$.

(b) For what values of a does the variance exist? For such a, find $V[X]$.

7.42 In a population, suppose personal income above $15,000 has a Pareto distribution with $a = 1.8$ (units are $10,000). Find the probability that a randomly chosen individual has income greater than $60,000.

7.43 In Newman [2005], the population of US cities larger than 40,000 is modeled with a Pareto distribution with $a = 2.30$. Find the probability that a random city's population is greater than $k = 3, 4, 5$, and 6 standard deviations above the mean.

7.44 Zipf's law is a discrete distribution related to the Pareto distribution. If X has a Zipf's law distribution with parameters $s > 0$ and $n \in \{1, 2, \ldots\}$, then

$$P(X = k) = \frac{1/k^s}{\sum_{i=1}^{n}(1/i^s)}, \quad \text{for } k = 1, \ldots, n.$$

The distribution is used to model frequencies of words in languages.

(a) Show that Zipf's law is scale-invariant.

(b) Assume there are a million words in the English language and word frequencies follow Zipf's law with $s = 1$. The three most frequently occurring words are, in order, "the," "of," and "and." What does Zipf's law predict for their relative frequencies?

7.45 The "99-10" rule on the Internet says that 99% of the content generated in Internet chat rooms is created by 10% of the users. If the amount of chat room content has a Pareto distribution, find the value of the parameter a.

Simulation and R

7.46 Conduct a simulation study to illustrate that sums of independent normal random variables are normal. In particular, let $X_1, \ldots, X_{30}$ be normally distributed with $\mu = 1$ and $\sigma^2 = 4$. Simulate $X_1 + \cdots + X_{30}$. Plot a histogram estimate of the distribution of the sum together with the exact density function of a normally distributed random variable with mean 30 and variance 120.

7.47 Suppose phone calls arrive at a Help Desk according to a Poisson process with parameter $\lambda = 10$. Show how to simulate the arrival of phone calls.

7.48 Perform a simulation to show that the sum of four i.i.d Exponential random variables has a Gamma distribution.

Chapter Review

Chapter review exercises are available through the text website. The URL is `www.wiley.com/go/wagaman/probability2e`.

8

DENSITIES OF FUNCTIONS OF RANDOM VARIABLES

Fortes fortuna iuvat. (*Fortune favors the brave.*)

Pliny the Elder, attributed by his nephew, Pliny the Younger

Learning Outcomes

1. Learn how to find densities for functions of random variables, including bivariate functions.
2. Solve problems involving minimums and maximums.
3. Apply the convolution formula to appropriate problems.
4. Use geometric thinking to solve problems.
5. (C) Simulate random variables using the inverse transform method.

Introduction. The diameter of a subatomic particle is modeled as a random variable. The volume of the particle is a function of that random variable. The rates of return for several stocks in a portfolio are modeled as random variables. The portfolio's maximum rate of return is a function of those random variables. In previous chapters, we have explored finding expectations and variances of functions of random variables, without finding the distribution for the new variable directly. We have also seen a few relationships between random variables illustrated via moment-generating functions (mgfs). There are other ways to examine these relationships, working with density functions directly. In this chapter, we explore ways of finding densities for functions of random variables via several methods and for several common functions such as minimums and maximums. Geometric

Probability: With Applications and R, Second Edition. Amy S. Wagaman and Robert P. Dobrow.

Companion Website: www.wiley.com/go/wagaman/probability2e

thinking is also introduced as a problem-solving tool. This material may prove challenging but is very powerful in terms of what you can accomplish working with it.

8.1 DENSITIES VIA CDFS

In this section, you will learn how to find the distribution of a function of a random variable. We start with an example which captures the essence of the general approach via cumulative distribution functions (cdfs).

Example 8.1 The radius R of a circle is uniformly distributed on $(0, 2)$. Let A be the area of the circle. Find the probability density function of A.

The area of the circle is a function of the radius. Write $A = g(R)$, where $g(r) = \pi r^2$. Our approach to finding the density of A will be to (i) find the cdf of A in terms of the cdf of R, and (ii) take derivatives to find the desired density function. This is called the cdf approach to finding the density.

As R takes values between 0 and 2, A takes values between 0 and 4π. For $0 < a < 4\pi$,

$$F_A(a) = P(A \leq a) = P(\pi R^2 \leq a) = P(R \leq \sqrt{a/\pi}) = F_R(\sqrt{a/\pi}).$$

Now take derivatives with respect to a. The left-hand side gives $f_A(a)$. The right-hand side, using the chain rule, gives

$$\begin{aligned}\frac{d}{da}F_R(\sqrt{a/\pi}) &= f_R(\sqrt{a/\pi})\left(\frac{d}{da}\sqrt{a/\pi}\right) = f_R(\sqrt{a/\pi})\frac{1}{2\sqrt{a\pi}}\\ &= \left(\frac{1}{2}\right)\frac{1}{2\sqrt{a\pi}} = \frac{1}{4\sqrt{a\pi}}.\end{aligned}$$

That is,

$$f_A(a) = \frac{1}{4\sqrt{a\pi}}, \quad \text{for } 0 < a < 4\pi,$$

and 0, otherwise. Note that we could have solved the problem using either our knowledge of the cdf of R (plugging it in before taking derivatives) or the density (as shown). ■

R: COMPARING THE EXACT DISTRIBUTION WITH A SIMULATION

We simulate $A = \pi R^2$. The theoretical density curve $f_A(a)$ is superimposed on the simulated histogram in Figure 8.1.

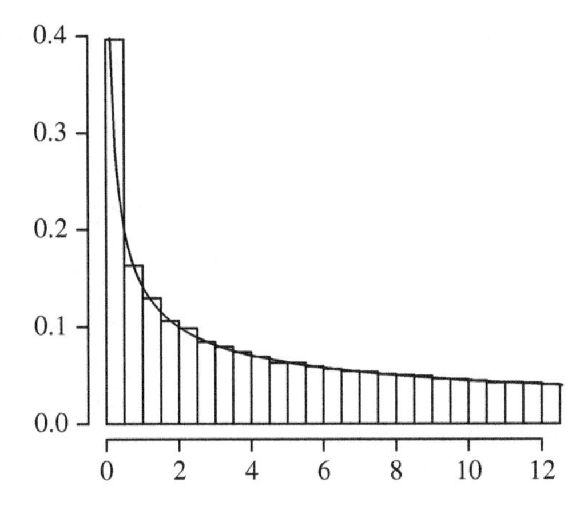

FIGURE 8.1: Simulated distribution of $A = \pi R^2$, where $R \sim \text{Unif}(0, 2)$. The curve is the density function $f(a) = 1/(4\sqrt{a\pi})$.

```
> n <- 10000
> simlist <- pi*runif(n, 0, 2)^2
> hist(simlist,prob=T)
> curve(1/(4*sqrt(x*pi)), 0, 4*pi, add = TRUE)
```

This example illustrates the general approach for finding the density of a function of a random variable. If $Y = g(X)$, start with the cdf of Y. Express the cdf of Y in terms of the cdf of X. Then take derivatives. Use what you know about X to simplify.

HOW TO FIND THE DENSITY OF $Y = g(X)$

1. Determine the possible values of Y based on the values of X and the function g.
2. Begin with the cdf $F_Y(y) = P(Y \leq y) = P(g(X) \leq y)$. Express the cdf in terms of the original random variable X.
3. From $P(g(X) \leq y)$ obtain an expression of the form $P(X \leq \cdots)$. The right-hand side of the expression will be a function of y. If g is invertible then $P(g(X) \leq y) = P(X \leq g^{-1}(y))$.
4. Differentiate with respect to y to obtain the density $f_Y(y)$.

Example 8.2 Suppose $X \sim \text{Unif}(0, 1)$, and $Y = \ln(X + 1)$. Find the density function of Y.

First, consider what values Y can take based on what values X can take. We see that $0 < y < \ln 2$. We also know the probability distribution function (pdf) of X is $f(x) = 1$, and the cdf is $F(x) = x$, for $0 < x < 1$. We now consider the cdf of Y and following the procedure outlined above, obtain

$$\begin{aligned} F_Y(y) &= P(Y \le y) = P(\log(X+1) \le y) \\ &= P(X + 1 \le e^y) = P(X \le e^y - 1) \\ &= e^y - 1. \end{aligned}$$

We differentiate with respect to Y to obtain the density function. Remember chain rule as applicable. Thus, the pdf of Y is

$$f_Y(y) = \frac{d}{dy}(e^y - 1) = e^y, \quad \text{for } 0 < y < \ln 2,$$

and 0, otherwise. ■

Example 8.3 Linear function of a uniform random variable. Suppose $X \sim \text{Unif}(0, 1)$. For $a < b$, let $Y = (b - a)X + a$. Find the density function of Y.

Because X takes values between 0 and 1, $Y = (b - a)X + a$ takes values between a and b. For $a < y < b$,

$$\begin{aligned} F_Y(y) &= P(Y \le y) = P((b-a)X + a \le y) \\ &= P\left(X \le \frac{y-a}{b-a}\right) = F_X\left(\frac{y-a}{b-a}\right) = \frac{y-a}{b-a}. \end{aligned}$$

To find the density of Y, differentiate to get

$$f_Y(y) = \frac{1}{b-a}, \quad \text{for } a < y < b.$$

The distribution of Y is uniform on (a, b).

Observe that we could have made this conclusion earlier by noting that the cdf of Y is the cdf of a uniform distribution on (a, b). ■

Example 8.4 Linear function of a random variable. Suppose X is a random variable with density f_X. Let $Y = aX + b$, where $a \neq 0$ and b are constants. Find the density of Y.

Suppose $a > 0$. The cdf of Y is

$$\begin{aligned} F_Y(y) &= P(Y \le y) = P(aX + b \le y) \\ &= P\left(X \le \frac{y-b}{a}\right) = F_X\left(\frac{y-b}{a}\right). \end{aligned}$$

Differentiating with respect to y gives

$$f_Y(y) = \frac{1}{a} f_X\left(\frac{y-b}{a}\right).$$

If $a < 0$, we have

$$F_y(y) = P\left(X \geq \frac{y-b}{a}\right) = 1 - F_X\left(\frac{y-b}{a}\right).$$

Differentiating gives

$$f_Y(y) = -\frac{1}{a} f_X\left(\frac{y-b}{a}\right).$$

In either case we have

$$f_Y(y) = \frac{1}{|a|} f_X\left(\frac{y-b}{a}\right).$$

Observe the relationship between the density functions of X and Y. The density of Y is obtained by translating the density of X b units to the right and stretching by a factor of a. Then compress by a factor of $|a|$ vertically. ■

Example 8.5 We illustrate the last result with a simple example. Suppose $X \sim \text{Exp}(4)$, and $Y = 6X + 11$. Find the density function of Y.

From the previous result, we have a linear function where $a = 6$ and $b = 11$. We also know that $f_X(x) = 4e^{-4x}$, for $x > 0$, and 0, otherwise. Applying the previous result, we have

$$\begin{aligned} f_Y(y) &= \frac{1}{|a|} f_X\left(\frac{y-b}{a}\right) \\ &= \frac{1}{6} e^{-4((y-11)/6)} = \frac{2}{3} e^{-2(y-11)/3}, \end{aligned}$$

for $y > 11$, and 0, otherwise. ■

Working through these examples, you might wonder why such approaches are needed when we have seen examples where mgfs can be used. While mgfs are useful in many cases, the resulting mgf may not always be identifiable as a distribution you know, and there are times you really want the density function itself.

Example 8.6 A lighthouse is one mile off the coast from the nearest point O on a straight, infinite beach. The lighthouse sends out pulses of light at random angles Θ uniformly distributed from $-\pi/2$ to $\pi/2$ (it rotates only 180°). Find the distribution

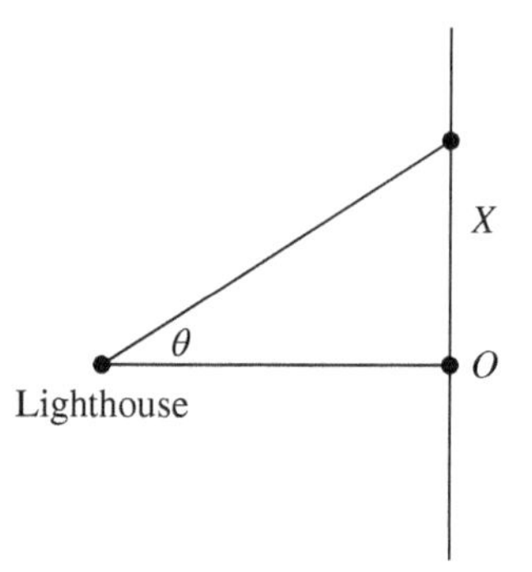

FIGURE 8.2: The geometry of the lighthouse problem.

of the distance X between where the light hits the shore and O. Also find the expected distance.

Figure 8.2 helps us to visualize the situation. Let $\Theta \sim \text{Unif}(-\pi/2, \pi/2)$ be the angle of the light to the shore. Let X be the distance from where the light hits the shore to O. By the geometry of the problem, $X = \tan\Theta$. For all real x,

$$F_X(x) = P(X \leq x) = P(\tan\Theta \leq x)$$
$$= P(\Theta \leq \tan^{-1}x) = \frac{\tan^{-1}x + \pi/2}{\pi}.$$

Taking derivatives with respect to x gives,

$$f_X(x) = \frac{1}{\pi(1+x^2)}, \quad -\infty < x < \infty.$$

This function is the density of the *Cauchy distribution*, also known as the Lorentz distribution in physics, where it is used to model energy states in quantum mechanics.

To find the expectation $E[X]$, we can set up the integral two ways.

1. Use the distribution of X.

$$E[X] = \int_{-\infty}^{\infty} x \frac{1}{\pi(1+x^2)}\, dx.$$

2. Use the distribution of Θ (and the law of the unconscious statistician).

$$E[X] = E[\tan\Theta] = \int_{-\pi/2}^{\pi/2} \frac{\tan\theta}{\pi}\, d\theta.$$

However, in either case, the integral does not converge. The Cauchy distribution has the property that its expectation does not exist.

What happens when one tries to simulate from a distribution with no expectation?

R: SIMULATING AN EXPECTATION THAT DOES NOT EXIST

The function noexp() simulates 100,000 trials of X, the distance from where the light hits the shore and O, and then computes the average.

```
> noexp <- function()
  mean(tan(runif(100000,-pi/2,pi/2)))}
```

Repeated simulations show no equilibrium. The averages are erratic; some are close to 0; others show large magnitudes.

```
> noexp()
[1] 0.4331702
> noexp()
[1] 3.168341
> noexp()
[1] 12.08399
> noexp()
[1] -1.447175
> noexp()
[1] 0.5758977
> noexp()
[1] 0.6160198
> noexp()
[1] -1.768458
> noexp()
[1] -0.8728917
```

Example 8.7 Darts. Stern [1997] collected data from 590 throws at a dart board, measuring the distance between the dart and the bullseye. To create a model of dart throws, it is assumed that the horizontal and vertical errors H and V are independent random variables normally distributed with mean 0 and variance σ^2. Let T be the distance from the dart to the bullseye. Then, $T = \sqrt{H^2 + V^2}$ is the radial distance. We find the distribution of T.

The radial distance is a function of two independent normals. The joint density of H and V is the product of their marginal densities. That is,

$$f(h, v) = \left(\frac{1}{\sqrt{2\pi}\sigma}e^{-h^2/2\sigma^2}\right)\left(\frac{1}{\sqrt{2\pi}\sigma}e^{-v^2/2\sigma^2}\right) = \frac{1}{2\pi\sigma^2}e^{-(h^2+v^2)/2\sigma^2}.$$

For $t > 0$,

$$P(T \le t) = P(\sqrt{H^2 + V^2} \le t) = P(H^2 + V^2 \le t^2)$$
$$= \int_{-t}^{t} \int_{-\sqrt{t^2-v^2}}^{\sqrt{t^2-v^2}} \frac{1}{2\pi\sigma^2} e^{-(h^2+v^2)/2\sigma^2}\, dh\, dv.$$

Changing to polar coordinates gives

$$P(T \le t) = \frac{1}{2\pi\sigma^2} \int_0^t \int_0^{2\pi} e^{-r^2/2\sigma^2} r\, d\theta\, dr$$
$$= \frac{1}{\sigma^2} \int_0^t re^{-r^2/2\sigma^2}\, dr = 1 - e^{-t^2/2\sigma^2}.$$

Differentiating with respect to t gives

$$f_T(t) = \frac{t}{\sigma^2} e^{-\frac{t^2}{2\sigma^2}}, \quad \text{for } t > 0.$$

This is the density of a *Weibull distribution* often used in reliability theory and industrial engineering.

In the darts model, the parameter σ is a measure of players' accuracy. Several statistical techniques can be used to estimate σ from the data. Estimates for two professional darts players were made at $\sigma \approx 13.3$. For this level of accuracy, the probability of missing the bullseye by more than 40 mm (about 1.5 inches) is

$$P(T > 40) = \int_{40}^{\infty} \frac{t}{(13.3)^2} e^{-\frac{t^2}{2(13.3)^2}} = 0.011.$$ ■

8.1.1 Simulating a Continuous Random Variable

A random variable X has density $f(x) = 2x$, for $0 < x < 1$, and 0, otherwise. Suppose we want to simulate observations from X (and we did not recognize the beta distribution). In this section, we present a simple and flexible method for simulating from a continuous distribution. It requires that the cdf of X is invertible.

INVERSE TRANSFORM METHOD

Suppose X is a continuous random variable with cdf F, where F is invertible with inverse function F^{-1}. Let $U \sim \text{Unif}(0, 1)$. Then the distribution of $F^{-1}(U)$ is equal to the distribution of X. To simulate X first simulate U and output $F^{-1}(U)$.

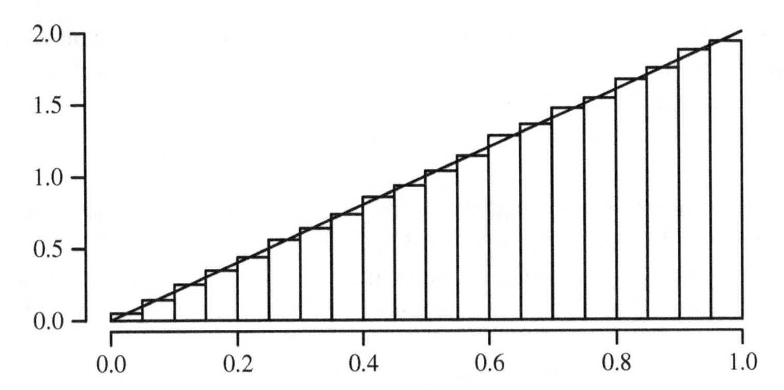

FIGURE 8.3: Simulating from density $f(x) = 2x$, for $0 < x < 1$ using the inverse transform method.

To illustrate with the initial example, the cdf of X is

$$F(x) = P(X \leq x) = \int_0^x 2t\,dt = x^2, \quad \text{for } 0 < x < 1.$$

On the interval $(0, 1)$ the function $F(x) = x^2$ is invertible and $F^{-1}(x) = \sqrt{x}$. The inverse transform method says that if $U \sim \text{Unif}(0, 1)$, then $F^{-1}(U) = \sqrt{U}$ has the same distribution as X. Thus to simulate X, generate $\sqrt{U}$.

R: IMPLEMENTING THE INVERSE TRANSFORM METHOD

The **R** commands

```
> n <- 10000
> simlist <- sqrt(runif(n))
> hist(simlist,prob=T,main="",xlab="")
> curve(2*x,0,1,add=T)
```

generate the histogram in Figure 8.3, along with the super-imposed curve of the theoretical density $f(x) = 2x$.

The proof of the inverse transform method is quick and easy. We need to show that $F^{-1}(U)$ has the same distribution as X. For x in the range of X,

$$P(F^{-1}(U) \leq x) = P(U \leq F(x)) = F(x) = P(X \leq x),$$

using the fact that the cdf of the uniform distribution on $(0, 1)$ is the identity function.

Example 8.8 We show how to simulate an exponential random variable using the inverse transform method. The cdf of the exponential distribution is $F(x) = 1 - e^{-\lambda x}$ with inverse function

$$F^{-1}(x) = \frac{\ln(1-x)}{-\lambda}.$$

To simulate an exponential random variable with parameter λ, simulate a uniform $(0, 1)$ variable U and generate $-\ln(1-U)/\lambda$.

Observe that if U is uniform on $(0, 1)$, then so is $1 - U$. Thus, an even simpler method for simulating an exponential random variable is to output $-\ln U/\lambda$. ■

Example 8.9 Simulating the Pareto distribution. **R** does not have a built-in function to simulate from the Pareto distribution. To simulate a Pareto random variable, we use the inverse transform method. Let $X \sim \text{Pareto}(m, a)$. Then

$$F_X(x) = 1 - \left(\frac{m}{x}\right)^a$$

with inverse function

$$F^{-1}(u) = \frac{m}{(1-u)^{1/a}}.$$

If $U \sim \text{Unif}(0, 1)$, then $m/(1-U)^{1/a}$ has the desired Pareto distribution. And we can even do a little better. Because $1 - U$ is also uniformly distributed on $(0, 1)$, simplify the simulation formula and use $m/U^{(1/a)}$ as an observation from a Pareto(m, a) distribution.

To illustrate the 80-20 rule for income discussed previously, we conduct a simulation study, simulating from a Pareto distribution with parameters $m = 1$ and $a = \log 5/\log 4$. See the script file **Pareto8020.R**.

R: SIMULATING THE 80-20 RULE

What percent of the population owns 80% of the income?

```
# Pareto8020.R
> m <- 1
> a <- log(5)/log(4)
> n <- 100000
> simlist <- m/runif(n)^(1/a)
> totalwealth <- sum(simlist)
> totalwealth80 <- 0.80*totalwealth  # 80% of wealth
> indx <- which(cumsum(simlist) > totalwealth80)[1]
> 1-indx/100000  # % of pop who own 80% of income
[1] 0.24548
```

■

8.1.2 Method of Transformations

Previously, you saw that it was possible to derive a result to find densities for linear functions of random variables in general, called the cdf method, which is applicable to any density. While the method of cdfs is effective, it is not always necessary to return to the cdf as the starting point for finding a pdf. In fact, some assumptions allow for a faster computation. This result is derived as a special case based on the cdf method.

Suppose $Y = g(X)$, for some function g. We further assume that g is invertible. Then by the cdf method,

$$F_Y(y) = P(Y \leq y) = P(g(X) \leq y) = P(X \leq g^{-1}(y)) = F_X(g^{-1}(y)).$$

Differentiating with respect to y, using the chain rule, gives

$$f_Y(y) = f_X(g^{-1}(y)) \left| \frac{d}{dy} g^{-1}(y) \right|,$$

where the absolute value results from pdfs needing to be nonnegative. You saw a similar absolute value appear in Example 8.4.

This approach has wide application, though you must check the invertibility of g. As a result, the approach may or may not work for g based on the values of X. For example, if $g(x) = x^2$, and $X \in (0, 4)$, the approach will work, as $g^{-1}(y) = \sqrt{y}$. However, if $X \in (-3, 3)$, then g is not invertible because not all values of Y map back to a single X.

Other texts equivalently describe the invertible condition as requiring g to be either an increasing or decreasing function of x for all x such that $f_X(x) > 0$. Frame the condition this way if it helps you, and be sure to check it!

We revisit two previous examples to show the applicability of this approach, often called the method of transformations.

Example 8.10 In Example 8.1, $R \sim \text{Unif}(0, 2)$, and $A = \pi R^2$. Thus, we know $f_R(r) = 1/2$, for $0 < r < 2$, and 0, otherwise. We find the inverse function is $R = \sqrt{A/\pi}$, with each value of A mapping back to a single value of R, so g is invertible. Applying the formula above, we find the density of A is

$$\begin{aligned} f_A(a) &= f_R(\sqrt{A/\pi}) \left| \frac{d}{dA} \left(\sqrt{\frac{A}{\pi}} \right) \right| \\ &= \frac{1}{2} \frac{1}{\pi} \frac{1}{2\sqrt{A}} = \frac{1}{4\sqrt{a\pi}}, \end{aligned}$$

for $0 < a < 4\pi$, and 0, otherwise. Note that you need to pay attention to chain rule. This matches our previous result. ■

Example 8.11 In Example 8.5, $X \sim \text{Exp}(4)$, and $Y = 6X + 11$, and we wanted the density function of Y. g is invertible as $X = \frac{Y-11}{6}$. Thus, the density of Y is

$$f_Y(y) = 4e^{4(y-11/6)} \left| \frac{d}{dy} \frac{y-11}{6} \right|$$
$$= \frac{2}{3} e^{-2(y-11)/3},$$

for $y > 11$, and 0, otherwise, as was found previously. ■

Next, we consider other common functions of random variables and ways to find their densities.

8.2 MAXIMUMS, MINIMUMS, AND ORDER STATISTICS

A scientist is monitoring four similar experiments and waiting for a chemical reaction to occur in each. Suppose the time until the reaction has some probability distribution and is modeled with random variables T_1, T_2, T_3, T_4, which represent the respective reaction time for each experiment. Of interest may be the time until the *first* reaction occurs. This is the *minimum* of the T_i's. The time until the *last* reaction is the *maximum* of the T_i's.

Maximums and minimums of collections of random variables arise frequently in applications. The key to working with them are the following algebraic relations.

INEQUALITIES FOR MAXIMUMS AND MINIMUMS

Let $x_1, \ldots, x_n, s$, and t be arbitrary numbers.

1. All of the x_k's are greater than or equal to s if and only if the *minimum* of the x_k's is greater than or equal to s. That is,

$$x_1 \geq s, \ldots, x_n \geq s \Leftrightarrow \min(x_1, \ldots, x_n) \geq s. \tag{8.1}$$

2. All of the x_k's are less than or equal to t if and only if the *maximum* of the x_k's is less than or equal to t. That is,

$$x_1 \leq t, \ldots, x_n \leq t \Leftrightarrow \max(x_1, \ldots, x_n) \leq t. \tag{8.2}$$

The same results hold if partial inequalities are replaced with strict inequalities.

Example 8.12 The reliability of three lab computers is being monitored. Let X_1, X_2, and X_3 be the respective times until their systems crash and need to reboot. The "crash times" are independent of each other and have exponential distributions with respective parameters $\lambda_1 = 2$, $\lambda_2 = 3$, and $\lambda_3 = 5$. Let M be the time that the first computer crashes. Find the distribution of M.

The time of the first crash is the minimum of X_1, X_2, and X_3. Using the above result for minimums, consider $P(M > m)$. For $m > 0$,

$$\begin{aligned}
P(M > m) &= P(\min(X_1, X_2, X_3) > m) \\
&= P(X_1 > m, X_2 > m, X_3 > m) \\
&= P(X_1 > m)P(X_2 > m)P(X_3 > m) \\
&= e^{-2m}e^{-3m}e^{-5m} = e^{-10m},
\end{aligned}$$

where the third equality is from independence of the X_i's. This gives $F_M(m) = P(M \le m) = 1 - e^{-10m}$, which is the cdf of an exponential random variable with parameter $\lambda = 10$. The minimum has an exponential distribution. See Figure 8.4 for a comparison of simulated distributions of X_1, X_2, X_3, and M. ■

This example illustrates a general result—the distribution of the minimum of independent exponential random variables is exponential.

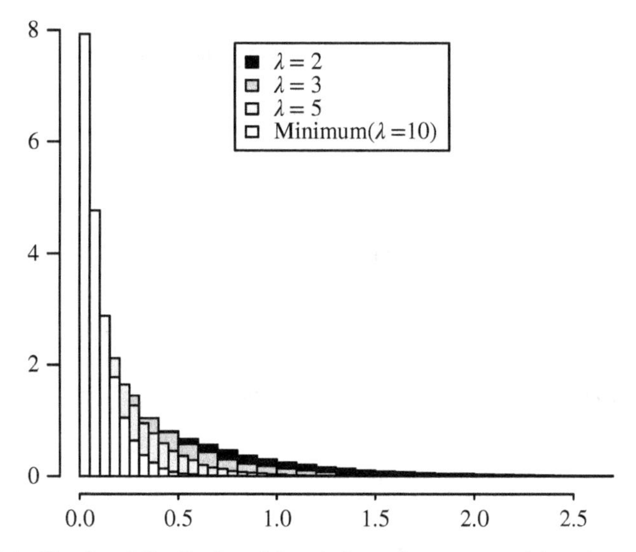

FIGURE 8.4: Simulated distribution of three independent exponential random variables and their minimum.

MINIMUM OF INDEPENDENT EXPONENTIAL DISTRIBUTIONS

Let $X_1, \ldots, X_n$ be independent random variables with $X_k \sim \text{Exp}(\lambda_k)$ for $k = 1, \ldots, n$. Then,

$$\min(X_1, \ldots, X_n) \sim \text{Exp}(\lambda_1 + \cdots + \lambda_n).$$

Example 8.13 After a severe storm, an insurance company expects many claims for damages. An actuary has modeled the company's payout per claim, in thousands of dollars, with the probability density function

$$f(x) = \frac{x^2}{9}, \quad 0 < x < 3.$$

Suppose $X_1, \ldots, X_n$ represent payouts from n independent claims. Find the expected value of the maximum payout.

Let $M = \max(X_1, \ldots, X_n)$. First find the cdf of M. For $0 < m < 3$,

$$\begin{aligned} F_M(m) &= P(M \leq m) = P(\max(X_1, \ldots, X_n) \leq m) \\ &= P(X_1 \leq m, \ldots, X_n \leq m) \\ &= P(X_1 \leq m) \cdots P(X_n \leq m) \\ &= [P(X_1 \leq m)]^n = [F(m)]^n. \end{aligned}$$

Differentiating with respect to m gives a general form of the pdf of a maximum as

$$f_M(m) = n[F(m)]^{n-1} f(m).$$

In this context, plugging in, we find that

$$\begin{aligned} f_M(m) &= n\left[\int_0^m \frac{t^2}{9}\, dt\right]^{n-1} \left[\frac{m^2}{9}\right] = n\left[\frac{m^3}{27}\right]^{n-1} \left[\frac{m^2}{9}\right] \\ &= \frac{n}{9(27^{n-1})} m^{3n-1} = \frac{3nm^{3n-1}}{27^n}, \quad \text{for } 0 < m < 3. \end{aligned}$$

The expectation with respect to this density is

$$\begin{aligned} E[M] &= \int_0^3 m \frac{3nm^{3n-1}}{27^n}\, dm = \frac{3n}{27^n} \int_0^3 m^{3n}\, dm \\ &= \frac{3n}{27^n} \frac{3^{3n+1}}{3n+1} = \frac{9n}{3n+1}. \end{aligned}$$

Observe that the expected maximum payout approaches 3, meaning \$3000, as the number of claims n gets large. ■

Example 8.14 Order statistics. Given a sequence of random variables $X_1, \ldots, X_n$, order the values from smallest to largest. Let $X_{(k)}$ denote the kth largest value. That is,

$$X_{(1)} \leq \cdots \leq X_{(n)}.$$

The minimum of the n variables is $X_{(1)}$ and the maximum is $X_{(n)}$. The random variables $(X_{(1)}, \ldots, X_{(n)})$ are called the *order statistics*.

For instance, take a sample $X_1, \ldots, X_5$, where the X_i's are uniformly distributed on (0, 1).

```
> runif(5)
[1] 0.1204 0.6064 0.5070 0.6805 0.1185
```

Then,

$$(X_{(1)}, X_{(2)}, X_{(3)}, X_{(4)}, X_{(5)}) = (0.1185, 0.1204, 0.5070, 0.6064, 0.6805).$$

Consider the order statistics for n i.i.d. random variables $X_1, \ldots, X_n$ uniformly distributed on (0, 1). We show that for each $k = 1, \ldots, n$, the kth order statistic $X_{(k)}$ has a beta distribution.

For $0 < x < 1$, $X_{(k)}$ is less than or equal to x if and only if at least k of the X_i's are less than or equal to x. As the X_i's are independent, and $P(X_i \leq x) = x$, the number of X_i's less than or equal to x has a binomial distribution with parameters n and $p = x$. Letting Y be a random variable with such a distribution gives

$$P(X_{(k)} \leq x) = P(Y \geq k) = \sum_{i=k}^{n} \binom{n}{i} x^i (1-x)^{n-i}.$$

Differentiate to find the density of $X_{(k)}$. Consider the case $k < n$.

$$\begin{aligned}
f_{X_{(k)}}(x) &= \sum_{i=k}^{n} \binom{n}{i} [ix^{i-1}(1-x)^{n-i} - (n-i)(1-x)^{n-i-1}x^i] \\
&= \sum_{i=k}^{n} \frac{n}{i!(n-i)!} ix^{i-1}(1-x)^{n-i} \\
&\quad - \sum_{i=k}^{n} \frac{n!}{i!(n-i)!}(n-i)x^i(1-x)^{n-i-1} \\
&= \sum_{i=k}^{n} \frac{n!}{(i-1)!(n-i)!} x^{i-1}(1-x)^{n-i} \\
&\quad - \sum_{i=k}^{n-1} \frac{n!}{i!(n-i-1)!} x^i(1-x)^{n-i-1}
\end{aligned}$$

$$= \sum_{i=k}^{n} \frac{n!}{(i-1)!(n-i)!} x^{i-1}(1-x)^{n-i}$$

$$- \sum_{i=k+1}^{n} \frac{n!}{(i-1)!(n-i)!} x^{i-1}(1-x)^{n-i}$$

$$= \frac{n!}{(k-1)!(n-k)!} x^{k-1}(1-x)^{n-k}.$$

This gives the density function of a beta distribution with parameters $a = k$ and $b = n - k + 1$.

For the case $k = n$, the order statistic is the maximum. The density function for the maximum of n independent uniforms is $f_{X_{(n)}}(x) = nx^{n-1}$, for $0 < x < 1$. This is a beta distribution with parameters $a = n$ and $b = 1$. In both cases, we have the following result.

DISTRIBUTION OF ORDER STATISTICS

Let $X_1, \ldots, X_n$ be an i.i.d. sequence uniformly distributed on $(0, 1)$. Let $X_{(1)}, \ldots, X_{(n)}$ be the corresponding order statistics. For each $k = 1, \ldots, n$,

$$X_{(k)} \sim \text{Beta}(k, n - k + 1).$$

■

Example 8.15 Simulating beta random variables. We show how to simulate beta random variables when the parameters a and b are integers using the previous result. Suppose $X_1, \ldots, X_{a+b-1}$ are i.i.d. uniform $(0, 1)$ random variables. Then

$$X_{(a)} \sim \text{Beta}(a, (b + a - 1) - a + 1) = \text{Beta}(a, b).$$

To simulate a Beta(a, b) random variable, generate $a + b - 1$ uniform $(0, 1)$ variables and choose the ath largest.

This is not the most efficient method for simulating beta variables, but it works fine for small and even moderate values of a and b. And it is easily coded. In **R**, type

```
> sort(runif(a+b-1))[a].
```

■

8.3 CONVOLUTION

We have already seen several applications of sums of independent random variables. Let X and Y be independent continuous variables with respective densities f and g. We give a general expression for the density function of $X + Y$. As $X \perp Y, f(x, y) = f(x)g(y)$, and the cdf of $X + Y$ is

$$P(X + Y \leq t) = \int_{-\infty}^{\infty} \int_{-\infty}^{t-y} f(x, y)\, dx\, dy$$
$$= \int_{-\infty}^{\infty} \left(\int_{-\infty}^{t-y} f(x)\, dx \right) g(y)\, dy.$$

Differentiating with respect to t and applying the fundamental theorem of calculus gives

$$f_{X+Y}(t) = \int_{-\infty}^{\infty} f(t - y)g(y)\, dy. \tag{8.3}$$

This is called the *convolution* of the densities of X and Y. Compare to the discrete formula for the probability mass function of a sum given in Equation 4.7.

Example 8.16 Once a web page goes on-line, suppose the times between successive hits received by the page are independent and exponentially distributed with parameter λ. Find the density function of the time that the second hit is received.

Let X be the time that the first hit is received. Let Y be the additional time until the second hit. The time that the second hit is received is $X + Y$. We showed this should be a Gamma distribution via mgfs previously, but now the density function itself is desired. We find the density of $X + Y$ two ways: (i) by using the convolution formula and (ii) from first principles and setting up the multiple integral.

(i) Suppose the density of X is f. The random variables X and Y have the same distribution and thus the convolution formula gives

$$f_{X+Y}(t) = \int_{-\infty}^{\infty} f(t - y)f(y)\, dy.$$

Consider the domain constraints on the densities and the limits of integration. As the exponential density is equal to 0 for negative values, the integrand expression $f(t - y)f(y)$ will be positive when $t - y > 0$ and $y > 0$. That is, $0 < y < t$. Hence for $t > 0$,

$$f_{X+Y}(t) = \int_{-\infty}^{\infty} f(t - y)f(y)\, dy$$
$$= \int_{0}^{t} (\lambda e^{-\lambda(t-y)})(\lambda e^{-\lambda y})\, dy$$

$$= \lambda^2 e^{-\lambda t} \int_0^t dy = \lambda^2 t e^{-\lambda t}.$$

(ii) First find the cdf of $X + Y$. For $t > 0$,

$$\begin{aligned}
P(X + Y \leq t) &= \iint_{\{(x,y):x+y\leq t\}} f(x, y)\, dx\, dy \\
&= \int_0^t \int_0^{t-y} \lambda e^{-\lambda x} \lambda e^{-\lambda y}\, dx\, dy \\
&= \int_0^t \lambda e^{-\lambda y}(1 - e^{-\lambda(t-y)})\, dy \\
&= \int_0^t \lambda(e^{-\lambda y} - e^{-\lambda t})\, dy \\
&= 1 - e^{-\lambda t} - \lambda t e^{-\lambda t}.
\end{aligned}$$

Differentiating with respect to t gives

$$f_{X+Y}(t) = \lambda e^{-\lambda t} - \lambda(e^{-\lambda t} - \lambda t e^{-\lambda t}) = \lambda^2 t e^{-\lambda t}, \text{for } t > 0.$$

As expected, the density function of $X + Y$ is the density of a gamma distribution, with parameters 2 and λ. ■

Example 8.17 Previously, we showed that the sum of two independent normal variables is also normal using mgfs. Now, we show it in the case of standard normal variables via the convolution formula, Equation 8.3.

Let X and Y be independent standard normal variables. Using the convolution formula to derive the density function of $X + Y$, for real t, we have

$$\begin{aligned}
f_{X+Y}(t) &= \int_{-\infty}^{\infty} f(t - y)f(y)\, dy \\
&= \int_{-\infty}^{\infty} \frac{1}{\sqrt{2\pi}} e^{-\frac{(t-y)^2}{2}} \frac{1}{\sqrt{2\pi}} e^{-\frac{y^2}{2}}\, dy \\
&= \frac{1}{2\pi} \int_{-\infty}^{\infty} e^{-\frac{(2y^2-2yt+t^2)}{2}}\, dy. \qquad (8.4)
\end{aligned}$$

Consider the exponent of e in the last integral. Write

$$2y^2 - 2yt + t^2 = 2\left(y^2 - yt + \frac{t^2}{4} + \frac{t^2}{4}\right) = 2\left(y - \frac{t}{2}\right)^2 + \frac{t^2}{2}.$$

This gives

$$
\begin{aligned}
f_{X+Y}(t) &= \frac{1}{2\pi}\int_{-\infty}^{\infty} e^{-\left(y-\frac{t}{2}\right)^2} e^{-\frac{t^2}{4}}\, dy \\
&= \frac{1}{\sqrt{2\pi}\sqrt{2}} e^{-\frac{t^2}{4}} \int_{-\infty}^{\infty} \frac{\sqrt{2}}{\sqrt{2\pi}} e^{-\left(y-\frac{t}{2}\right)^2}\, dy \\
&= \frac{1}{\sqrt{2\pi}\sqrt{2}} e^{-t^2/4},
\end{aligned}
$$

where the last equality follows because the integrand is the density of a normal distribution with mean $t/2$ and variance $1/2$, and thus integrates to one. The final expression is the density of a normal distribution with mean 0 and variance 2. That is, $X + Y \sim \text{Norm}(0, 2)$. ■

Example 8.18 Sum of independent uniforms. Let X and Y be i.i.d. random variables uniformly distributed on $(0, 1)$. Find the density of $X + Y$.

Let f be the density function for the uniform distribution. We use the convolution formula. As f is equal to 0 outside the interval $(0, 1)$, the integrand expression $f(t - y)f(y)$ will be positive when both factors are positive giving $0 < t - y < 1$ and $0 < y < 1$, or $t - 1 < y < t$ and $0 < y < 1$. We can write these two conditions as $\max(0, t - 1) < y < \min(1, t)$.

Because $X + Y$ takes values between 0 and 2, we consider two cases:

(i) For $0 < t < 1$,

$$\max(0, t - 1) = 0 < y < \min(1, t) = t$$

and thus $0 < y < t$. In that case,

$$f_{X+Y}(t) = \int_0^t f(t - y)f(y)\, dy = \int_0^t dy = t.$$

(ii) For $1 \leq t \leq 2$,

$$\max(0, t - 1) = t - 1 < y < \min(1, t) = 1$$

and thus $t - 1 < y < 1$. In that case,

$$f_{X+Y}(t) = \int_{t-1}^1 f(t - y)f(y)\, dy = \int_{t-1}^1 dy = 2 - t.$$

Together, we get the *triangular density*

$$f_{X+Y}(t) = \begin{cases} t, & \text{if } 0 < t \leq 1 \\ 2 - t, & \text{if } 1 < t \leq 2, \end{cases}$$

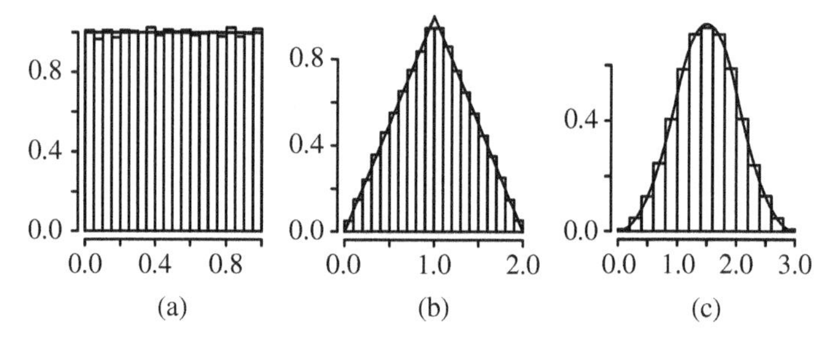

FIGURE 8.5: Distributions from (a) to (c): uniform, sum of two independent uniforms, sum of three independent uniforms. Histogram is from 100,000 trials. Curve is the theoretical density.

and 0, otherwise. See Figure 8.5 for the graph of the triangular density and the density of the sum of three independent uniform random variables. ■

There are several equivalent ways to tackle this problem. A different approach using geometry is shown in the next section.

8.4 GEOMETRIC PROBABILITY

In this section, we use geometry to solve probability problems. Geometric methods are a powerful tool for working with independent uniform random variables.

If X and Y are independent and each uniformly distributed on a bounded interval, then (X, Y) is uniformly distributed on a rectangle. Often problems involving two independent uniform random variables can be recast in two-dimensions where they can be approached geometrically.

Example 8.19 When Angel meets Lisa. Angel and Lisa are planning to meet at a coffee shop for lunch. Let M be the event that they meet. Each will arrive at the coffee shop at some time uniformly distributed between 1:00 and 1:30 p.m. independently of each other. When each arrives they will wait for the other person for 5 minutes, but then leave if the other person does not show up. What is the probability that they will meet?

Let A and L denote their respective arrival times in minutes from one o'clock. Then both A and L are uniformly distributed on $(0, 30)$. They will meet if they both arrive within 5 minutes of each other. Let M be the event that they meet. Then $M = \{|A - L| < 5\}$ and

$$P(M) = P(|A - L| < 5) = P(-5 < A - L < 5).$$

The analytic approach to solving this probability involves a tricky double integral and working with the joint density function. We take a geometric approach.

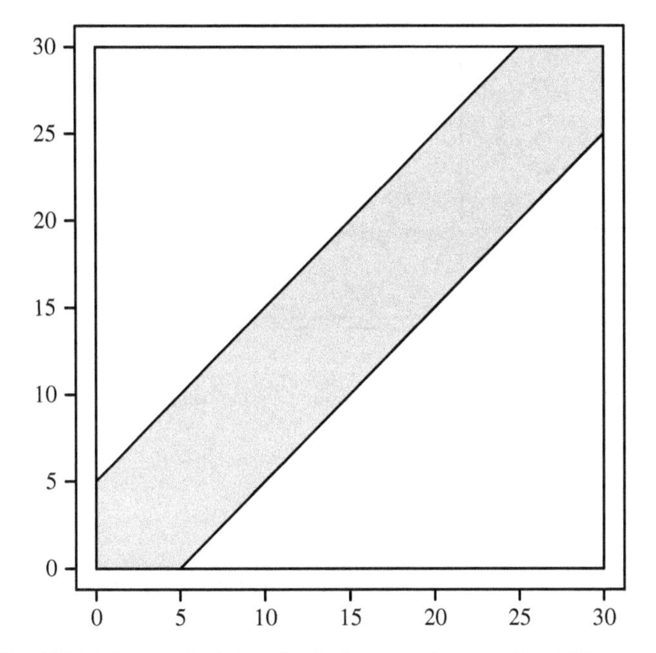

FIGURE 8.6: The shaded region is the event that Angel and Lisa meet.

Represent the arrival times A and L as a point (A, L) in a $[0, 30] \times [0, 30]$ square in the a-l plane. As A and L are independent, the point (A, L) is uniformly distributed on the square. The event that Angel and Lisa meet corresponds to the region five units above and below the line $L = A$. See Figure 8.6.

The probability $P(M)$ is the area of the shaded region as a proportion of the area of the square. The area of the shaded region is best found by computing the area of the non-shaded region, which consists of two 25×25 triangles for an area of $2 \times 25 \times 25/2 = 625$. As the area of the square is $30 \times 30 = 900$, this gives

$$P(M) = \frac{900 - 625}{900} = \frac{11}{36} = 0.306. \quad \blacksquare$$

We can use geometric methods not only to find probabilities but to derive cdfs and density functions as well.

Example 8.20 Let X and Y be independent and uniformly distributed on $(0, a)$. Let $W = |X - Y|$ be their absolute difference. Find the density and expectation of W.

The cdf of W is $F(w) = P(W \leq w) = P(|X - Y| \leq w)$, for $0 < w < a$. Observe that in the last example we found this probability for Angel and Lisa with $a = 30$ and $w = 5$. Generalizing the results there with a $[0, a] \times [0, a]$ square gives

$$P(|X - Y| \leq w) = \frac{a^2 - (a - w)^2}{a^2}.$$

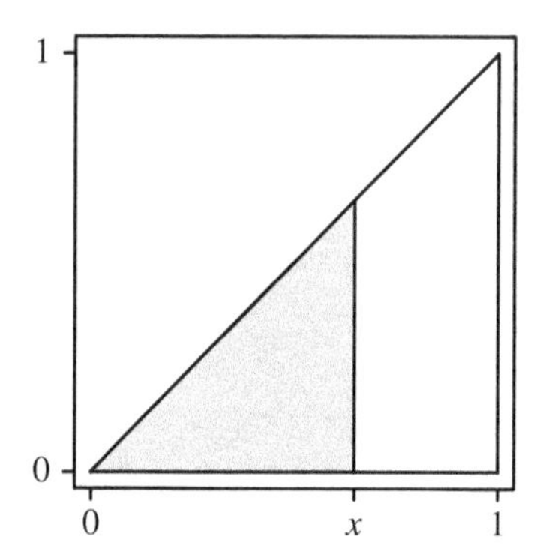

FIGURE 8.7: Let (X, Y) be uniformly distributed on the triangle. The shaded region is the event $\{X \leq x\}$.

Differentiating gives the density of W

$$f_W(w) = \frac{2(a-w)}{a^2}, \quad \text{for } 0 < w < a.$$

The expectation of the absolute difference of X and Y is

$$E[W] = \int_0^a w\frac{2(a-w)}{a^2}\,dw = \frac{2}{a^2}\int_0^a (aw - w^2)\,dw = \frac{2}{a^2}\left(\frac{aw^2}{2} - \frac{w^3}{3}\right)\Bigg|_0^a = \frac{a}{3}.$$

■

Example 8.21 A random point (X, Y) is uniformly distributed on the triangle with vertices at $(0, 0)$, $(1, 0)$, and $(1, 1)$. Find the distribution of the X-coordinate.

In Example 6.27, we gave the analytic solution using multiple integrals. Here we use geometry. First draw the picture (see Fig. 8.7). For $0 < x < 1$, the event $\{X \leq x\}$ consists of the shaded region, with area $x^2/2$. The area of the large triangle is 1/2. Thus

$$P(X \leq x) = \frac{x^2/2}{1/2} = x^2.$$

Differentiating gives $f(x) = 2x$, for $0 < x < 1$.

Note that the marginal distribution of X is *not* uniform as the x-coordinate of the triangle is more likely to be large than small. ■

Example 8.22 Sum of uniforms revisited. The distribution of $X + Y$ for independent random variables uniformly distributed on $(0, 1)$ was derived with the convolution formula in Example 8.18. Here we do it geometrically.

For $0 < t < 2$, consider $P(X + Y \leq t)$. The region $x + y \leq t$ in the $[0, 1] \times [0, 1]$ square is the region below the line $y = t - x$. The y-intercept of the line is $y = t$,

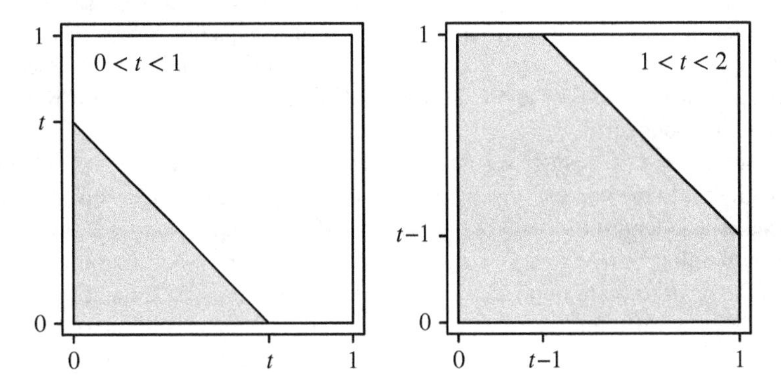

FIGURE 8.8: Region $\{X + Y \leq t\}$ for $0 < t < 1$ and $1 < t < 2$.

which is inside the square, if $t < 1$, and outside the square, if $t > 1$ (see Fig. 8.8). The shaded region is the event $\{X + Y \leq t\}$ for different choices of t.

For $0 < t < 1$, the area of the shaded region is $t^2/2$. For $1 < t < 2$, the area of the shaded region is $1 - (2 - t)^2/2$. Thus,

$$P(X + Y \leq t) = \begin{cases} 0, & \text{if } t \leq 0 \\ t^2/2, & \text{if } 0 < t \leq 1 \\ 1 - (2 - t)^2/2, & \text{if } 1 < t \leq 2 \\ 1, & \text{if } t \geq 2. \end{cases}$$

Differentiating gives

$$f_{X+Y}(t) = \begin{cases} t, & \text{if } 0 < t \leq 1 \\ 2 - t, & \text{if } 1 < t \leq 2, \end{cases}$$

and 0, otherwise. ■

Example 8.23 Buffon's needle*. The most famous problem in geometric probability is Buffon's needle problem, introduced in 1733 by Georges-Louis Leclerc, Comte de Buffon, who posed the question:

> Suppose we have a floor made of parallel strips of wood, each the same width, and we drop a needle onto the floor. What is the probability that the needle will lie across a line between strips?

Assume that the lines between strips of wood are one unit apart, and the length of the needle also has length one. It should be clear that we only need to consider what happens on one strip of wood.

Parameterize the needle's position with two numbers. Let D be the distance between the center of the needle and the closest line. Let Θ be the angle that the needle makes with that line. Then all positions of the needle can be described by (D, Θ), with $0 < D < 1/2$ and $0 < \Theta < \pi$. A probability model is obtained for a random position of the needle by letting $D \sim \text{Unif}(0, 1/2)$ and $\Theta \sim \text{Unif}(0, \pi)$, with D and Θ independent.

Suppose $\Theta = \theta$. Consider a right triangle whose hypotenuse is the half of the needle closest to the nearest line and with two vertices at the center and endpoint of the needle. The length of the hypotenuse is 1/2. The triangle's angle at the needle's center will either be θ, if $0 < \theta < \pi/2$, or $\pi - \theta$, if $\pi/2 < \theta < \pi$. In either case, the side of the right triangle opposite that angle has length $(\sin\theta)/2$. See Figure 8.9.

Considering different cases shows that the needle intersects a line if and only if the distance of that side is greater than D. Thus, the event that the needle crosses a line is equal to the event that $\sin(\Theta)/2 > D$ and

$$P(\text{Needle crosses line}) = P\left(\frac{\sin\Theta}{2} > D\right).$$

The needle's position (D, θ) can be expressed as a point in the $[0, \pi] \times [0, 1/2]$ rectangle in the θ-d plane. The event $\{\sin\Theta/2 > D\}$ is the region under the curve $d = (\sin\theta)/2$ (see Fig. 8.10).

The area of this region is

$$\int_0^{\pi} \frac{\sin\theta}{2}\, d\theta = \left.\frac{-\cos\theta}{2}\right|_0^{\pi} = 1.$$

As a fraction of the area of the rectangle $\pi/2$ this gives

$$P(\text{Needle crosses line}) = \frac{1}{\pi/2} = \frac{2}{\pi} = 0.6366\ldots$$

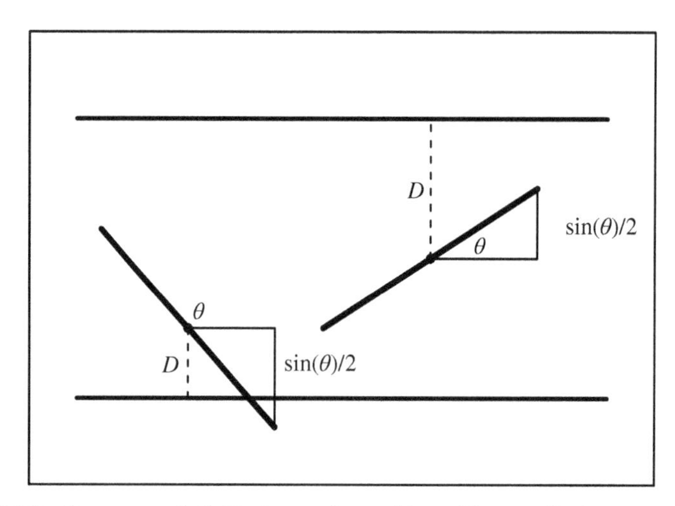

FIGURE 8.9: Geometry of Buffon's needle problem. The needle intersects a line if $\sin(\theta)/2 > D$.

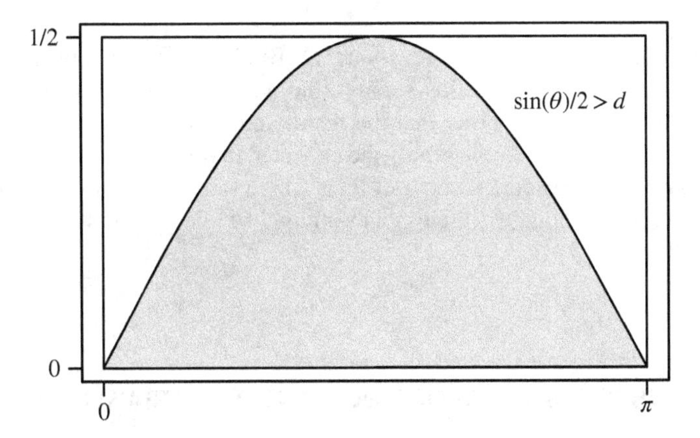

FIGURE 8.10: Buffon's needle problem is solved by finding the area under the curve $d = \sin(\theta)/2$ as a proportion of the area of the $[0, \pi] \times [0, 1/2]$ rectangle.

Although the number π, pervasive throughout mathematics, is not a physical constant, here we have a physical method to simulate π: drop needles on a wooden floor and count the number of times a line is crossed. For large n,

$$P(\text{Needle crosses line}) \approx \frac{\text{Number of needle crosses}}{n},$$

which gives

$$\pi = \frac{2}{P(\text{Needles crosses line})} \approx \frac{2n}{\text{Number of needle crosses}}.$$

■

Ants, Fish, and Noodles

- In *Ants estimate area using Buffon's needle*, Mallon and Franks [2000] show a remarkable connection between the Buffon needle problem and how ants measure the size of their potential nest sites. Size assessment is made by individual ant scouts who are able to detect intersections of their tracks along potential nest sites. The authors argue that the ants use a "Buffon needle algorithm" to assess nest areas.
- Power plants use large volumes of water for cooling. Water is usually drawn into the plant by means of intake pumps which have large impeller blades. Sometimes the water contains small larval fish. In *Larval fish, power plants and Buffon's needle*, Ebey and Beauchamp [1977] extend the classic Buffon needle problem to estimate the probability of a larval fish being killed by an impeller blade of a pump in a power plant.

- Among the many generalizations of Buffon's needle, our favorite is "Buffon's noodle." In this version, the needle is allowed to be a curve. Although the probability that the noodle crosses the line depends on the shape of the curve, remarkably the expected number of crossings does not. That is, the expected number of lines which the noodle crosses is equal to the expected number of lines that a straight needle will cross (see Ramaley [1969]).

8.5 TRANSFORMATIONS OF TWO RANDOM VARIABLES*

Most of our work finding densities revolved around working with a single variable or special functions of several random variables (sums, maximums, etc.). What if we wanted to find the density of $V = XY$, where X and Y are continuous? We can use a method like that introduced for a single variable to find the density, appropriately adapted to a setting with two variables.

Suppose X and Y are continuous random variables with joint density $f_{X,Y}$. Given functions of two variables $g_1(x, y)$ and $g_2(x, y)$, let $V = g_1(X, Y)$ and $W = g_2(X, Y)$. For example, if $g_1(x, y) = x + y$ and $g_2(x, y) = x - y$, then $V = X + Y$ and $W = X - Y$. In this section, we show how to find the joint density $f_{V,W}$ of V and W from the joint density $f_{X,Y}$ of X and Y.

The method for finding the joint density of a function of two random variables extends how we found the density of a function of one variable. In one dimension, if $Y = g(X)$, for some function g, and g is invertible, then we found earlier in the chapter that

$$f_Y(y) = f_X(g^{-1}(y)) \left| \frac{d}{dy} g^{-1}(y) \right|.$$

We see that the density of Y involves an "inverse piece" and a "derivative piece."

For two variables, suppose g_1 and g_2 are "invertible" in the sense that $v = g_1(x, y)$ and $w = g_2(x, y)$ can be solved uniquely for x and y with $x = h_1(v, w)$ and $y = h_2(v, w)$. This is equivalent to saying that g_1 and g_2 define a one-to-one transformation in the plane.

For instance, if $g_1(x, y) = x + y$ and $g_2(x, y) = x - y$, then solving

$$v = x + y \quad \text{and} \quad w = x - y$$

gives

$$x = \frac{v + w}{2} \quad \text{and} \quad y = \frac{v - w}{2},$$

and thus $h_1(v, w) = (v + w)/2$ and $h_2(v, w) = (v - w)/2$.

If h_1 and h_2 have continuous partial derivatives, let J denote the Jacobian consisting of the determinant of partial derivatives

$$J = \begin{vmatrix} \frac{\partial h_1}{\partial v} & \frac{\partial h_1}{\partial w} \\ \frac{\partial h_2}{\partial v} & \frac{\partial h_2}{\partial w} \end{vmatrix} = \frac{\partial h_1}{\partial v}\frac{\partial h_2}{\partial w} - \frac{\partial h_2}{\partial v}\frac{\partial h_1}{\partial w}.$$

Assume $J \neq 0$. The foregoing gives the joint density function of V and W.

JOINT DENSITY OF V AND W

Under the aforementioned assumptions,

$$f_{V,W}(v, w) = f_{X,Y}(h_1(v, w), h_2(v, w))|J|. \tag{8.5}$$

The factor $f_{X,Y}(h_1(v, w), h_2(v, w))$ is the "inverse piece," and the Jacobian is the "derivative piece." We do not give the proof of the joint density formula. It follows directly from the change of variable formula for multiple integrals in multivariable calculus. Most multivariable calculus textbooks contain a derivation.

In our working example, with $h_1(v, w) = (v + w)/2$ and $h_2(v, w) = (v - w)/2$, the Jacobian is

$$J = \begin{vmatrix} \frac{\partial h_1}{\partial v} & \frac{\partial h_1}{\partial w} \\ \frac{\partial h_2}{\partial v} & \frac{\partial h_2}{\partial w} \end{vmatrix} = \begin{vmatrix} \frac{\partial}{\partial v}\frac{v+w}{2} & \frac{\partial}{\partial w}\frac{v+w}{2} \\ \frac{\partial}{\partial v}\frac{v-w}{2} & \frac{\partial}{\partial w}\frac{v-w}{2} \end{vmatrix}$$

$$= \begin{vmatrix} 1/2 & 1/2 \\ 1/2 & -1/2 \end{vmatrix} = (-1/4) - (1/4) = -1/2.$$

This gives the joint density of V and W in terms of the joint density of X and Y:

$$f_{V,W}(v, w) = \frac{1}{2} f_{X,Y}\left(\frac{v+w}{2}, \frac{v-w}{2}\right).$$

Example 8.24 Suppose X and Y are independent standard normal random variables. Then from the result just derived, the joint distribution of $V = X + Y$ and $W = X - Y$ is

$$\begin{aligned} f_{V,W}(v, w) &= \frac{1}{2} f_{X,Y}\left(\frac{v+w}{2}, \frac{v-w}{2}\right) \\ &= \frac{1}{2} f\left(\frac{v+w}{2}\right) f\left(\frac{v-w}{2}\right) \\ &= \frac{1}{4\pi} e^{-(v+w)^2/8} e^{-(v-w)^2/8} \end{aligned}$$

$$= \frac{1}{4\pi} e^{-(v^2+w^2)/4}$$

$$= \left(\frac{1}{2\sqrt{\pi}} e^{-v^2/4} \right) \left(\frac{1}{2\sqrt{\pi}} e^{-w^2/4} \right).$$

This is a product of two normal densities, each with mean $\mu = 0$ and variance $\sigma^2 = 2$. In addition to finding the joint density of $X + Y$ and $X - Y$, we have also established an interesting result for normal random variables. If X and Y are independent, then so are $X + Y$ and $X - Y$. ■

Example 8.25 Sometimes a one-dimensional problem can be approached more easily by working in two-dimensions. Let X and Y be independent exponential random variables with parameter λ. Find the density of $X/(X + Y)$.

Let $V = X/(X + Y)$. We consider a two-dimensional problem letting $W = X + Y$. Then $V = g_1(X, Y)$, where $g_1(x, y) = x/(x + y)$. And $W = g_2(X, Y)$, where $g_2(x, y) = x + y$. Solving $v = x/(x + y)$ and $w = x + y$ for x and y gives $x = vw$ and $y = w - vw$. Thus,

$$h_1(v, w) = vw \quad \text{and} \quad h_2(v, w) = w - vw.$$

The Jacobian is

$$J = \begin{vmatrix} w & v \\ -w & 1 - v \end{vmatrix} = w - wv + wv = w.$$

The joint density of X and Y is

$$f_{X,Y}(x, y) = \lambda^2 e^{\lambda(x+y)}, \ x > 0, y > 0.$$

The joint density of V and W is thus

$$f_{V,W}(v, w) = f_{X,Y}(vw, w - vw)w = \lambda^2 w e^{-\lambda(vw+(w-vw))} = \lambda^2 w e^{-\lambda w},$$

for $0 < v < 1$ and $w > 0$.

To find the density of $V = X/(X + Y)$, integrate out the w variable, giving

$$f_V(v) = \int_0^\infty \lambda^2 w e^{-\lambda w}\, dw = \lambda \left(\frac{1}{\lambda} \right) = 1, \quad \text{for } 0 < v < 1.$$

That is, the distribution of $X/(X + Y)$ is uniform on $(0, 1)$. ■

Example 8.26 We consider a final example where working in two-dimensions is beneficial. Suppose a scientist is working with a compound whose active component is chemical X, which can be colored red or blue.

Suppose the density function for X_1, the proportion of a sample of the compound that is chemical X, is given by $f_{X_1}(x_1) = 6x_1(1 - x_1)$, for $0 \le x_1 \le 1$, and 0, otherwise. Further suppose that the density function for X_2, the proportion of red colored chemical X out of the chemical X in the sample, is given by $f_{X_2}(x_2) = 3x_2^2$, for $0 \leq x_2 \leq 1$, and 0, otherwise, and also that $X_1$and X_2 are independent.

The variable $Y = X_1X_2$ represents the proportion of the sample that is red colored chemical X. Find the probability density function for Y.

To make this two-dimensional, let $W = X_1$. Then $Y = g_1(X_1, X_2)$, where $g_1(x_1, x_2) = x_1x_2$. And $W = g_2(X_1, X_2)$, where $g_2(x_1, x_2) = x_1$. Solving for the inverse functions gives $x_1 = w$ and $x_2 = y/w$. Thus,

$$h_1(y, w) = w \quad \text{and} \quad h_2(y, w) = y/w.$$

The Jacobian is

$$J = \begin{vmatrix} 0 & 1 \\ 1/w & -y/w^2 \end{vmatrix} = 0 - 1/w = -1/w.$$

By independence, the joint density of X_1 and X_2 is

$$f_{X_1,X_2}(x_1, x_2) = 18x_1(1 - x_1)x_2^2, \ 0 \leq x_1 \leq 1, 0 \leq x_2 \leq 1.$$

The joint density of V and W is thus

$$f_{Y,W}(y, w) = f_{X_1,X_2}(w, y/w)(1/w) = 18w(1 - w)\left(\frac{y}{w}\right)^2(1/w) = 18y^2\frac{1 - w}{w^2},$$

for $0 \leq y \leq w \leq 1$.

For the bounds, these are obtained working from the inverse functions. Both X_1 and X_2 range from 0 to 1, so this gives $0 \leq w \leq 1$ and $0 \leq y/w \leq 1$. Rearranging gives $0 \leq y \leq w \leq 1$.

To find the density of $Y = X_1X_2$, integrate out the w variable, giving

$$f_Y(y) = \int_y^1 18y^2\frac{1 - w}{w^2}\, dw = 18y^2(\ln y + (1/y) - 1), \quad \text{for } 0 \leq y \leq 1.$$

We leave the reader to verify this integration. Equivalently, the density of Y could have been obtained via a similar process if we had let $W = X_2$. ■

Results for transformations of more than two random variables extend naturally from the two-variable case. Let $(X_1, \ldots, X_k)$ have joint density function $f_{X_1,\ldots,X_k}$. Given functions of k variables $g_1, \ldots, g_k$, let

$$V_i = g_i(X_1, \ldots, X_k), \ \text{for } i = 1, \ldots, k.$$

Assume the functions $g_1, \ldots, g_k$ define a one-to-one transformation in k dimensions with inverse functions $h_1, \ldots, h_k$ as in the two-variable case. Then the joint density of $(V_1, \ldots, V_k)$ is

$$f_{V_1,\ldots,V_k}(v_1, \ldots, v_k) = f_{X_1,\ldots,X_k}(h_1(v_1, \ldots, v_k), \ldots, h_k(v_1, \ldots, v_k))|J|,$$

where J is the determinant of the $k \times k$ matrix of partial derivatives whose (i,j)th entry is $\frac{\partial h_i}{\partial v_j}$, for $i, j = 1, \ldots, k$.

8.6 SUMMARY

This chapter focuses on finding densities for functions of random variables. One method relies on the density function for continuous random variables being the derivative of the cdf. Special functions of random variables are also explored, including maximums and minimums, and sums of independent variables.

When X and Y are independent and uniformly distributed, the random pair (X, Y) is uniformly distributed on a rectangle in the plane. Often, such problems can be treated geometrically. The chapter includes several examples of geometric probability.

Transformations of two random variables are presented in the final section. The formula for the probability density function is a consequence of the change of variable formula in multivariable calculus.

- **Densities of functions of random variables:** Suppose $Y = g(X)$. To find the density of Y, start with the cdf $P(Y \leq y) = P(g(X) \leq y)$. Obtain an expression of the form $P(X \leq h(y))$ for some function h. Differentiate to obtain the desired density. Most likely you will need to apply the chain rule.
- **Inverse transform method:** If the cdf F of a random variable X is invertible, and $U \sim \text{Unif}(0, 1)$, then $F^{-1}(U)$ has the same distribution as X. This gives a method for simulating X.
- **Convolution:** If X and Y are continuous and independent, with respective marginal densities f_X and f_Y, then the density of $X + Y$ is given by the convolution formula
$$f_{X+Y}(t) = \int_{-\infty}^{\infty} f_X(t - y) f_Y(y)\, dy.$$
- **Geometric probability:** Geometric probability is a powerful method for solving problems involving independent uniform random variables. If $X \sim \text{Unif}(a, b)$ and $Y \sim \text{Unif}(c, d)$ are independent, then (X, Y) is uniformly distributed on the rectangle $[a, b] \times [c, d]$. Finding uniform probabilities in two-dimensions reduces to finding areas. Try to find a geometrically based solution for these types of problems.
- **Transformation of two random variables—joint density:** Let X and Y have joint density $f_{X,Y}$. Suppose $V = g_1(X, Y)$ and $W = g_2(X, Y)$ with $X = h_1(V, W)$ and $Y = h_2(V, W)$. Let J be the Jacobian $J = \frac{\partial h_1}{\partial v}\frac{\partial h_2}{\partial w} - \frac{\partial h_2}{\partial v}\frac{\partial h_1}{\partial w}$. Then
$$f_{V,W}(v, w) = f_{X,Y}(h_1(v, w), h_2(v, w))|J|.$$

EXERCISES

Practice with Finding Densities

8.1 Suppose $X \sim \text{Exp}(\lambda)$. Find the density of $Y = cX$ for $c > 0$. Describe the distribution of Y.

8.2 Suppose X has the pdf given by $f(x) = \frac{x}{\theta^2} e^{-x^2/(2\theta^2)}$, for $x > 0$. Find the pdf for $U = X^2$. (Note the distribution of X here is called the Rayleigh density.)

8.3 Suppose that X has pdf given by $f(x) = (3/2)x^2$ on $-1 \le x \le 1$.

(a) Find the pdf of $Y = 2X$.

(b) Find the pdf of $W = X + 4$.

8.4 Suppose $U \sim \text{Unif}(0, 1)$. Find the density of

$$Y = \tan\left(\pi U - \frac{\pi}{2}\right).$$

8.5 Let $X \sim \text{Unif}(0, 1)$. Find $E[e^X]$ two ways:

(a) By finding the density of e^X and then computing the expectation with respect to that distribution.

(b) By using the law of the unconscious statistician.

8.6 The density of a random variable X is given by

$$f(x) = \begin{cases} 3x^2, & \text{for } 0 < x < 1 \\ 0, & \text{otherwise.} \end{cases}$$

Let $Y = e^X$.

(a) Find the density function of Y.

(b) Find $E[Y]$ two ways: (i) using the density of Y and (ii) using the density of X.

8.7 Let X have a Cauchy distribution. That is, the density of X is

$$f(x) = \frac{1}{\pi(1 + x^2)}, -\infty < x < \infty.$$

Show that $1/X$ has a Cauchy distribution.

8.8 Let X have an exponential distribution conditioned to be greater than 1. That is, for $t > 1$, $P(X \le t) = P(Y \le t | Y > 1)$, where $Y \sim \text{Exp}(\lambda)$.

(a) Find the density of X.

(b) Find $E[X]$.

8.9 Let $X \sim \text{Unif}(a, b)$. Suppose Y is a linear function of X. That is $Y = mX + n$, where m and n are constants. Assume also that $m > 0$. Show that Y is uniformly distributed on the interval $(ma + n, mb + n)$.

8.10 Suppose X has density function

$$f(x) = \frac{e^2}{2(e^2 - 1)}|x|e^{-x^2/2}, \quad \text{for } -2 < x < 2.$$

Find the density of $Y = X^2$.

8.11 If r is a real number, the *ceiling of* r, denoted $\lceil r \rceil$, is the smallest integer not less than r. For instance, $\lceil 0.25 \rceil = 1$ and $\lceil 4 \rceil = 4$. Suppose $X \sim \text{Exp}(\lambda)$. Let $Y = \lceil X \rceil$. Show that Y has a geometric distribution.

8.12 Your friend missed probability class today. Explain to your friend, in simple language, how the *inverse transform* method works.

8.13 Suppose $X \sim \text{Exp}(\lambda)$. Show how to use the inverse transform method to simulate X.

8.14 Suppose X has density function

$$f(x) = \frac{1}{(1 + x)^2}, \quad \text{for } x > 0.$$

Show how to use the inverse transform method to simulate X.

8.15 Let X be a continuous random variable with cdf F. As $F(x)$ is a function of x, $F(X)$ is a random variable which is a function of X. Suppose F is invertible. Find the distribution of $F(X)$.

8.16 Let $X \sim \text{Norm}(\mu, \sigma^2)$. Suppose $a \neq 0$ and b are constants. Show that $Y = aX + b$ is normally distributed using the method of cdfs. (This was solved using mgfs in Chapter 7.)

8.17 If Z has a standard normal distribution, find the density of Z^2 without using mgfs.

8.18 If X has a gamma distribution and c is a positive constant, show that cX has a gamma distribution without using mgfs. Find the parameters.

8.19 Let $X \sim \text{Beta}(a, b)$. For $s < t$, let $Y = (t - s)X + s$. Then Y has an extended beta distribution on (s, t). Find the density function of Y.

8.20 Let $X \sim \text{Exp}(a)$. Let $Y = me^X$. Show that $Y \sim \text{Pareto}(m, a)$.

8.21 Let $X \sim \text{Beta}(a, b)$. Find the distribution of $Y = 1 - X$.

8.22 Let $X \sim \text{Beta}(a, 1)$. Show that $Y = 1/X$ has a Pareto distribution.

8.23 Suppose X and Y are i.i.d. exponential random variables with $\lambda = 1$. Find the density of X/Y and use it to compute $P(X/Y < 1)$.

8.24 Let X_1 and X_2 be independent exponential random variables with parameter λ. Show that $Y = |X_1 - X_2|$ is also exponentially distributed with parameter λ.

8.25 Consider the following attempt at generating a point uniformly distributed in the circle of radius 1 centered at the origin. In polar coordinates, pick R uniformly at random on $(0, 1)$. Pick Θ uniformly at random on $(0, 2\pi)$, independently of R. Show that this method does *not* work. That is show (R, Θ) is not uniformly distributed on the circle.

8.26 Let $A = UV$, where U and V are independent and uniformly distributed on $(0, 1)$.

(a) Find the density of A.

(b) Find $E[A]$ two ways: (i) using the density of A and (ii) not using the density of A.

Maxs, Mins, and Convolution

8.27 In Example 8.16 we found the distribution of the sum of two i.i.d. exponential variables with parameter λ. Call the sum X. Let Y be a third independent exponential variable with parameter λ. Use the convolution formula 8.3 to find the sum of three independent exponential random variables by finding the distribution of $X + Y$.

8.28 Your friend missed probability class today. Explain to your friend, in simple language, how the *convolution* formula works.

8.29 Jakob and Kayla each pick uniformly random real numbers between 0 and 1. Find the expected value of the smaller number.

8.30 Let $X_1, \ldots, X_n$ be an i.i.d. sequence of Uniform $(0, 1)$ random variables. Let $M = \max(X_1, \ldots, X_n)$.

(a) Find the density function of M.

(b) Find $E[M]$ and $V[M]$.

8.31 Perdita moved into a new apartment and put new identical lightbulbs in her four lamps. Suppose the time to failure for each lightbulb is exponentially distributed with $\lambda = 1/10$, and that lightbulbs fail independently. Perdita will replace all the bulbs when the first fails.

(a) Find the density function of M, the time that the first bulb fails.

(b) Find $E[M]$ and $V[M]$.

8.32 Let X, Y, and Z be i.i.d. random variables uniformly distributed on $(0, 1)$. Find the density of $X + Y + Z$.

8.33 Let U_1, U_2 be independent draws from a Uniform distribution on $(0, \theta)$.

(a) Find the density function of M, the maximum of U_1 and U_2.

(b) Find $E(M)$.

(c) Suppose M is observed to be 6. What is a good guess for θ?

8.34 **Extreme value distribution.** Suppose $X_1, \ldots, X_n$ is an independent sequence of exponential random variables with parameter $\lambda = 1$. Let

$$Z = \max(X_1, \ldots, X_n) - \log n.$$

(a) Show that the cdf of Z is

$$F_Z(z) = \left(1 - \frac{e^{-z}}{n}\right)^n, z > -\log n.$$

(b) Show that for all z,

$$F_Z(z) \to e^{-e^{-z}} \text{as } n \to \infty.$$

The limit is a probability distribution called an *extreme value distribution*. It is used in many fields which model extreme values, such as hydrology (intense rainfall), actuarial science, and reliability theory.

(c) Suppose the times between heavy rainfalls are independent and have an exponential distribution with mean 1 month. Find the probability that in the next 10 years, the maximum time between heavy rainfalls is greater than 3 months in duration.

8.35 **Order statistics.** Suppose $X_1, \ldots, X_{100}$ are independent and uniformly distributed on $(0, 1)$.

(a) Find the probability the 25th smallest variable is less than 0.20.

(b) Find $E[X_{(95)}]$ and $V[X_{(95)}]$.

(c) The *range* of a set of numbers is the difference between the maximum and the minimum. Find the expected range.

8.36 If n is odd, the *median* of a list of n numbers is the middle value. Suppose a sample of size $n = 13$ is taken from a uniform distribution on $(0, 1)$. Let M be the median. Find $P(M > 0.55)$.

Geometric Probability

8.37 Suppose X and Y are independent random variables, each uniformly distributed on $(0, 2)$.

(a) Find $P(X^2 < Y)$.

(b) Find $P(X^2 < Y | X + Y < 2)$.

8.38 Suppose (X, Y) is uniformly distributed on the region in the plane between the curves $y = \sin x$ and $y = \cos x$, for $0 < x < \pi/2$. Find $P(Y > 1/2)$.

8.39 Suppose (X, Y, Z) is uniformly distributed on the sphere of radius 1 centered at the origin. Find the probability that (X, Y, Z) is contained in the inscribed cube.

8.40 Solve Buffon's needle problem for a "short" needle. That is, suppose the length of the needle is $x < 1$.

8.41 Suppose you use Buffon's needle problem to simulate π. Let n be the number of needles you drop on the floor. Let X be the number of needles that cross a line. Find the distribution, expectation and variance of X.

8.42 Suppose X and Y are independent random variables uniformly distributed on $(0, 1)$. Use geometric arguments to find the density of $Z = X/Y$.

8.43 A *spatial Poisson process* is a model for the distribution of points in two-dimensional space. For a set $A \subseteq \Re^2$, let N_A denote the number of points in A. The two defining properties of a spatial Poisson process with parameter λ are

1. If A and B are disjoint sets, then N_A and N_B are independent random variables.
2. For all $A \subseteq \Re^2$, N_A has a Poisson distribution with parameter $\lambda|A|$ for some $\lambda > 0$, where $|A|$ denotes the area of A. That is,

$$P(N_A = k) = \frac{e^{-\lambda|A|}(\lambda|A|)^k}{k!}, k = 0, 1, \ldots$$

Consider a spatial Poisson process with parameter λ. Let x be a fixed point in the plane.

(a) Find the probability that there are no points of the spatial process that are within two units distance from x. (Draw the picture.)

(b) Let X be the distance between x and the nearest point of the spatial process. Find the density of X. (Hint: Find $P(X > x)$.)

Bivariate Transformations

8.44 A candy mix includes three different types of candy. Let X and Y be the proportions of the first two types of candy in the mixture. Suppose that the joint pdf of X and Y is uniform over the region where $0 \leq x \leq 1$, $0 \leq y \leq 1$, and $0 \leq x + y \leq 1$, and is 0, elsewhere.

(a) Find the pdf of $W = X + Y$, the proportion of the first two types of candy in the mixture.

(b) How likely is it that W will be greater than 0.5?

8.45 Suppose X and Y are independent exponential random variables with parameter λ. Find the joint density of $V = X/Y$ and $W = X + Y$. Use the joint density to find the marginal distributions.

8.46 Suppose that a point (X, Y) is chosen in the unit square with probability governed by the joint pdf $f(x, y) = x + y$, $0 < x < 1, 0 < y < 1$. What is the pdf of W, the area of the rectangle formed by the points $(0, 0)$, $(x, 0)$, $(0, y)$, and (x, y)?

8.47 Suppose X and Y have joint density

$$f(x, y) = 4xy, \quad \text{for } 0 < x < 1, 0 < y < 1.$$

Find the joint density of $V = X$ and $W = XY$. Find the marginal density of W.

8.48 Let X and Y be jointly continuous with density $f_{X,Y}$. Let (R, Θ) be the polar coordinates of (X, Y).

(a) Give a general expression for the joint density of R and Θ.

(b) Suppose X and Y are independent with common density function $f(x) = 2x$, for $0 < x < 1$. Use your result to find the probability that (X, Y) lies inside the circle of radius one centered at the origin.

8.49 Recall that the density function of the Cauchy distribution is

$$f(x) = \frac{1}{\pi(1 + x^2)}, \qquad \text{for all } x.$$

Show that the ratio of two independent standard normal random variables has a Cauchy distribution by finding a suitable transformation of two variables.

8.50 Pairs of standard normal random variables can be generated from a pair of independent uniforms. To investigate, we work in reverse. Let X and Y be independent standard normal random variables. Let $V = X^2 + Y^2$ and $W = \tan^{-1}(Y/X)$.

(a) Show that V and W are independent with $V \sim \text{Exp}(1/2)$ and $W \sim \text{Unif}(0, 2\pi)$. (Hint: $x = \sqrt{v}\cos w$ and $y = \sqrt{v}\sin w$.)

(b) Now let $U_1, U_2 \sim \text{Unif}(0, 1)$. Show that $V = -2\ln U_1$ and $W = 2\pi U_2$ are independent with distributions described above.

(c) Implement this method and plot 1000 pairs of points to show that pairs of independent standard normal random variables can be generated from pairs of independent uniforms. In other words, show that $(U_1, U_2) \to (V, W) \to (X, Y)$ yields independent standard normal random variables.

8.51 (This exercise requires knowledge of three-dimensional determinants.) Let (X, Y, Z) be independent standard normal random variables. Let (Φ, Θ, R) be the corresponding spherical coordinates. The correspondence between rectangular and spherical coordinates is given by

$$x = r \sin \phi \cos \theta, \quad y = r \sin \phi \sin \theta, \quad z = r \cos \phi,$$

for $0 \leq \phi \leq \pi$, $0 \leq \theta \leq 2\pi$, and $r > 0$. Find the joint density of (Φ, Θ, R) and the marginal density of R.

Simulation and R

8.51 Let $R \sim \text{Unif}(1, 4)$. Let A be the area of the circle of radius R. Use **R** to simulate R. Simulate the mean and pdf of A and compare to the exact results. Create one graph with both the theoretical density and the simulated distribution.

8.52 Write an **R** script to estimate π using Buffon's needle problem. How many simulation iterations do you need to perform to be reasonably confident that your estimation is good within two significant digits? That is, $\pi \approx 3.14$.

8.53 Let X be a random variable with density function $f(x) = 4/(3x^2)$, for $1 < x < 4$, and 0, otherwise. Simulate $E[X]$ using the inverse transform method. Compare to the exact value.

8.54 Make up your own example to demonstrate the inverse transform method. State the pdf of the random variable of interest. Implement the method to simulate $E(X)$ and compare to the exact value that you calculate.

8.55 Let $X_1, \ldots, X_n$ be independent random variables each uniformly distributed on $[-1, 1]$. Let $p_n = P(X_1^2 + \cdots + X_n^2 < 1)$. Conduct a simulation study to approximate p_n for increasing values of n. For $n = 2$, p_2 is the probability that a point uniformly distributed on the square $[-1, 1] \times [-1, 1]$ falls in the inscribed circle of radius 1 centered at the origin. For $n = 3$, p_3 is the probability that a point uniformly distributed on the cube $[-1, 1] \times [-1, 1] \times [-1, 1]$ falls in the inscribed sphere of radius 1 centered at the origin. For $n > 3$, you are in higher dimensions estimating the probability that a point in a "hypercube" falls within the inscribed "hypersphere." What happens when n gets large?

Chapter Review

Chapter review exercises are available through the text website. The URL is `www.wiley.com/go/wagaman/probability2e`.

9

CONDITIONAL DISTRIBUTION, EXPECTATION, AND VARIANCE

Conditioning is a must for martial artists.

—Bruce Lee

Learning Outcomes

1. Define the terms: conditional distribution, conditional expectation, and conditional variance.
2. Find conditional distributions using relationships between joint and marginal distributions.
3. Compute conditional expectation and variance using appropriate rules.
4. Understand the application of the conditional concepts in the bivariate normal setting.
5. (C) Simulate conditional distributions.

INTRODUCTION

At this crossroads, we bring together several important ideas in both discrete and continuous probability. We focus on conditional distributions and conditional expectation, briefly introduced previously. We will introduce conditional density functions, extend the law of total probability, present problems with both discrete and continuous components, and expand our available tools for computing probabilities. Finally, we explore many of these concepts via the bivariate normal density.

Probability: With Applications and R, Second Edition. Amy S. Wagaman and Robert P. Dobrow.

Companion Website: www.wiley.com/go/wagaman/probability2e

9.1 CONDITIONAL DISTRIBUTIONS

Conditional distributions for discrete variables were introduced in Chapter 4. For continuous variables, the conditional (probability) density function plays the analogous role to the conditional probability mass function.

CONDITIONAL DENSITY FUNCTION

If X and Y are jointly continuous random variables, the conditional density of Y given $X = x$ is

$$f_{Y|X}(y|x) = \frac{f(x, y)}{f_X(x)}, \tag{9.1}$$

for $f_X(x) > 0$.

The conditional density function is a valid probability density function, as it is nonnegative and integrates to 1. For a given x,

$$\int_{-\infty}^{\infty} f_{Y|X}(y|x)\, dy = \int_{-\infty}^{\infty} \frac{f(x, y)}{f_X(x)}\, dy = \frac{1}{f_X(x)} \int_{-\infty}^{\infty} f(x, y)\, dy = \frac{f_X(x)}{f_X(x)} = 1.$$

Geometrically, the conditional density of Y given $X = x$ is a one-dimensional "slice" of the two-dimensional joint density along the line $X = x$, but "renormalized" so that the resulting curve integrates to 1 (see Fig. 9.1).

Conditional densities are used to compute conditional probabilities. For $A \subseteq \Re$,

$$P(Y \in A|X = x) = \int_A f_{Y|X}(y|x)\, dy.$$

When working with conditional density functions $f_{Y|X}(y|x)$, it is most important to remember that the conditioning variable x is treated as *fixed*. The conditional density function is a function of its first argument y.

■ **Example 9.1** Random variables X and Y have joint density

$$f(x, y) = \frac{y}{4x}, \quad \text{for } 0 < y < x < 4,$$

and 0, otherwise. Find $P(Y < 1|X = x)$, for $x \geq 1$, and $P(Y < 1|X = 2)$.

As you tackle this problem, think carefully about the steps you need to follow to find the desired probabilities, as outlined below. Note that if $x < 1$, $P(Y < 1|X = x) = 1$.

By definition, with $x \geq 1$, the first desired probability is

$$P(Y < 1|X = x) = \int_0^1 f_{Y|X}(y|x)\, dy.$$

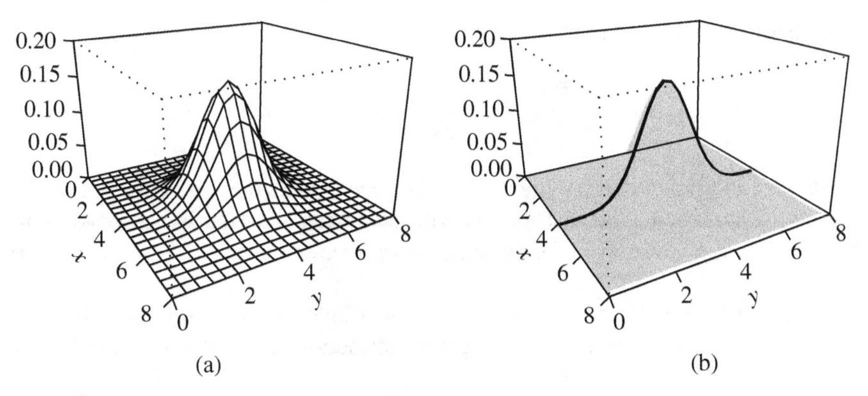

FIGURE 9.1: (a) Graph of a joint density function. (b) Conditional density of Y given $X = 4$.

To find the conditional density of Y given $X = x$, we first need the marginal density of X. Paying attention to the range of Y, we find the marginal of X as

$$f_X(x) = \int_0^x \frac{y}{4x}\, dy = \frac{1}{4x}\left(\frac{y^2}{2}\right)\Bigg|_0^x = \frac{x}{8}, \quad \text{for } 0 < x < 4.$$

The conditional density is

$$f_{Y|X}(y|x) = \frac{f(x,y)}{f_X(x)} = \frac{y/(4x)}{x/8} = \frac{2y}{x^2}, \quad \text{for } 0 < y < x < 4.$$

Now we can find the desired probability as

$$P(Y < 1|X = x) = \int_0^1 \frac{2y}{x^2}\, dy = \frac{1}{x^2},$$

assuming $x \geq 1$, and by plugging in $x = 2$, we find $P(Y < 1|X = 2) = 1/4$. ■

Example 9.2 Random variables X and Y have joint density function

$$f(x,y) = e^{-x^2 y}, \quad \text{for } x > 1, y > 0.$$

Find and describe the conditional distribution of Y given $X = x$.

The marginal density of X is

$$f_X(x) = \int_0^\infty f(x,y)\, dy = \int_0^\infty e^{-x^2 y}\, dy = \frac{1}{x^2}, \quad \text{for } x > 1.$$

The conditional density function is

$$f_{Y|X}(y|x) = \frac{f(x,y)}{f_X(x)} = \frac{e^{-x^2 y}}{1/x^2} = x^2 e^{-x^2 y}, \quad \text{for } y > 0.$$

In the conditional density function, x^2 is treated as a constant. Thus, this is the density of an exponential distribution with parameter x^2. That is, the conditional distribution of Y given $X = x$ is exponential with parameter x^2. ■

Example 9.3 Let X and Y be uniformly distributed in the circle of radius one centered at the origin. Find (i) the marginal distribution of X and (ii) the conditional distribution of Y given $X = x$.

(i) The area of the circle is π. The equation of the circle is $x^2 + y^2 = 1$. The joint density of X and Y is

$$f(x,y) = \frac{1}{\pi}, \quad \text{for } -1 \leq x \leq 1, -\sqrt{1-x^2} \leq y \leq \sqrt{1-x^2},$$

and 0, otherwise. Integrating out the y term gives the marginal density of X as

$$f_X(x) = \int_{-\sqrt{1-x^2}}^{\sqrt{1-x^2}} \frac{1}{\pi}\, dy = \frac{2\sqrt{1-x^2}}{\pi}, \quad \text{for } -1 \leq x \leq 1.$$

(ii) The conditional density of Y given $X = x$ is

$$f_{Y|X}(y|x) = \frac{f(x,y)}{f_X(x)} = \frac{1/\pi}{2\sqrt{1-x^2}/\pi}$$
$$= \frac{1}{2\sqrt{1-x^2}}, \quad \text{for } -\sqrt{1-x^2} < y < \sqrt{1-x^2}.$$

The function $f_{Y|X}(y|x)$ does not depend on y. As x is treated as a constant, the conditional distribution is uniform on the interval $(-\sqrt{1-x^2}, \sqrt{1-x^2})$.

Observe that while the conditional distribution of Y given $X = x$ is uniform, the marginal distribution of X is not. The marginal distribution of X is the distribution of the X-coordinate of a point (X, Y) in the circle. Points tend to be closer to the origin than to the outside of the circle as described by the half-circular marginal density. ■

Example 9.4 Lillian is working on a project that has many tasks to complete, including tasks A and B. Let X be the proportion of the project time she spends on task A. Let Y be the proportion of time she spends on B. The joint density of X and Y is

$$f(x,y) = 24xy, \quad \text{for } 0 < x < 1, 0 < y < 1 - x,$$

and 0, otherwise. If the fraction of the time Lillian spends on task A is x, find the probability that she spends at least half the project time on task B.

The total fraction of the time that Lillian spends on both tasks A and B must be less than 1. There are two cases based on the given value of x. For $x < 1/2$, the desired probability is $P(Y > 1/2|X = x)$. If $x \geq 1/2$, then $P(Y > 1/2|X = x) = 0$. We focus on the first case.

The marginal density of X is

$$f_X(x) = \int_0^{1-x} 24xy\,dy = 12x(1-x)^2, \quad \text{for } 0 < x < 1.$$

You should recognize this as a beta distribution. The conditional density of Y given $X = x$ is

$$f_{Y|X}(y|x) = \frac{f(x,y)}{f_X(x)} = \frac{24xy}{12x(1-x)^2} = \frac{2y}{(1-x)^2}, \quad \text{for } 0 < y < 1-x.$$

Thus, for $0 < x < 1/2$,

$$P(Y > 1/2|X = x) = \int_{1/2}^{1-x} \frac{2y}{(1-x)^2}\,dy = 1 - \frac{1}{4(1-x)^2}.$$

■

Many random experiments are performed in stages. Consider the following two-stage, *hierarchical* model. Riley picks a number X uniformly distributed in $(0, 1)$. Riley shows her number $X = x$ to Miguel, who picks a number Y uniformly distributed in $(0, x)$. The conditional distribution of Y given $X = x$ is uniform on $(0, x)$ and thus the conditional density is

$$f_{Y|X}(y|x) = \frac{1}{x}, \quad \text{for } 0 < y < x < 1.$$

If we know that Riley picked 2/3, then the probability that Miguel's number is greater than $1/3$ is

$$P(Y > 1/3|X = 2/3) = \int_{1/3}^{2/3} \frac{3}{2}\,dy = \frac{1}{2}.$$

On the other hand, suppose we only see the second stage of the experiment. If Miguel's number is 1/3, what is the probability that Riley's original number is greater than 2/3? The desired probability is $P(X > 2/3|Y = 1/3)$. This requires the conditional density of X given $Y = y$.

We are given the conditional density $f_{Y|X}(y|x)$ and we want to find the "inverse" conditional density $f_{X|Y}(x|y)$. This has the flavor of Bayes formula used to invert conditional probabilities. We appeal to the continuous version of Bayes formula. By rearranging the conditional density formula,

$$f_{X|Y}(x|y) = \frac{f(x,y)}{f_Y(y)} = \frac{f_{Y|X}(y|x)f_X(x)}{f_Y(y)}$$

with denominator equal to

$$f_Y(y) = \int_{-\infty}^{\infty} f(t,y)\,dt = \int_{-\infty}^{\infty} f_{Y|X}(y|t)f_X(t)\,dt.$$

This gives the continuous version of Bayes formula.

BAYES FORMULA

Let X and Y be jointly distributed continuous random variables. Then

$$f_{X|Y}(x|y) = \frac{f_{Y|X}(y|x)f_X(x)}{\int_{-\infty}^{\infty} f_{Y|X}(y|t)f_X(t)\,dt}. \tag{9.2}$$

Using Bayes formula on Riley and Miguel's problem, we find the conditional density of X given $Y = y$

$$f_{X|Y}(x|y) = \frac{(1/x)(1)}{\int_y^1 (1/t)(1)\,dt} = \frac{-1}{x\ln y}, \text{ for } 0 < y < x < 1.$$

The desired probability is

$$P(X > 2/3|Y = 1/3) = \int_{2/3}^{1} \frac{-1}{x\ln 1/3}\,dx = (\ln 3)(\ln 3/2) = 0.445.$$

Example 9.5 The time between successive tsunamis in the Caribbean is modeled with an exponential distribution. See Parsons and Geist [2008] for the use of probability in predicting tsunamis. The parameter value of the exponential distribution is unknown and is itself modeled as a random variable Λ uniformly distributed on (0, 1). Suppose the time X between the last two consecutive tsunamis was 2 years. Find the conditional density of Λ given $X = 2$. (We use the notation Λ since it is the Greek capital letter lambda.)

The conditional distribution of X given $\Lambda = \lambda$ is exponential with parameter λ. That is, $f_{X|\Lambda}(x|\lambda) = \lambda e^{-\lambda x}$, for $x > 0$. The (unconditional) density of Λ is $f_\Lambda(\lambda) = 1$, for $0 < \lambda < 1$. By Bayes formula,

$$\begin{aligned} f_{\Lambda|X}(\lambda|x) &= \frac{f_{X|\Lambda}(x|\lambda)f_\Lambda(\lambda)}{\int_{-\infty}^{\infty} f_{X|\Lambda}(x|t)f_\Lambda(t)\,dt} \\ &= \frac{\lambda e^{-\lambda x}}{\int_0^1 te^{-tx}\,dt} \\ &= \frac{\lambda e^{-\lambda x}}{(1 - e^{-x} - xe^{-x})/x^2}, \text{ for } 0 < \lambda < 1. \end{aligned}$$

The conditional density of Λ given $X = 2$ is

$$\begin{aligned} f_{\Lambda|X}(\lambda|2) &= \frac{\lambda e^{-2\lambda}}{(1 - e^{-2} - 2e^{-2})/4} \\ &= \frac{4\lambda e^{-2\lambda}}{1 - 3e^{-2}}, \text{ for } 0 < \lambda < 1. \end{aligned}$$

Of interest to researchers is estimating λ given the observed data $X = 2$. One approach is to find the value of λ that maximizes the conditional density function (see Fig. 9.2). The density is maximized at $\lambda = 1/2$.

In hindsight, this is not surprising. A natural estimate for λ is 2 given that the mean of an exponential distribution with parameter $\lambda = 1/2$ is 2, the value of X. ■

Motivating the definition. For jointly continuous random variables X and Y, and $A \subseteq R$, the probability $P(Y \in A|X = x)$ is found by integrating the conditional density function. Although the probability is conditional, we *cannot* use the conditional

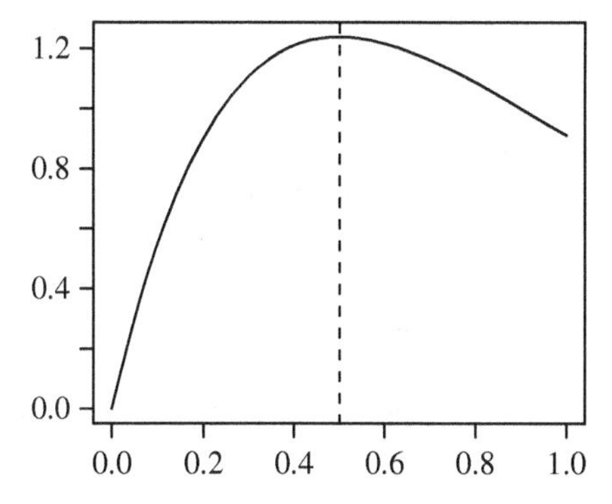

FIGURE 9.2: Graph of conditional density of Λ given $X = 2$. The density function takes its largest value at $\lambda = 1/2$.

probability formula

$$P(Y \in A|X = x) = \frac{P(Y \in A, X = x)}{P(X = x)},$$

because if X is continuous both numerator and denominator are equal to 0, and the conditional probability is undefined.

However, when we try to divide 0 by 0, there often is a limit lurking in the background. To motivate the definition of the conditional density function in Equation 9.1, we have

$$\begin{aligned}
P(Y \in A|X = x) &= \lim_{\epsilon \to 0} P(Y \in A|x \le X \le x + \epsilon) \\
&= \lim_{\epsilon \to 0} \frac{P(Y \in A, x \le X \le x + \epsilon)}{P(x \le X \le x + \epsilon)} \\
&= \lim_{\epsilon \to 0} \frac{\int_x^{x+\epsilon} \int_A f(t, y)\, dy\, dt}{\int_x^{x+\epsilon} f_X(t)\, dt} \\
&= \frac{\lim_{\epsilon \to 0} \frac{1}{\epsilon} \int_x^{x+\epsilon} \int_A f(t, y)\, dy\, dt}{\lim_{\epsilon \to 0} \frac{1}{\epsilon} \int_x^{x+\epsilon} f_X(t)\, dt} \\
&= \frac{\int_A f(x, y)\, dy}{f_X(x)} = \int_A \left[\frac{f(x, y)}{f_X(x)}\right] dy \\
&= \int_A f_{Y|X}(y|x)\, dy, \text{ for all } A \subseteq \mathfrak{R}.
\end{aligned}$$

9.2 DISCRETE AND CONTINUOUS: MIXING IT UP

Hopefully the reader has appreciated the close similarity between results for discrete and continuous random variables. In an advanced probability course, using the tools of real analysis and measure theory, the two worlds are unified. A new type of integral—called a Lebesgue integral—is used for *both* continuous and discrete random variables, and for random variables that exhibit properties of both. There is no need to treat discrete and continuous problems as two separate categories.

Such a discussion is beyond the scope of this book. However, we can, and will, define joint and conditional distributions of random variables where one variable is discrete and the other is continuous. Some common pairings are illustrated in the examples below.

To illustrate, here is a "continuous-discrete" two-stage random experiment.

Example 9.6 Pick a number U uniformly at random between 0 and 1. Given $U = u$, consider a "biased" coin whose probability of heads is equal to u. Flip such a coin n times and let H be the number of heads.

The conditional distribution of H given $U = u$ is a binomial distribution, with parameters n and u. The conditional probability mass function of H given $U = u$ is

$$P(H = h|U = u) = \binom{n}{h} u^h(1 - u)^{n-h}, \quad h = 0, \ldots, n.$$

We write this as $H|U = u \sim \text{Binom}(n, u)$. Now suppose we want to find the *unconditional* distribution of H.

The joint density of H and U is

$$\begin{aligned} f(h, u) &= P(H = h|U = u)f_U(u) \\ &= \binom{n}{h} u^h(1 - u)^{n-h}, \quad h = 0, \ldots, n, 0 < u < 1. \end{aligned} \tag{9.3}$$

The function contains a discrete component and a continuous component. To compute probabilities, we sum *and* integrate respectively:

$$P(H \in A, U \in B) = \sum_{h \in A} \int_B f(h, u)\, du = \sum_{h \in A} \int_B \binom{n}{h} u^h(1 - u)^{n-h}\, du.$$

The marginal distribution of H is obtained by integrating out u in the joint density, giving

$$P(H = h) = \int_0^1 \binom{n}{h} u^h(1 - u)^{n-h}\, du = \binom{n}{h} \int_0^1 u^h(1 - u)^{n-h}\, du.$$

The integral can be solved using integration by parts, or you can recognize that the integrand is proportional to a beta density with parameters $a = h + 1$ and $b = n - h + 1$. Hence, the integral is equal to

$$\frac{\Gamma(h + 1)\Gamma(n - h + 1)}{\Gamma(n + 2)} = \frac{h!(n - h)!}{(n + 1)!}.$$

This gives

$$P(H = h) = \binom{n}{h} \frac{h!(n - h)!}{(n + 1)!} = \frac{1}{n + 1}, \text{ for } h = 0, \ldots, n.$$

We see that H is uniformly distributed on the integers $\{0, \ldots, n\}$.

Another quantity of interest in this example is the conditional density of U given $H = h$. For $h = 0, \ldots, n$,

$$\begin{aligned} f_{U|H}(u|h) = \frac{f(h, u)}{f_H(h)} &= \frac{\binom{n}{h} u^h(1 - u)^{n-h}}{1/(n + 1)} \\ &= \frac{(n + 1)!}{h!(n - h)!} u^h(1 - u)^{n-h}, \ 0 < u < 1. \end{aligned}$$

The density almost looks like a binomial probability expression. But h is fixed and the density is a continuous function of u. The conditional distribution of U given $H = h$ is a beta distribution with parameters $a = h + 1$ and $b = n - h + 1$. ■

The example we have used to motivate this discussion is a case of the *beta-binomial model*, an important model in Bayesian statistics and numerous applied fields.

Example 9.7 Will the sun rise tomorrow? Suppose an event, such as the daily rising of the sun, has occurred n times without fail. What is the probability that it will occur again?

In 1774, Laplace formulated his *law of succession* to answer this question. In modern notation, let S be the unknown probability that the sun rises on any day. And let X be the number of days the sun has risen among the past n days. If $S = s$, and we assume successive days are independent, then the conditional distribution of X is binomial with parameters n and s, and $P(X = n|S = s) = s^n$. Laplace assumed in the absence of any other information that the "unknown" sun rising probability S was uniformly distributed on (0, 1).

Using Bayes formula, the conditional density of the sun rising probability, given that the sun has risen for the past n days, is

$$f_{S|X}(s|n) = \frac{P(X = n|S = s)f_S(s)}{\int_0^1 P(X = n|S = t)f_S(t)\,dt} = \frac{s^n}{\int_0^1 t^n\,dt} - (n + 1)s^n.$$

Laplace computed the mean sun rising probability with respect to this conditional density

$$\int_0^1 s(n + 1)s^n\,ds = \frac{n + 1}{n + 2}.$$

He argued that the sun has risen for the past 5000 years, or 1,826,213 days. And thus the probability that the sun will rise tomorrow is

$$\frac{1{,}826{,}214}{1{,}826{,}215} = 0.9999994524\ldots.$$

With such "certainty," hopefully you will sleep better tonight! ■

Example 9.8 The following two-stage "exponential-Poisson" setting has been used to model traffic flow on networks and highways. Consider e-mail traffic during a 1-hour interval. Suppose the unknown rate of e-mail traffic Λ has an exponential distribution with parameter one. Further suppose that if $\Lambda = \lambda$, the number of e-mail messages M that arrive during the hour has a Poisson distribution with parameter 100λ. Find the probability mass function of M.

The conditional distribution of M given $\Lambda = \lambda$ is Poisson with parameter 100λ. The joint density of M and Λ is

$$f(m, \lambda) = f_{M|\Lambda}(m|\lambda)f_{\Lambda}(\lambda) = \frac{e^{-100\lambda}(100\lambda)^m}{m!}e^{-\lambda}$$
$$= \frac{e^{-101\lambda}(100\lambda)^m}{m!}, \quad \text{for } \lambda > 0 \text{ and } m = 0, 1, 2, \ldots.$$

Integrating the "mixed" joint density with respect to λ gives the discrete probability mass function

$$P(M = m) = \frac{1}{m!}\int_0^\infty e^{-101\lambda}(100\lambda)^m \, d\lambda = \frac{1}{m!}\left(\frac{m!}{101}\left(\frac{100}{101}\right)^m\right)$$
$$= \left(1 - \frac{1}{101}\right)^{m-1}\frac{1}{101}, \quad \text{for } m = 1, 2, \ldots.$$

Written this way, we see that the number of e-mail messages that arrive during the hour has a geometric distribution with parameter $p = 1/101$. ■

R: SIMULATING EXPONENTIAL-POISSON TRAFFIC FLOW MODEL

The following code simulates the joint distribution of Λ and M in the two-stage traffic flow model. Letting n be the number of trials in the simulation, the data are stored in an $n \times 2$ matrix `simarray`. Each row consists of an outcome of (Λ, M).

```
> n <- 1000
> simarray <- matrix(0, n, 2)
> for (i in 1:n) {
simarray[i, 1] <- rexp(1, 1)
simarray[i, 2] <- rpois(1, 100*simarray[i, 1]) }
```

Marginal distributions of Λ and M are simulated by simply taking the respective first and second columns of the `simarray` matrix.

See Figure 9.3 for graphs of the simulated joint distribution of (Λ, M) and the marginal distributions. Compare the outcomes of M with the geometric distribution with parameter $p = 1/101$. The simulation and R commands for generating the graphs can be found in the script file **Traffic.R**.

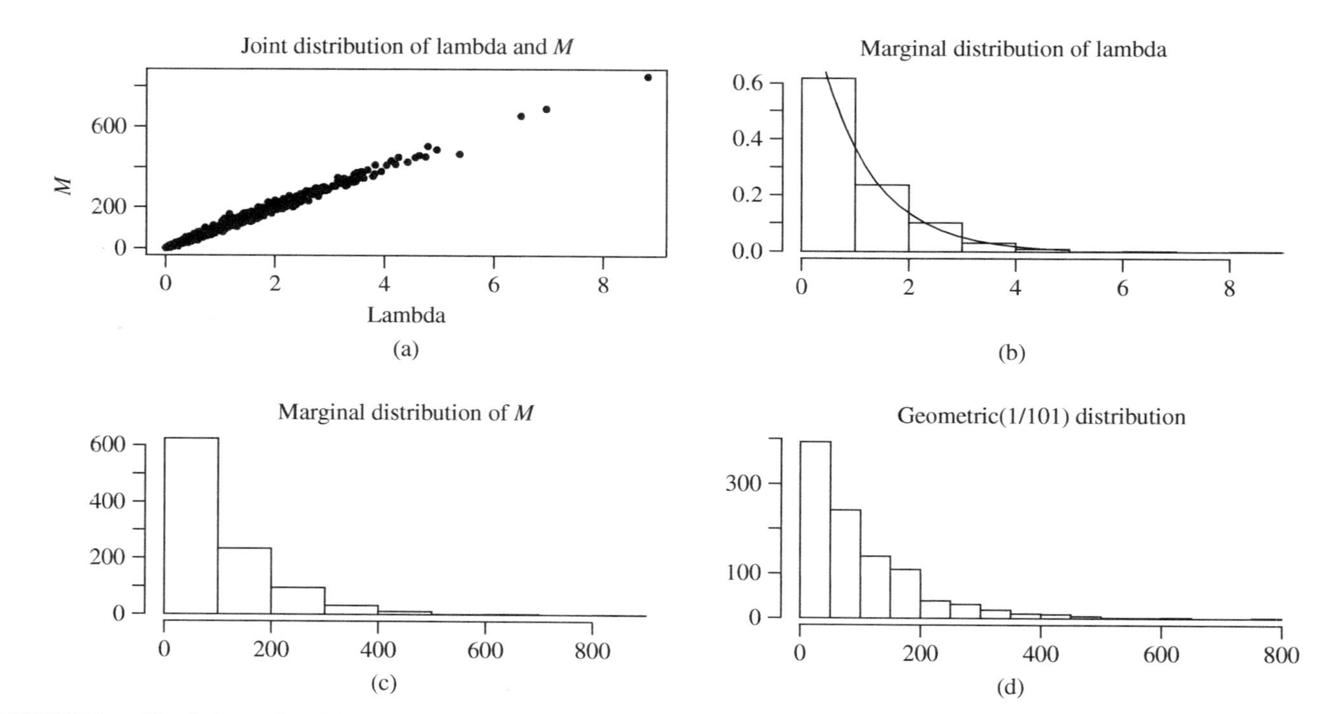

FIGURE 9.3: Simulations of (a) joint distribution of (Λ, M), (b) marginal distribution of Λ together with exponential density curve ($\lambda = 1$), (c) marginal distribution of M, and (d) geometric distribution ($p = 1/101$).

9.3 CONDITIONAL EXPECTATION

A *conditional expectation* is an expectation computed with respect to a conditional distribution. We write $E[Y|X = x]$ for the *conditional expectation of* Y *given* $X = x$. Our treatment of conditional expectation combines both discrete and continuous settings.

CONDITIONAL EXPECTATION OF Y GIVEN $X = x$

$$E[Y|X = x] = \begin{cases} \sum_y yP(Y = y|X = x), & \text{discrete} \\ \int_y y f_{Y|X}(y|x)\, dy, & \text{continuous.} \end{cases}$$

Most important is that $E[Y|X = x]$ is a *function* of x.

Example 9.9 Random variables X and Y have joint probability mass function such that $P(X = x_1, Y = 1) = 0.1, P(X = x_1, Y = 2) = 0.2, P(X = x_2, Y = 1) = 0.3$, and $P(X = x_2, Y = 2) = 0.4$. Find $E[Y|X = x]$.

We treat $X = x_1$ and $X = x_2$ separately. The conditional probability mass function of Y given $X = x_1$, is

$$P(Y = k|X = x_1) = \begin{cases} 1/3, & \text{if } k = 1 \\ 2/3, & \text{if } k = 2. \end{cases}$$

Thus,

$$E[Y|X = x_1] = 1\left(\frac{1}{3}\right) + 2\left(\frac{2}{3}\right) = \frac{5}{3}.$$

Similarly, the conditional pmf of Y given $X = x_2$ is,

$$P(Y = k|X = x_2) = \begin{cases} 3/7, & \text{if } k = 1 \\ 4/7, & \text{if } k = 2, \end{cases}$$

which gives

$$E[Y|X = x_2] = 1\left(\frac{3}{7}\right) + 2\left(\frac{4}{7}\right) = \frac{11}{7}.$$

Writing the conditional expectation $E[Y|X = x]$ as a function of x gives

$$E[Y|X = x] = \begin{cases} 5/3, & \text{if } x = x_1 \\ 11/7, & \text{if } x = x_2. \end{cases}$$

■

Example 9.10 Suppose X and Y have joint density

$$f(x, y) = \frac{2}{xy}, \text{ for } 1 < y < x < e,$$

and 0, otherwise. Find $E[Y|X = x]$.

The marginal density of X is

$$f_X(x) = \int_1^x \frac{2}{xy}\, dy = \frac{2 \ln x}{x}, \quad \text{for } 1 < x < e.$$

The conditional density of Y given $X = x$ is

$$f_{Y|X}(y|x) = \frac{f(x, y)}{f_X(x)} = \frac{2/xy}{2 \ln x/x} = \frac{1}{y \ln x}, \quad \text{for } 1 < y < x < e.$$

The conditional expectation is

$$E[Y|X = x] = \int_1^x y f_{Y|X}(y|x)\, dy = \int_1^x \frac{y}{y \ln x}\, dy = \frac{x - 1}{\ln x}.$$

■

Example 9.11 Suppose X and Y have joint density function

$$f(x, y) = e^{-y}, 0 < x < y < \infty.$$

Find the conditional expectations (i) $E[X|Y = y]$ and (ii) $E[Y|X = x]$. Think carefully about the steps needed to find these expectations as you follow the solution below.

(i) The marginal density of Y is

$$f_Y(y) = \int_{-\infty}^{\infty} f(x, y)\, dx = \int_0^y e^{-y}\, dx = ye^{-y}, y > 0.$$

This gives

$$f_{X|Y}(x|y) = \frac{f(x, y)}{f_Y(y)} = \frac{e^{-y}}{ye^{-y}} = \frac{1}{y}, 0 < x < y.$$

Remember that y is fixed. The conditional distribution of X given $Y = y$ is uniform on the interval $(0, y)$. It immediately follows that the conditional expectation of X given $Y = y$ is the midpoint of the interval $(0, y)$. That is, $E[X|Y = y] = y/2$.

(ii) The marginal density of X is

$$f_X(x) = \int_x^{\infty} e^{-y}\, dy = e^{-x}, x > 0.$$

This gives the conditional density function

$$f_{Y|X}(y|x) = \frac{f(x,y)}{f_X(x)} = \frac{e^{-y}}{e^{-x}} = e^{-(y-x)}, y > x. \tag{9.4}$$

The function looks like an exponential density, but it is "shifted over" x units. Let W be an exponential random variable with parameter $\lambda = 1$. Then the conditional distribution of Y given $X = x$ is the same distribution as that of $W + x$. In particular, for $y > x$,

$$P(W + x \le y) = P(W \le y - x) = 1 - e^{-(y-x)}.$$

Differentiating with respect to y shows that

$$f_{W+x}(y) = e^{-(y-x)}, \text{ for } y > x.$$

Because the two distributions are the same, their expectations are the same. This gives

$$E[Y|X = x] = E[W + x] = E[W] + x = 1 + x.$$

Or, if you prefer the integral,

$$E[Y|X = x] = \int_{y=x}^{\infty} ye^{-(y-x)}\, dy = 1 + x.$$

■

9.3.1 From Function to Random Variable

We are about to take a big leap in our treatment of conditional expectation. Please make sure that your seat belt is securely fastened.

In the previous section, we emphasized that $E[Y|X = x]$ is a function of x. Temporarily write this function with functional notation as $g(x) = E[Y|X = x]$. As g is a function, we can define a random variable $g(X)$. What exactly is $g(X)$? When X takes the value x, the random variable $g(X)$ takes the value $g(x) = E[Y|X = x]$.

We give a new name to $g(X)$ and call it $E[Y|X]$, the *conditional expectation of Y given X*.

CONDITIONAL EXPECTATION $E[Y|X]$

For jointly distributed random variables X and Y, the *conditional expectation of Y given X*, denoted $E[Y|X]$, is a random variable, which

1. Is a function of X, and
2. Is equal to $E[Y|X = x]$ when $X = x$.

Example 9.12 Remember Riley and Miguel? Riley picks a random number X uniformly distributed on (0, 1). If Riley picks x, she shows it to Miguel who picks a number Y uniformly distributed on $(0, x)$. Find the conditional expectation of Y given X.

The conditional distribution of Y given $X = x$ is provided explicitly. As the conditional distribution is uniform on $(0, x)$, it follows immediately that $E[Y|X = x] = x/2$, the midpoint of the interval. This holds for all $0 < x < 1$, and thus $E[Y|X] = X/2$. ■

There is much to be learned from this simple, two-stage experiment. We will return to it again.

As much as they look the same, there is a fundamental difference between the conditional expectations $E[Y|X = x]$ and $E[Y|X]$. The former is a function of x. Its domain is a set of real numbers. The function can be evaluated and graphed. For instance, in the last example, $E[Y|X = x] = x/2$ is a linear function of x with slope 1/2. On the other hand, $E[Y|X]$ is a random variable. As such, it has a probability distribution. And thus it makes sense to take *its* expectation with respect to that distribution.

The expectation of a conditional expectation might be a lot to chew on. But it leads to one of the most important results in probability.

LAW OF TOTAL EXPECTATION

For a random variable Y and any random variable X defined jointly with Y, the expectation of Y is equal to the expectation of the conditional expectation of Y given X. That is,

$$E[Y] = E[E[Y|X]]. \quad (9.5)$$

We prove this for the discrete case and leave the continuous case as an exercise. As $E[Y|X]$ is a random variable that is a function of X, it will be helpful to write it explicitly as $E[Y|X] = g(X)$, where $g(x) = E[Y|X = x]$. When we take the expectation $E[E[Y|X]] = E[g(X)]$ of this function of a random variable, we use the law of the unconscious statistician:

$$\begin{aligned}E[E[Y|X]] = E[g(X)] &= \sum_x g(x)P(X = x)\\ &= \sum_x E[Y|X = x]P(X = x)\\ &= \sum_x \left(\sum_y yP(Y = y|X = x)\right) P(X = x)\\ &= \sum_y y \sum_x P(Y = y|X = x)P(X = x)\\ &= \sum_y y \sum_x P(X = x, Y = y)\\ &= \sum_y yP(Y = y) = E[Y].\end{aligned}$$

The fifth equality is achieved by changing the order of the double summation.

Example 9.13 Ayesha will harvest T tomatoes in her vegetable garden, where T has a Poisson distribution with parameter λ. Each tomato is checked for defects. The chance that a tomato has defects is p. Assume that having defects or not is independent from tomato to tomato. Find the expected number of defective tomatoes.

Let X be the number of defective tomatoes. Intuitively, the expected number of defective tomatoes is $E[X] = pE[T] = p\lambda$.

Rigorously, observe that the conditional distribution of X given $T = n$ is binomial with parameters n and p. Thus, $E[X|T = n] = pn$. As this is true for all n, $E[X|T] = pT$. By the law of total expectation,

$$E[X] = E[E[X|T]] = E[pT] = pE[T] = p\lambda. \qquad \blacksquare$$

Example 9.14 When Trinity goes to the gym, she will either run, bicycle, or row. She will choose one of the aerobic activities with respective probabilities 0.5, 0.3, and 0.2. And having chosen an activity the amount of time (in minutes) she spends exercising is exponentially distributed with respective parameters 0.05, 0.025, and 0.01. Find the expectation of Trinity's exercise time.

Let T be her exercise time, and let A be a random variable that takes values 1, 2, and 3 corresponding to her choice of running, bicycling, and rowing. The conditional distribution of exercise time given her choice is exponentially distributed. Thus, the conditional expectation of exercise time given her choice is the reciprocal of the corresponding parameter value. By the law of total expectation,

$$E[T] = E[E[T|A]] = \sum_{a=1}^{3} E[T|A = a]P(A = a)$$
$$= \frac{1}{0.05}(0.5) + \frac{1}{0.025}(0.3) + \frac{1}{0.01}(0.2) = 42 \text{ minutes.}$$

Trinity's expected exercise time is 42 minutes. See **Exercise.R** for a simulation. ■

Example 9.15 At the gym, continued. When Teagan goes to the gym, he has similar habits as Trinity, except, whenever he chooses rowing he only rows for 10 minutes, stops to take a drink of water, and starts all over, choosing one of the three activities at random as if he had just walked in the door. Find the expectation of Teagan's exercise time.

The conditional expectation of Teagan's exercise time given that he chooses running or bicycling is the same as Trinity's. The conditional expectation given that he picks rowing is $E[T|A = 3] = 10 + E[T]$ since after 10 minutes of rowing, Teagan's subsequent exercise time has the same distribution as if he had just walked into the gym and started anew. His expected exercise time is

$$\begin{aligned} E[T] &= E[E[T|A]] \\ &= E[T|A = 1]P(A = 1) + E[T|A = 2]P(A = 2) \\ &\quad + E[T|A = 3]P(A = 3) \\ &= \frac{0.5}{0.05} + \frac{0.3}{0.025} + (E[T] + 10)(0.2) \\ &= 24 + E[T](0.2). \end{aligned}$$

Solving for $E[T]$ gives $E[T] = 24/(0.8) = 30$ minutes. See **Exercise2.R** for a simulation. ■

In the last example, observe carefully the difference between $E[T]$, $E[T|A = a]$, and $E[T|A]$. The first is a number, the second is a function of a, and the third is a random variable.

Example 9.16 Recall in Example 9.12 with Riley and Miguel, that we found $E[Y|X = x] = x/2$ and $E[Y|X] = X/2$. What is $E[Y]$?

By the law of total expectation,

$$E[Y] = E[E[Y|X]] = E\left[\frac{X}{2}\right] = \frac{1}{2}E[X] = \frac{1}{4}.$$

As in the previous example, note the difference between $E[Y]$, $E[Y|X = x]$, and $E[Y|X]$. The first is a number, the second is a function of x, and the third is a random variable. ■

As conditional expectations *are* expectations, they have the same properties as regular expectations, such as linearity. The law of the unconscious statistician also applies. In addition, we highlight two new properties specific to conditional expectation.

PROPERTIES OF CONDITIONAL EXPECTATION

1. (Linearity) For constants a, b, and random variables X, Y, and W,

$$E[aW + bY|X] = aE[W|X] + bE[Y|X].$$

2. (Law of the unconscious statistician) If g is a function, then

$$E[g(Y)|X = x] = \begin{cases} \sum_y g(y)P(Y = y|X = x), & \text{discrete} \\ \int_y g(y)f_{Y|X}(y|x)\, dy, & \text{continuous.} \end{cases}$$

3. (Independence) If X and Y are independent, then

$$E[Y|X] = E[Y].$$

4. (Y function of X) If $Y = g(X)$ is a function of X, then

$$E[Y|X] = E[g(X)|X] = g(X) = Y. \tag{9.6}$$

We prove the last two properties. For Property 3, we show the discrete case, and leave the similar continuous case for the reader. If X and Y are independent,

$$E[Y|X = x] = \sum_y yP(Y = y|X = x) = \sum_y yP(Y = y) = E[Y],$$

for all x. Thus, $E[Y|X] = E[Y]$.

For Property 4, let $Y = g(X)$ for some function g, then

$$E[Y|X = x] = E[g(X)|X = x] = E[g(x)|X = x] = g(x),$$

where the last equality follows because the expectation of a constant is that constant. As $E[Y|X = x] = g(x)$ for all x, we have that $E[Y|X] = g(X) = Y$.

Example 9.17 Let X and Y be independent Poisson random variables with respective parameters $\lambda_X > \lambda_Y$. Let U be uniformly distributed on $(0, 1)$ and independent of X and Y. The conditional expectation $E[UX + (1 - U)Y|U]$ is a random variable. Find its distribution, mean, and variance.

We have

$$\begin{aligned} E[UX + (1-U)Y|U] &= E[UX|U] + E[(1-U)Y|U] \\ &= UE[X|U] + (1-U)E[Y|U] \\ &= UE[X] + (1-U)E[Y] \\ &= U\lambda_X + (1-U)\lambda_Y \\ &= \lambda_Y + (\lambda_X - \lambda_Y)U, \end{aligned}$$

where the first equality uses linearity, the second equality uses Property 4, and the third equality uses independence. As U is uniformly distributed on (0, 1), it follows that the conditional expectation is uniformly distributed on (λ_Y, λ_X).

We could use the law of total expectation for the mean, but this is not necessary. Appealing to results for the uniform distribution, the mean of the conditional expectation is the midpoint $(\lambda_X + \lambda_Y)/2$, and the variance is $(\lambda_X - \lambda_Y)^2/12$. ■

Example 9.18 Suppose X and Y have joint density

$$f(x, y) = xe^{-y}, \ 0 < x < y < \infty.$$

Find $E[e^{-Y}|X]$. Think carefully about the steps you need to take to find this expectation.

We will need the conditional density of Y given $X = x$. The marginal density of X is

$$f_X(x) = \int_x^\infty xe^{-y}\, dy = xe^{-x}, \text{ for } x > 0.$$

The conditional density of Y given $X = x$ is

$$f_{Y|X}(y|x) = \frac{f(x,y)}{f_X(x)} = \frac{xe^{-y}}{xe^{-x}} = e^{-(y-x)}, \text{ for } y > x.$$

This gives

$$\begin{aligned} E[e^{-Y}|X = x] &= \int_{-\infty}^\infty e^{-y} f_{Y|X}(y|x)\, dy = \int_x^\infty e^{-y}e^{-(y-x)}\, dy \\ &= e^x \int_x^\infty e^{-2y}\, dy = e^x\left(\frac{e^{-2x}}{2}\right) = \frac{e^{-x}}{2}, \end{aligned}$$

for all $x > 0$. Thus, $E[e^{-Y}|X] = e^{-X}/2$. ■

Example 9.19 Let $X_1, X_2, \ldots$ be an i.i.d. sequence of random variables with common mean μ. Let $S_n = X_1 + \cdots + X_n$, for each $n = 1, 2, \ldots$. Find $E[S_m|S_n]$ for (i) $m \leq n$, and (ii) $m > n$.

(i) For $m \leq n$,

$$E[S_m|S_n] = E[X_1 + \cdots + X_m|S_n]$$
$$= \sum_{i=1}^{m} E[X_i|S_n] = mE[X_1|S_n], \quad (9.7)$$

where the second equality is from linearity of conditional expectation, and the last equality is because of symmetry, because all the X_i's are identically distributed. As S_n is, of course, a function of S_n, we have that

$$S_n = E[S_n|S_n] = E[X_1 + \cdots + X_n|S_n] = \sum_{i=1}^{n} E[X_i|S_n] = nE[X_1|S_n],$$

where X_1 is chosen for convenience. Thus, $E[X_1|S_n] = S_n/n$. With Equation 9.7 this gives

$$E[S_m|S_n] = mE[X_1|S_n] = \left(\frac{m}{n}\right) S_n.$$

(ii) For $m > n$,

$$E[S_m|S_n] = E[S_n + X_{n+1} + \cdots + X_m|S_n]$$
$$= E[S_n|S_n] + \sum_{i=n+1}^{m} E[X_i|S_n]$$
$$= S_n + \sum_{i=n+1}^{m} E[X_i] = S_n + (m - n)\mu,$$

where μ is the common mean of the X_i's. The last equality is because for $i > n$, X_i is independent of $(X_1, \ldots, X_n)$and thus X_i is independent of S_n.

Let us apply these results. Suppose the amounts that a waiter earns every day in tips form an i.i.d. sequence with mean \$50. Given that he earns \$1400 in tips during 1 month (30 days), what is his expected earnings via tips for the first week? By (i),

$$E[S_7|S_{30} = 1400] = (7/30)(1400) = \$326.67.$$

On the other hand, if the waiter makes \$400 in tips in the next week, how much can he expect to earn in tips in the next month? By (ii),

$$E[S_{30}|S_7 = 400] = 400 + (30 - 7)\mu = 400 + 23(50) = \$1550.$$ ■

9.3.2 Random Sum of Random Variables

Sums of random variables where the *number* of summands is also a random variable arise in numerous applications. Van Der Laan and Louter [1986] use such a model to study the total cost of damage from traffic accidents.

Let X_k be the amount of damage from an individual's kth traffic accident. It is assumed that $X_1, X_2, \ldots$ is an i.i.d. sequence of random variables with common mean μ. Furthermore, the number of accidents N for an individual driver is a random variable with mean λ. It is often assumed that N has a Poisson distribution.

The total cost of damages is

$$X_1 + \cdots + X_N = \sum_{k=1}^{N} X_k.$$

The number of summands, N, is random. To find the expected total cost, it is *not* correct to write $E\left[\sum_{k=1}^{N} X_k\right] = \sum_{k=1}^{N} E[X_k]$, assuming that linearity of expectation applies. Linearity of expectation does not apply here because the number of summands is random, not fixed. (Observe that the equation does not even make sense as the left-hand side is a number and the right-hand side is a random variable.)

To find the expectation of a random sum, condition on the number of summands N. Let $T = X_1 + \cdots + X_N$. By the law of total expectation, $E[T] = E[E[T|N]]$. To find $E[T|N]$, consider

$$\begin{aligned} E[T|N = n] &= E\left[\sum_{k=1}^{N} X_k | N = n\right] = E\left[\sum_{k=1}^{n} X_k | N = n\right] \\ &= E\left[\sum_{k=1}^{n} X_k\right] = \sum_{k=1}^{n} E[X_k] = n\mu. \end{aligned}$$

The third equality follows because N is independent of the X_i's. The equality holds for all n, thus $E[T|N] = N\mu$. By the law of total expectation,

$$E[T] = E[E[T|N]] = E[N\mu] = \mu E[N] = \mu\lambda.$$

The result is intuitive. The expected total cost is the product of the expected number of accidents times the expected cost per accident.

9.4 COMPUTING PROBABILITIES BY CONDITIONING

When we first introduced indicator variables, we showed that probabilities can actually be treated as expectations, since for any event A, $P(A) = E[I_A]$, where I_A is the associated indicator random variable. Applying the law of total expectation gives

$$P(A) = E[I_A] = E[E[I_A|X]].$$

If X is continuous with density f_X,

$$P(A) = E[E[I_A|X]] = \int_{-\infty}^{\infty} E[I_A|X = x]\, f_X(x)\, dx$$
$$= \int_{-\infty}^{\infty} P(A|X = x) f_X(x)\, dx.$$

This is the continuous version of the law of total probability, a powerful tool for finding probabilities by conditioning.

For instance, consider $P(X < Y)$, where X and Y are continuous random variables. Treating $\{X < Y\}$ as the event A, and conditioning on Y, gives

$$P(X < Y) = \int_{-\infty}^{\infty} P(X < Y|Y = y) f_Y(y)\, dy = \int_{-\infty}^{\infty} P(X < y|Y = y) f_Y(y)\, dy.$$

If X and Y are independent,

$$P(X < Y) = \int_{-\infty}^{\infty} P(X < y|Y = y) f_Y(y)\, dy$$
$$= \int_{-\infty}^{\infty} P(X < y) f_Y(y)\, dy$$
$$= \int_{-\infty}^{\infty} F_X(y) f_Y(y)\, dy.$$

We apply these ideas in the following examples.

Example 9.20 The times X and Y that Lief and Haley arrive to class have exponential distributions with respective parameters λ_X and λ_Y. If their arrival times are independent, the probability that Haley arrives before Lief is

$$P(Y < X) = \int_0^{\infty} (1 - e^{-\lambda_Y w}) \lambda_X e^{-\lambda_X w}\, dw$$
$$= 1 - \lambda_X \int_0^{\infty} e^{-w(\lambda_Y + \lambda_X)}\, dw$$
$$= 1 - \frac{\lambda_X}{\lambda_Y + \lambda_X} = \frac{\lambda_Y}{\lambda_Y + \lambda_X}.$$

■

Example 9.21 The density function of X is $f(x) = xe^{-x}$, for $x > 0$. Given $X = x$, Y is uniformly distributed on $(0, x)$. Find $P(Y < 2)$.

Observe that

$$P(Y < 2|X = x) = \begin{cases} 1, & \text{if } 0 < x \le 2 \\ 2/x, & \text{if } x > 2. \end{cases}$$

By conditioning on X,

$$\begin{aligned}
P(Y < 2) &= \int_0^{\infty} P(Y < 2|X = x)f_X(x)\,dx \\
&= \int_0^{2} P(Y < 2|X = x)xe^{-x}\,dx + \int_2^{\infty} P(Y < 2|X = x)xe^{-x}\,dx \\
&= \int_0^{2} xe^{-x}\,dx + \int_2^{\infty} \left(\frac{2}{x}\right) xe^{-x}\,dx \\
&= (1 - 3e^{-2}) + 2e^{-2} = 1 - e^{-2} = 0.8647.
\end{aligned}$$

■

Example 9.22 We demonstrate the convolution formula Equation 8.3 for the sum of independent random variables $X + Y$ by conditioning on one of the variables.

$$\begin{aligned}
P(X + Y \leq t) &= \int_{-\infty}^{\infty} P(X + Y \leq t|Y = y)f_Y(y)\,dy \\
&= \int_{-\infty}^{\infty} P(X + y \leq t|Y = y)f_Y(y)\,dy \\
&= \int_{-\infty}^{\infty} P(X \leq t - y)f_Y(y)\,dy,
\end{aligned}$$

where the last equality is because of independence. Differentiating with respect to t gives the density of $X + Y$

$$f_{X+Y}(t) = \int_{-\infty}^{\infty} f(t - y)f(y)\,dy.$$

■

The following example treats a random variable that has both discrete and continuous components. It arises naturally in many applications, including modeling insurance claims.

Example 9.23 Mohammed's insurance will pay for a medical expense subject to a \$100 deductible. Suppose the amount of the expense is exponentially distributed with parameter λ. Find (i) the distribution of the amount of the insurance company's payment and (ii) the expected payout.

(i) Let M be the amount of the medical expense and let X be the company's payout. Then

$$X = \begin{cases} M - 100, & \text{if } M > 100 \\ 0, & \text{if } M \leq 100, \end{cases}$$

where $M \sim \text{Exp}(\lambda)$.

The random variable X has both discrete and continuous components. Observe that $X = 0$ if and only if $M \leq 100$. Thus,

$$P(X = 0) = P(M \leq 100) = \int_0^{100} \lambda e^{-\lambda m}\, dm = 1 - e^{-\lambda 100}.$$

For $x > 0$, we find the cdf of X by conditioning on M. Thus,

$$\begin{aligned}
P(X \leq x) &= \int_0^{\infty} P(X \leq x | M = m)\lambda e^{-\lambda m}\, dm \\
&= \int_0^{100} P(0 \leq x | M = m)\lambda e^{-\lambda m}\, dm \\
&\quad + \int_{100}^{\infty} P(M - 100 \leq x | M = m)\lambda e^{-\lambda m}\, dm \\
&= \int_0^{100} \lambda e^{-\lambda m}\, dm + \int_{100}^{\infty} P(M \leq x + 100)\lambda e^{-\lambda m}\, dm \\
&= 1 - e^{-\lambda 100} + \int_{100}^{x+100} \lambda e^{-\lambda m}\, dm \\
&= 1 - e^{-\lambda 100} + (e^{-100\lambda} - e^{-\lambda(100+x)}) \\
&= 1 - e^{-\lambda(100+x)}.
\end{aligned}$$

Thus, the cdf of X is

$$P(X \leq x) = \begin{cases} 0, & \text{if } x < 0 \\ 1 - e^{-\lambda(100+x)}, & \text{if } x \geq 0. \end{cases}$$

The cdf is not continuous and has a jump discontinuity at $x = 0$ (see Fig. 9.4 for the graph of the cdf).

(ii) For the expected payout $E[X]$, we apply the law of total expectation, giving

$$\begin{aligned}
E[X] = E[E[X|M]] &= \int_0^{\infty} E[X|M = m]\lambda e^{-\lambda m}\, dm \\
&= \int_{100}^{\infty} E[M - 100|M = m]\lambda e^{-\lambda m}\, dm \\
&= \int_{100}^{\infty} (m - 100)\lambda e^{-\lambda m}\, dm = \frac{e^{-100\lambda}}{\lambda}.
\end{aligned}$$

■

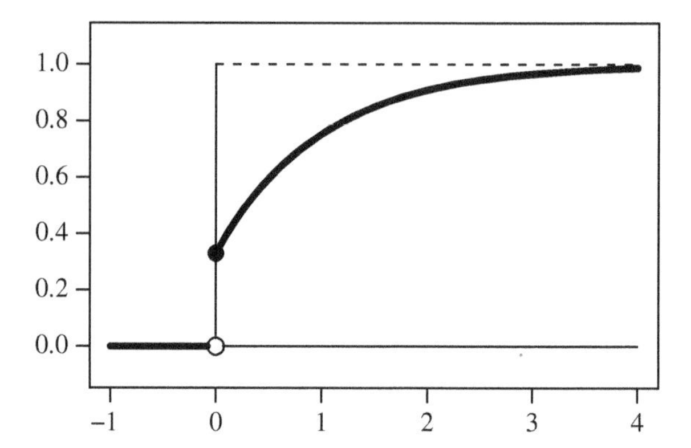

FIGURE 9.4: Cumulative distribution function of insurance payout has jump discontinuity at $x = 0$. The distribution has both discrete and continuous components.

9.5 CONDITIONAL VARIANCE

Conceptually, the conditional variance is derived in a way similar to that of the conditional expectation. It is a variance taken with respect to a conditional distribution. The conditional variance $V[Y|X = x]$ is defined as

CONDITIONAL VARIANCE OF Y GIVEN $X = x$

$$V[Y|X = x] = \begin{cases} \sum_y (y - E[Y|X = x])^2 P(Y = y|X = x), & Y \text{ is discrete} \\ \int_y (y - E[Y|X = x])^2 f_{Y|X}(y|x)\, dy, & Y \text{ is continuous.} \end{cases}$$

Compare with the unconditional variance formula. Note the conditional expectation $E[Y|X = x]$ takes the place of the unconditional expectation $E[Y]$ in the usual variance formula.

Example 9.24 Recall the two-stage uniform model in Example 9.12. Let $X \sim$ Unif(0, 1). Conditional on $X = x$, let $Y \sim$ Unif$(0, x)$. Find the conditional variance $V[Y|X = x]$.

From the defining formula,

$$V[Y|X = x] = \int_0^x (y - E[Y|X = x])^2 \frac{1}{x}\, dy = \int_0^x \frac{(y - x/2)^2}{x}\, dy = \frac{x^2}{12}. \quad ■$$

Properties for the regular variance transfer to the conditional variance.

PROPERTIES OF CONDITIONAL VARIANCE

1. $$V[Y|X = x] = E[Y^2|X = x] - (E[Y|X = x])^2.$$
2. For constants a and b,
$$V[aY + b|X = x] = a^2V[Y|X = x].$$
3. If W and Y are independent, then
$$V[W + Y|X = x] = V[W|X = x] + V[Y|X = x].$$

As with the development of conditional expectation, we define the *conditional variance*, $V[Y|X]$, as the random variable that is a function of X which takes the value $V[Y|X = x]$ when $X = x$. For instance, in Example 9.24, $V[Y|X = x] = x^2/12$, for $0 < x < 1$. And thus $V[Y|X] = X^2/12$.

While the conditional expectation formula $E[Y] = E[E[Y|X]]$ may get credit for one of the most important formulas in this book, the following variance formula is perhaps the most aesthetically pleasing.

LAW OF TOTAL VARIANCE

$$V[Y] = E[V[Y|X]] + V[E[Y|X]].$$

The proof is easier than you might think. We start with the summands on the right hand side. We have that

$$\begin{aligned}E[V[Y|X]] &= E[E[Y^2|X] - (E[Y|X])^2]\\ &= E[E[Y^2|X]] - E[(E[Y|X])^2]\\ &= E[Y^2] - E[(E[Y|X])^2].\end{aligned}$$

And

$$\begin{aligned}V[E[Y|X]] &= E[(E[Y|X])^2] - (E[E[Y|X]])^2\\ &= E[(E[Y|X])^2] - (E[Y])^2.\end{aligned}$$

Thus,

$$\begin{aligned}&E[V[Y|X]] + V[E[Y|X]]\\ &\quad = (E[Y^2] - E[(E[Y|X])^2]) + (E[(E[Y|X])^2] - (E[Y])^2)\\ &\quad = E[Y^2] - (E[Y])^2 = V[Y].\end{aligned}$$

Example 9.25 We continue with the two-stage random experiment of first picking X uniformly in $(0, 1)$ and, if $X = x$, then picking Y uniformly in $(0, x)$. Find the variance of Y.

Recall what we already know in this setting. In Examples 9.12, 9.16, and 9.24, we found that $E[Y|X] = X/2$, $E[Y] = 1/4$, and $V[Y|X] = X^2/12$. By the law of total variance,

$$\begin{aligned} V[Y] &= E[V[Y|X]] + V[E[Y|X]] \\ &= E\left[\frac{X^2}{12}\right] + V\left[\frac{X}{2}\right] \\ &= \frac{1}{12}E[X^2] + \frac{1}{4}V[X] \\ &= \frac{1}{12}\left(\frac{1}{3}\right) + \frac{1}{4}\left(\frac{1}{12}\right) = \frac{7}{144} = 0.04861. \end{aligned}$$

The nature of the hierarchical experiment makes it easy to simulate.

R: SIMULATION OF TWO-STAGE UNIFORM EXPERIMENT

```
> n <- 100000
> simlist <- replicate(n, runif(1, 0, runif(1, 0, 1)))
> mean(simlist)
[1] 0.2495062
> var(simlist)
[1] 0.04868646
```

■

Example 9.26 During her 4 years at college, Hayley takes N exams, where N has a Poisson distribution with parameter λ. On each exam, she scores an A with probability p, independently of any other test. Let Y be the number of A's she receives. Find the correlation between N and Y.

We first find the covariance $\text{Cov}(N, Y) = E[NY] - E[N]E[Y]$. Conditional on $N = n$, the number of A's Hayley receives has a binomial distribution with parameters n and p. Thus, $E[Y|N = n] = np$ and $E[Y|N] = Np$. This gives

$$E[Y] = E[E[Y|N]] = E[Np] = pE[N] = p\lambda.$$

By conditioning on N,

$$\begin{aligned} E[NY] &= E[E[NY|N]] \\ &= E[NE[Y|N]] = E[N(pN)] \\ &= pE[N^2] = p(\lambda + \lambda^2). \end{aligned}$$

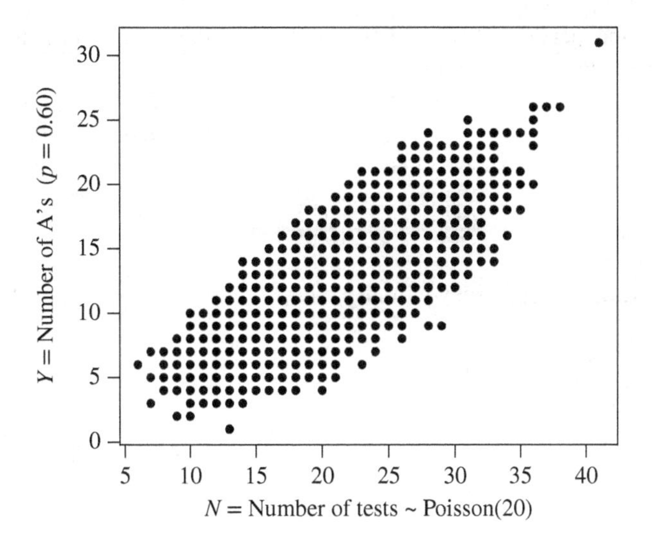

FIGURE 9.5: Number of exams versus number of A's. Correlation is 0.775.

The second equality is because of Property 4 in Equation 9.6. We have

$$\text{Cov}(N, Y) = E[NY] - E[N]E[Y] = p(\lambda + \lambda^2) - (p\lambda)\lambda = p\lambda.$$

To find the correlation, we need the standard deviations. The conditional variance is $V[Z|N = n] = np(1 - p)$ and thus $V[Z|N] = Np(1 - p)$. By the law of total variance,

$$\begin{aligned} V[Y] &= E[V[Y|N]] + V[E[Y|N]] \\ &= E[p(1-p)N] + V[pN] = p(1-p)E[N] + p^2 V[N] \\ &= p(1-p)\lambda + p^2\lambda = p\lambda. \end{aligned}$$

Thus, $\text{SD}[N]\ \text{SD}[Y] = \sqrt{\lambda}\sqrt{p\lambda} = \lambda\sqrt{p}$. This gives

$$\text{Corr}(N, Y) = \frac{\text{Cov}(N, Y)}{\text{SD}[N]\ \text{SD}[Y]} = \frac{\lambda p}{\lambda\sqrt{p}} = \sqrt{p}.$$

See the script file **CorrTest.R** for a simulation. The graph in Figure 9.5 was created with parameters $\lambda = 20$ and $p = 0.60$. The correlation between N and Y is $\sqrt{0.60} = 0.775$. ■

Example 9.27 Random sums continued. The number of customers N who come in every day to Alice's Restaurant has mean and variance μ_N and σ_N^2, respectively. Customers each spend on average μ_C dollars with variance σ_C^2. Customers' spending is independent of each other and of N. Find the mean and standard deviation of customers' total spending.

Let $C_1, C_2, \ldots$ be the amounts each customer spends at the restaurant. Then the total spending is $T = C_1 + \cdots + C_N$, a random sum of random variables.

In Section 9.3.2, we showed that

$$E[T] = E[C_1]E[N] = \mu_C \mu_N.$$

For the variance, condition on N. By the law of total variance,

$$\begin{aligned} V[T] &= V\left[\sum_{k=1}^{N} C_k\right] \\ &= E\left[V\left[\sum_{k=1}^{N} C_k \,\middle|\, N\right]\right] + V\left[E\left[\sum_{k=1}^{N} C_k \,\middle|\, N\right]\right]. \end{aligned}$$

We have that

$$\begin{aligned} V\left[\sum_{k=1}^{N} C_k \,\middle|\, N = n\right] &= V\left[\sum_{k=1}^{n} C_k \,\middle|\, N = n\right] \\ &= V\left[\sum_{k=1}^{n} C_k\right] = \sum_{k=1}^{n} V[C_k] \\ &= n\sigma_C^2. \end{aligned}$$

The second equality is because N and the C_k's are independent. The third equality is because all of the C_k's are independent. This gives

$$V\left[\sum_{k=1}^{N} C_k \,\middle|\, N\right] = \sigma_C^2 N.$$

From results for conditional expectation, we have that

$$E\left[\sum_{k=1}^{N} C_k \,\middle|\, N\right] = E[C_1]N = \mu_C N.$$

This gives

$$V[T] = E\left[V\left[\sum_{k=1}^{N} C_k \,\middle|\, N\right]\right] + V\left[E\left[\sum_{k=1}^{N} C_k \,\middle|\, N\right]\right]$$

$$= E[\sigma_C^2 N] + V[\mu_C N]$$
$$= \sigma_C^2 E[N] + (\mu_C)^2 V[N]$$
$$= \sigma_C^2 \mu_N + (\mu_C)^2 \sigma_N^2.$$

Suppose, on average, $\lambda = 100$ customers arrive each day. Customers each spend on average \$14 with standard deviation \$2. Then total spending at Alice's Restaurant has mean $E[T] = (100)(14) = \$1400$ and standard deviation

$$\text{SD}[T] = \sqrt{(2^2)(100) + (14^2)(100)} = \sqrt{20{,}000} = \$141.42. \qquad \blacksquare$$

R: TOTAL SPENDING AT ALICE'S RESTAURANT

```
# Alice.R
> n <- 100000
> simlist <- numeric(n)
> for (i in 1:n) {
    N <- rpois(1,100) # Number of customers
    cust <- rnorm(N, 14, 2)
    total <- sum(cust)
    simlist[i] <- total  }
> mean(simlist)
[1] 1400.268
> sd(simlist)
[1] 141.0764
```

9.6 BIVARIATE NORMAL DISTRIBUTION*

The multivariate normal distribution for random variables $(X_1, \ldots, X_n)$ generalizes the one-dimensional normal distribution to n dimensions. Here we introduce the two-dimensional bivariate normal distribution for X and Y. We can explore many of the conditional results from this chapter with this distribution.

The bivariate normal distribution is widely used in science and statistics. In Hollowed et al. [2011], a bivariate normal distribution is used to model the habitat of arrowtooth flounder in Alaska fisheries (see Fig. 9.6).

The distribution is specified by five parameters: $\mu_X, \mu_Y, \sigma_X^2, \sigma_Y^2$, and ρ: the means and variances of X and Y and their correlation. If $\mu_X = \mu_Y = 0$ and $\sigma_X^2 = \sigma_Y^2 = 1$, this gives the bivariate standard normal distribution, with correlation ρ. See Figure 9.7 for example contour graphs of bivariate standard normal densities with varying correlations, including one with correlation 0, and Figure 9.8 for a 3-D visual.

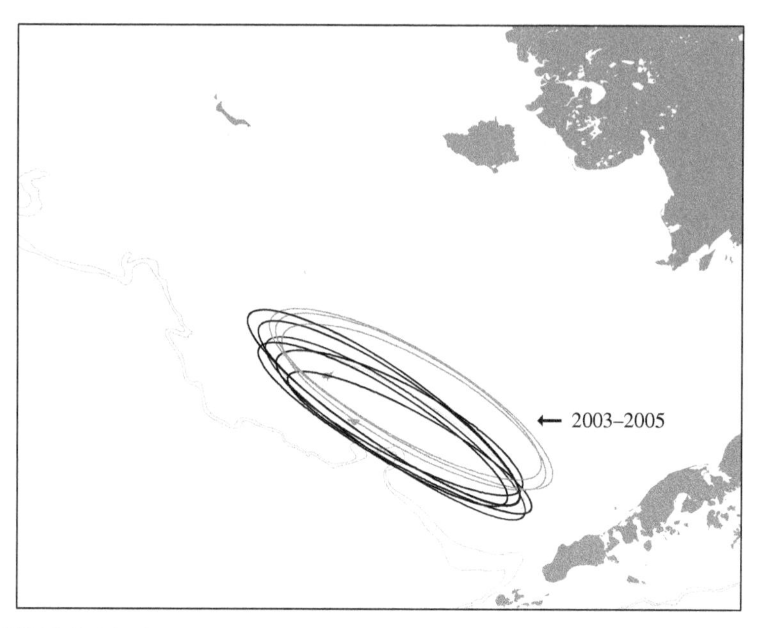

FIGURE 9.6: Distribution of arrowtooth flounder in Alaskan fisheries. Ellipses representing 30% probability contours of bivariate normal distribution fit to EBS survey CPUE data for arrowtooth flounder for the five coldest (black; 1994, 1999, 2008–2010) and warmest (gray; 1996, 1998, 2003–2005) years from 1982 to 2010. Source: Hollowed et al. [2011].

BIVARIATE STANDARD NORMAL DISTRIBUTION

Random variables X and Y have a *bivariate standard normal distribution with correlation* ρ if the joint density function of (X, Y) is

$$f(x,y) = \frac{1}{2\pi\sqrt{1-\rho^2}} e^{-\frac{x^2-2\rho xy+y^2}{2(1-\rho^2)}}, \tag{9.8}$$

for $-\infty < x, y < \infty$, where $-1 < \rho < 1$.

Derivation. Here we derive the bivariate normal distribution and its joint density function.

The derivation starts with a pair of independent standard normal random variables Z_1 and Z_2. Let $-1 < \rho < 1$. We transform (Z_1, Z_2) into a pair of random variables (X, Y) such that (i) marginally X and Y each have standard normal distributions and (ii) the correlation between X and Y is ρ. We then derive the joint density function of X and Y.

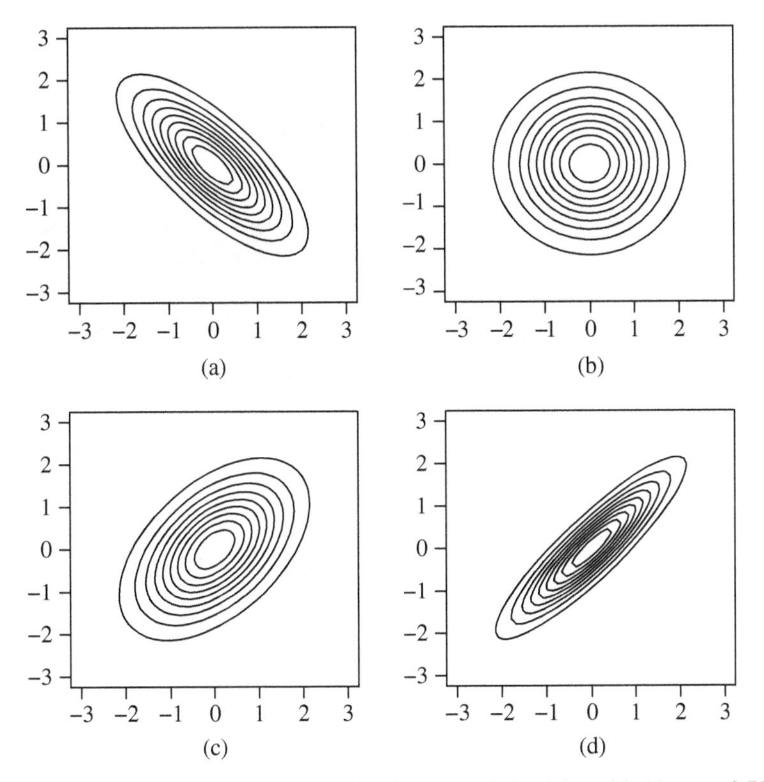

FIGURE 9.7: Contour plots of standard bivariate normal densities with (a) $\rho = -0.75$, (b) $\rho = 0$, (c) $\rho = 0.5$, and (d) $\rho = 0.9$.

Let $X = Z_1$ and $Y = \rho Z_1 + \sqrt{1-\rho^2} Z_2$. Trivially, $X \sim$ Norm(0, 1). As Y is the sum of two independent normal variables, it follows that Y is normally distributed with mean

$$E[Y] = \rho E[Z_1] + \sqrt{1-\rho^2} E[Z_2] = 0$$

and variance

$$V[Y] = \rho^2 V[Z_1] + (1-\rho^2) V[Z_2] = \rho^2 + 1 - \rho^2 = 1.$$

Because both X and Y have mean 0 and variance 1,

$$\begin{aligned} \text{Corr}(X, Y) = E[XY] &= E[Z_1(\rho Z_1 + \sqrt{1-\rho^2} Z_2)] \\ &= \rho E[Z_1^2] + \sqrt{1-\rho^2} E[Z_1 Z_2] = \rho. \end{aligned}$$

We now show that the joint density function of X and Y is the bivariate standard normal density using the method of transformations for two variables, Equation 8.5.

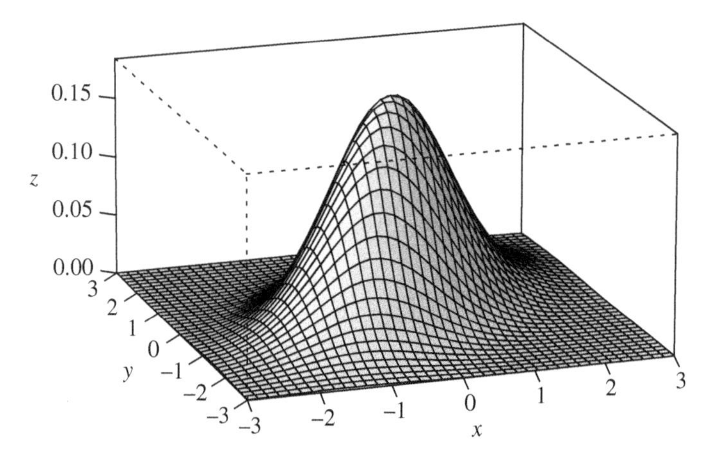

FIGURE 9.8: Bivariate standard normal distribution.

Let $g_1(z_1, z_2) = z_1$ and $g_2(z_1, z_2) = \rho z_1 + \sqrt{1-\rho^2} z_2$. Solving for the inverse functions give

$$h_1(x, y) = x \quad \text{and} \quad h_2(x, y) = \frac{y - \rho x}{\sqrt{1-\rho^2}}.$$

The Jacobian is

$$J = \begin{vmatrix} 1 & 0 \\ -\rho/\sqrt{1-\rho^2} & 1/\sqrt{1-\rho^2} \end{vmatrix} = \frac{1}{\sqrt{1-\rho^2}}.$$

The joint density of X and Y is thus

$$\begin{aligned} f_{X,Y}(x, y) &= \frac{1}{\sqrt{1-\rho^2}} f_{Z_1,Z_2}(x, (y - \rho x)/\sqrt{1-\rho^2}) \\ &= \frac{1}{2\pi\sqrt{1-\rho^2}} e^{-(x^2/2+(y-\rho x)^2/(1-\rho^2))} \\ &= \frac{1}{2\pi\sqrt{1-\rho^2}} e^{-(x^2-2\rho xy+y^2)/2(1-\rho^2)}. \end{aligned}$$

■

The joint density function of the general bivariate distribution is complicated. Please, do not memorize it! Often it will suffice to work with the standard normal bivariate density. As in the univariate case, X and Y have a bivariate normal distribution with parameters $\mu_X, \mu_Y, \sigma_X^2, \sigma_Y^2, \rho$ if and only if the standardized variables $(X - \mu_X)/\sigma_X$ and $(Y - \mu_Y)/\sigma_Y$ have a bivariate standard normal distribution.

For completeness, we give the general expression for the bivariate normal density.

BIVARIATE NORMAL DENSITY

Random variables X and Y have a joint bivariate normal distribution with parameters μ_X, μ_Y, σ_X^2, σ_Y^2, and ρ, if the joint density of X and Y is

$$f(x, y) = \frac{1}{\sigma_X \sigma_Y 2\pi \sqrt{1-\rho^2}} \exp\left(-\frac{d(x, y)}{2(1-\rho^2)}\right), \tag{9.9}$$

where

$$d(x, y) = \left(\frac{x-\mu_X}{\sigma_X}\right)^2 - 2\rho\left(\frac{x-\mu_X}{\sigma_X}\right)\left(\frac{y-\mu_Y}{\sigma_Y}\right) + \left(\frac{y-\mu_Y}{\sigma_Y}\right)^2.$$

The parameter constraints are $\sigma_X > 0$, $\sigma_Y > 0$, and $-1 < \rho < 1$.

Example 9.28 Fathers and sons. Sir Francis Galton, one of the "founding fathers" of statistics, introduced the concept of correlation in the late nineteenth century in part based on his study of the relationship between heights of fathers and their adult sons. Galton took 1078 measurements of father–son pairs. From his data, the mean height of fathers is 69 inches, the mean height of their sons is 70 inches, and the standard deviation of height is 2 inches for both fathers and son. The correlation is 0.5. Galton's data are well fit by a bivariate normal distribution (see Fig. 9.9). ■

The bivariate normal distribution has many remarkable properties, including the fact that both marginal and conditional distributions are normal. We summarize the main properties for the bivariate standard normal distribution next. Results extend naturally to the general case.

PROPERTIES OF BIVARIATE STANDARD NORMAL DISTRIBUTION

Suppose random variables X and Y have a bivariate standard normal distribution with correlation ρ. Then the following properties hold.

1. **Marginal distribution:** The marginal distributions of X and Y are each standard normal.
2. **Conditional distribution:** The conditional distribution of X given $Y = y$ is normally distributed with mean ρy and variance $1 - \rho^2$. That is, $E[X|Y = y] = \rho y$ and variance $V[X|Y = y] = 1 - \rho^2$. Similarly, the conditional distribution of Y given $X = x$ is normal with $E[Y|X = x] = \rho x$ and $V[Y|X = x] = 1 - \rho^2$.

3. **Correlation and independence:** If $\rho = 0$, that is, X and Y are uncorrelated, then X and Y are independent random variables.
4. **Transforming X and Y to independent random variables:** Let $Z_1 = X$ and $Z_2 = (Y - \rho X)/\sqrt{1-\rho^2}$. Then Z_1 and Z_2 are independent standard normal random variables.
5. **Linear functions of X and Y:** For nonzero constants a and b, $aX + bY$ is normally distributed with mean 0 and variance $a^2 + b^2 + 2ab\rho$.

We remark briefly on each of the five properties.

1. **Marginals:** The marginal density of Y is found by a straightforward calculation, which we leave to the reader. The key step in the derivation is completing the square in the exponent of e, writing $x^2 - 2\rho xy + y^2 = (x - \rho y)^2 + (1-\rho^2)y^2$. Similarly for the marginal density of X.
2. **Conditional distributions:** The conditional density of Y given $X = x$ is

$$f_{Y|X}(y|x) = \frac{\frac{1}{2\pi\sqrt{1-\rho^2}} \exp\left(-\frac{x^2-2\rho xy+y^2}{2(1-\rho^2)}\right)}{\frac{1}{\sqrt{2\pi}} \exp\left(-\frac{x^2}{2}\right)}$$

$$= \frac{1}{\sqrt{2\pi}\sqrt{1-\rho^2}} \exp\left(-\frac{y^2 - 2\rho xy - \rho^2 x^2}{2(1-\rho^2)}\right)$$

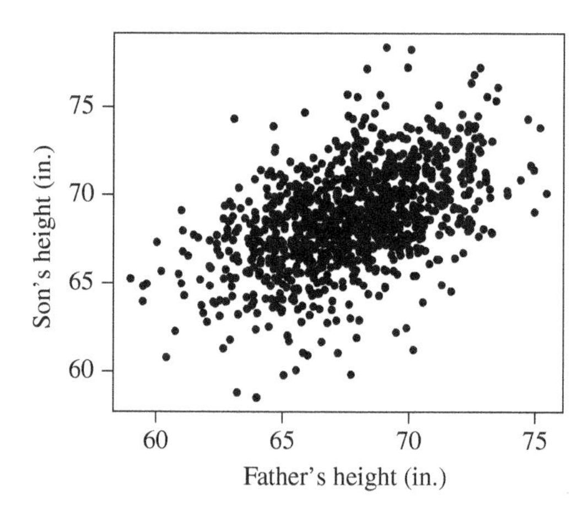

FIGURE 9.9: Galton's height data for fathers and sons are well fit by a bivariate normal distribution with parameters $(\mu_F, \mu_S, \sigma_F^2, \sigma_S^2, \rho) = (69, 70, 2^2, 2^2, 0.5)$.

$$= \frac{1}{\sqrt{2\pi}\sqrt{1-\rho^2}} \exp\left(-\frac{(y-\rho x)^2}{2(1-\rho^2)}\right),$$

which is the density function of a normal distribution with mean ρx and variance $1 - \rho^2$.

3. **Correlation and independence:** Independent random variables always have correlation equal to 0. However, the converse is generally not true. For normal random variables, however, it is. Let $\rho = 0$ in the bivariate joint density function. The density is equal to

$$f(x, y) = \left(\frac{1}{\sqrt{2\pi}} e^{-x^2/2}\right)\left(\frac{1}{\sqrt{2\pi}} e^{-y^2/2}\right),$$

which is the product of the marginal densities of X and Y. Thus, X and Y are independent.

It follows that if X and Y are standard normal random variables that have a joint bivariate normal distribution, and if $E[XY] = E[X]E[Y]$, then X and Y are independent.

4. **Transforming X and Y:** This property follows from considering the derivation of the standard normal bivariate density in reverse, as shown previously.
5. **Linear functions of X and Y:** Property 4 tells us that we can write $X = Z_1$ and $Y = \sqrt{1-\rho^2}Z_2 + \rho Z_1$, where Z_1 and Z_2 are independent standard normals. Thus,

$$aX + bY = aZ_1 + b(\sqrt{1-\rho^2}Z_2 + \rho Z_1) = (a + b\rho)Z_1 + b\sqrt{1-\rho^2}Z_2.$$

That is, we can write $aX + bY$ as a sum of independent random variables. The result follows.

R: SIMULATING BIVARIATE NORMAL RANDOM VARIABLES

We generate 1000 observations from a bivariate standard normal distribution with correlation $\rho = -0.75$. The resulting plot is shown in Figure 9.10.

```
> n <- 1000
> rho <- -0.75
> xlist <- numeric(n)
> ylist <- numeric(n)
> for (i in 1:n) {
    z1 <- rnorm(1)
    z2 <- rnorm(1)
xlist[i] <- z1
```

```
ylist[i] <- rho*z1 + sqrt(1-rho^2)*z2
   }
> plot(cbind(xlist,ylist))
```

The properties and results for the bivariate standard normal distribution extend to the general bivariate normal distribution. We summarize the conditional distribution results for the general case.

CONDITIONAL DISTRIBUTION OF Y GIVEN $X = x$

Suppose the distribution of X and Y is bivariate normal with parameters $\mu_X, \mu_Y, \sigma_X^2, \sigma_Y^2, \rho$. Then the conditional distribution of Y given $X = x$ is normal with conditional mean

$$E[Y|X = x] = \mu_Y + \rho\frac{\sigma_Y}{\sigma_X}(x - \mu_X)$$

and conditional variance

$$V[Y|X = x] = \sigma_Y^2(1 - \rho^2).$$

The result is derived with the assistance of Property 4. Write $X = \sigma_X Z_1 + \mu_X$, where $Z_1 \sim$ Norm(0, 1). Similarly, write $Y = \sigma_Y Z_2 + \mu_Y$, where $Z_2 \sim$ Norm(0, 1).

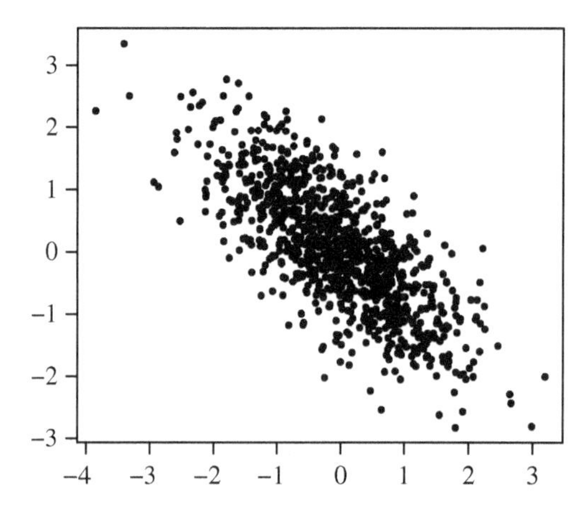

FIGURE 9.10: Plot of 1000 observations from bivariate standard normal distribution with $\rho = -0.75$.

Then

$$\begin{aligned}E[Y|X=x] &= E[\mu_Y+\sigma_Y Z_2|\mu_X+\sigma_X Z_1=x]\\ &= E[\mu_Y+\sigma_Y Z_2|Z_1=(x-\mu_X)/\sigma_X]\\ &= \mu_Y+\sigma_Y E[Z_2|Z_1=(x-\mu_X)/\sigma_X]\\ &= \mu_Y+\rho\frac{\sigma_Y}{\sigma_X}(x-\mu_X)\end{aligned}$$

and

$$\begin{aligned}V[Y|X=x] &= V[\mu_Y+\sigma_Y Z_2|\mu_X+\sigma_X Z_1=x]\\ &= V[\mu_Y+\sigma_Y Z_2|Z_1=(x-\mu_X)/\sigma_X]\\ &= \sigma_Y^2 V[Z_2|Z_1=(x-\mu_X)/\sigma_X]\\ &= \sigma_Y^2(1-\rho^2).\end{aligned}$$

Example 9.29 Fathers and sons continued. We use the Galton dataset as the basis of a model for heights for fathers and their adult sons. For a father–son pair, let F denote father's height and S denote son's height. We assume (F,S) has a bivariate normal distribution with a parameters $\mu_F=69$, $\mu_S=70$, $\sigma_F=\sigma_S=2$, and $\rho=0.50$. (i) Find the probability that a son is taller than his father. (ii) Suppose a father is 67 inches tall. What is the probability that his son will be over 6 feet tall?

(i) The desired probability is $P(S>F)=P(S-F>0)$. By Property 5, $S-F$ is normally distributed with mean

$$E[S-F]=\mu_S-\mu_F=70-69=1$$

and variance

$$\begin{aligned}V[S-F] &= V[S]+V[F]-2\,\mathrm{Cov}(S,F)\\ &= \sigma_S^2+\sigma_F^2-2\rho\sigma_S\sigma_F\\ &= 4+4-2(0.5)(2)(2)=4.\end{aligned}$$

Thus $S-F\sim$ Norm(1, 4). The desired probability $P(S>F)=P(S-F>0)$ is found in **R**.

```
> 1-pnorm(0,1,2)
[1] 0.6914625
```

(ii) The question asks for $P(S>72|F=67)$. The conditional distribution of S given $F=f$ is normal with mean

$$E[S|F=f] = 70 + (0.5)\frac{2}{2}(f-69) = 70 + \frac{f-69}{2}.$$

At $f = 67$, the conditional mean is $70 + (67 - 69)/2 = 69$. The conditional variance is

$$V[S|F=67] = (1-(0.5)^2)4 = 3.$$

The desired probability is obtained in **R**.

```
> 1- pnorm(72,69,sqrt(3))
[1] 0.04163226
```

■

9.7 SUMMARY

Conditional distribution, expectation, and variance are the focus of this chapter. Many results are given for both discrete and continuous settings. For jointly distributed continuous random variables, the conditional density function is a density function defined with respect to a conditional distribution. The continuous version of Bayes formula is presented. Conditional distributions arise naturally in two-stage, hierarchical models, with several related examples given throughout the chapter.

Conditional expectation is given an extensive treatment. A conditional expectation is an expectation with respect to a conditional distribution. Similarly for conditional variance, we first focus on the conditional expectation of Y given $X = x$ $E[Y|X = x]$, which is a function of x. We then introduce $E[Y|X]$, the conditional expectation of Y given a random variable X. The law of total expectation, a central result in probability, is presented. From the law of total expectation, we show how to compute continuous probabilities by conditioning on random variables. The last section presents the related law of total variance.

The bivariate normal distribution is introduced as an optional topic. Conditional results are illustrated with the distribution. A derivation of the joint density function of the bivariate normal distribution is also included.

- **Conditional density function:** The conditional density of Y given $X = x$ is

$$f_{Y|X}(y|x) = \frac{f(x,y)}{f_X(x)},$$

for $f_X(x) > 0$.

- **Continuous Bayes formula:**

$$f_{X|Y}(x|y) = \frac{f_{Y|X}(y|x)f_X(x)}{\int_{-\infty}^{\infty} f_{Y|X}(y|t)f_X(t)\,dt}.$$

- **Conditional expectation of Y given $X = x$:**

$$E[Y|X = x] = \begin{cases} \int_y y f_{Y|X}(y|x)\, dy, & \text{continuous} \\ \sum_y y P(Y = y|X = x), & \text{discrete.} \end{cases}$$

- **Conditional expectation of Y given X:** $E[Y|X]$ is a random variable and a function of X. When $X = x$, the random variable $E[Y|X]$ takes the value $E[Y|X = x]$.
- **Law of total expectation:** $E[Y] = E[E[Y|X]]$.
- **Properties of conditional expectation:**
  1. For constants a and b,

$$E[aY + bZ|X] = aE[Y|X] + bE[Z|X].$$

  2. If g is a function,

$$E[g(Y)|X = x] = \begin{cases} \sum_y g(y)P(Y = y|X = x), & \text{discrete} \\ \int_y g(y) f_{Y|X}(y|x)\, dy, & \text{continuous.} \end{cases}$$

  3. If X and Y are independent, then $E[Y|X] = E[Y]$.
  4. If $Y = g(X)$ is a function of X, then $E[Y|X] = Y$.
- **Law of total probability (continuous version):**

$$P(A) = \int_{-\infty}^{\infty} P(A|X = x) f_X(x)\, dx.$$

- **Conditional variance:**

$$V[Y|X = x] = \begin{cases} \sum_y (y - E[Y|X = x])^2 P(Y = y|X = x), & \text{discrete} \\ \int_y (y - E[Y|X = x])^2 f_{Y|X}(y|x)\, dy, & \text{continuous.} \end{cases}$$

- **Properties of conditional variance:**
  1. $V[Y|X = x] = E[Y^2|X = x] - (E[Y|X = x])^2$.
  2. For constants a and b, $V[aY + b|X = x] = a^2 V[Y|X = x]$.
  3. If Y and Z are independent,

$$V[Y + Z|X = x] = V[Y|X = x] + V[Z|X = x].$$

- **Law of total variance:** $V[Y] = E[V[Y|X]] + V[E[Y|X]]$.

- **Bivariate standard normal density:** X and Y have a bivariate standard normal distribution with correlation ρ, if the joint density function is

$$f(x, y) = \frac{1}{2\pi\sqrt{1-\rho^2}} e^{-\frac{x^2-2\rho xy+y^2}{2(1-\rho^2)}}.$$

- **Properties of bivariate standard normal distribution:** If X and Y have a bivariate standard normal distribution with correlation ρ, then
  1. Marginal distributions of X and Y are standard normal.
  2. Conditional distribution of X given $Y = y$ is normal with mean ρy and variance $1 - \rho^2$. Similarly for Y given $X = x$.
  3. If $\rho = 0$, X and Y are independent.
  4. Let $Z_1 = X$ and $Z_2 = (Y - \rho X)/\sqrt{1-\rho^2}$. Then Z_1 and Z_2 are independent standard normal variables.
  5. For nonzero constants a and b, $aX + bY$ is normally distributed with mean 0 and variance $a^2 + b^2 + 2ab\rho$.

EXERCISES

Conditional Distributions

9.1 Let X and Y have joint density

$$f(x, y) = 12x(1 - x), \quad \text{for } 0 < x < y < 1.$$

(a) Find the marginal densities of X and Y.

(b) Find the conditional density of Y given $X = x$. Describe the conditional distribution.

(c) Find $P(X < 1/4 | Y = 0.5)$.

9.2 Let $X \sim$ Unif$(0, 2)$. If $X = x$, let Y be uniformly distributed on $(0, x)$.

(a) Find the joint density of X and Y.

(b) Find the marginal densities of X and Y.

(c) Find the conditional densities of X and Y.

9.3 Let X and Y have joint density

$$f(x, y) = 4e^{-2x}, \quad \text{for } 0 < y < x < \infty.$$

(a) Describe the steps needed to find the conditional density of Y given X.

(b) Find the conditional density of Y given X.

(c) Find $P(1 < Y < 2 | X = 3)$.

9.4 Let X and Y be uniformly distributed on the triangle with vertices $(0, 0)$, $(1, 0)$, and $(1, 1)$.

(a) Find the joint and marginal densities of X and Y.

(b) Find the conditional density of Y given $X = x$. Describe the conditional distribution.

9.5 Let X and Y be i.i.d. exponential random variables with parameter λ. Find the conditional density function of $X + Y$ given $X = x$. Describe the conditional distribution.

9.6 Let X and Y have joint density

$$f(x, y) = \sqrt{\frac{2}{\pi}} y e^{-xy} e^{-y^2/2}, \quad \text{for } x > 0, y > 0.$$

(a) By examining the joint density function can you guess the conditional density of X given $Y = y$? (Treat y as a constant.) Confirm your guess.

(b) Find $P(X < 1|Y = 1)$.

9.7 Let $Z \sim$ Gamma(a, λ), where a is an integer. Conditional on $Z = z$, let $X \sim$ Pois(z). Show that X has a negative binomial distribution with the following interpretation: for $k = 0, 1, \ldots$, $P(X = k)$ is equal to the number of failures before a successes in a sequence of i.i.d. Bernoulli trials with success parameter $\lambda/(1 + \lambda)$.

Conditional Expectation

9.8 Explain carefully the difference between $E[Y|X]$ and $E[Y|X = x]$.

9.9 Show that $E[E[Y|X]] = E[Y]$ in the continuous case.

9.10 On one block of "Eat Street" in downtown Minneapolis there are 10 restaurants to choose from. The waiting time for each restaurant is exponentially distributed with respective parameters $\lambda_1, \ldots, \lambda_{10}$. Luka will decide to eat at restaurant i with probability p_i for $i = 1, \ldots, 10$. (Note: $p_1 + \cdots + p_{10} = 1$). What is Luka's expected waiting time?

9.11 Let X and Y have joint density function

$$f(x, y) = e^{-x(y+1)}, \; x > 0, 0 < y < e - 1.$$

(a) Find and describe the conditional distribution of X given $Y = y$.

(b) Find $E[X|Y = y]$ and $E[X|Y]$.

(c) Find $E[X]$ in two ways: (i) using the law of total expectation; (ii) using the distribution of X.

9.12 Let X and Y have joint density

$$f(x, y) = x + y, \ 0 < x < 1, 0 < y < 1.$$

Find $E[X|Y = y]$. (It may help you to outline the steps needed to compute the expectation.)

9.13 Let X and Y have joint density

$$f(x, y) = (3/160)(x + 2y), 0 < x < y < 4.$$

(a) Find the marginal pdf of Y.

(b) Find the conditional pdf of X given $Y = y$.

(c) Find $E(X|Y = 2)$.

9.14 Let X and Y have joint density

$$f(x, y) = 24xy, \ 0 < x, y < 1, 0 < x + y < 1.$$

(a) Find the conditional pdf of Y given $X = x$.

(b) Find $E(Y|X)$.

(c) Find $E(Y)$ using the law of total expectation.

9.15 Let $P(X = 0, Y = 0) = 0.1$, $P(X = 0, Y = 1) = 0.2$, $P(X = 1, Y = 0) = 0.3$, and $P(X = 1, Y = 1) = 0.4$. Show that

$$E[X|Y] = \frac{9 - Y}{12}.$$

9.16 Let A and B be events such that $P(A) = 0.3$, $P(B) = 0.5$, and $P(AB) = 0.2$.

(a) Find $E[I_A|I_B = 0]$ and $E[I_A|I_B = 1]$.

(b) Show that $E[E[I_A|I_B]] = E[I_A]$.

9.17 Let X and Y be independent and uniformly distributed on $(0, 1)$. Let $M = \min(X, Y)$ and $N = \max(X, Y)$.

(a) Find the joint density of M and N. (Hint: For $0 < m < n < 1$, show that $M \leq m$ and $N > n$ if and only if either $\{X \leq m \text{ and } Y > n\}$ or $\{Y \leq m \text{ and } X > n\}$.)

(b) Find the conditional density of N given $M = m$ and describe the conditional distribution.

9.18 Given an event A, define the conditional expectation of Y given A as

$$E[Y|A] = \frac{E[YI_A]}{P(A)}, \tag{9.10}$$

where I_A is the indicator random variable.

(a) Let $Y \sim Exp(\lambda)$. Find $E[Y|Y > 1]$.

(b) An insurance company has a \$250 deductible on a claim. Suppose C is the amount of damages claimed by a customer. Let X be the amount that the insurance company will pay on the claim. Suppose C has an exponential distribution with mean 300. That is,

$$X = \begin{cases} 0, & \text{if } C \leq 250 \\ C - 250, & \text{if } C > 250. \end{cases}$$

Find $E[X]$, the expected payout by the insurance company.

9.19 Let X_1, X_2 be the rolls of two four-sided tetrahedral dice. Let $S = X_1 + X_2$ be the sum of the dice. Let $M = \max(X_1, X_2)$ be the largest of the two numbers rolled. Find the following:

(a) $E[X_1|X_2]$.

(b) $E[X_1|S]$.

(c) $E[M|X_1 = x]$.

(d) $E[X_1X_2|X_1]$.

9.20 A *random walker* starts at one vertex of a triangle, moving left or right with probability 1/2 at each step. The triangle is *covered* when the walker visits all three vertices. Find the expected number of steps for the walker to cover the triangle.

9.21 Repeat the last exercise. Only this time at each step the walker either moves left, moves right, or stays put, each with probability 1/3. Staying put counts as one step.

Computing Probabilities with Conditioning

9.22 Let $X \sim$ Unif(0, 1). If $X = x$, then $Y \sim Exp(x)$. Find $P(Y > 1)$ by conditioning on X.

9.23 Let $X \sim$ Unif(0, 1). If $X = x$, then $Y \sim$ Unif$(0, x)$.

(a) Find $P(Y < 1/4)$ by conditioning on X.

(b) Find $P(Y < 1/4)$ by using the marginal density of Y.

(c) Approximate $P(Y < 1/4)$ by simulation.

9.24 Suppose the density of X is proportional to $x^2(1 - x)$ for $0 < x < 1$. If $X = x$, then $Y \sim$ Binom$(10, x)$. Find $P(Y = 6)$ by conditioning on X.

9.25 Suppose that X is uniform on (0, 3), and that $Y|X = x$ is uniform on the interval $(0, x^2)$. Find $P(Y < 4)$ by conditioning on X.

9.26 Suppose that X has a Gamma distribution with parameters $a = 4, \lambda = 1/5$, and that $Y|X = x$ is uniform on the interval $(0, x^2)$. Find $P(Y > 16)$ by conditioning on X.

9.27 Suppose X and Y are independent and positive random variables with density functions f_X and f_Y, respectively. Use conditioning to find a general expression for the density function of

(a) XY.

(b) X/Y.

(c) $X - Y$.

9.28 Let X and Y be independent uniform random variables on $(0, 1)$. Find the density function of $Z = X/Y$. Show that the mean of Z does not exist.

Conditional Variance

9.29 A biased coin has heads probability p. Let $N \sim$ Pois(λ). If $N = n$, we will flip the coin n times. Let X be the number of heads.

(a) Use the law of total expectation to find $E[X]$.

(b) Use the law of total variance to find $V[X]$.

9.30 Suppose Λ is an exponential random variable with mean 1. Conditional on $\Lambda = \lambda$, N is a Poisson random variable with parameter λ.

(a) Find $E[N|\Lambda]$ and $V[N|\Lambda]$.

(b) Use the law of total expectation to find $E[N]$.

(c) Use the law of total variance to find $V[N]$.

(d) Find the probability mass function of N.

(e) Find $E[N]$ and $V[N]$ again using the pmf of N.

9.31 Madison tosses 100 coins. Let H be the number of heads she gets. For each head that she tosses, Madison will get a reward. The amount of each reward is normally distributed with $\mu = 5$ and $\sigma^2 = 1$. (The units are in dollars, and Madison might get a negative "reward.") Individual rewards are independent of each other and independent of H. Find the expectation and standard deviation of Madison's total reward.

9.32 The joint density of X and Y is

$$f(x, y) = xe^{-3xy}, \quad 1 < x < 4, y > 0.$$

(a) Describe the marginal distribution of X.

(b) Describe the conditional distribution of Y given $X = x$.

(c) Find $E[Y|X]$.

(d) Use the law of total expectation to show $E[Y] = (\ln 4)/9$.

(e) Find $E[XY]$ by conditioning on X.

(f) Find $\text{Cov}(X, Y)$.

9.33 The number of deaths by horsekick per army corps has a Poisson distribution with mean λ. However, λ varies from corps unit to corps unit and can be thought of as a random variable. Determine the mean and variance of the number of deaths by horsekick when $\Lambda \sim \text{Unif}(0, 3)$.

9.34 Revisit Exercise 9.15. Find $V[X|Y = 0]$ and $V[X|Y = 1]$. Find a general expression for $V[X|Y]$ as a function of Y.

9.35 Let $X_1, \ldots, X_n$ be i.i.d. random variables with mean μ and variance σ^2. Let $\overline{X} = (X_1 + \cdots + X_n)/n$ be the average.

(a) Find $E[\overline{X}|X_1]$.

(b) Find $V[\overline{X}|X_1]$.

9.36 If X and Y are independent, does $V[Y|X] = V[Y]$? If not, what is the relationship?

9.37 If $Y = g(X)$ is a function of X, what is $V[Y|X]$?

9.38 Suppose X and Y have joint pdf given by

$$f(x, y) = 2(x + y), \ 0 < x < y < 1.$$

(a) Find the conditional pdf of X given $Y = y$.

(b) Find $V(X|Y)$.

(c) Find $V(X)$ using the law of total variance.

9.39 An arborist is modeling heights of trees of different species. He has decided to model the height of the trees, X, using different means for each species, but a constant variance, so he sets up $X|\mu \sim N(\mu, \sigma^2)$, where $\mu \sim \text{Gamma}(\alpha, \beta)$, and the Gamma distribution changes by species.

(a) Explain any concerns you have regarding the arborist's tree height modeling distribution choices.

(b) Using the arborist's setup, find $E(X)$.

(c) Using the arborist's setup, find $V(X)$.

9.40 Suppose X and Y have joint pdf given by

$$f(x, y) = e^{-y}, \ 0 < x < y < \infty.$$

(a) Find the conditional pdf of X given $Y = y$.

(b) Find $E(X|Y)$ and $V(X|Y)$. Hint: You should recognize the distribution.

(c) Find $E(X)$ and $V(X)$ using the law of total expectation and law of total variance, respectively.

Bivariate Normal Distribution

9.41 Let X and Y be independent and identically distributed normal random variables. Show that $X + Y$ and $X - Y$ are independent.

9.42 Let U and V have a bivariate standard normal distribution with correlation ρ.

(a) Find $E[UV]$.

(b) Now suppose X and Y have a bivariate normal distribution with parameters $\mu_X, \mu_Y, \sigma_X^2, \sigma_Y^2, \rho$. Find $E[XY]$.

9.43 Suppose that math and reading SAT scores have a bivariate normal distribution with the mean of both scores 500, the standard deviation of both scores 100, and correlation 0.70. For someone who scores 650 on the reading SAT, find the probability that they score over 700 on the math SAT.

9.44 Let (X, Y) have a bivariate standard normal distribution with correlation ρ. Using results for the conditional distribution of Y given $X = x$, illustrate the law of total variance and find $V[Y]$ with a bare minimum of calculation.

9.45 Let X and Y have joint density

$$f(x, y) = \frac{1}{\pi\sqrt{3}} e^{-\frac{2}{3}(x^2 - xy + y^2)},$$

for real x, y.

(a) Identify the distribution of X and Y and parameters.

(b) Identify the conditional distribution of X given $Y = y$.

(c) Use **R** to find $P(X > 1 | Y = 0.5)$.

9.46 Let X and Y have a bivariate normal distribution with parameters $\mu_X = -1$, $\mu_y = 4$, $\sigma_X^2 = 1$, $\sigma_Y^2 = 25$, and $\rho = -0.75$.

(a) Find $P(3 < Y < 6 | X = 0)$.

(b) Find $P(3 < Y < 6)$.

9.47 Let X and Y have a bivariate standard normal distribution with correlation $\rho = 0$. That is, X and Y are independent. Let (x, y) be a point in the plane. The rotation of (x, y) about the origin by angle θ gives the point

$$(u, v) = (x\cos\theta - y\sin\theta, x\sin\theta + y\cos\theta).$$

Show that the joint density of X and Y has *rotational symmetry* about the origin. That is, show that $f(x, y) = f(u, v)$.

9.48 Let X and Y have a bivariate standard normal distribution with correlation ρ. Find $P(X > 0, Y > 0)$ by the following steps.

(a) Write $X = Z_1$ and $Y = \rho Z_1 + \sqrt{1-\rho^2}Z_2$, where (Z_1, Z_2) are independent standard normal random variables. Rewrite the probability in terms of Z_1 and Z_2.

(b) Think geometrically. Express the event as a region in the plane.

(c) See the last Exercise 9.47. Use rotational symmetry and conclude that

$$P(X > 0, Y > 0) = \frac{1}{4} + \frac{\sin^{-1}\rho}{2\pi}.$$

9.49 If X and Y have a joint bivariate normal distribution, show that

$$\rho^2 = \frac{V[E[Y|X]]}{V[Y]}.$$

Simulation and R

9.50 Suppose X has a Poisson distribution whose parameter value is the outcome of an independent exponential random variable with parameter μ. (i) Write an **R** function `rexppois(k,μ)` for simulating k copies of such a random variable. (ii) Use your function to estimate $E[X]$ for the case $\mu = 1$ and show it matches up with the result of Example 9.8.

9.51 Let N be a Poisson random variable with parameter λ. Write **R** commands for simulating $Z = \sum_{i=1}^{N} X_i$, where $X_1, X_2, \ldots$ are i.i.d. normal random variables with parameters μ and σ^2. Then show how to write a command for estimating $E[Z]$.

9.52 See the previous exercise. Suppose $\lambda = 10$ and $\mu = \sigma = 1$. Find the mean and variance of Z following the logic of the Alice's restaurant example. Simulate the distribution of Z and superimpose a normal density curve with your mean and variance.

Chapter Review

Chapter review exercises are available through the text website. The URL is `www.wiley.com/go/wagaman/probability2e`.

10

LIMITS

We must learn our limits.

—Blaise Pascal

Learning Outcomes

1. Explain what the following results say: Markov and Chebyshev inequalities, the weak and strong laws of large numbers, and the Central Limit Theorem.
2. Describe the different types of convergence expressed in the WLLN, SLLN, and CLT.
3. Apply the method of moments.*
4. Solve problems using Monte Carlo Integration and the Central Limit Theorem.
5. (C) Simulate to solve problems involving limits.

Introduction. We study limits in probability to understand the long-term behavior of random processes and sequences of random variables. Limits can lead to simplified formulas for otherwise intractable probability models, and they may give insight into complex problems. We have already seen some limit results—the normal approximation, and the Poisson approximation, of the binomial distribution.

The use of simulation to approximate the probability of an event A is justified by one of the most important limit results in probability—the law of large numbers. A consequence of the law of the large numbers is that in repeated trials of a random experiment the proportion of trials in which A occurs converges to $P(A)$, as the

Probability: With Applications and R, Second Edition. Amy S. Wagaman and Robert P. Dobrow.

Companion Website: www.wiley.com/go/wagaman/probability2e

number of trials goes to infinity. This is often described as the relative frequency interpretation of probability.

The earliest version of the law of large numbers was discovered at the beginning of the eighteenth century by James Bernoulli, who called it his "golden theorem."

Let $X_1, X_2, \ldots$ be an independent and identically distributed sequence of random variables with finite expectation μ. For $n = 1, 2, \ldots$, let

$$S_n = X_1 + \cdots + X_n.$$

The law of large numbers says that the average S_n/n converges to μ, as $n \to \infty$.

In statistics, where $X_1, \ldots, X_n$ might represent data from a random sample taken from a population, this says that the sample mean converges to the population mean.

Recall mathematician John Kerrich's experiment tossing 10,000 coins when interned in a prisoner of war camp during World War II, described in Section 7.1. His results are shown in Figure 10.1, which graphs the proportion of heads he got for 10–10,000 tosses. The data, given in Freedman et al. [2007], illustrate the law of large numbers. Wide fluctuations in possible outcomes at the beginning of the trials eventually settle down and reach an equilibrium at the mean of the distribution. In this example, $X_1, X_2, \ldots, X_{10,000}$ represents independent coin tosses, and S_n/n is the proportion of heads in n tosses, with $\mu = 1/2$.

To see how the law of large numbers leads to the relative frequency interpretation of probability, consider a random experiment and some event A. Repeat the experiment many times, obtaining a sequence of outcomes, and let

$$X_k = \begin{cases} 1, & \text{if } A \text{ occurs on the } k\text{th trial of the experiment} \\ 0, & \text{otherwise,} \end{cases}$$

for $k = 1, 2, \ldots$. The X_k's form an i.i.d. sequence with mean $\mu = E[X_k] = P(A)$. The average $S_n/n = (X_1 + \cdots + X_n)/n$ is the proportion of n trials in which A occurs.

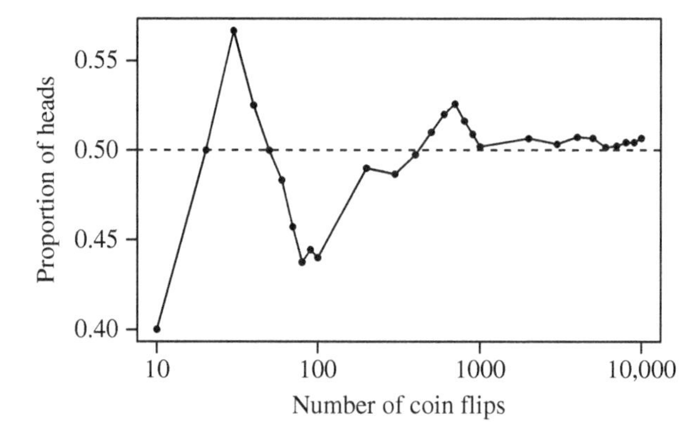

FIGURE 10.1: Mathematician John Kerrich tossed 10,000 coins when he was interned in a prisoner of war camp during World War II. His results illustrate the law of large numbers.

The law of large numbers says that $S_n/n \to \mu = P(A)$, as $n \to \infty$. That is, the proportion of n trials in which A occurs converges to $P(A)$.

James Bernoulli's discovery marked a historic shift in the development of probability theory. Previously most probability problems that were studied involved games of chance where outcomes were equally likely. The law of large numbers gave rigorous justification to the fact that *any* probability could be computed, or at least estimated, with a large enough sample size of repeated trials.

Today there are two versions of the law of large numbers, each based on different ways to define what it means for a sequence of random variables to *converge*. What we now know as the *weak law of large numbers* (WLLN) is the "golden theorem" discovered by Bernoulli some 300 years ago.

THE "LAW OF AVERAGES" AND A RUN OF BLACK AT THE CASINO

The law of large numbers is sometimes confused with the so-called non-existent "law of averages," also called the "gambler's fallacy." This says that outcomes of a random experiment will "even out" within a small sample.

Darrel Huff, in his excellent and amusing book *How to take a chance* [1964], tells the story of a run on black at roulette in a Monte Carlo casino in 1913:

> Black came up a record 26 times in succession. Except for the question of the house limit, if a player had made a one-louis ($4) bet when the run started and pyramided for precisely the length of the run on black, he could have taken away 268 million dollars. What actually happened was a near-panicky rush to bet on red, beginning about the time black had come up a phenomenal 15 times... Players doubled and tripled their stakes [believing] that there was not a chance in a million of another repeat. In the end the unusual run enriched the Casino by some millions of franc.

10.1 WEAK LAW OF LARGE NUMBERS

The WLLN says that for any $\epsilon > 0$ the sequence of probabilities

$$P\left(\left|\frac{S_n}{n} - \mu\right| < \epsilon\right) \to 1, \quad \text{as } n \to \infty.$$

That is, the probability that S_n/n is arbitrarily close to μ converges to 1. Equivalently,

$$P\left(\left|\frac{S_n}{n} - \mu\right| \geq \epsilon\right) \to 0, \quad \text{as } n \to \infty.$$

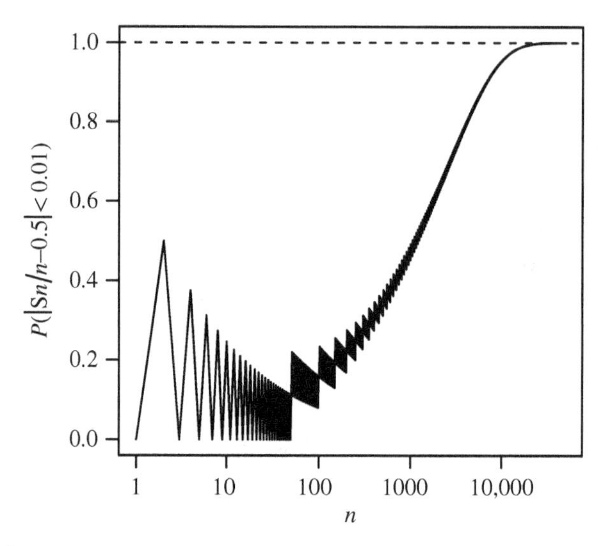

FIGURE 10.2: An illustration of the weak law of large numbers for Bernoulli coin flips with $\epsilon = 0.01$.

To understand the result, it is helpful to work out a specific example in detail. Consider i.i.d. coin flips, that is, Bernoulli trials with $p = \mu = 1/2$. Let $\epsilon > 0$. Then

$$P\left(\left|\frac{S_n}{n} - \frac{1}{2}\right| < \epsilon\right) = P\left(-\epsilon < \frac{S_n}{n} - \frac{1}{2} < \epsilon\right)$$
$$= P\left(\frac{n}{2} - n\epsilon < S_n < \frac{n}{2} + n\epsilon\right).$$

The sum $S_n = X_1 + \cdots + X_n$ has a binomial distribution with parameters n and $1/2$. We find the probability $P(|S_n/n - \mu| < \epsilon)$ in **R**. An illustration of the limiting behavior, with $\epsilon = 0.01$, is shown in Figure 10.2. For low n, the probability fluctuates, but as n increases (i.e., as a limit), the probability goes to one, as the WLLN states.

R: WEAK LAW OF LARGE NUMBERS

The function `wlln(n,ε)` computes the probability $P(|S_n/n - \mu| < \epsilon)$ for $\mu = p = 1/2$.

```
> wlln <- function(n,eps){
    p <- 1/2
    pbinom(n*p+n*eps, n, p)-pbinom(n*p-n*eps, n, p)}
> wlln(100, 0.01)
```

```
[1] 0.1576179
> wlln(1000, 0.01)
[1] 0.4726836
> wlln(10000, 0.01)
[1] 0.9544943
```

10.1.1 Markov and Chebyshev Inequalities

Bernoulli's original proof of the weak law of large numbers is fairly complicated and technical. A much simpler proof was discovered in the mid-1800s based on what is now called Chebyshev's inequality by way of Markov's. The use of inequalities to bound probabilities is a topic of independent interest.

Markov's inequality. Let X be a nonnegative random variable with finite expectation. Then for all $\epsilon > 0$,

$$P(X \geq \epsilon) \leq \frac{E[X]}{\epsilon}.$$

Proof: We give the proof for the case when X is continuous with density function f, and invite the reader to show it for the discrete case. We have

$$E[X] = \int_0^\infty xf(x)\,dx \geq \int_\epsilon^\infty xf(x)\,dx \geq \int_\epsilon^\infty \epsilon f(x)\,dx = \epsilon P(X \geq \epsilon).$$

The first inequality is a consequence of the fact that the integrand is nonnegative and $\epsilon > 0$. Rearranging gives the result. □

Example 10.1 Let $\epsilon = kE[X] = k\mu$ in Markov's inequality for positive integer k. Then

$$P(X \geq k\mu) \leq \frac{\mu}{k\mu} = \frac{1}{k}.$$

For instance, the probability that a nonnegative random variable is at least twice its mean is at most 1/2. ■

Corollary. If g is an increasing positive function, then

$$P(X \geq \epsilon) = P(g(X) \geq g(\epsilon)) \leq \frac{E[g(X)]}{g(\epsilon)}.$$

By careful choice of the function g, one can often improve the Markov inequality upper bound.

Chebyshev's inequality. Let X be a random variable (not necessarily positive) with finite mean μ and variance σ^2. Then for all $\epsilon > 0$,

$$P(|X - \mu| \geq \epsilon) \leq \frac{\sigma^2}{\epsilon^2}.$$

Proof: Let $g(x) = x^2$ on $(0, \infty)$. By our Corollary, applied to the nonnegative random variable $|X - \mu|$,

$$P(|X - \mu| \geq \epsilon) = P(|X - \mu|^2 \geq \epsilon^2) \leq \frac{E[(X - \mu)^2]}{\epsilon^2} = \frac{\sigma^2}{\epsilon^2}. \qquad \square$$

At times, it may make sense to consider an equivalent expression using our understanding of complements, giving

$$P(|X - \mu| < \epsilon) > 1 - \frac{\sigma^2}{\epsilon^2}.$$

Example 10.2 Let X be an exponential random variable with mean and variance equal to 1. Consider $P(X \geq 4)$. By Markov's inequality,

$$P(X \geq 4) \leq \frac{1}{4} = 0.25.$$

To bound $P(X \geq 4)$ using Chebyshev's inequality, we have

$$P(X \geq 4) = P(X - 1 \geq 3) = P(|X - 1| \geq 3) \leq \frac{1}{9} = 0.111. \qquad (10.1)$$

We see the improvement of Chebyshev's bound over Markov's bound.

In fact, $P(X \geq 4) = e^{-4} = 0.0183$. So both bounds are fairly crude. However, the power of Markov's and Chebyshev's inequalities is that they apply without regard to distribution of the random variable, so long as their requirements are satisfied.

As a note, observe that the second equality in Equation 10.1 holds because

$$\begin{aligned}\{|X - 1| \geq 3\} &= \{X - 1 \geq 3 \text{ or } X - 1 \leq -3\} \\ &= \{X - 1 \geq 3 \text{ or } X \leq -2\} = \{X - 1 \geq 3\},\end{aligned}$$

as X is a positive random variable. However, in general, for a random variable Y and constant c,

$$P(Y \geq c) \leq P(|Y| \geq c),$$

because $\{Y \geq c\}$ implies $\{|Y| \geq c\}$. ■

Example 10.3 Suppose X is a random variable with finite mean and variance. Let $\epsilon = k\sigma$ in Chebyshev's inequality for positive integer k. The probability that X is within k standard deviations of the mean is

$$P(|X - \mu| < k\sigma) = 1 - P(|X - \mu| \geq k\sigma) \geq 1 - \frac{\sigma^2}{k^2\sigma^2} = 1 - \frac{1}{k^2}.$$

With $k = 2$, the probability that *any* random variable is within two standard deviations from the mean is at least 75%. The probability that any random variable is within $k = 3$ standard deviations from the mean is at least 88.89%.

For a normally distributed random variable, these probabilities are, respectively, 95 and 99.7%. But again, the utility of the inequalities is that they apply to *all* random variables, regardless of distribution, if their conditions are met. ■

Example 10.4 Can Chebyshev's inequality, as a general bound for all random variables with finite mean and variance, be improved upon? For $k \geq 1$, define a random variable X that takes values -1, 0, and 1, with

$$P(X = -1) = P(X = 1) = \frac{1}{2k^2} \quad \text{and} \quad P(X = 0) = 1 - \frac{1}{k^2}.$$

Observe that $\mu = E[X] = 0$ and $\sigma^2 = V[X] = 1/k^2$. Chebyshev's inequality gives

$$P(|X - \mu| \geq k\sigma) \leq \frac{1/k^2}{k^2\sigma^2} = \frac{1}{k^2}.$$

An exact calculation finds

$$P(|X - \mu| \geq k\sigma) = P(|X| \geq 1) = P(X = 1) + P(X = -1) = \frac{1}{k^2}.$$

Thus, this random variable X achieves the Chebyshev bound. This shows that for a general bound that applies to *all* distributions, Chebyshev's bound is the best possible; it cannot be improved upon. ■

The proof of the WLLN is remarkably easy with Chebyshev's inequality.

WEAK LAW OF LARGE NUMBERS

Let $X_1, X_2, \ldots$ be an i.i.d. sequence of random variables with finite mean μ and variance σ^2. For $n = 1, 2, \ldots$, let $S_n = X_1 + \cdots + X_n$. Then

$$P\left(\left|\frac{S_n}{n} - \mu\right| \geq \epsilon\right) \to 0,$$

as $n \to \infty$.

Proof: We have

$$E\left[\frac{S_n}{n}\right] = \mu \quad \text{and} \quad V\left[\frac{S_n}{n}\right] = \frac{\sigma^2}{n}.$$

Let $\epsilon > 0$. By Chebyshev's inequality,

$$P\left(\left|\frac{S_n}{n} - \mu\right| \geq \epsilon\right) \leq \frac{\sigma^2}{n\epsilon^2} \to 0,$$

as $n \to \infty$. □

Remarks:

1. The requirement that the X_i's have finite variance $\sigma^2 < \infty$ is not necessary for the WLLN to hold. We include it only to simplify the proof.
2. The type of convergence stated in the WLLN is known as *convergence in probability*. More generally, say that a sequence of random variables $X_1, X_2, \ldots$ converges in probability to a random variable X, if, for all $\epsilon > 0$,

$$P(|X_n - X| \geq \epsilon) \to 0,$$

   as $n \to \infty$. We write $X_n \xrightarrow{p} X$. Thus, the weak law of large numbers says that S_n/n converges in probability to μ.
3. In statistics, a sequence of estimators $\widehat{p}_1, \widehat{p}_2, \ldots$ for an unknown parameter p is called *consistent* if $\widehat{p}_n \xrightarrow{p} p$, as $n \to \infty$. For instance, the sample mean $\overline{X}_n = (X_1 + \cdots + X_n)/n$ is a consistent estimator for the population mean μ.

The WLLN says that for large n, the average S_n/n is with high probability close to μ. For instance, in a million coin flips we find, using the **R** function `wlln (n, ϵ)` given earlier, that the probability that the proportion of heads is within one-one thousandth of the mean is about 0.954.

```
> wlln(1000000, 0.001)
[1] 0.9544997
```

However, the weak law does *not* say that as you continue flipping more coins your sequence of coin flips will *stay* close to 1/2. You might get a run of "bad luck" with a long sequence of tails, which will temporarily push the average S_n/n below 1/2, further away than your ϵ.

Recall that if a sequence of numbers $x_1, x_2, \ldots$ converges to a limit x, then eventually, for n sufficiently large, the terms $x_n, x_{n+1}, x_{n+2} \cdots$ will *all* be arbitrarily close to x. That is, for any $\epsilon > 0$, there is some index N such that $|x_n - x| \leq \epsilon$ for all $n \geq N$.

The WLLN, however, does not say that the sequence of averages $S_1/1, S_2/2, S_3/3, \ldots$ behaves like this. It does not say that having come close to μ with high probability the sequence of averages will always *stay* close to μ.

It might seem that because the terms of the sequence $S_1/1, S_2/2, S_3/3, \ldots$ are random variables it would be unlikely to guarantee such strong limiting behavior. And yet the remarkable strong law of large numbers (SLLN) says exactly that.

10.2 STRONG LAW OF LARGE NUMBERS

The Strong Hotel has infinitely many rooms. In each room, a guest is flipping coins—forever. Each guest generates an infinite sequence of zeros and ones. We are interested in the limiting behavior of the sequences in each room. In six of the rooms, we find the following outcomes:

Room 1: 0, 1, 1, 0, 0, 0, 0, 0, 0, 0, 0, 1, 0, 0, 0, 0, 0, 1, 1, 0, 1, 0, 1, 0, 1, 1, 0, ...
Room 2: 1, 0, 0, 0, 0, 1, 0, 1, 1, 1, 1, 0, 1, 1, 1, 0, 1, 1, 1, 0, 0, 1, 0, 0, 0, 1, 0, ...
Room 3: 0, 1, 1, 0, 1, 0, 1, 1, 0, 1, 1, 1, 1, 0, 1, 1, 1, 1, 0, 0, 0, 1, 0, 0, 0, 1, 1, ...
Room 4: 1, 1, 1, 0, 0, 0, 1, 0, 1, 1, 1, 0, 1, 1, 0, 0, 1, 0, 1, 1, 1, 1, 1, 0, 1, 1, 1, ...
Room 5: 1, 0, 0, 1, 0, 0, 1, 0, 0, 1, 0, 0, 1, 0, 0, 1, 0, 0, 1, 0, 0, 1, 0, 0, 1, 0, 0, ...
Room 6: 1, 1, 1, 1, 1, 1, 1, 1, 1, 1, 1, 1, 1, 1, 1, 1, 1, 1, 1, 1, 1, 1, 1, 1, 1, 1, 1, ...

While the sequences in the first four rooms do not reveal any obvious pattern, in room 5, heads appears once every three flips. And in room 6, the guest seems to be continually flipping heads. The sequence of partial averages appears to be converging to 1/3 in room 5, and to 1 in room 6. One can imagine many rooms in which the guest will create sequences whose partial averages do not converge to 1/2.

However, the strong law of large numbers says that in *virtually every room of the hotel* the sequence of averages will converge to 1/2. And not only will these averages get arbitrarily close to 1/2 after a very long time, but each will stay close to 1/2 for all the remaining terms of the sequence. Those sequences whose averages converge to 1/2 constitute a set of "probability 1." And those sequences whose averages do not converge to 1/2 constitute a set of "probability 0."

STRONG LAW OF LARGE NUMBERS

Let $X_1, X_2, \ldots$ be an i.i.d. sequence of random variables with finite mean μ. For $n = 1, 2, \ldots$, let $S_n = X_1 + \cdots + X_n$. Then

$$P\left(\lim_{n\to\infty} \frac{S_n}{n} = \mu\right) = 1. \tag{10.2}$$

We say that S_n/n *converges to* μ *with probability* 1.

The proof of the strong law is beyond the scope of this book. However, we will spend some time explaining what it says and means, as well as giving examples of its use.

Implicit in the statement of the theorem is the fact that the set

$$\left\{ \lim_{n\to\infty} \frac{S_n}{n} = \mu \right\}$$

is an event on some sample space. Otherwise, it would not make sense to take its probability. In order to understand the strong law, we need to understand the sample space and the probability function defined on that sample space.

In the case where $X_1, X_2, \ldots$ is an i.i.d. sequence of Bernoulli trials representing coin flips, the sample space Ω is the set of all infinite sequences of zeros and ones. A simple outcome $\omega \in \Omega$ is an infinite sequence.

It can be shown (usually in a more advanced analysis course) that Ω is uncountable. The set can be identified (put in one-to-one correspondence) with the set of real numbers in the interval (0, 1). Every real number r between 0 and 1 has a binary base two expansion of the form

$$r = \frac{x_1}{2} + \frac{x_2}{4} + \cdots + \frac{x_k}{2^k} + \cdots,$$

where the x_k's are 0 or 1. For every $r \in (0, 1)$, the correspondence yields a 0–1 sequence $(x_1, x_2, \ldots) \in \Omega$. Conversely, every coin-flipping sequence of zeros and ones $(x_1, x_2, \ldots)$ yields a real number $r \in (0, 1)$.

The construction of the "right" probability function on Ω is also beyond the scope of this book, and one of the important topics in an advanced probability course based on measure theory. But it turns out that the desired probability function is equivalent to the uniform distribution on (0, 1).

As an illustration, consider the probability that in an infinite sequence of coin flips the first two outcomes are both tails. With our notation, we have that $X_1 = X_2 = 0$. Sequences of zeros and ones in which the first two terms are zero yield binary expansions of the form

$$\frac{0}{2} + \frac{0}{4} + \frac{x_3}{8} + \frac{x_4}{16} + \frac{x_5}{32} + \cdots,$$

where the x_k's are 0 or 1 for $k \geq 3$. The set of all such resulting real numbers gives the interval $(0, 1/4)$. And if $U \sim \text{Unif}(0, 1)$, then $P(0 < U < 1/4) = 1/4$, the probability that the first two coins come up tails.

For a 0–1 sequence $\omega \in \Omega$, write $\omega = (\omega_1, \omega_2, \ldots)$. The set $\{S_n/n \to \mu\}$ is the set of all sequences ω with the property that $(\omega_1 + \cdots + \omega_n)/n \to \mu$, as $n \to \infty$. This set is in fact an event, that is, it is contained in Ω, and thus we can take its probability. And the SLLN says that this probability is equal to 1.

Remarks:

1. The type of convergence described in the SLLN is known as *almost sure convergence*. More generally, we say that a sequence of random variables $X_1, X_2, \ldots$ converges almost surely to a random variable X if $P(X_n \to X) = 1$. We write $X_n \xrightarrow{a.s.} X$. The strong law says that S_n/n converges almost surely to μ.
2. Notice that the SLLN talks about the probability of a limit, while the WLLN is about a limit of a probability.
3. Almost sure convergence is a stronger form of convergence than convergence in probability. Sequences of random variables that converge almost surely also converge in probability. However, the converse is not necessarily true.
4. The set of 0–1 sequences whose partial averages do not converge to 1/2 is very large. In fact, it is uncountable. Nevertheless, the SLLN asserts that such sequences constitute a set of probability 0. In that sense, the set of such sequences is very small!

The SLLN is illustrated in Figure 10.3, which shows the characteristic convergence of S_n/n for four sequences of 1000 coin flips. Wide fluctuations at the beginning of the sequence settle down to an equilibrium very close toμ.

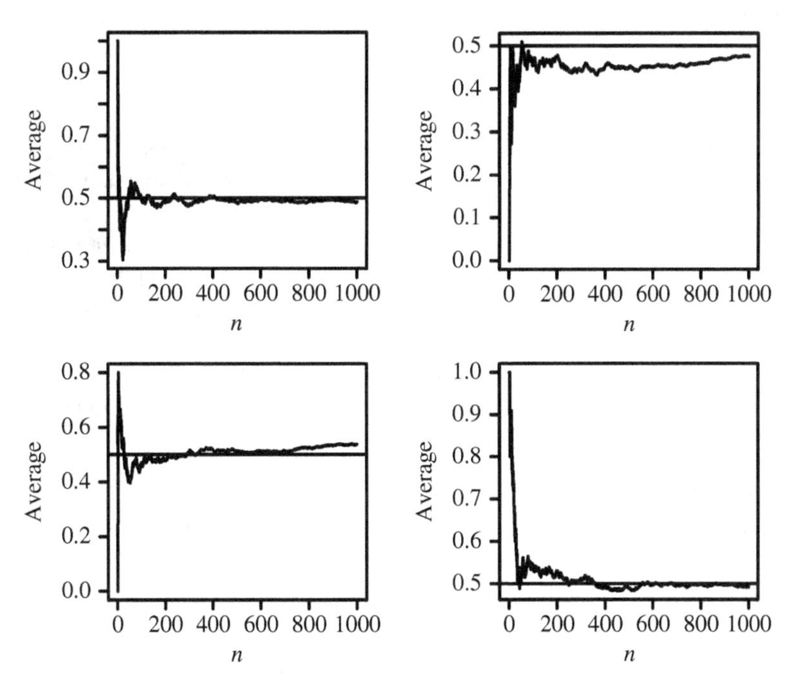

FIGURE 10.3: Four realizations of convergence of S_n/n in 1000 fair coin flips.

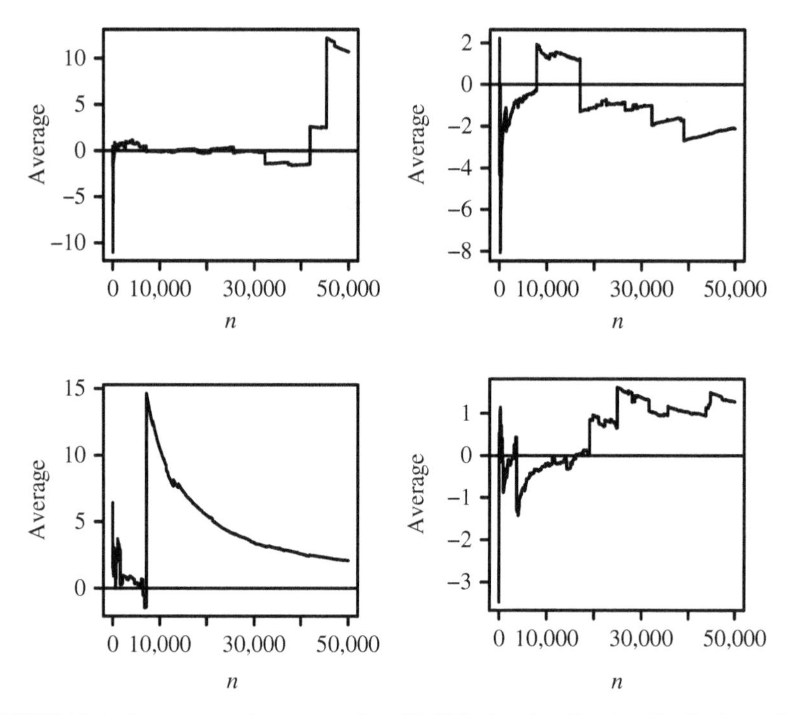

FIGURE 10.4: Sequences of averages ($n = 50{,}000$) for the Cauchy distribution whose expectation does not exist. Observe the erratic behavior.

It is interesting to compare such convergence with what happens for a sequence of random variables in which the expectation does not exist, and for which the SLLN does not apply. The Cauchy distribution, with density $f(x) = 1/(\pi(1 + x^2))$, for all real x, provides such an example. The distribution is symmetric about zero, but does not have finite expectation. Observe the much different behavior of the sequence of averages in Figure 10.4. Several sequences display erratic behavior with no settling down to an equilibrium as in the finite expectation case.

See Exercise 10.29 for the **R** code for creating the graphs in Figures 10.3 and 10.4. The code is easily modified to show the strong law for i.i.d. sequences with other distributions.

Example 10.5 The **Weierstrass approximation theorem** says that a continuous function g on $[0, 1]$ can be approximated arbitrarily closely by polynomials. In particular, for $0 \leq p \leq 1$,

$$\sum_{k=0}^{n} g\left(\frac{k}{n}\right)\binom{n}{k}p^k(1-p)^{n-k} \to g(p),$$

as $n \to \infty$. The nth degree polynomial function of p on the left-hand side is known as the *Bernstein polynomial.*

The result seems to have no connection to probability. However, there is a probabilistic proof based on the law of large numbers. We give the broad strokes. ■

Proof outline. Let $X_1, X_2, \ldots$ be an independent sequence of Bernoulli random variables with parameter p. Then by the SLLN,

$$\frac{X_1 + \cdots + X_n}{n} \to p, \quad \text{as } n \to \infty,$$

with probability 1. If g is a continuous function, then with probability 1,

$$g\left(\frac{X_1 + \cdots + X_n}{n}\right) \to g(p).$$

It also follows that

$$E\left[g\left(\frac{X_1 + \cdots + X_n}{n}\right)\right] \to E[g(p)] = g(p), \quad \text{as } n \to \infty.$$

As $X_1 + \cdots + X_n \sim \text{Binom}(n, p)$, the left-hand side is

$$E\left[g\left(\frac{X_1 + \cdots + X_n}{n}\right)\right] = \sum_{k=0}^{n} g\left(\frac{k}{n}\right)\binom{n}{k} p^k (1-p)^{n-k},$$

which gives the result. □

Consider the problem of estimating the area of a "complicated" set. Here is a Monte Carlo method based on the accept–reject method introduced in Section 6.7.1.

Let S be a bounded set in the plane. Suppose R is a rectangle that encloses S. Let $X_1, \ldots, X_n$ be i.i.d. points uniformly distributed in R. Define

$$I_k = \begin{cases} 1, & \text{if } X_k \in S \\ 0, & \text{otherwise,} \end{cases}$$

for $k = 1, \ldots, n$. The I_k's form an i.i.d. sequence. Their common expectation is

$$E[I_k] = P(X_k \in S) = \frac{\text{Area }(S)}{\text{Area }(R)}.$$

The proportion of the n points that lie in S is

$$\frac{I_1 + \cdots + I_n}{n} \to \frac{\text{Area }(S)}{\text{Area }(R)}, \quad \text{as } n \to \infty,$$

with probability 1. Thus for large n,

$$\text{Area }(S) \approx \left(\frac{I_1 + \cdots + I_n}{n}\right) \text{Area }(R).$$

The area of S is approximately equal to the area of the rectangle R times the proportion of the n points that fall in S.

Example 10.6 Estimating the area of the United States. What is the area of the continental United States? The distance from a line parallel to the northernmost point in the continental United States (Lake of the Woods, Minnesota) to a line parallel to the southernmost point (Ballast Key, Florida) is 1720 miles. From the westernmost point (Cape Alava, Washington) to the easternmost point (West Quoddy Head, Maine) is 3045 miles. The continental United States fits inside a rectangle that is $1720 \times 3045 = 5{,}237{,}400$ square miles.

We enclosed a map of the continental United States inside a comparable rectangle. One thousand points uniformly distributed in the rectangle were generated and we counted 723 in the map (see Fig. 10.5). The area of the continental United States is estimated as $(0.723)(5{,}237{,}400) = 3{,}786{,}640$ square miles.

The area of the continental United States is in fact 3,718,710 square miles. ■

The SLLN has some other applications, including the method of moments and Monte Carlo integration, explored in the next sections. It is not our last result for limits though, as you will see with the central limit theorem (CLT).

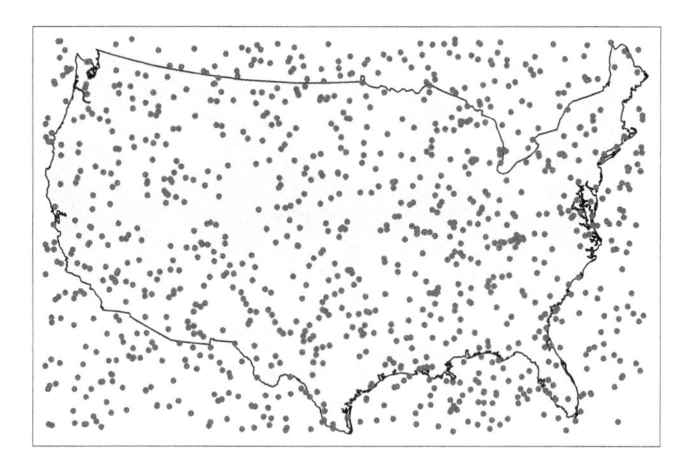

FIGURE 10.5: Using probability and random numbers to estimate the area of the United States.

10.3 METHOD OF MOMENTS*

The *method of moments* is a statistical technique for using data to estimate the unknown parameters of a probability distribution. If you continue to study statistics, you are likely to see a maximum likelihood approach as well. We explore the method of moments briefly due to its connection to the SLLN.

Recall that the kth moment of a random variable X is $E[X^k]$. We will also call this the *kth theoretical moment.*

Let $X_1, \ldots, X_n$ be an i.i.d. sample from a probability distribution with finite moments. Think of $X_1, \ldots, X_n$ as representing data from a random sample. The kth *sample moment* is defined as

$$\frac{1}{n}\sum_{i=1}^{n} X_i^k.$$

In a typical statistical context, the values of the X_i's are known (they are the observation values in the data set), and the parameters of the underlying probability distribution are unknown.

In the method of moments, one sets up equations that equate sample moments with corresponding theoretical moments. The equation(s) are solved for the unknown parameter(s) of interest. The method is reasonable because if X is a random variable from the probability distribution of interest then by the SLLN, with probability 1,

$$\frac{1}{n}\sum_{i=1}^{n} X_i^k \to E[X^k], \quad \text{as } n \to \infty,$$

and thus for large n,

$$\frac{1}{n}\sum_{i=1}^{n} X_i^k \approx E[X^k].$$

For example, suppose a biased die has some unknown probability of coming up 5. An experiment is performed whereby the die is rolled repeatedly until a 5 occurs. Data are kept on the number of rolls until a 5 occurs. The experiment is repeated 10 times. The following data are generated:

$$3 \quad 1 \quad 4 \quad 4 \quad 1 \quad 6 \quad 4 \quad 2 \quad 9 \quad 2.$$

(Thus on the first experiment, the first 5 occurred on the third roll of the die. On the second experiment, a 5 occurred on the first roll of the die, etc.)

The data are modeled as outcomes from a geometric distribution with unknown success parameter p, where p is the probability of rolling a 5. We consider the sample as the outcome of i.i.d. random variables $X_1, \ldots, X_{10}$, where $X_i \sim \text{Geom}(p)$.

Let $X \sim \text{Geom}(p)$. The first theoretical moment is $E[X] = 1/p$. The first sample moment is

$$\frac{1}{10}\sum_{i=1}^{10} X_i = \frac{3+1+4+4+1+6+4+2+9+2}{10} = 3.6.$$

Set the theoretical and sample moments equal to each other. Solving $1/p = 3.6$ gives $p = 1/3.6 = 0.278$, which is the method of moments estimate for p.

Example 10.7 Hoffman [2003] suggests a negative binomial distribution to model water bacteria counts for samples taken from a water purification system. The goal is to calculate an upper control limit for the number of bacteria in future water samples. Hoffman illustrates the approach for their data, and we illustrate the method on simulated data to have a slightly larger sample size.

Suppose that 20 water samples were taken from a normally functioning water purification system. The (simulated) data of bacteria counts per milliliter are:

1 6 3 6 28 8 4 1 8 13 14 3 11 8 6 7 8 13 11 5.

Recall that negative binomial distribution is a flexible family of discrete distributions with two parameters r and p. It is widely used to model data in the form of counts. The variation of the distribution used by **R**, takes nonnegative integer values starting at 0, counting the number of failures before the rth success. An even more general formulation of the distribution allows the parameter r to be real.

For a random variable $X \sim \text{NegBin}(r, p)$, using the variation of the distribution described above, the first two theoretical moments are

$$E[X] = \frac{r(1-p)}{p}$$

and

$$E[X^2] = V[X] + E[X]^2 = \frac{r(1-p)}{p^2} + \left(\frac{r(1-p)}{p}\right)^2.$$

The first and second sample moments for these data are

$$\frac{1}{20}\sum_{i=1}^{20} X_i = 8.2 \quad \text{and} \quad \frac{1}{20}\sum_{i=1}^{20} X_i^2 = 101.7.$$

The two methods of moments equations are thus

$$\frac{r(1-p)}{p} = 8.2$$

and

$$\frac{r(1-p)}{p^2} + \left(\frac{r(1-p)}{p}\right)^2 = 101.7.$$

Substituting the first equality into the second equation gives

$$101.7 = \frac{8.2}{p} + (8.2)^2.$$

Solving gives $p = 0.238$. Back substituting then finds $r = 2.05$.

In reality, the data was simulated from a distribution with $r = 2$ and $p = 0.2$. The method of moments estimators are reasonable.

How could we use our estimators? Well, suppose for quality control purposes, engineers wanted to use the fitted model to set an upper limit of bacteria counts for future samples. The top 5% of the distribution might be considered "unusual." That is, we might want to find the *95th percentile* of the distribution, and state that bacteria counts greater than that cutoff could be a warning sign that the purification system is not functioning normally. Here, we use **R** to find that cutoff as 17, so we would flag counts of 17 or higher as unusual. Under the theoretical distribution, the cutoff is 20, so our estimators have us being slightly more cautious than we might need to be. ■

Method of moments estimators can be found for continuous distribution parameters as well, as shown in the next example.

Example 10.8 Suppose we have i.i.d. random variables $X_1, \ldots, X_n$, where $X_i \sim \text{Unif}(0, \theta)$, and we want to estimate θ. Find the method of moments estimator for θ.

From our previous results, we know that $E(X_i) = \theta/2$. The first sample moment is $\overline{X}$, and the idea in method of moments is that this is an approximation for $E(X_i)$ for large n. Thus, if we solve for θ, we find that the estimator is $\theta = 2\overline{X}$.

Suppose $n = 10$ values were generated from a Uniform distribution with unknown θ, with the following values observed:

8.28 3.99 7.03 12.48 3.69 9.32 13.74 8.86 1.97 2.52.

We find the sample mean, $\overline{X} = 7.188$. Thus, the method of moments estimate of θ is twice that or 14.376.

To demonstrate the performance of the estimator, you can choose a θ value and perform simulations. (More theoretical considerations of estimators await readers who continue further study in statistics.) See **MoM.R** for an example simulation in this setting. ■

10.4 MONTE CARLO INTEGRATION

We consider *Monte Carlo integration* as another application of the SLLN. Let g be a continuous function on $(0, 1)$. Consider solving the integral

$$I = \int_0^1 g(x)\,dx,$$

assuming the integral converges. If g does not have a known antiderivative, solving the integral may be hard and require numerical approximation. Although the problem does not seem to have any connection with probability, observe that we can write the integral as the expectation of a random variable. In particular,

$$\int_0^1 g(x)\,dx = E[g(X)],$$

where X is uniformly distributed on $(0, 1)$. Writing the integral as an expectation allows for using the law of large numbers to approximate the integral in an approach called Monte Carlo integration.

Let $X_1, X_2, \ldots$ be an i.i.d. sequence of uniform $(0, 1)$ random variables. Then $g(X_1), g(X_2), \ldots$ is also an i.i.d. sequence with expectation $E[g(X)]$. By the SLLN, with probability 1,

$$\frac{g(X_1) + \cdots + g(X_n)}{n} \to E[g(X)] = I, \quad \text{as } n \to \infty.$$

This gives a recipe for a Monte Carlo approximation of the integral.

MONTE CARLO INTEGRATION ON $(0, 1)$

Let g be a continuous function on $(0, 1)$. The following Monte Carlo algorithm gives an approximation for the integral

$$I = \int_0^1 g(x)\,dx.$$

1. Generate n uniform $(0, 1)$ random variables $X_1, \ldots, X_n$.
2. Evaluate g at each X_i, giving $g(X_1), \ldots, g(X_n)$.
3. Take the average as a Monte Carlo approximation of the integral. That is,

$$I \approx \frac{g(X_1) + \cdots + g(X_n)}{n}.$$

Example 10.9 Solve

$$\int_0^1 (\sin x)^{\cos x}\, dx.$$

Here $g(x) = (\sin x)^{\cos x}$. The integral has no simple antiderivative. We use Monte Carlo approximation with $n = 10{,}000$.

R: MONTE CARLO INTEGRATION

```
> g <- function(x) sin(x)^cos(x)
> n <- 10000
> simlist <- g(runif(n))
> mean(simlist)
[1] 0.5032993
```

The solution can be "checked" using numerical integration. The **R** command `integrate` numerically solves the integral within a margin of error.

```
> integrate(g, 0, 1)
0.5013249 with absolute error < 1.1e-09
```

■

The method we have given can be used, in principle, to approximate *any* convergent integral on (0, 1) using uniform random variables. However, the Monte Carlo method can be used in a wider context, with different ranges of integration, types of distributions, and with either integrals or sums.

Example 10.10 Solve

$$\int_0^\infty x^{-x}\, dx.$$

As the range of integration is $(0, \infty)$, we cannot express the integral as a function of a uniform random variable on (0, 1). However, we can write

$$\int_0^\infty x^{-x}\, dx = \int_0^\infty x^{-x}e^{x}e^{-x}\, dx = \int_0^\infty \left(\frac{e}{x}\right)^x e^{-x}\, dx.$$

Letting $g(x) = (e/x)^x$, the integral is equal to $E[g(X)]$, where X has an exponential distribution with parameter $\lambda = 1$. For Monte Carlo integration, generate exponential random variables, evaluate using g, and take the resulting average.

R: MONTE CARLO INTEGRATION

```
> g <- function(x) (exp(1)/x)^x
> n <- 10000
> simlist <- g(rexp(n, 1))
> mean(simlist)
[1] 1.998325
```

Here is a check by numerical integration:

```
> h <- function(x) x^(-x)
> integrate(h, 0, Inf)
1.995456 with absolute error < 0.00016
```

■

As mentioned above, Monte Carlo approximation is not restricted to integrals. Here we approximate the sum of a series using our knowledge of discrete random variables.

Example 10.11 Solve

$$\sum_{k=1}^{\infty} \frac{\log k}{3^k}.$$

Write

$$\sum_{k=1}^{\infty} \frac{\log k}{3^k} = \sum_{k=1}^{\infty} \frac{\log k}{2} \left(\frac{1}{3}\right)^{k-1} \left(\frac{2}{3}\right) = \sum_{k=1}^{\infty} \frac{\log k}{2} P(X = k),$$

where X has a geometric distribution with parameter $p = 2/3$. The last expression is equal to $E[g(X)]$, where $g(x) = (\log x)/2$.

R: MONTE CARLO SUMMATION

```
> g <- function(x) (log(x))/2
> n <- 10000
> simlist <- g(rgeom(n, 2/3) + 1)
> mean(simlist)
[1] 0.144658
```

The exact sum of the series, to seven decimal places, is 0.1452795. ■

Example 10.12 Solve

$$I = \int_{-\infty}^{\infty} \int_{-\infty}^{\infty} \int_{-\infty}^{\infty} \sin(x^2 + 2y - z) e^{-(x^2+y^2+z^2)/2} \, dx \, dy \, dz.$$

The multiple integral can be expressed as an expectation with respect to a joint distribution. Let (X, Y, Z) be independent standard normal random variables. The joint density of (X, Y, Z) is

$$f(x, y, z) = \frac{1}{2\pi\sqrt{2\pi}} e^{-(x^2+y^2+z^2)/2}.$$

Write the multiple integral as $I = 2\pi\sqrt{2\pi} E[g(X, Y, Z)]$, where $g(x, y, z) = \sin(x^2 + 2y - z)$. Let $(X_i, Y_i, Z_i), i = 1, \ldots, n$ be an i.i.d. sample from the joint normal distribution. Then

$$I = 2\pi\sqrt{2\pi} E[g(X, Y, Z)] \approx \frac{2\pi\sqrt{2\pi}}{n} \sum_{i=1}^{n} g(X_i, Y_i, Z_i).$$

R: MONTE CARLO INTEGRATION

```
> n <- 1000000
> simlist <- numeric(n)
> for (i in 1:n) {
  vect <- rnorm(3)
  x <- vect[1]
  y <- vect[2]
  z <- vect[3]
  simlist[i] <- sin(x^2+2*y-z) }
> 2*pi*sqrt(2*pi)*mean(simlist)
[1] 0.4701659
```

A symbolic mathematical software system took 3 minutes to solve the integral giving the exact answer

$$I = 2\pi^{3/2} e^{-5/2} \sqrt{\frac{1}{\sqrt{5}} - \frac{1}{5}} = 0.454522\ldots \qquad \blacksquare$$

Is Monte Carlo integration a practical method for solving integrals? For a single integral, probably not, as there are many deterministic numerical integration methods that give excellent approximations to desired levels of accuracy.

However in high dimensions, with large multiple integrals, numerical techniques break down and are not able to give accurate results. It is not uncommon in statistics, biology, physics, and many fields to encounter intractable integrals involving hundreds, even thousands, of variables. For such problems, randomized methods are the only way to go. Monte Carlo integration, often using sophisticated methods involving random processes called Markov chains, is the only practical way in which such integrals can be solved. According to the prominent mathematician and probabilist Diaconis [2008], the use of simulation for such problems has "revolutionized applied mathematics." We introduce this revolutionary methodology, called Markov chain Monte Carlo, in Section 11.3.

10.5 CENTRAL LIMIT THEOREM

> Everyone believes in the law of errors, the experimenters because they think it is a mathematical theorem, the mathematicians because they think it is an experimental fact.
>
> —Gabriel Lipman in conversation with Henri Poincaré

The next theorem rivals the law of large numbers in importance. It gives insight into the behavior of sums of random variables, it is fundamental to much of statistical inference, and it quantifies the size of the error in using Monte Carlo methods to approximate integrals, expectations, and probabilities.

CENTRAL LIMIT THEOREM (CLT)

Let $X_1, X_2, \ldots$ be an i.i.d. sequence of random variables with finite mean μ and variance σ^2. For $n = 1, 2, \ldots$, let $S_n = X_1 + \cdots + X_n$. Then the distribution of the standardized random variable $(S_n/n - \mu)/(\sigma/\sqrt{n})$ converges to a standard normal distribution in the following sense. For all t,

$$P\left(\frac{S_n/n - \mu}{\sigma/\sqrt{n}} \leq t\right) \to P(Z \leq t), \text{ as } n \to \infty,$$

where $Z \sim \text{Norm}(0, 1)$.

We first illustrate the CLT with a simulation experiment. Let $X_1, X_2, \ldots,$ denote i.i.d. Bernoulli trials with $p = 1/2$ (e.g., fair coin flips). The mean and variance of the X_i's are, respectively, $\mu = 1/2$ and $\sigma^2 = 1/4$. Consider the sequence

$$\frac{S_n/n - \mu}{\sigma/\sqrt{n}} = \frac{S_n/n - 1/2}{1/(2\sqrt{n})} = 2\sqrt{n}\left(\frac{S_n}{n} - \frac{1}{2}\right), \; n = 1, 2, \ldots . \tag{10.3}$$

Observe that by the law of large numbers, $(S_n/n - 1/2) \to 0$, as $n \to \infty$, because $S_n/n \to 1/2$. Thus, on the right-hand side of Equation 10.3, we take a sequence that converges to 0 and multiply it by $2\sqrt{n}$, a sequence that diverges to infinity. This gives the indeterminate form $\infty \cdot 0$, and *a priori* we do not know whether the resulting sequence will converge to 0, tend to infinity, or do something "in between" like converge to a constant. Furthermore, the terms of the sequence are random variables.

In the simulation code given next, the function `cltsequence(n)` simulates terms of this sequence. Choose a large value of n, enter values repeatedly at your keyboard, and it becomes apparent that the sequence does not converge to 0, nor diverge to infinity, but does "something else." When we repeat many times, graph the distribution of outcomes with a histogram, and superimpose a normal density curve on top of the graph, we obtain an excellent fit to a standard normal distribution, as seen in Figure 10.6.

R: SIMULATION EXPERIMENT

```
> cltsequence <- function(n)
   (mean(rbinom(n,1,1/2))-1/2)*(2*sqrt(n))
> cltsequence(1000)
```

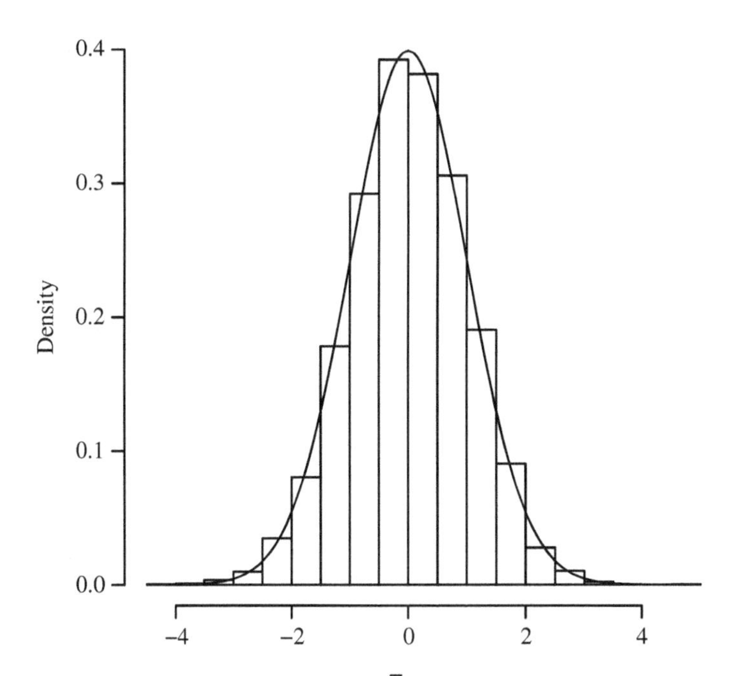

FIGURE 10.6: Histogram of $(S_n/n - \mu)/(\sigma/\sqrt{n})$ from an underlying sequence of $n = 1000$ Bernoulli trials. The density curve is the standard normal.

```
[1] 0.8221922
> cltsequence(1000)
[1] 1.70763
> cltsequence(1000)
[1] 0.8221922
> cltsequence(1000)
[1] 0.1264911
> simlist <- replicate(10000, cltsequence(1000))
> hist(simlist, prob = T)
> curve(dnorm(x),-4, 4, add = T)
```

The astute reader may notice that because a sum of Bernoulli trials has a binomial distribution, what we have illustrated is just the normal approximation of the binomial distribution, discussed in Section 7.1.2.

True enough. However, the power of the CLT is that it applies to *any* population distribution with finite mean and variance. In Figure 10.7, observe the results from

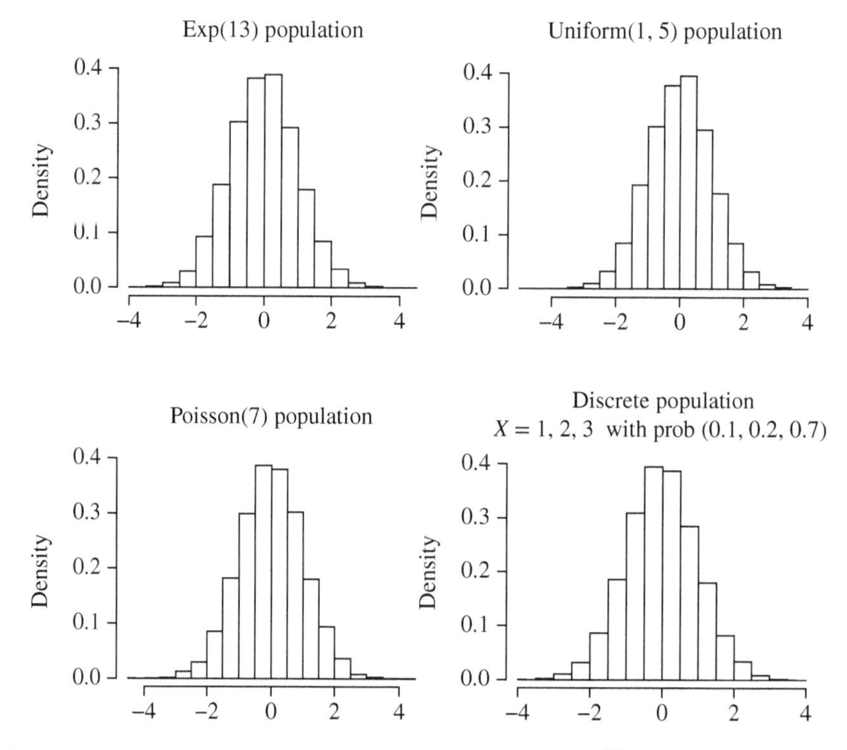

FIGURE 10.7: The simulated distribution of $(S_n/n - \mu)/(\sigma/\sqrt{n})$ for four population distributions ($n = 1000$).

simulation experiments with four different population distributions: (i) exponential with $\lambda = 13$; (ii) uniform on $(1, 5)$; (iii) Poisson with $\lambda = 7$; and (iv) a discrete distribution that takes values 1, 2, and 3, with respective probabilities 0.1, 0.2, and 0.7. In every case, the distribution of the standardized random variable $(S_n/n - \mu)/(\sigma/\sqrt{n})$ tends to a standard normal distribution as n tends to infinity.

Remarks:

1. If the distribution of the X_i's is normal, then by results for sums of independent normal random variables, the distributions of S_n and S_n/n are *exactly* normal. The specialness of the CLT is that it applies to *any* distribution of the X_i's with finite mean and variance.
2. There are several equivalent ways to formulate the CLT. For large n, the sum $S_n = X_1 + \cdots + X_n$ is approximately normal with mean $n\mu$ and variance $n\sigma^2$. Also, the average S_n/n is approximately normal with mean μ and variance σ^2/n.

EQUIVALENT EXPRESSIONS FOR THE CENTRAL LIMIT THEOREM

If the sequence of random variables $X_1, X_2, \ldots$ satisfies the assumptions of the CLT, then for large n,

$$X_1 + \cdots + X_n \approx \text{Norm}(n\mu, n\sigma^2)$$

and

$$\overline{X}_n = \frac{X_1 + \cdots + X_n}{n} \approx \text{Norm}(\mu, \sigma^2/n).$$

3. The type of convergence described in the CLT is called *convergence in distribution*. We say that random variables $X_1, X_2, \ldots$ *converge in distribution to* X, if for all t, $P(X_n \leq t) \to P(X \leq t)$, as $n \to \infty$.

Example 10.13 Customers at a popular restaurant are waiting to be served. Waiting times are independent and exponentially distributed with mean $1/\lambda = 30$ minutes. If 16 customers are waiting what is the probability that their average wait is less than 25 minutes?

The average waiting time is $S_{16}/16$. As waiting time is exponentially distributed, the mean and standard deviation of individual waiting time is $\mu = \sigma = 30$. By the CLT,

$$P(S_{16}/16 < 25) = P\left(\frac{S_{16}/16 - \mu}{\sigma/\sqrt{n}} < \frac{25 - 30}{7.5}\right)$$
$$\approx P(Z < -0.667) = 0.252.$$

TABLE 10.1. Grade distribution for AP exams

1	2	3	4	5
0.214	0.211	0.236	0.195	0.144

For this example, we can compare the central limit approximation with the exact probability. The distribution of S_{16}, the sum of independent exponentials, is a gamma distribution with parameters 16 and $1/30$. This gives $P(S_{16}/16 < 25) = P(S_{16} < 400)$. In **R**,

```
> pgamma(400,16,1/30)
[1] 0.2666045
> pnorm(-0.6667)
[1] 0.2524819
```

■

Example 10.14 More than three million high school students took an AP exam in 2011. The grade distribution for the exams is given in Table 10.1.

In one high school, a sample of 30 AP exam scores was taken. What is the probability that the average score is above 3?

Assume scores are independent. The desired probability is $P(S_{30}/30 > 3)$. The expectation of exam score is

$$1(0.214) + 2(0.211) + 3(0.236) + 4(0.195) + 5(0.144) = 2.844.$$

The mean of the squared scores is

$$1(0.214)^2 + 2(0.211)^2 + 3(0.236)^2 + 4(0.195)^2 + 5(0.144)^2 = 9.902,$$

giving the standard deviation of exam score to be $\sqrt{9.902 - (2.844)^2} = 1.347$. By the CLT,

$$\begin{aligned} P(S_{30}/30 > 3) &= P\left(\frac{S_{30}/30 - \mu}{\sigma/\sqrt{n}} > \frac{3 - 2.844}{1.347/\sqrt{30}}\right) \\ &\approx P(Z > 0.634) = 0.263. \end{aligned}$$

■

Example 10.15 Random walks. Random walks are fundamental models in physics, ecology, and numerous other fields. They have been popularized in finance with applications to the stock market. A particle starts at the origin on the integer number line. At each step, the particle moves left or right with probability 1/2. Find the expectation and standard deviation of the distance of the walk from the origin after n steps.

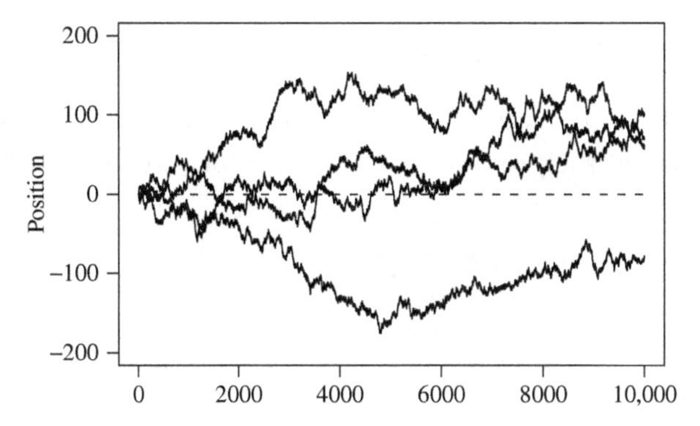

FIGURE 10.8: Four random walk paths of 10,000 steps. The horizontal axis represents the number of steps n. The vertical axis is position.

A random walk process is constructed as follows. Let $X_1, X_2, \ldots$ be an independent sequence of random variables taking values ± 1 with probability 1/2 each. The X_i's represent the individual steps of the random walk. For $n = 1, 2, \ldots$, let $S_n = X_1 + \cdots + X_n$ be the position of the walk after n steps. The random walk process is the sequence $(S_1, S_2, S_3, \ldots)$. Four "paths" for such a process, each taken for 10,000 steps, are shown in Figure 10.8.

The X_i's have mean 0 and variance 1. Thus, $E[S_n] = 0$ and $V[S_n] = n$. By the CLT, for large n the distribution of S_n is approximately normal with mean 0 and variance n.

After n steps, the random walk's distance from the origin is $|S_n|$. Using the normal approximation, the expected distance from the origin is

$$E[|S_n|] \approx \int_{-\infty}^{\infty} |t| \frac{1}{\sqrt{2\pi n}} e^{-t^2/2n} \, dt = \frac{2}{\sqrt{2\pi n}} \int_0^{\infty} te^{-t^2/2n} \, dt$$

$$= \frac{2n}{\sqrt{2\pi n}} = \sqrt{\frac{2}{\pi}}\sqrt{n} \approx (0.80)\sqrt{n}.$$

For the standard deviation of distance, $E[|S_n|^2] = E[S_n^2] = n$. Thus,

$$V[|S_n|] = E[|S_n^2|] - E[|S_n|]^2 \approx n - \frac{2n}{\pi} = n\left(\frac{\pi - 2}{\pi}\right),$$

giving

$$\text{SD}[|S_n|] \approx \sqrt{\frac{\pi - 2}{\pi}}\sqrt{n} \approx (0.60)\sqrt{n}.$$

For instance, after 10,000 steps, a random walk is about $(0.80)\sqrt{10{,}000} = 80$ steps from the origin give or take about $(0.60)\sqrt{10{,}000} = 60$ steps. ■

R: RANDOM WALK DISTANCE FROM ORIGIN

In **RandomWalk.R**, we simulate the expectation and standard deviation of a random walk's distance from the origin. Also included is the code for generating the random walk graphs in Figure 10.8.

```
> reps <- 5000
> simlist <- numeric(reps)
> for (i in 1:reps) {
     rw <- sample(c(-1, 1), 10000, replace = T)
     simlist[i] <- abs(tail(cumsum(rw), 1))  }
> mean(simlist)
[1] 79.1296
> sd(simlist)
[1] 59.46567

# random walk path
> steps <- 10000
> rw <- sample(c(-1, 1), steps, replace = T)
> plot(c(1, 1), type = "n", xlim = c(0, steps),
  ylim = c(-200, 200), xlab = "", ylab = "Position")
> lines(cumsum(rw), type = "l")
```

The CLT requires that n be "large" for the normal approximation to hold. But how large? In statistics, empirical evidence based on the behavior of many variables in actual datasets suggests that $n \approx 30$–40 is usually "good enough" for the CLT to take effect. Heavily skewed distributions require larger n. However, if the population distribution is fairly symmetric with no extreme values, even relatively small values of n may work reasonably well.

Example 10.16 With just six dice and the CLT one can obtain a surprisingly effective method for simulating normal random variables.

Let $S_6 = X_1 + \cdots + X_6$ be the sum of six dice throws. Then $E[S_6] = 6(3.5) = 21$. And $V[S_6] = 6V[X_1] = 6(35/12) = 35/2$, with $\text{SD}[S_6] = \sqrt{35/2} \approx 4.1833$.

Roll six dice and take $(S_6 - 21)/4.1833$ as a simulation of one standard normal random variable. Here are some results.

R: SUM OF SIX DICE ARE CLOSE TO NORMAL

```
> (sum(sample(1:6, 6, rep = T))-21)/4.1833
[1] 0.2390457
> (sum(sample(1:6, 6, rep = T))-21)/4.1833
[1] 0.4780915
```

```
> (sum(sample(1:6, 6, rep = T))-21)/4.1833
[1]  1.434274
> (sum(sample(1:6, 6, rep = T))-21)/4.1833
[1] -0.4780915
> (sum(sample(1:6, 6, rep = T))-21)/4.1833
[1] 0.9561829
```

To explore the distribution of the standardized sum, we simulate 10,000 replications. The normal distribution satisfies the "68-95-99.7 rule." Counting the number of observations within 1, 2, and 3 standard deviations, respectively, gives fairly close agreement with the normal distribution to within about 0.02.

```
> reps <- 10000
> simlist <- replicate(reps,
   (sum(sample(1:6, 6, rep = T))-21)/4.1833)
> sum(-1 <= simlist & simlist <= 1)/reps
[1] 0.7195
> sum(-2 <= simlist & simlist <= 2)/reps
[1] 0.9625
> sum(-3 <= simlist & simlist <= 3)/reps
[1] 0.9991
```

See the comparison in Figure 10.9 between the distribution of the standardized sum of six dice throws and the standard normal density curve.

■

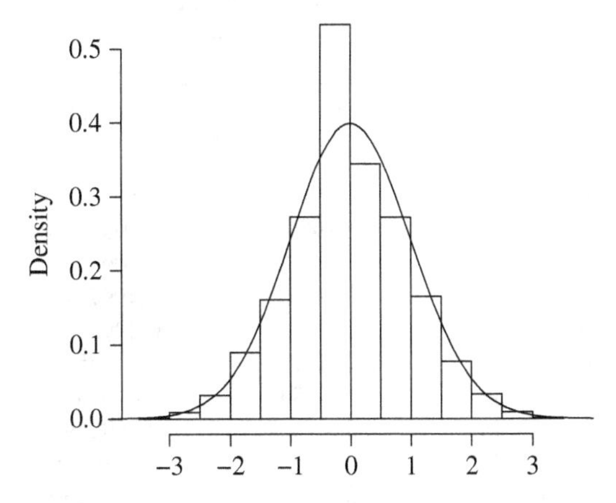

FIGURE 10.9: The normalized sum of just six dice throws comes close to the normal distribution.

10.5.1 Central Limit Theorem and Monte Carlo

In broad strokes, the law of large numbers asserts that $S_n/n \approx \mu$, when n is large. The CLT states that for large n, $(S_n/n - \mu)/(\sigma/\sqrt{n}) \approx Z$, where Z is a standard normal random variable. Equivalently,

$$\frac{S_n}{n} \approx \mu + \frac{\sigma}{\sqrt{n}} Z.$$

Thus, the CLT can be seen as giving the second-order term in the approximation of μ by S_n/n. This suggests that the error $|S_n/n - \mu|$ in the approximation decreases on the order of $1/\sqrt{n}$, as $n \to \infty$.

What this means is that increasing the number of trials in a simulation by a factor of $n = 100$ will roughly decrease the error by a factor of $1/\sqrt{n} = 1/10$. In other words, every 100-fold increase in the number of Monte Carlo trials improves accuracy in the simulation by about one significant digit.

To illustrate, we use Monte Carlo to approximate the mean μ of a uniform (0, 1) distribution, taking samples of size n from the distribution and using the average to approximate $\mu = 1/2$.

See Table 10.2 to observe the accuracy of Monte Carlo approximation for $n = 10^1, 10^3, 10^5$, and 10^7. For each value of n, the simulation is repeated 12 times. With just 10 trials, the results are not uniformly accurate for even one digit. With 1000 trials, the results are uniformly accurate in the first digit. That is, each approximation is equal to 0.5 when rounded to one digit. With 100,000 trials, the results are precise to two digits. And with 10 million trials, all of the approximations round to 0.500 to three digits.

TABLE 10.2. Monte Carlo approximation of the mean of a uniform $(0, 1)$ distribution. Compare the precision for $n = 10^1, 10^3, 10^5, 10^7$. Each simulation is repeated 12 times

	Number of trials			
Outcome	10	1000	100,000	10,000,000
1	0.7999731	0.4887204	0.5000470	0.4999549
2	0.6176654	0.4922006	0.5004918	0.5000157
3	0.5858701	0.5034127	0.4999026	0.4999761
4	0.5304252	0.4914003	0.5003041	0.5000184
5	0.5602126	0.5070693	0.4997541	0.4999820
6	0.4448233	0.5097890	0.4992946	0.4998743
7	0.3839213	0.4936749	0.5024522	0.5001255
8	0.5434252	0.5010844	0.4997956	0.5000897
9	0.3328748	0.5147627	0.4988693	0.4999183
10	0.5452109	0.4917534	0.5016036	0.5001748
11	0.4943713	0.5178478	0.4999336	0.5001578
12	0.5573252	0.5001346	0.5025030	0.4998666

10.6 A PROOF OF THE CENTRAL LIMIT THEOREM

Recall that the moment-generating function for a random variable X is $m_X(t) = E[e^{tX}]$. Previously, we used moment-generating functions to find moments, identify distributions, and examine relationships between random variables. Moment-generating functions play a significant role in establishing limit theorems for random variables because of the following result, which we state without proof.

CONTINUITY THEOREM

Let $X_1, X_2, \ldots$ be a sequence of random variables with corresponding mgfs $m_{X_1}(t), m_{X_2}(t), \ldots$ Further suppose that for all t,

$$m_{X_n}(t) \to m_X(t), \quad \text{as } n \to \infty,$$

where $m_X(t)$ is the mgf of a random variable X. Then

$$P(X_n \le x) \to P(X \le x), \quad \text{as } n \to \infty,$$

at each x where $P(X \le x)$ is continuous.

Thus, convergence of the mgfs corresponds to convergence in distribution for random variables.

Here we prove the CLT using moment generating functions. We show that the cumulative distribution function of $(S_n - n\mu)/(\sigma\sqrt{n})$ converges to the cdf of the standard normal distribution.

Proof. Let $X_1, X_2, \ldots$ be an i.i.d. sequence of random variables with finite mean μ and variance σ^2. First suppose that $\mu = 0$ and $\sigma^2 = 1$. Then $S_n/n = (X_1 + \cdots + X_n)/n$ has mean 0 and variance $1/n$. We need to show that the mgf of $(S_n/n)/(1/\sqrt{n}) = S_n/\sqrt{n}$ converges to the mgf of Z, where $Z \sim \text{Norm}(0, 1)$.

Let m be the common mgf of the X_i's. By the properties of the mgfs,

$$m_{S_n/\sqrt{n}}(t) = \left[m\left(\frac{t}{\sqrt{n}}\right)\right]^n.$$

The mgf of the standard normal distribution is $m_Z(t) = e^{t^2/2}$. We need to show that for all t,

$$\left[m\left(\frac{t}{\sqrt{n}}\right)\right]^n \to e^{t^2/2}, \quad \text{as } n \to \infty.$$

Equivalently, we take logarithms and show that

$$\ln\left(m\left(\frac{t}{\sqrt{n}}\right)^n\right) \to \frac{t^2}{2}, \quad \text{as } n \to \infty.$$

Consider the limit, as $n \to \infty$, of $\ln\,[m(t/\sqrt{n})]^n = n \ln m(t/\sqrt{n})$. Assuming continuity of the mgf,

$$\lim_{n\to\infty} \ln\left(m\left(\frac{t}{\sqrt{n}}\right)\right) = \ln\left(m\left(\lim_{n\to\infty}\frac{t}{\sqrt{n}}\right)\right) = \ln m(0) = \ln 1 = 0.$$

Thus, $\lim_{n\to\infty} n \ln m(t/\sqrt{n})$ is an indeterminate form of type $\infty \cdot 0$. Apply l'Hôpital's rule two times, but first make a change of variables letting $\epsilon = t/\sqrt{n}$, so $n = t^2/\epsilon^2$, giving

$$\begin{aligned}
\lim_{n\to\infty} n \ln m(t/\sqrt{n}) &= t^2 \lim_{\epsilon\to 0} \frac{\ln m(\epsilon)}{\epsilon^2} \\
&= t^2 \lim_{\epsilon\to 0} \frac{m'(\epsilon)}{m(\epsilon)2\epsilon} \\
&= \lim_{\epsilon\to 0} t^2 \frac{m''(\epsilon)}{m'(\epsilon)2\epsilon + 2m(\epsilon)} \\
&= t^2 \frac{m''(0)}{2m(0)} = \frac{t^2}{2}.
\end{aligned}$$

This proves the theorem for the case when $\mu = 0$ and $\sigma^2 = 1$.

For the general case, let $X_i^* = (X_i - \mu)/\sigma$. Then $X_i = \sigma X_i^* + \mu$ and

$$S_n = \sum_{i=1}^{n} X_i = \sum_{i=1}^{n} (\sigma X_i^* + \mu) = \sigma S_n^* + n\mu,$$

where $S_n^* = X_1^* + \cdots + X_n^*$. This gives

$$\frac{S_n/n - \mu}{\sigma/\sqrt{n}} = \frac{S_n^*}{\sqrt{n}}.$$

Observe that S_n^* is the sum of i.i.d. random variables with mean 0 and variance 1. Thus, the mgf of $S_n^*/\sqrt{n}$ converges to the standard normal mgf, and the result is shown. □

There are proofs for the CLT that do not use mgfs, but they are beyond the scope of this text. There are other CLT proofs that use mgfs that rely more on Taylor series results.

10.7 SUMMARY

This chapter introduces the important limit theorems of probability and two inequalities for bounding general probabilities. For i.i.d. random variables $X_1, X_2, \ldots$, with finite mean μ, the law of large numbers says that the sequence of averages $S_n/n = (X_1 + \cdots + X_n)/n$ converges to μ, as n tends to infinity. There are two versions of the limit result, the weak law and the strong law, based on different ways in which a sequence of random variables can converge to a limit. The proof of the WLLN uses Markov's and Chebyshev's inequalities, which bound probabilities of the form $P(X \geq x)$. The SLLN asserts that S_n/n converges to μ "almost surely," or with probability 1.

The method of moments is introduced to show how estimation of parameters in distributions may be done. The use of Monte Carlo simulation to approximate integrals and sums is shown. Both approaches are applications of the SLLN.

The CLT is discussed and proved in this chapter. The theorem says that for large n, the standardized averages $(S_n/n - \mu)/(\sigma/\sqrt{n})$ have an approximate standard normal distribution. The result is remarkable in that it applies to all population distributions with finite mean and variance.

The last section of this chapter uses moment-generating functions to prove the CLT.

For the results given next, let $X_1, X_2, \ldots$ be an i.i.d. sequence of random variables with finite mean μ and variance σ^2. Let $S_n = X_1 + \cdots + X_n$.

- **Weak law of large numbers:** For all $\epsilon > 0$,

$$P(|S_n/n - \mu| < \epsilon) \to 1, \text{ as } n \to \infty.$$

- **Markov's inequality:** For nonnegative random variable X with finite expectation, $P(X \geq \epsilon) \leq E[X]/\epsilon$.
- **Chebyshev's inequality:** If the mean μ and variance σ^2 of X are finite, then $P(|X - \mu| \geq \epsilon) \leq \sigma^2/\epsilon^2$.
- **Strong law of large numbers:** $P(\lim_{n\to\infty} S_n/n = \mu) = 1$.
- **Method of moments:** A technique for estimating the unknown parameters of a probability distribution. For sufficiently many k, equate sample moments with theoretical moments $E[X^k]$ and solve for unknown parameters.
- **Sample moments:** Given i.i.d. random variables $X_1, \ldots, X_n$, the kth sample moment is $(1/n)\sum_{i=1}^{n} X_i^k$.
- **Monte Carlo integration:** An integral such as $I = \int_0^1 g(x)\,dx$ can be written as the expectation of a random variable $E[g(X)]$, where $X \sim \text{Unif}(0, 1)$. If $X_1, X_2, \ldots$ are i.i.d. random variables uniformly distributed on $(0, 1)$, then

$$\frac{g(X_1) + \cdots + g(X_n)}{n} \approx E[g(X)] = I.$$

- **Central limit theorem:** For all t,

$$P\left(\frac{S_n/n - \mu}{\sigma/\sqrt{n}} \le t\right) \to P(Z \le t), \text{ as } n \to \infty,$$

where $Z \sim \text{Norm}(0, 1)$.

- **Continuity theorem:** Let $X_1, X_2, \ldots$ be a sequence of random variables. If $m_{X_n}(t) \to m_X(t)$, as $n \to \infty$, where $m_X(t)$ is the mgf of X, then $P(X_n \le x) \to P(X \le x)$, as $n \to \infty$. This result for moment-generating functions is used to prove the CLT.

EXERCISES

Law of Large Numbers

10.1 Describe in your own words the law of large numbers.

10.2 Your roommate missed probability class again. Explain to him/her the difference between the weak and strong laws of large numbers.

10.3 Let S be the sum of 100 fair dice rolls. Use (i) Markov's inequality and (ii) Chebyshev's inequality to bound $P(S \ge 380)$.

10.4 Find the best value of c so that $P(X \ge 5) \le c$ using Markov's and Chebyshev's inequalities, filling in the subsequent table. Compare with the exact probabilities.

Distribution	Markov	Chebyshev	Exact probability
Pois(2)			
Exp(1/2)			
Norm(2, 4)			
Geom(1/2)			

10.5 Let X be a positive random variable with $\mu = 50$ and $\sigma^2 = 25$.

(a) What can you say about $P(X \ge 60)$ using Markov's inequality?

(b) What can you say about $P(X \ge 60)$ using Chebyshev's inequality?

10.6 Prove Markov's inequality for the discrete case.

10.7 Let X be a positive random variable. Show that for all c,

$$P(\log X \ge c) \le \mu e^{-c}.$$

10.8 The expected sum of two fair dice is 7; the variance is 35/6. Let X be the sum after rolling n pairs of dice. Use Chebyshev's inequality to find z such that

$$P(|X - 7n| < z) \geq 0.95.$$

In 10,000 rolls of two dice there is at least a 95% chance that the sum will be between what two numbers?

Applications: Method of Moments and Monte Carlo Integration

10.9 Following are data from an i.i.d. sample taken from a Geometric distribution with unknown parameter p.

$$1 \quad 3 \quad 2 \quad 3 \quad 1 \quad 2 \quad 2 \quad 1 \quad 4 \quad 2.$$

Find the method of moments estimate for p. Explain the method of moments logic as part of your solution.

10.10 Following are data from an i.i.d. sample taken from a Poisson distribution with unknown parameter λ.

$$2 \quad 3 \quad 0 \quad 7 \quad 2 \quad 2 \quad 3 \quad 5 \quad 2 \quad 2 \quad 2 \quad 0.$$

Find the method of moments estimate for λ.

10.11 Let $X_1, \ldots, X_{25}$ be an i.i.d. sample from a binomial distribution with parameters n and p. Suppose n and p are unknown. Write down the method of moments equations that would need to be solved to estimate n and p.

10.12 Let $X_1, \ldots, X_n$ be an i.i.d. sample from a normal distribution with mean μ and variance σ^2. Find the general method of moments estimators for μ and σ^2.

10.13 Describe how to use Monte Carlo techniques to approximate the following integrals and sums. State clearly what the distribution and necessary function $g(X)$ are.

(a)

$$I = \int_0^1 \sin(x)e^{-x^2}\, dx.$$

(b)

$$I = \int_0^\infty \sin(x)e^{-x^2}\, dx.$$

(c)

$$I = \int_{-\infty}^\infty \log(x^2)e^{-x^2}\, dx.$$

(d)

$$S = \sum_{k=1}^{\infty} \frac{\sin k}{2^k}.$$

(e)

$$S = \sum_{k=0}^{\infty} \frac{\cos \cos k}{k!}.$$

10.14 Make up your own "hard" integral to solve using Monte Carlo approximation. Do the same with a "hard" sum.

Central Limit Theorem

10.15 The local farm packs its tomatoes in crates. Individual tomatoes have mean weight of 10 ounces and standard deviation 3 ounces. Find the probability that a crate of 50 tomatoes weighs between 480 and 510 ounces.

10.16 The waiting time on the cashier's line at the school cafeteria is exponentially distributed with mean 2 minutes. Use the central limit theorem to find the approximate probability that the average waiting time is more than 2.5 minutes for a group of 20 people. Use **R** and compare with the exact probability.

10.17 Recall the game of roulette and the casino's fortunes when a player places a "red" bet (see Example 4.29). For one $1 red bet, let G be the casino's gain. Then $P(G = 1) = 20/38$ and $P(G = -1) = 18/38$. Suppose in 1 month, one million red bets are placed. Let T be the casino's total gain. Find $P(50{,}000 < T < 55{,}000)$.

10.18 A baseball player has a batting average of 0.328. Let X be the number of hits the player gets during 20 times at bat. Use the central limit theorem to find the approximate probability $P(X \leq k)$ for $k = 1, 3, 6$. Compare with the exact probability for each k.

10.19 Let $X_1, \ldots, X_{30}$ be i.i.d. random variables with density

$$f(x) = 3x^2, \text{ if } 0 < x < 1.$$

Use the central limit theorem to approximate

$$P(22 < X_1 + \cdots + X_{30} < 23).$$

10.20 Let $X_1, \ldots, X_n$ be an i.i.d. sample from a population with unknown mean μ and standard deviation σ. We take the sample mean $\overline{X} = (X_1 + \cdots + X_n)/n$ as an estimate for μ.

(a) According to Chebyshev's inequality, how large should the sample size n be so that with probability 0.99 the error $|\overline{X} - \mu|$ is less than two standard deviations?

(b) According to the central limit theorem, how large should n be so that with probability 0.99 the error $|\overline{X} - \mu|$ is less than two standard deviations?

10.21 Let $X_1, \ldots, X_{10}$ be independent Poisson random variables with $\lambda = 1$. Consider $P(\sum_{i=1}^{10} X_i \geq 14)$.

(a) What does Markov's inequality say about this probability?

(b) What does Chebyshev's inequality say?

(c) What does the central limit theorem say?

(d) What does the central limit theorem say with the continuity correction?

(e) Find the exact probability.

10.22 Consider a random walk as described in Example 10.15. After one million steps, find the probability that the walk is within 500 steps of the origin.

10.23 Let $X \sim \text{Gamma}(a, \lambda)$, where a is a large integer. Without doing any calculations, explain why $X \approx \text{Norm}(a/\lambda, a/\lambda^2)$.

10.24 Show that

$$\lim_{n\to\infty} \int_0^n \frac{e^{-x}x^{n-1}}{(n-1)!}\, dx = \frac{1}{2}.$$

Hint: Consider an independent sum of n Exponential(1) random variables and apply the central limit theorem.

10.25 A random variable Y is said to have a *lognormal distribution* if $\log Y$ has a normal distribution. Equivalently, we can write $Y = e^X$, where X has a normal distribution.

(a) If $X_1, X_2, \ldots$ is an independent sequence of uniform $(0, 1)$ variables, show that the product $Y = \prod_{i=1}^n X_i$ has an approximate lognormal distribution. Show that the mean and variance of $\log Y$ are, respectively, $-n$ and n.

(b) If $Y = e^X$, with $X \sim \text{Norm}(\mu, \sigma^2)$, it can be shown that

$$E[Y] = e^{\mu+\sigma^2/2} \quad \text{and} \quad V[Y] = (e^{\sigma^2-1})e^{2\mu+\sigma^2}.$$

Let $X_1, \ldots, X_{100}$ be an independent sequence of uniform (0, 1) variables. Estimate

$$P\left(3^{-100} \leq \prod_{i=1}^{100} X_i \leq 2^{-100}\right).$$

(c) Verify the aforementioned results with a simulation experiment in **R**.

10.26 Consider a *biased* random walk that starts at the origin and that is twice as likely to move to the right as it is to move to the left. After how many steps will the probability be greater than 99% that the walk is to the right of the origin?

10.27 Let X be a random variable, not necessarily positive.

(a) Using Markov's inequality, show that for $x > 0$ and $t > 0$,

$$P(X \geq x) \leq \frac{E[e^{tX}]}{e^{tx}} = e^{-tx}m(t), \tag{10.4}$$

assuming that $E[e^{tX}]$ exists, where m is the mgf of X.

(b) For the case when X has a standard normal distribution, give the upper bound in Equation 10.4. Note that the bound holds for all $t > 0$.

(c) Find the value of t that minimizes your upper bound. If $Z \sim \text{Norm}(0, 1)$, show that for $z > 0$,

$$P(Z \geq z) \leq e^{-z^2/2}. \tag{10.5}$$

The upper bounds in Equations 10.4 and 10.5 are called *Chernoff bounds*.

10.28 Use the mgfs to show that the binomial distribution converges to the Poisson distribution. The convergence is taken so that $pn \to \lambda > 0$. (Hint: Write the p in the binomial distribution as λ/n.)

Simulation and R

10.29 Code similar to the following was used to generate the graphs in Figures 10.3 and 10.4. Modify the code to illustrate the SLLN for an i.i.d. sequence with the following distributions: (i) Pois($\lambda = 5$), (ii) Norm($-4, 4$), (iii) Exp($\lambda = 0.01$). Be sure your solution is reproducible.

```
par(mfrow = c(2, 2))
n <- 1000
p <- 1/2
for (i in 1:4){
  seq <- rbinom(n, 1, p)  # Distribution
  avgs <- cumsum(seq)/(1:n)
```

```
plot(avgs, type = "l", xlab = "n",
ylab = "Average")
abline(h = p)}
```

10.30 See the code in Example 10.15 for generating a simple random walk. We aim to simulate a biased random walk where the probability of moving left and right is p and $1 - p$, respectively. Graph your function obtaining pictures like Figure 10.8. What do you notice about the behavior of the random walk when $p = 0.60, 0.55, 0.505$?

10.31 (i) Write a function to simulate a random walk in the plane that moves up, down, left, and right with equal probability. Use your function to estimate the average distance from the origin after $n = 1000$ steps. (ii) Modify your function to simulate a three-dimensional random walk that moves in one of six directions with equal probability. Estimate the average distance from the origin after $n = 1000$ steps.

Chapter Review

Chapter review exercises are available through the text website. The URL is `www.wiley.com/go/wagaman/probability2e`.

11

BEYOND RANDOM WALKS AND MARKOV CHAINS

Not what we have but what we enjoy, constitutes our abundance.

—Epicurus

Learning Outcomes

1. Define key terms: random walk, transition matrix, stationary distribution.
2. Solve problems such as finding a stationary distribution.
3. Explain what MCMC is and describe the Metropolis–Hastings and Gibbs sampler algorithms.
4. (C) Explore simple MCMC applications such as the decoding example.

In this chapter, we explore additional topics on random walks and Markov chains, leading up to a discussion of Markov chain Monte Carlo methods.

11.1 RANDOM WALKS ON GRAPHS

A *graph* consists of a set of vertices and a set of edges. The graph in Figure 11.1 has four vertices and four edges. We say that two vertices are *neighbors* if there is an edge joining them. Thus, vertex c has two neighbors b and d. And vertex a has only one neighbor b.

Imagine the vertices as lily-pads on a pond, with a frog sitting on one lily-pad. At each discrete unit of time, the frog hops to a neighboring lily-pad with probability proportional to the number of its neighbors.

Probability: With Applications and R, Second Edition. Amy S. Wagaman and Robert P. Dobrow.

Companion Website: www.wiley.com/go/wagaman/probability2e

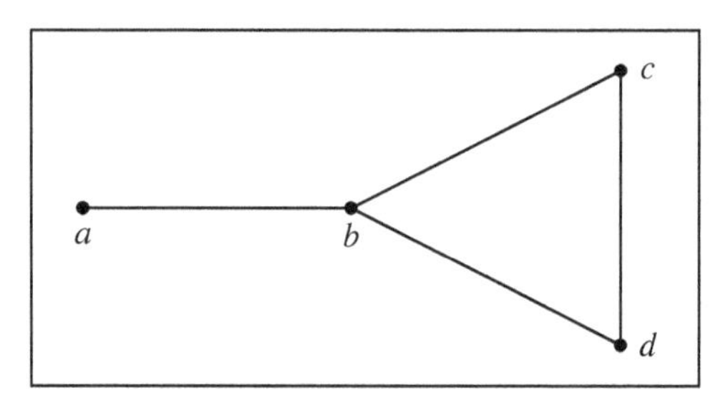

FIGURE 11.1: Graph on four vertices.

For instance, if the frog is on lily-pad c, it jumps to b or d with probability 1/2 each. If the frog is on a, it always jumps to b. And if the frog is on b it jumps to a, c, or d, with probability 1/3 each.

As the frog hops from lily-pad to lily-pad, the frog's successive locations form a random walk on the graph. For instance, if the frog starts at a, the random walk sequence might look like $a, b, a, b, c, b, d, c, d, b, a, b, \ldots$

Define a sequence of random variables $X_0, X_1, X_2, \ldots$, taking values in the vertex set of the graph. Let X_0 be the frog's initial lily-pad (e.g., vertex). For each $n = 1, 2, \ldots$, let X_n be the frog's position after n hops. The sequence $X_0, X_1, X_2, \ldots$ forms a *random walk on the graph*.

(Note that we have quietly extended our definition of random variable to allow the variables to take values in the vertex set, not just real numbers. Thus if the frog's random walk starts off as $a, b, a, b, c, \ldots$, then $X_0 = a$, $X_1 = b$, $X_2 = a$, etc.)

For vertices i and j, write $i \sim j$ if i and j are neighbors, that is, if there is an edge joining them. The *degree* of vertex i, written $\deg(i)$, is the number of neighbors of i. In the graph in Figure 11.1 (hereafter called the *frog graph*), $\deg(a) = 1$, $\deg(b) = 3$, and $\deg(c) = \deg(d) = 2$.

A *simple random walk* moves from a vertex to its neighbor with probability proportional to the degree of the vertex. That is, if the frog is on vertex i, then the probability that it hops to vertex j is $1/\deg(i)$, if $i \sim j$, and 0, otherwise. This is true at any time n. That is, for vertices i and j,

$$P(X_{n+1} = j | X_n = i) = \begin{cases} 1/\deg(i), & \text{if } i \sim j \\ 0, & \text{otherwise,} \end{cases} \tag{11.1}$$

for $n = 0, 1, 2 \ldots$ This *transition probability* describes the basic mechanism of the random walk process. The transition probability does not depend on n and thus for all n,

$$P(X_{n+1} = j | X_n = i) = P(X_1 = j | X_0 = i).$$

The transition probabilities can be represented as a matrix T whose ijth entry is the probability of moving from vertex i to vertex j in one step. This matrix is called the *transition matrix* of the random walk. Write T_{ij} for the ijth entry of T. For a random walk on a graph with k vertices labeled $\{1, \ldots, k\}$, T will be a $k \times k$ matrix with

$$T_{ij} = P(X_1 = j | X_0 = i), \text{ for } i, j = 1, \ldots, k.$$

For the frog's random walk, the transition matrix is

$$T = \begin{matrix} & a & b & c & d \\ a & 0 & 1 & 0 & 0 \\ b & 1/3 & 0 & 1/3 & 1/3 \\ c & 0 & 1/2 & 0 & 1/2 \\ d & 0 & 1/2 & 1/2 & 0 \end{matrix}.$$

Observe that each row of a transition matrix sums to 1. If the random walk is at vertex i, then row i gives the probability distribution for the next step of the walk. That is, row i is the conditional distribution of the walk's next position given that it is at vertex i. For each i, the sum of the ith row is

$$\sum_{j=1}^{k} T_{ij} = \sum_{j=1}^{k} P(X_1 = j | X_0 = i)$$

$$= \sum_{j=1}^{k} \frac{P(X_0 = i, X_1 = j)}{P(X_0 = i)} = \frac{P(X_0 = i)}{P(X_0 = i)} = 1.$$

Example 11.1 The cycle and complete graphs. Examples of the *cycle graph* and *complete graph* are shown in Figure 11.2. A simple random walk on the cycle graph moves left or right with probability 1/2. Each vertex has degree two. The transition matrix is described using "clock arithmetic":

$$T_{ij} = \begin{cases} 1/2, & \text{if } j = i \pm 1 \\ 0, & \text{otherwise.} \end{cases}$$

In the complete graph, every pair of vertices is joined by an edge. The complete graph on k vertices has $\binom{k}{2}$ edges. Each vertex has degree $k - 1$. The entries of the transition matrix are

$$T_{ij} = \begin{cases} 1/(k-1), & \text{if } i \neq j \\ 0, & \text{if } i = j. \end{cases}$$

■

Example 11.2 Card shuffling by random transpositions. Random walks are used to model methods of shuffling cards. Consider the following shuffle. Given a deck of cards, pick two distinct positions in the deck at random. Switch the two cards at those positions. This is known as the "random transpositions" shuffle. Successive random transpositions form a random walk on a graph whose vertices are the set

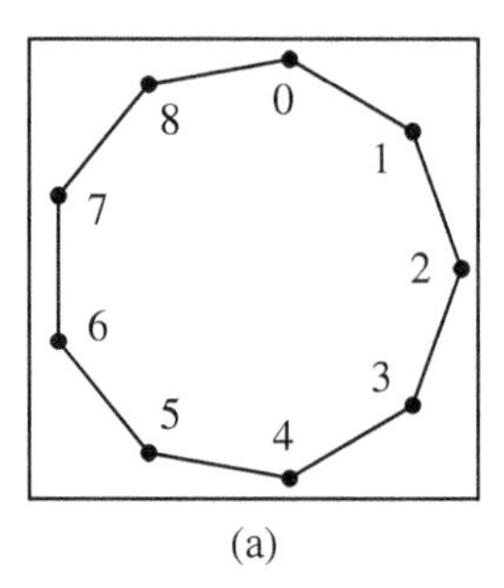

(a)

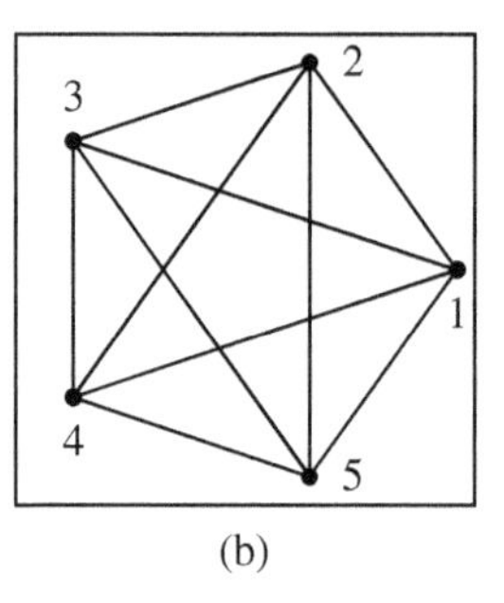

(b)

FIGURE 11.2: (a) Cycle graph on $k = 9$ vertices. (b) Complete graph on $k = 5$ vertices.

of orderings (permutations) of the deck of cards. For a deck of k cards, this gives a graph with $k!$ vertices. Because there are $\binom{k}{2}$ ways to select two positions in the deck, the degree of each vertex is $\binom{k}{2}$. Letting i and j denote orderings of the deck of cards, the transition matrix is defined by

$$T_{ij} = \begin{cases} 1/\binom{k}{2}, & \text{if } i \text{ differs from } j \text{ in exactly two locations} \\ 0, & \text{otherwise.} \end{cases}$$

Observe that the matrix T is symmetric because if the walk can move from i to j by switching two cards, it can move from j to i by switching those same two cards. Here is the transition matrix for random transpositions on a three-card deck with cards labeled 1, 2, and 3.

$$T = \begin{matrix} & \begin{matrix} 123 & 132 & 213 & 231 & 312 & 321 \end{matrix} \\ \begin{matrix} 123 \\ 132 \\ 213 \\ 231 \\ 312 \\ 321 \end{matrix} & \begin{pmatrix} 0 & 1/3 & 1/3 & 0 & 0 & 1/3 \\ 1/3 & 0 & 0 & 1/3 & 1/3 & 0 \\ 1/3 & 0 & 0 & 1/3 & 1/3 & 0 \\ 0 & 1/3 & 1/3 & 0 & 0 & 1/3 \\ 0 & 1/3 & 1/3 & 0 & 0 & 1/3 \\ 1/3 & 0 & 0 & 1/3 & 1/3 & 0 \end{pmatrix} \end{matrix}.$$ ■

Example 11.3 Hypercube. The *k-hypercube graph* has vertex set consisting of all k-element sequences of zeros and ones. Two vertices (sequences) are connected by an edge if they differ in exactly one coordinate. The graph has 2^k vertices and $\binom{k}{2}$ edges. Each vertex has degree k (see Fig. 11.3).

A random walk on the hypercube can be described as follows. Given a k-element 0–1 sequence, pick one of the k coordinates uniformly at random. Then "flip the bit"

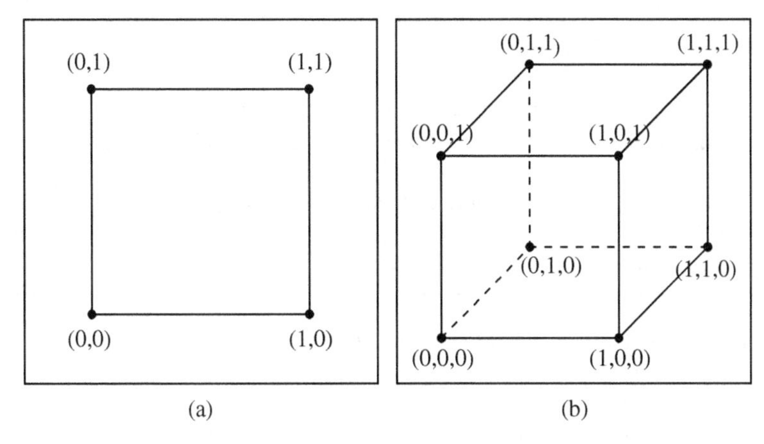

FIGURE 11.3: The k-hypercube graph for $k = 2$ (a) and $k = 3$ (b).

at that coordinate. That is, switch the number from 0 to 1, or from 1 to 0. Here is the transition matrix for the 3-hypercube.

$$T = \begin{array}{c} \\ 000 \\ 100 \\ 010 \\ 110 \\ 001 \\ 101 \\ 110 \\ 111 \end{array} \begin{array}{c} \begin{array}{cccccccc} 000 & 100 & 010 & 110 & 001 & 101 & 011 & 111 \end{array} \\ \begin{pmatrix} 0 & 1/3 & 1/3 & 0 & 1/3 & 0 & 0 & 0 \\ 1/3 & 0 & 0 & 1/3 & 0 & 1/3 & 0 & 0 \\ 1/3 & 0 & 0 & 1/3 & 0 & 0 & 1/3 & 0 \\ 0 & 1/3 & 1/3 & 0 & 0 & 0 & 0 & 1/3 \\ 1/3 & 0 & 0 & 0 & 0 & 1/3 & 1/3 & 0 \\ 0 & 1/3 & 0 & 0 & 1/3 & 0 & 0 & 1/3 \\ 0 & 0 & 1/3 & 0 & 1/3 & 0 & 0 & 1/3 \\ 0 & 0 & 0 & 1/3 & 0 & 1/3 & 1/3 & 0 \end{pmatrix} \end{array}.$$

■

11.1.1 Long-Term Behavior

Of particular interest in the study of random walk is the long-term behavior of the process. The *limiting distribution* describes this long-term behavior.

For notation in this chapter, we will write finite probability distributions as vectors. Thus if $\mu = (\mu_1, \ldots, \mu_k)$ is the distribution of a random variable X taking values $\{1, \ldots, k\}$, then $P(X = i) = \mu_i$, for $i = 1, \ldots, k$.

The *initial distribution* of a random walk is the distribution of the starting vertex X_0. A random walk is described by its initial distribution and transition matrix.

If a random walk has a limiting distribution, then as the walk evolves over time the distribution of the walk's position reaches an "equilibrium," which is independent of where the walk began. That is, the effect of the initial distribution "wears off" as the random walk evolves.

LIMITING DISTRIBUTION

Let $X_0, X_1, X_2, \ldots$ be a random walk on a graph. A probability distribution $\pi = (\pi_1, \ldots, \pi_k)$ is the *limiting distribution* of the random walk if for all vertices j,

$$P(X_n = j|X_0 = i) \rightarrow \pi_j, \text{ as } n \rightarrow \infty,$$

for all initial vertices i.

A consequence of this definition is that $P(X_n = j) \rightarrow \pi_j$, as $n \rightarrow \infty$, as

$$\begin{aligned}\lim_{n\to\infty} P(X_n = j) &= \lim_{n\to\infty} \sum_{i=1}^{k} P(X_n = j|X_0 = i)P(X_0 = i)\\ &= \sum_{i=1}^{k} \lim_{n\to\infty} P(X_n = j|X_0 = i)P(X_0 = i)\\ &= \sum_{i=1}^{k} \pi_j P(X_0 = i) = \pi_j.\end{aligned}$$

To illustrate the limiting distribution, imagine a frog hopping on the cycle graph on nine vertices as in Figure 11.2. Suppose the frog starts at vertex 0. After just a few hops, the frog will tend to be relatively close to 0. However, after the frog has been hopping around for a long time, intuitively the frog's position will tend to be uniformly distributed around the cycle, independent of the frog's starting vertex.

Observe the behavior of random walk on the cycle graph on $k = 9$ vertices in Table 11.1. We simulated the random walk for a fixed number of steps n, and then repeated 100,000 times to simulate the distribution of X_n. The walk starts at vertex 0. After five steps, the walk is close to its starting position with high probability. The probability is almost 60% that $X_5 \in \{8, 0, 1\}$. The distribution of X_5 is strongly dependent on the initial vertex. After 10 steps, there is still bias toward vertex 0 but

TABLE 11.1. Simple random walk for the cycle graph on nine vertices after n steps. Simulation of X_n for $n = 5, 10, 50, 100$.

Steps versus	Vertices								
vertex	5	6	7	8	0	1	2	3	4
5	0.025	0.06	0.125	0.185	0.21	0.185	0.125	0.06	0.025
10	0.075	0.09	0.12	0.14	0.15	0.14	0.12	0.09	0.075
50	0.11	0.11	0.12	0.10	0.12	0.10	0.12	0.10	0.11
100	0.11	0.11	0.11	0.11	0.11	0.11	0.11	0.11	0.11

not as much. Still the random walk is almost twice as likely to be close to vertex 0 than it is to be at the opposite side of the cycle near vertex 4 or 5. By 50 steps, however, the dependency on the initial vertex has almost worn off, and the distribution of X_{50} is close to uniform. By 100 steps, the distribution is uniform to two decimal places. The table suggests that the limiting distribution of the random walk is uniform on the set of vertices.

For the cycle graph, the limiting distribution is uniform. What about for the "frog graph" in Figure 11.1? If the frog has been hopping from lily-pad to lily-pad on this graph for a long time what is the probability that it is now on, say, vertex a?

Based on the structure of the graph, it is reasonable to think that after a long time, the frog is least likely to be at vertex a and most likely to be at vertex b as there are few edges into a and many edges into b.

The limiting distribution for random walk on this graph, and in general, has a nice intuitive description. The long-term probability that the walk is at a particular vertex is proportional to the degree of the vertex.

LIMITING DISTRIBUTION FOR A SIMPLE RANDOM WALK ON A GRAPH

Let π be the limiting distribution for a simple random walk on a graph. Then

$$\pi_j = \frac{\deg(j)}{\sum_{i=1}^{k} \deg(i)} = \frac{\deg(j)}{2e}, \quad \text{for all vertices } j, \tag{11.2}$$

where e is the number of edges in the graph.

For the frog-hopping random walk, the limiting distribution is

$$\pi = (\pi_a, \pi_b, \pi_c, \pi_d) = (1/8, 3/8, 2/8, 2/8) = (0.125, 0.375, 0.25, 0.25).$$

See the following simulation using Frog.R.

R: RANDOM WALK ON A GRAPH

The script file **Frog.R** simulates random walk on the frog graph for $n = 200$ steps starting from a uniformly random vertex. The 200-step walk is repeated 10,000 times, keeping track of the last position X_{200} each time.

```
> n <- 200
> trials <- 10000
> simlist <- numeric(trials)
> for (i in 1:trials) {
```

```
    k <- 0
    pos <- sample(1:4, 1) # initial position
    while (k < n) {
    k <- k+1
    if (pos==1) {pos <- 2; next}
    if (pos==2) {pos <- sample(c(1, 3, 4), 1); next}
    if (pos==3) {pos <- sample(c(2, 4), 1); next}
    if (pos==4) pos <- sample(c(2, 3), 1); }
    simlist[i] <- pos  }
> table(simlist)/trials
simlist
     1      2      3      4
0.1204 0.3805 0.2463 0.2528
```

Observe how closely the simulated distribution matches the exact limiting distribution $\pi = (1/8, 3/8, 1/4, 1/4)$.

Remarks:

1. For any graph, the sum of the vertex degrees is equal to twice the number of edges as every edge contributes two to the sum of the degrees, one for each endpoint.
2. Not every graph has a limiting distribution. Whether or not a random walk has a limiting distribution depends on the structure of the graph. There are two necessary conditions:
   (i) The graph must be *connected*. This means that for every pair of vertices there is a sequence of edges that forms a path connecting the two vertices.
   (ii) The graph must be *aperiodic*. To understand this condition, consider a random walk on the square with vertices labeled (clockwise) 0, 1, 2, and 3. If the walk starts at vertex 0, then after an even number of steps it will always be on either 0 or 2. And after an odd number of steps it will always be on 1 or 3. Thus, the position of the walk depends on the starting state and there is no limiting distribution. The square is an example of a *bipartite graph*. In such a graph, one can color the vertices of the graph with two colors black and white so that every edge has one endpoint that is black and the other endpoint that is white. Random walks on a bipartite graph give rise to periodic behavior and no limiting distribution exists.

   Every graph that is connected and nonbipartite has a unique limiting distribution. All the graphs that we discuss will meet this condition.
3. Observe from the limiting distribution formula in Equation 11.2 that if all the vertex degrees of a graph are equal, then the limiting distribution is uniform. A graph with all degrees the same is called *regular*. The complete graph,

hypercube graph, and random transpositions graph are all regular and thus the simple random walks on these graphs have uniform limiting distributions. The cycle graph is also regular but is bipartite if the number of vertices is even. The cycle graph has a uniform limiting distribution when the number of vertices is odd.

4. A classic problem in probability asks: How many shuffles does it take to mix up a deck of cards? The problem can be studied by means of random walks as in Example 11.2. If the uniform distribution is the limiting distribution of a card-shuffling random walk, then the problem is equivalent to asking: How many steps of a card-shuffling random walk are necessary to get close to the limiting distribution, when the deck of cards is "mixed up?" Such questions study the rate of convergence of random walk to the limiting distribution and form an active area of modern research in probability.

11.2 RANDOM WALKS ON WEIGHTED GRAPHS AND MARKOV CHAINS

> Let us finish the article and the whole book with a good example of dependent trials, which approximately can be regarded as a simple chain.
>
> —Alexei Andreyevich Markov (Masharin et al. [2004])

Random walks on graphs are a special case of Markov chains. A Markov chain is a sequence of random variables $X_0, X_1, X_2, \ldots$, with the property that for all n, the conditional distribution of X_{n+1} given the past history $X_0, \ldots, X_n$ is equal to the conditional distribution of X_{n+1} given X_n. This is sometimes stated as the distribution of the future given the past only depends on the present. The set of values of the Markov chain is called the *state space*.

A simple random walk on a graph is a Markov chain because the distribution of the walk's position at any fixed time only depends on the last vertex visited and not on the previous locations of the walk. The state space is the vertex set of the graph, leading to a discrete state space. Extensions exist to continuous state spaces.

As in the case of random walk on a graph, a Markov chain can be described by its transition matrix T and an initial distribution. The concept of a limiting distribution introduced in the last section extends naturally to general Markov chains.

Markov chains are remarkably useful models. They are used extensively in virtually every applied field to model random processes that exhibit some dependency structure between successive outcomes. A recent Google search of "Markov chain" returned over 29 million hits.

Example 11.4 Consider a Markov chain model for winter weather in the state of North Dakota. Suppose on any day the weather can be in one of three states: clear, rain, and snow. A meteorologist suggests the following transition matrix:

$$T = \begin{array}{c} \\ c \\ r \\ s \end{array} \begin{array}{c} \begin{array}{ccc} c & r & s \end{array} \\ \begin{pmatrix} 1/6 & 1/3 & 1/2 \\ 1/8 & 1/8 & 3/4 \\ 1/3 & 1/6 & 1/2 \end{pmatrix} \end{array}.$$

The script file **Markov.R** contains the function `markov(mat, start, n)` for simulating n steps of a Markov chain with a given transition matrix (mat) and starting state (start). The function takes the transition matrix as an input, so that must be entered into **R**. In the transition matrix entered below called weather, note that the states are coded as 1 = clear, 2 = rain, and 3 = snow.

What is the long-term behavior of the Markov chain? That is, what is the long-term weather according to this model? Suppose we start on a clear day. We ran the Markov chain for $n = 200$ steps (representing days) and repeated 10,000 times with the following results. After that, we also observed the chain's behavior after 300 steps (days).

R: WEATHER MARKOV CHAIN

```
> weather <- matrix(c(1/6, 1/3, 1/2, 1/8, 1/8, 3/4,
          1/3, 1/6, 1/2), nrow = 3, byrow = T)
> reps <- 10000
> simlist <- replicate(reps, markov(weather, 1, 200))
> table(simlist)/reps
simlist
     1      2      3
0.2533 0.1977 0.5490
> simlist <- replicate(reps, markov(weather, 1, 300))
> table(simlist)/reps
simlist
     1      2      3
0.2472 0.1967 0.5561
```

There is little difference in the distribution of X_{200} and X_{300}. The simulated limiting distribution suggests that the long-term weather forecast is approximately 25% chance of clear skies, 20% chance of rain, and 55% chance of snow. ■

General Markov chains can be studied in terms of random walks on graphs if we extend our notions of graph. A *weighted graph* is a graph with a positive number, or weight, assigned to each edge. A *directed graph* is a graph where edges have direction associated with them. Thus for every pair of vertices i and j, one can have

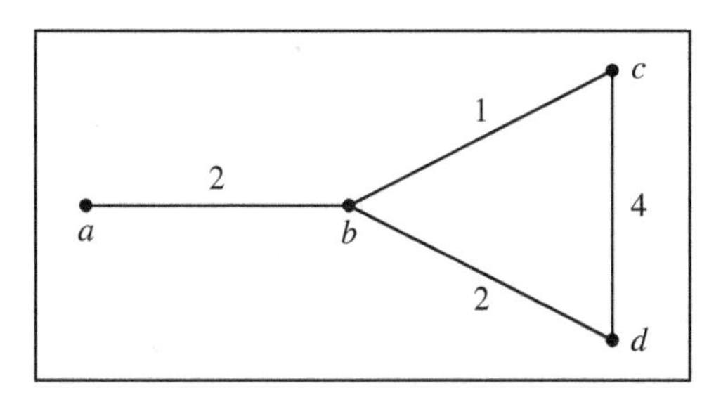

FIGURE 11.4: Weighted graph.

an edge from i to j and an edge from j to i. An example of a weighted, but undirected, graph is shown in Figure 11.4, referred to as the weighted frog graph.

Example 11.5 PageRank. The PageRank algorithm used by the search engine Google was presented in the book's introduction. When you make an inquiry using Google, it returns an ordered list of sites by assigning a rank to each page. The rank it assigns is essentially the limiting distribution of a Markov chain.

This Markov chain can be described as a random walk on the *web graph*. In the web graph, vertices represent web pages, edges represent hyperlinks. A directed edge joins page i to page j if page i links to page j. Imagine a random walker that visits web pages according to this model moving from a page with probability proportional to the number of "out-links" from that page. The long-term probability that the random walker is at page i is precisely the PageRank of page i.

Write $i \vec{\sim} j$ if there is a directed edge from i to j. Let link(i) be the number of directed edges from i. Transition probabilities are defined as follows:

$$T_{ij} = \begin{cases} P(X_1 = j | X_0 = i) = 1/\text{link}(i), & \text{if } i \vec{\sim} j \\ 0, & \text{otherwise.} \end{cases}$$

There are over one billion websites on the Internet (the exact number fluctuates). Computing the PageRank distribution amounts to finding the limiting distribution for a matrix that has hundreds of millions of rows and columns. Using the tools of probability and linear algebra, the computation is remarkably fast and efficient. For a detailed treatment, see *The $25,000,000,000 eigenvector* by Bryan and Leise [2006].

For a random walk on a weighted graph, transition probabilities are proportional to the sum of the weights. If vertices i and j are neighbors, let $w(i,j)$ denote the weight of the edge joining i and j. Let $w(i) = \sum_{i \sim j} w(i,j)$ be the sum of the weights on all the edges joining i to its neighbors. The transition probabilities, and the transition matrix, for random walk on a weighted graph are defined as

$$T_{ij} = P(X_1 = j | X_0 = i) = \begin{cases} w(i,j)/w(i), & \text{if } i \sim j \\ 0, & \text{otherwise.} \end{cases}$$

For the weighted frog graph, $w(a) = 2$, $w(b) = w(c) = 5$, and $w(d) = 6$. The transition matrix is

$$T = \begin{array}{c} \\ a \\ b \\ c \\ d \end{array} \begin{array}{c} \begin{array}{cccc} a & b & c & d \end{array} \\ \begin{pmatrix} 0 & 1 & 0 & 0 \\ 2/5 & 0 & 1/5 & 2/5 \\ 0 & 1/5 & 0 & 4/5 \\ 0 & 1/3 & 0 & 2/3 \end{pmatrix} \end{array}.$$

We will also allow a random walk to transition from a vertex back to itself in one step. This is done by adding an edge called a "loop" at a vertex. If a vertex has a loop, then it is a neighbor to itself. In the weighted frog graph, if we add a loop of weight one at every vertex, then $w(a) = 3$, $w(b) = w(c) = 6$, and $w(d) = 7$. The new transition matrix is

$$T = \begin{array}{c} \\ a \\ b \\ c \\ d \end{array} \begin{array}{c} \begin{array}{cccc} a & b & c & d \end{array} \\ \begin{pmatrix} 1/3 & 2/3 & 0 & 0 \\ 1/3 & 1/6 & 1/6 & 1/3 \\ 0 & 1/6 & 1/6 & 2/3 \\ 0 & 2/7 & 4/7 & 1/7 \end{pmatrix} \end{array}.$$

Every discrete Markov chain can be described as a random walk on a directed, weighted graph. Given a Markov chain with transition matrix T_{ij}, form a graph whose vertex set is the state space of the Markov chain. Then define weights $w(i,j) = T_{ij}$ for all i and j. This gives a directed, weighted graph. Similarly, given a directed, weighted graph with weight function $w(i,j)$ define transition probabilities by $T_{ij} = w(i,j)/w(i)$ for all i and j to obtain the transition matrix for a Markov chain. ■

11.2.1 Stationary Distribution

Stationary distributions play an important role in the study of Markov chains with intimate connections to limiting distributions.

STATIONARY DISTRIBUTION

Given a Markov chain with transition matrix T, a probability distribution $\mu = (\mu_1, \ldots, \mu_k)$ is a stationary distribution of the Markov chain if for all states j,

$$\mu_j = \sum_{i=1}^{k} \mu_i T_{ij}. \tag{11.3}$$

The reason for the name "stationary" is because a Markov chain that starts in its stationary distribution stays in that distribution. Suppose μ is both a stationary distribution and the initial distribution of a Markov chain. That is, $P(X_0 = j) = \mu_j$, for all j. Consider the distribution of X_1. By conditioning on X_0,

$$P(X_1 = j) = \sum_{i=1}^{k} P(X_1 = j|X_0 = i)P(X_0 = i) = \sum_{i=1}^{k} T_{ij}\mu_i = \mu_j,$$

for all j. Thus, the distribution of X_1 is also given by μ. And if $X_1 \sim \mu$, by the same argument $X_2 \sim \mu$, and so on. Hence, $X_n \sim \mu$ for all n. Thus, the sequence $X_0, X_1, X_2, \ldots$ is identically distributed with common distribution μ. We refer to the *stationary Markov chain* when the process is started in its stationary distribution.

Example 11.6 For the weather chain of Example 11.4, the distribution $\mu = (1/4, 1/5, 11/20)$ is a stationary distribution. We verify that it is so by showing that it satisfies the defining property. Checking for each $j = 1, 2, 3$ gives

$$\sum_{i=1}^{3} \mu_i T_{i1} = (1/4)(1/6) + (1/5)(1/8) + (11/20)(1/3) = 1/4 = \mu_1,$$

$$\sum_{i=1}^{3} \mu_i T_{i2} = (1/4)(1/3) + (1/5)(1/8) + (11/20)(1/6) = 1/5 = \mu_2,$$

$$\sum_{i=1}^{3} \mu_i T_{i3} = (1/4)(1/2) + (1/5)(3/4) + (11/20)(1/2) = 11/20 = \mu_3.$$

■

There is an intimate connection between the stationary distribution of a Markov chain and the limiting distribution. In fact, for a large class of Markov chains they are equal. We show that every limiting distribution is also a stationary distribution.

Suppose the limiting distribution of a Markov chain is π. Then for all states j,

$$\begin{aligned}\sum_{i=1}^{k} T_{ij}\pi_i &= \sum_{i=1}^{k} P(X_{n+1} = j|X_n = i) \left[\lim_{n\to\infty} P(X_n = i)\right]\\ &= \lim_{n\to\infty} \sum_{i=1}^{k} P(X_{n+1} = j|X_n = i)P(X_n = i)\\ &= \lim_{n\to\infty} \sum_{i=1}^{k} P(X_{n+1} = j, X_n = i)\\ &= \lim_{n\to\infty} P(X_{n+1} = j) = \pi_j.\end{aligned}$$

Thus, π is a stationary distribution.

A limiting distribution of a Markov chain is always a stationary distribution. The converse, however, is not necessarily true. Some Markov chains can have many stationary distributions. And some Markov chains with stationary distributions do not have limiting distributions. However, a random walk on a weighted graph that is connected and nonbipartite admits a unique positive stationary distribution, which is also the limiting distribution. This result is a fundamental theorem of Markov chains, which we restate without proof.

STATIONARY, LIMITING DISTRIBUTION FOR RANDOM WALK ON WEIGHTED GRAPHS

Suppose a weighted graph G is connected and nonbipartite. Then a random walk on the graph has a unique, positive stationary distribution π, which is also the limiting distribution.

The following result is extremely useful for finding stationary distributions.

DETAILED BALANCE CONDITION

Given a Markov chain with transition matrix T, a probability distribution $\mu = (\mu_1, \ldots, \mu_k)$ is said to satisfy the *detailed balance condition* if

$$\mu_i T_{ij} = \mu_j T_{ji}, \text{ for all } i \text{ and } j. \tag{11.4}$$

If a probability distribution satisfies the detailed balance condition, then it is a stationary distribution of the Markov chain.

Proof: Suppose μ satisfies the detailed balance condition. Then for all j,

$$\sum_{i=1}^{k} \mu_i T_{ij} = \sum_{i=1}^{k} \mu_j T_{ji} = \mu_j \sum_{i=1}^{k} T_{ji} = \mu_j.$$

Thus, μ is a stationary distribution. The first equality is because of the detailed balance condition. The last equality is because the jth row of the transition matrix sums to 1. □

We apply this result to find the stationary distribution for random walk on a weighted graph. We search for a candidate π that satisfies the detailed balance condition. Let T be the transition matrix. Then for all i and j, $\pi_i T_{ij} = \pi_j T_{ji}$ and thus

$$\pi_i \frac{w(i,j)}{w(i)} = \pi_j \frac{w(j,i)}{w(j)}.$$

As $w(i,j) = w(j,i)$ for all i and j, we have that $\pi_i/w(i) = \pi_j/w(j)$ for all vertices. Call this constant c. Hence $\pi_i = cw(i)$ for all i. Summing over i gives $1 = c\sum_{i=1}^{k} w(i)$. And thus $c = 1/\sum_{i=1}^{k} w(i)$. We find that

$$\pi_j = \frac{w(j)}{\sum_{i=1}^{k} w(i)}$$

is the stationary distribution. It is also the limiting distribution.

Observe that the simple random walk on a graph can be considered a random walk on a weighted graph where all the weights are equal to 1. In that case, $w(j) = \deg(j)$ and $\pi_j = \deg(j)/\sum_{i=1}^{k} \deg(i)$.

For the random walk on the graph in Figure 11.4, the stationary distribution is

$$\pi = (\pi_a, \pi_b, \pi_c, \pi_d) = (2/18, 5/18, 5/18, 6/18).$$

Example 11.7 Two-state Markov chain and Alexander Pushkin. The general two-state Markov chain on states a and b has transition matrix

$$T = \begin{array}{c} \\ a \\ b \end{array} \begin{array}{c} \begin{array}{cc} a & b \end{array} \\ \begin{pmatrix} 1-p & p \\ q & 1-q \end{pmatrix} \end{array},$$

where $0 \leq p, q, \leq 1$. If p and q are neither 0 nor 1, the Markov chain can be cast as a random walk on a weighted graph with loops as shown in Figure 11.5. This gives $w(a) = q$ and $w(b) = p$.

The stationary distribution probabilities are

$$\pi_a = \frac{w(a)}{w(a)+w(b)} = \frac{q}{p+q} \quad \text{and} \quad \pi_b = \frac{w(b)}{w(a)+w(b)} = \frac{p}{p+q}.$$

Andrei Andreyevich Markov introduced Markov chains 100 years ago. He first applied a two-state chain to analyze the successive vowels and consonants

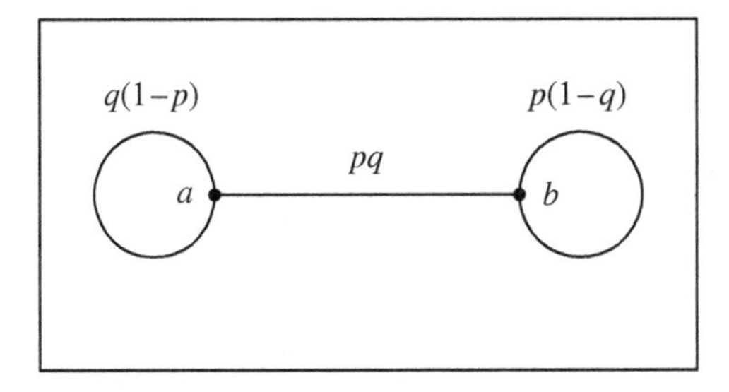

FIGURE 11.5: Weighted graph for the general two-state Markov chain.

in Alexander Puskin's poem *Eugéne Onégin*. In 20,000 letters of the poems, Markov found that there are 8638 vowels and 11,362 consonants with 1104 vowel–vowel pairs, 7534 vowel–consonant and consonant–vowel pairs, and 3828 consonant–consonant pairs. In modern notation, the transition matrix is

$$T = \begin{array}{c} \\ v \\ c \end{array}\begin{array}{c} \begin{array}{cc} \quad v \qquad\qquad & \qquad c \end{array} \\ \begin{pmatrix} 1104/8638 & 7534/8638 \\ 7534/11{,}362 & 3828/8638 \end{pmatrix} \end{array} = \begin{array}{c} \\ v \\ c \end{array}\begin{array}{c} \begin{array}{cc} v \qquad & \qquad c \end{array} \\ \begin{pmatrix} 0.128 & 0.872 \\ 0.663 & 0.337 \end{pmatrix} \end{array},$$

with stationary distribution

$$\begin{aligned} \pi = (\pi_v, \pi_c) &= \left(\frac{0.663}{0.872 + 0.663}, \frac{0.872}{0.872 + 0.663} \right) \\ &= (0.432, 0.568), \end{aligned}$$

which gives the proportion of vowels and consonants, respectively, in the poem. ■

A Markov chain whose stationary distribution π satisfies the detailed balance condition is said to be *time-reversible*. The descriptive language is used because a time-reversible stationary chain "looks the same" going forward in time as it does going backward.

Suppose a stationary Markov chain is time-reversible with stationary distribution π. For all states i and j,

$$\begin{aligned} P(X_0 = i, X_1 = j) &= P(X_1 = j | X_0 = i) P(X_0 = i) \\ &= T_{ij}\pi_i = \pi_j T_{ji} \\ &= P(X_1 = i, X_0 = j). \end{aligned}$$

More generally, it can be shown that for all n,

$$P(X_0 = i_0, X_1 = i_1, \ldots, X_n = i_n) = P(X_n = i_0, X_{n-1} = i_1, \ldots, X_0 = i_n),$$

for all $i_0, i_1, \ldots, i_n$.

11.3 FROM MARKOV CHAIN TO MARKOV CHAIN MONTE CARLO

> The impact of Gibbs sampling and MCMC was to change our entire method of thinking and attacking problems, representing a *paradigm shift*. Now the collection of real problems that we could solve grew almost without bound. Markov chain Monte Carlo changed our emphasis from "closed form" solutions to algorithms, expanded our impact to solving "real" applied problems and to improving numerical algorithms using statistical ideas, and led us into a world where "exact" now means "simulated." This has truly been a quantum leap in the evolution of the field of statistics, and the

evidence is that there are no signs of slowing down .... The size of the data sets, and of the models, for example, in genomics or climatology, is something that could not have been conceived 60 years ago, when Ulam and von Neumann invented the Monte Carlo method.

—Robert and Casella [2011], pg. 110, 111

We are given a large graph in which the number of vertices is unknown. However, the graph can be described *locally* in the sense that from any vertex i we can easily find the neighbors of i. Thus, we are able to run a simple random walk on the graph. Let V be the number of vertices in the graph. The goal is to sample from the uniform distribution on the vertex set. That is, from the distribution $P(i) = 1/V$, for all i. But, as we said, V is unknown.

At first sight, it seems incredible that such a problem can be solved. Yet we will show—using Markov chains—that it can.

An example of the type of graph we have in mind is the web graph introduced in the context of the PageRank algorithm in Example 11.5. The number of vertices in the graph, the total number of web pages on the Internet, is extremely hard to compute and, for all practical purposes, unknown. Yet from any vertex in the graph, from any web page, the number of neighbors is just the number of hyperlinks on that page. We would like to simulate an observation from the uniform distribution on the set of all web pages.

Why might such a simulation be useful? Suppose we would like to know the average number of hyperlinks on a web page on the Internet. Call this μ. One approach to finding μ is to look at every web page on the Internet, count the number of hyperlinks, and take the average. My computer science colleagues tell me that even with a lot of processors it would take between a week and several months to crawl the Internet and collect these data from *every* web page.

On the other hand, suppose we are able to simulate i.i.d. observations $X_1, \ldots, X_n$ from the uniform distribution on the web graph, where X_i represents a uniformly random web page (e.g., vertex). Given a web page i, let $h(i)$ be the number of hyperlinks on i. Then, $h(X_k)$ is the number of hyperlinks on a uniformly random web page. A Monte Carlo estimate of μ is

$$\mu \approx \frac{h(X_1) + \cdots + h(X_n)}{n}, \text{ for large } n.$$

But in order to make this estimate, we need i.i.d. observations from the uniform distribution.

Markov chain Monte Carlo is a powerful method for simulating observations from an unknown probability distribution. The method constructs a Markov chain whose limiting distribution is the unknown distribution of interest.

Let π be the "target" probability distribution of interest. The basic problem is to simulate from π, that is, to generate a random variable X whose distribution is π. Markov chain Monte Carlo constructs a Markov chain sequence $X_0, X_1, X_2, \ldots$

whose limiting distribution is π. Having constructed such a sequence, we can then take $X = X_n$, for n sufficiently large, as an approximate sample from π.

The MCMC algorithm for constructing a Markov chain with a given limiting distribution is called the Metropolis–Hastings algorithm. It is named after Nicholas Metropolis, a physicist, and W. Keith Hastings, a statistician. Metropolis co-authored a 1953 paper that first proposed the algorithm. Hastings later extended the work in 1970.

Metropolis–Hastings algorithm. For simplicity assume that π is finite, taking values $\{1, \ldots, k\}$. Let T be the transition matrix of *any* Markov chain on $\{1, \ldots, k\}$. We assume that we know how to sample from this chain. In particular, if the chain is at state i, we can simulate the next step of the chain according to the ith row of T. The T matrix is called the *proposal* matrix.

We construct a new Markov chain $X_0, X_1, X_2, \ldots$ by describing its transition mechanism. Suppose after m steps, $X_m = i$. The next state X_{m+1} is determined by a two-step procedure.

1. **Propose:** Choose a *proposal state* according to the ith row of T. That is, state j is chosen with probability $T_{ij} = P(X_{m+1} = j|X_m = i)$.
2. **Accept:** Decide whether or not to *accept* the proposal state. Suppose the proposal state is j. Compute the *acceptance function*

$$a(i,j) = \frac{\pi_j T_{ji}}{\pi_i T_{ij}}.$$

Let U be a random variable uniformly distributed on (0, 1). Then the next state of the chain is

$$X_{m+1} = \begin{cases} j, & \text{if } U \leq a(i,j) \\ i, & \text{if } U > a(i,j). \end{cases}$$

MCMC: METROPOLIS–HASTINGS ALGORITHM

Let X_0 be an arbitrary initial state. The sequence of random variables $X_0, X_1, X_2, \ldots$ constructed by the aforementioned algorithm is a time-reversible Markov chain whose limiting distribution is π.

Before proving this theorem consider the following toy example. Suppose $\pi = (0.1, 0.2, 0.3, 0.4)$ is the desired target distribution. Let T be the transition matrix for simple random walk on the square, labeled 1, 2, 3, 4. From any vertex, the random walk moves left or right with probability 1/2 each.

The algorithm proceeds as follows. Let X_0 be any vertex. Start a simple random walk on the square. If the current state is $X_m = i \in \{1, 2, 3, 4\}$, choose the proposal $j = i \pm 1$ (in clock arithmetic) with probability 1/2 each (e.g., if the walk is at state 1, then choose 4 or 2 with probability 1/2 each). Compute the acceptance function

$$a(i,j) = \frac{\pi_j T_{ji}}{\pi_i T_{ij}} = \frac{\pi_j}{\pi_i}, \quad \text{if } i \sim j,$$

because $T_{ij} = T_{ji} = 1/2$, for $i \sim j$. Let $U \sim \text{Unif}(0, 1)$. Then, the next step of the walk is set to be j, if $U \leq \pi_j/\pi_i$, and i, otherwise.

Briefly, consider how the acceptance function works in the context of the toy example. If $i = 1$, then for $j = 2$, the acceptance function is $a(1, 2) = 0.2/0.1 = 2$. Clearly, the value of U will always be less than 2, so $j = 2$ will always be accepted. With $i = 1$ and $j = 4$, a similar effect occurs. When might we not accept a new proposal? Consider $i = 2$ and $j = 1$. Then the acceptance function is $a(2, 1) = 0.1/0.2 = 0.5$. So, we would only accept $j = 1$ moving from $i = 2$ about 50% of the time. Note this will help us accumulate fewer 1's than 2's, as desired. We still need more 3's than 2's though, so note for $i = 2$ and $j = 3$, the acceptance function is $a(2, 3) = 0.3/0.2 = 1.5$, so we would always move from 2 to 3 when 3 is proposed.

The following is a simple implementation to demonstrate the algorithm (see also **MCMCtoy.R**).

R: MCMC—A TOY EXAMPLE

The target distribution is $\pi = (0.1, 0.2, 0.3, 0.4)$, with state space $\{1, 2, 3, 4\}$. The function `mcmc(n)` simulates X_n from a Markov chain constructed according to the Metropolis–Hastings algorithm. After 100 steps, X_{100} is output, and this process is repeated 10,000 times. The simulated distribution of X_{100} serves as an approximation of π.

```
> pi <- c(0.1, 0.2, 0.3, 0.4)
> mcmc <- function(n) {
   current <- 1
   for (i in 1:n) {
     proposal <- (current + sample(c(-1, 1), 1)) %% 4
     if(proposal == 0) proposal <- 4
     accept <- pi[proposal] / pi[current]
     if(runif(1) < accept) current <- proposal}
   current }
> replicate(20,mcmc(100))
 [1] 3 2 4 4 4 3 4 3 3 4 2 2 2 2 2 3 4 1 2 1
> trials <- 10000
> simlist <- replicate(trials,mcmc(100))
> table(simlist)/trials
simlist
     1      2      3      4
0.0977 0.2008 0.3089 0.3926
```

Proof of Metropolis–Hastings algorithm. Let $X_0, X_1, \ldots$ be the sequence constructed by the Metropolis–Hastings algorithm with transition matrix P. By construction, for each m, X_m only depends on the previous state X_{m-1} and not on the full past history of the sequence, and thus the sequence is in fact a Markov chain. Let P be the transition matrix. To prove that the limiting distribution is π, we show that the detailed balance condition is satisfied.

For $i \neq j$, consider $P(X_1 = j|X_0 = i)$, the ijth entry of P. If $X_0 = i$, in order to transition to j, (i) state j must be a proposal state and (ii) j must be accepted. The first event (i) occurs with probability T_{ij}. And (ii) occurs if $U \leq a(i,j)$, where $U \sim$ Unif(0, 1). Observe that

$$P(U \leq a(i,j)) = \begin{cases} a(i,j), & \text{if } \pi_j T_{ji} \leq \pi_i T_{ij} \\ 1, & \text{if } \pi_j T_{ji} > \pi_i T_{ij}. \end{cases}$$

Thus, the transition matrix P is given by

$$P_{ij} = P(X_1 = j|X_0 = i) = \begin{cases} T_{ij}a(i,j), & \text{if } \pi_j T_{ji} \leq \pi_i T_{ij} \\ T_{ij}, & \text{if } \pi_j T_{ji} > \pi_i T_{ij}, \end{cases}$$

for $i \neq j$. The diagonal entries P_{ii} are found so that the rows sum to 1. That is, $P_{ii} = 1 - \sum_{j \neq i} P_{ij}$.

To show that π satisfies the detailed balance condition that $\pi_i T_{ij} = \pi_j T_{ji}$, let $i \neq j$ and suppose $\pi_j T_{ji} / \pi_i T_{ij} \leq 1$. Then necessarily, the reciprocal $\pi_i T_{ij} / \pi_j T_{ji} > 1$ and

$$\pi_i P_{ij} = \pi_i T_{ij} a(i,j) = \pi_i T_{ij} \frac{\pi_j T_{ji}}{\pi_i T_{ij}} = \pi_j T_{ji} = \pi_j P_{ji}.$$

The result holds similarly for the case $\pi_j T_{ji} / \pi_i T_{ij} > 1$.

We have shown that $X_0, X_1, \ldots$ is a time-reversible Markov chain with limiting distribution π. □

Remarks:

1. The algorithm requires the computation of ratios of the form π_i / π_j. Often, the distribution π is specified up to a proportionality constant, which may be unknown. As only ratios are required in the algorithm, the proportionality constant is not needed.
2. In the original version of the algorithm, the matrix T was taken to be symmetric, with $T_{ij} = T_{ji}$. In that case, the acceptance function simplifies to $a(i,j) = \pi_j / \pi_i$.
3. There are several accounts of the rich history of Markov chain Monte Carlo. We suggest the papers by Richey [2010] and Robert and Casella [2011].

Example 11.8 We revisit the general problem introduced at the beginning of this section. Let G be a graph where the number of vertices V is both very large and unknown. However, given any vertex i, we are able to compute the degree of i and perform a simple random walk on the graph. Our goal is to sample from the uniform distribution on the set of vertices. To implement MCMC, let T be the transition matrix for simple random walk on the graph. Using T as the proposal matrix, the acceptance function is

$$a(i,j) = \frac{\pi_j T_{ji}}{\pi_i T_{ij}} = \frac{(1/V)(1/\deg(j))}{(1/V)(1/\deg(i))} = \frac{\deg(i)}{\deg(j)}, \quad \text{for } i \sim j.$$

Thus, an MCMC algorithm to simulate from the uniform distribution is based on modifying a simple random walk on G: From vertex i, propose vertex j according to such a random walk. Then compute $a(i,j) = \deg(i)/\deg(j)$. Let $U \sim \text{Unif}(0,1)$. If $U \leq a(i,j)$, then accept j as the next step in the chain. Otherwise, stay at i.

To implement the algorithm on the web graph in order to generate a uniformly random web page, the basic transition mechanism is as follows: from web page i choose a hyperlink uniformly at random to reach a proposal page j. The acceptance function is then

$$a(i,j) = \frac{\text{Number of hyperlinks on page } i}{\text{Number of hyperlinks on page } j}.$$

Let $U \sim \text{Unif}(0,1)$. The Markov chain moves to j if $U < a(i,j)$, or stays at i for the next step of the sequence. Run the chain a "long time" and output X_n as an approximate sample from the uniform distribution. ■

The reader will no doubt wonder after the last example: What is a "long time"? This is perhaps the biggest open question in the study of Markov chain Monte Carlo. In practice there are numerous empirical methods for estimating the number of steps required in order to implement MCMC effectively. There are theoretical results for Markov chains on graphs that exhibit a lot of symmetry, such as the cycle or complete graphs. But these types of highly structured chains are typically not encountered in real-life applications.

Example 11.9 Cryptography

> ahicainqcaqx ic zqcqwbl bwq zwqbj xjustlicz tlhamx ic jyq kbr ho jybj albxx ho jyicmqwx kyh ybgq tqqc qnuabjqn jh mchk chjyicz ho jyq jyqhwr ho dwhtbtilijiqx jybj jyqhwr jh kyiay jyq shxj zlhwihux htpqajx ho yusbc wqxqbway bwq icnqtjqn ohw jyq shxj zlhwihux ho illuxjwbjihcx
>
> —qnzbw bllqc dhq jyq suwnqwx ic jyq wuq shwzuq

This example is based on Diaconis [2008], who gives a compelling demonstration of the power and breadth of MCMC with an application to cryptography.

The coded message above was formed by a simple substitution cipher—each letter standing for one letter in the alphabet. The hidden message can be thought of as a *coding function* from the letters of the coded message to the regular alphabet. For example, given an encrypted message *xoaaoaaoggo*, the function that maps x to m, o to i, a to s, and g to p decodes the message to *mississippi*.

If one keeps tracks of all 26 letters and spaces, then there are $27! \approx 10^{28}$ possible coding functions. The goal is to find the one that decodes the message.

A colleague, Jack Goldfeather, assisted in downloading the complete works of Jane Austen (about four million characters) from Project Gutenberg [2013] and recorded the number of transitions of consecutive text symbols. For simplicity, he ignored case and only kept track of spaces and the letters a to z. The counts are kept in a 27×27 matrix M of transitions indexed by $(a, b, \ldots, z, \text{[space]})$. For example, there are 6669 places in Austen's work where b follows a and thus $M_{12} = 6669$.

The encoded message has 320 characters, denoted $(c_1, \ldots, c_{320})$. For each coding function g associate a *score function*

$$\text{score}(g) = \prod_{i=1}^{319} M_{g(c_1),g(c_{i+1})}.$$

The score is a product over all successive pairs of letters in the decrypted text $(g(c_i), g(c_{i+1}))$ of the number of occurrences of that pair in the reference Austen text. The score is higher when successive pair frequencies in the decrypted message match those of the reference text. Coding functions with high scores are good candidates for decryption. The goal is to find the coding function of maximum score.

A probability distribution proportional to the scores is obtained by letting

$$\pi_g = \frac{\text{score}(g)}{\sum_h \text{score}(h)}. \tag{11.5}$$

From a Monte Carlo perspective, we want to sample from π, with the idea that a sample is most likely to return a value of maximum probability. Of course, the denominator in Equation 11.5 is intractable, being the sum of 27! terms. But the beauty of Metropolis–Hastings is that the denominator is not needed as the algorithm relies on ratios of the form π_f / π_g.

The MCMC implementation runs a random walk on the set of coding functions. Given a coding function g, the transition to a proposal function g^* is made by picking two letters at random and switching the values that g assigns to these two symbols. This method of "random transpositions" gives a symmetric proposal matrix T, simplifying the acceptance function

$$a(g, g^*) = \frac{\pi_{g^*} T_{g^*,g}}{\pi_g T_{g,g^*}} = \frac{\pi_{g^*}}{\pi_g} = \frac{\text{score}(g^*)}{\text{score}(g)}.$$

The algorithm is as follows:

1. Start with any g. For convenience we use the identity function.
2. Pick two letters uniformly at random and switch the values that g assigns to these two symbols. Call this new proposal function g^*.
3. Compute the acceptance function $a(g, g^*) = \text{score}(g^*)/\text{score}(g)$.
4. Let $U \sim \text{Unif}(0, 1)$. If $U < a$, accept g^*. Otherwise, stay with g.

We ran this algorithm on the coded message. See the script file **Decode.R**. The results are shown next. Before iteration 2700, the message was decoded.

DECODING THE MESSAGE

[0] ahicainqcaqx ic zqcqwbl bwq zwqbj xjustlicz tlhamx ic jyq kbr ho jybj albxx ho jyicmqwx kyh ybgq tqqc qnuabjqn jh mchk chjyicz ho jyq jyqhwr ho dwhtbtilijiqx jybj jyqhwr jh kyiay jyq shxj zlhwihux htpqajx ho yusbc wqxqbway bwq icnqtjqn ohw jyq shxj zlhwihux ho illuxjwbjihcx qnzbw bllqc dhq jyq suwnqwx ic jyq wuq shwzuq

[100] goiegimsegsn ie asesrbl brs arsbh nhuktliea tlogdn ie hys pbc of hybh glbnn of hyiedsrn pyo ybvs tsse smugbhsm ho deop eohyiea of hys hysorc of wrotbtil-ihisn hybh hysorc ho pyigy hys konh alorioun otxsghn of yukbe rsnsbrgy brs iemsthsm for hys konh alorioun of illunhrbhioen smabr bllse wos hys kurmsrn ie hys rus koraus

[500] gosegsviegin se lieirap ari lriat ntucmpsel mpogyn se thi wad of that gpann of thseyirn who haxi miie ivugativ to yeow eothsel of thi thiord of bromamspstsin that thiord to whsgh thi cont lporsoun omkigtn of hucae riniargh ari sevimtiv for thi cont lporsoun of sppuntratsoen ivlar appie boi thi curvirn se thi rui corlui

[1000] goingidenges in meneral are mreat stucplinm plogys in the wak of that glass of thinyers who have peen edugated to ynow nothinm of the theork of bropapilities that theork to whigh the cost mlorious opzegts of hucan researgh are indepted for the cost mlorious of illustrations edmar allen boe the curders in the rue cormue

[1500] goingidenges in meneral are mreat stucplinm plogks in the way of that glass of thinkers who have peen edugated to know nothinm of the theory of bropapilities that theory to whigh the cost mlorious opxegts of hucan researgh are indepted for the cost mlorious of illustrations edmar allen boe the curders in the rue cormue

[2000] coincidences in beneral are breat stumplinb plocks in the way of that class of thinkers who have peen educated to know nothinb of the theory of gropa-pilities that theory to which the most blorious opjects of human research are indepted for the most blorious of illustrations edbar allen goe the murders in the rue morbue

[2500] coincidences in general are great stumpling plocks in the way of that class of thinkers who have peen educated to know nothing of the theory of bropapilities that theory to which the most glorious opjects of human research are indepted for the most glorious of illustrations edgar allen boe the murders in the rue morgue

[2700] coincidences in general are great stumbling blocks in the way of that class of thinkers who have been educated to know nothing of the theory of probabilities that theory to which the most glorious objects of human research are indebted for the most glorious of illustrations edgar allen poe the murders in the rue morgue

■

Gibbs sampler. The original MCMC algorithm was developed in 1953 motivated by problems in physics. In 1984, a landmark paper by Geman and Geman [1984] showed how the algorithm could be adapted for the high-dimensional problems that arise in Bayesian statistics. The name "Gibbs sampling" was coined in that paper because of connections with Gibbs fields and the work of physicist Josiah Gibbs.

In order to understand Gibbs sampling, we need to broaden our notion of a Markov chain and transition matrix to allow for infinite and continuous state spaces. In particular, the transition matrix T in the Metropolis–Hastings algorithm is replaced by a transition *function* $T(i, j)$ that, for fixed i, is a conditional density function, as opposed to a conditional pmf in the discrete case.

In the Gibbs sampler, the target distribution is a joint distribution

$$\pi(\boldsymbol{x}) = \pi(x_1, \ldots, x_k).$$

A Markov chain is constructed whose limiting distribution is π and that takes values in a k-dimensional space. The algorithm generates elements of the form

$$\left(X_1^{(0)}, \ldots, X_k^{(0)}\right), \left(X_1^{(1)}, \ldots, X_k^{(1)}\right), \left(X_1^{(2)}, \ldots, X_k^{(2)}\right), \ldots$$

eventually generating $(X_1^{(n)}, \ldots, X_k^{(n)})$ for large n as a sample from π.

Chain elements are simulated by sampling from conditional distributions. We illustrate with a simple example: simulating from a bivariate standard normal distribution with correlation ρ. Recall that if (X, Y) has a bivariate standard normal distribution, then the conditional distribution of X given $Y = y$ is normal with mean ρy and variance $1 - \rho^2$. Similarly, the conditional distribution of Y given $X = x$ is normal with mean ρx and variance $1 - \rho^2$.

At each step of the Gibbs sampler, each of the two coordinates of the joint distribution are updated by sampling from the conditional distribution of one given the other back and forth as described in the following implementation:

1. Initiate $(X_0, Y_0) = (0, 0)$.
2. At step m, having already simulated (X_{m-1}, Y_{m-1}),
   (a) take X_m as a sample from the conditional distribution of X_m given Y_{m-1}. That is, if $Y_{m-1} = y$, then simulate X_m from a normal distribution with mean ρy and variance $1 - \rho^2$, Then,
   (b) take Y_m as a sample from the conditional distribution of Y_m given X_m. That is, having just simulated $X_m = x$, now simulate Y_m from a normal distribution with mean ρx and variance $1 - \rho^2$.
3. For large n, output (X_n, Y_n) as a sample from the desired bivariate distribution.

R: SIMULATION OF BIVARIATE STANDARD NORMAL DISTRIBUTION

This code simulates 1000 observations of a bivariate standard normal distribution with $\rho = -0.5$ using the Gibbs sampler. We take $n = 100$ iterations and output (X_{100}, Y_{100}). See Figure 11.6 for a graph of example output, and the script **Gibbs.R** for code.

```
> gibbsnormal <- function(n, rho) {
  x <- 0
  y <- 0
sd <- sqrt(1-rho^2)
for (i in 1:n) {
    x <- rnorm(1, rho*y, sd)
    y <- rnorm(1, rho*x, sd) }
return(c(x, y))}
> simlist <- replicate(1000, gibbsnormal(100, -0.5)
> plot(t(simlist))
```

Why does this work? The Gibbs sampler is actually a special case of the Metropolis–Hastings algorithm, where transitions are based on conditional distributions. Consider one step of the Gibbs sampler. Suppose $\boldsymbol{i} = (x, y)$ is the current state and we are proposing an update for the first coordinate from x to x^*. Let $\boldsymbol{j} = (x^*, y)$.

The proposal x^* is obtained from the conditional distribution of X given $Y = y$. Let $f_{X|Y}(x|y)$ denote the conditional density of X given $Y = y$. Let $f_Y(y)$ denote the

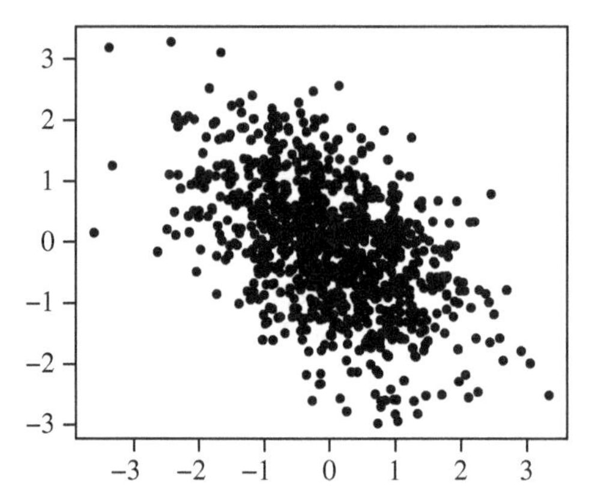

FIGURE 11.6: Gibbs sampler simulation of bivariate standard normal with $\rho = -0.5$.

marginal density of Y. The acceptance function is thus

$$a(\boldsymbol{i},\boldsymbol{j}) = \frac{\pi_j T_{ji}}{\pi_i T_{ij}} = \frac{\pi(x^*, y) f_{X|Y}(x|y)}{\pi(x, y) f_{X^*|Y}(x^*|y)} = \frac{\pi(x^*, y)\pi(x, y) f_Y(y)}{\pi(x, y)\pi(x^*, y) f_Y(y)} = 1.$$

Because the acceptance function is equal to one, we always accept a transition. Similarly, when the second coordinate is updated from (x^*, y). Thus, Gibbs sampling is a special case of the Metropolis–Hastings algorithm in which we always accept proposals.

More generally, suppose we want to sample from a multivariate probability distribution

$$\pi(\boldsymbol{x}) = \pi(x_1, \ldots, x_k) = P(X_1 = x_1, \ldots, X_k = x_k).$$

We assume that for each i, we can sample from the conditional distribution

$$P(X_i = x_i | X_j = x_j, j \neq i).$$

That is, we can simulate from the conditional distribution of each coordinate of $(X_1, \ldots, X_k)$ given the other $k - 1$ coordinates. As shown earlier, the acceptance function is always equal to 1, and the algorithm proceeds by iteratively choosing coordinates to update, simulating from the conditional distribution given the remaining coordinates.

Example 11.10 The following three-dimensional example is based on Casella and George [1992]. Consider the mixed joint density

$$\pi(x, p, n) \propto \binom{n}{x} p^x (1 - p)^{n-x} e^{-4} \frac{4^n}{n!}, \tag{11.6}$$

for $x = 0, 1, \ldots, n, 0 < p < 1, n = 0, 1, \ldots$. The p variable is continuous; variables x and n are discrete. The distribution arises from the following application. Conditional on $N = n$ and $P = p$, let X represent the number of successful hatchings from n insect eggs, where each egg has success probability p. Both N and P are random and vary across insects. We seek the expectation and standard deviation of X, the number of successful hatchings among all insects.

There is no simple closed form expression for the marginal distribution of X. Gibbs sampling is used to simulate from the joint density of (X, P, N) and then extract the marginal information.

The key to using the Gibbs sampler is our ability to sample easily from the conditional distributions of each coordinate variable given the other coordinate values. For fixed p and n in Equation 11.6, the conditional distribution of X given $P = p$ and $N = n$ is binomial with parameters n and p. (To see this, observe that anything in the expression that does not involve x can be treated as a constant and thus absorbed in the constant of proportionality.) Similarly, the conditional distribution of P given $X = x$ and $N = n$ is a beta distribution with parameters $x + 1$ and $n - x + 1$. And the conditional pmf of N given $X = x$ and $P = p$, up to proportionality, is

$$\begin{aligned} P(N = n|P = p, X = x) &\propto \frac{1}{(n-x)!}(1-p)^{n-x}4^n \\ &\propto e^{-4(1-p)}\frac{(4(1-p))^{n-x}}{(n-x)!}, \end{aligned}$$

for $n = x, x + 1, \ldots$ This is a "shifted" Poisson distribution. It is the distribution of $Y + x$, where $Y \sim \text{Pois}(4(1-p))$.

As we can simulate from the three conditional distributions, the Gibbs sampler proceeds by cycling through each of the three coordinates of the joint distribution and updating each coordinate with the conditional distribution of that coordinate given the other two coordinate values.

R: GIBBS SIMULATION OF TRIVARIATE DISTRIBUTION

The Gibbs sampler is run for 500 iterations and outputs $(X_{500}, P_{500}, N_{500})$ as an approximate sample from the joint distribution. We then repeat 10,000 times storing the output in the 3×500 matrix `simmat`. The first row `simmat[1,]` is taken as a sample from the marginal distribution of X. See the histogram of the distribution in Figure 11.7. Code is provided in the script **Gibbs2.R**.

```
> gibbsthree <- function(trials) {
  x <- 1
  p <- 1/2
  n <- 2
> for (i in 1:trials) {
```

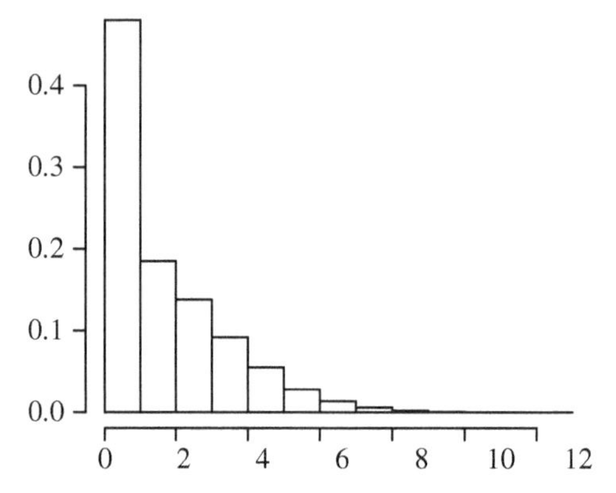

FIGURE 11.7: Distribution of the marginal distribution of X from 10,000 runs of the Gibbs sampler. Each sampler is run for 500 iterations.

```
    x <- rbinom(1, n, p)
    p <- rbeta(1, x+1, n-x+1)
    n <- x + rpois(1, 4*(1-p)) }
    return(c(x, p, n))  }
> simmat <- replicate(10000, gibbsthree(500))
> marginal <- simmat[1,]
> mean(marginal)
[1] 1.995
> sd(marginal)
[1] 1.826392
> hist(marginal)
```

■

In these examples, we have barely scratched the surface of the many uses of MCMC methods. We encourage the reader to continue their studies and investigate further on their own.

11.4 SUMMARY

The final chapter of the book is devoted to random walks on graphs and Markov chains, culminating in the final treatment of Markov chain Monte Carlo (MCMC). The simple random walk and the random walk on weighted graphs are introduced. The treatment gives a gentle introduction to Markov chains, which does not require knowledge of linear algebra. Several examples of the power and scope of MCMC are presented. Both the Metropolis–Hastings algorithm and the Gibbs sampler are introduced.

- **Graph:** A set of vertices together with a set of edges.
- **Neighbors and degrees:** For vertices i and j, we say i is a neighbor of j if there is an edge between them. The degree of a vertex is the number of its neighbors.
- **Simple random walk on a graph:** The random walk moves from a given vertex by choosing one of its neighbors uniformly at random.
- **Transition matrix:** For a graph on k vertices, this is a $k \times k$ matrix whose ijth entry $P(X_1 = j|X_0 = i)$ is the probability of moving to vertex j given that the walk is on vertex i.
- **Initial distribution:** The distribution of X_0, the initial vertex of the random walk.
- **Limiting distribution:** A probability distribution π on the vertex set with the property that for all initial vertices i, $P(X_n = j|X_0 = i) \to \pi_j$, as $n \to \infty$.
- **Limiting distribution for a simple random walk on a graph:** Under suitable conditions on the graph, $\pi_j = \deg(j)/\sum_{i=1}^{k} \deg(i)$.
- **Weighted graph:** A graph with a positive number assigned to each edge.
- **Random walk on weighted graph:** The random walk moves from a given vertex i by choosing a neighbor j with probability $w(i,j)/w(i)$, where $w(i,j)$ is the weight of the edge joining i and j, and $w(i)$ is the sum of the weights on the edges that join i to all of its neighbors.
- **Markov chain:** A sequence of $X_0, X_1, \ldots$ with the property that for all n, the conditional distribution of X_{n+1} (the "future") given $X_0, \ldots, X_n$ (the "past") only depends on X_n (the "present").
- **Stationary distribution:** A probability distribution $\mu = (\mu_1, \ldots, \mu_k)$ on the states of a Markov chain that satisfies $\mu_j = \sum_{i=1}^{k} \mu_i T_{ij}$ for all j.
- **Detailed balance condition:** A probability distribution μ satisfies this condition if $\mu_i T_{ij} = \mu_j T_{ji}$ for all i and j. If a probability distribution satisfies this condition, it is a stationary distribution of the Markov chain.
- **Markov chain Monte Carlo:** A technique for simulating from a given probability distribution by running a Markov chain whose limiting distribution is the distribution of interest.
- **Metropolis–Hastings algorithm:** Given a probability distribution π, this algorithm constructs a Markov chain whose limiting distribution is π.
- **Gibbs sampler:** An MCMC method for simulating from a multivariate joint distribution $\pi = (\pi_1, \ldots, \pi_k)$. The algorithm is based on the ability to sample from the conditional distribution of each variable given the other variables.

EXERCISES

Random Walk on Graphs and Markov Chains

11.1 The number of umbrellas in a college student's room at the start of each day is described by a Markov chain with transition matrix

$$U = \begin{array}{c} \\ 0 \\ 1 \\ 2 \end{array} \begin{array}{c} \begin{array}{ccc} 0 & 1 & 2 \end{array} \\ \begin{pmatrix} 0.6 & 0.4 & 0 \\ 0.24 & 0.52 & 0.24 \\ 0 & 0.24 & 0.76 \end{pmatrix} \end{array}.$$

(a) What is the probability there will be no umbrellas available in the student's room at the start of the day after tomorrow if there is only one available today?

(b) Find the stationary distribution.

11.2 Suppose you decide to construct a music composition according to a Markov chain with three states—the notes, of A, C sharp, and E flat. Suppose you choose the transition matrix

$$T = \begin{array}{c} \\ \text{A} \\ \text{Csharp} \\ \text{Eflat} \end{array} \begin{array}{c} \begin{array}{ccc} \text{A} & \text{Csharp} & \text{Eflat} \end{array} \\ \begin{pmatrix} 0.1 & 0.6 & 0.3 \\ 0.25 & 0.05 & 0.7 \\ 0.7 & 0.3 & 0 \end{pmatrix} \end{array}.$$

(a) What is the probability A is the note that is played two notes from now assuming you just played E flat?

(b) If you are equally likely to start with any note, what is the probability A is the third note played? (i.e., you play an initial note, one more note, and then A).

(c) In the long-run (assuming a stationary distribution exists) what fraction of notes will be C sharps?

11.3 The star graph on k vertices contains one center vertex and $k - 1$ other vertices called leaves. Between each leaf and the center vertex there is one edge. Thus the graph has $k - 1$ edges. Find the stationary distribution of the random walk on the star graph.

11.4 The *lollipop graph* on $2k - 1$ vertices is defined as follows: a complete graph on k vertices is joined with a path on k vertices by identifying one of the endpoints of the path with one of the vertices of the complete graph (see Fig. 11.8). Find the limiting distribution for a simple random walk on the lollipop graph.

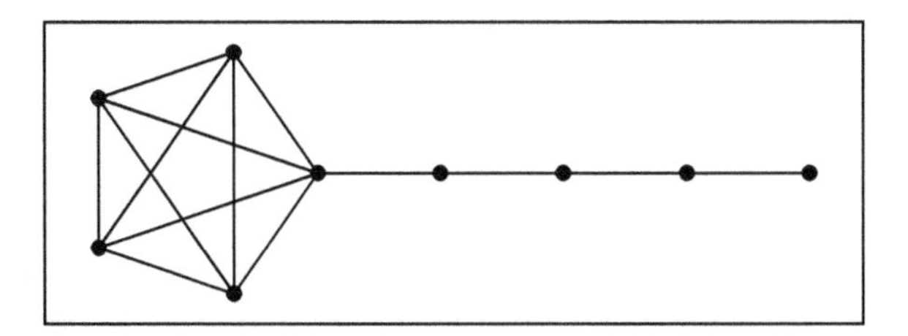

FIGURE 11.8: Lollipop graph on nine vertices.

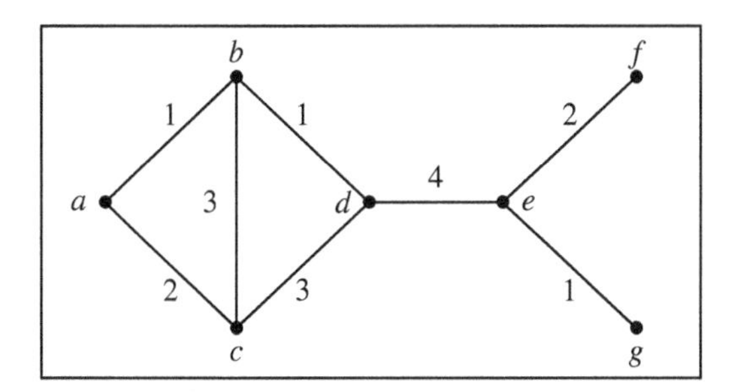

FIGURE 11.9: Weighted graph.

11.5 Find the stationary distribution for a random walk on the weighted graph in Figure 11.9.

11.6 The rows of a Markov chain transition matrix sum to one. A matrix is called *doubly stochastic* if its columns also sum to one. If a Markov chain has a doubly stochastic transition matrix, show that its stationary distribution is uniform.

11.7 Show that if a transition matrix for a Markov chain is symmetric, that is, if $T_{ij} = T_{ji}$ for all i and j, then the Markov chain is time-reversible.

11.8 A lone king on a chessboard conducts a random walk by moving to a neighboring square with probability proportional to the number of neighbors. The walk defines a simple random walk on a graph consisting of 64 vertices. Find the value of the stationary distribution at a corner square of the chessboard.

11.9 The weather Markov chain of Example 11.4 has stationary distribution is $\pi = (1/4, 1/5, 11/20)$. Determine whether or not the Markov chain is time-reversible.

11.10 Suppose a time-reversible Markov chain has transition matrix P and stationary distribution π. Show that the Markov chain can be regarded as a random walk on a weighted graph with edge weights $w(i,j) = \pi_i T_{ij}$ for all states i and j.

11.11 A Markov chain has transition matrix

$$T = \begin{array}{c} \\ a \\ b \\ c \\ d \end{array} \begin{array}{c} \begin{array}{cccc} a & b & c & d \end{array} \\ \begin{pmatrix} 0 & 1/4 & 1/4 & 1/2 \\ 1/4 & 0 & 0 & 3/4 \\ 1/2 & 0 & 0 & 1/2 \\ 1/3 & 1/2 & 1/6 & 0 \end{pmatrix} \end{array}.$$

(a) Show that the stationary distribution is $\pi = (1/4, 1/4, 1/8, 3/8)$.

(b) The Markov chain can be regarded as a random walk on a weighted graph. Determine the graph and the weights.

11.12 Suppose a Markov chain with unique positive stationary distribution π starts at state i. The expected number of steps until the chain revisits i is called the *expected return time of state i*. We state without proof that the expected return time of state i is equal to $1/\pi_i$. Find the expected return time of the center vertex in the lollipop graph of Figure 11.8.

11.13 See Exercise 11.12. A lone knight performs a random walk on a chessboard. From any square, the knight looks at the squares that it can legally move to in chess, and picks one uniformly at random to move to. If the knight starts this random walk at one of the corner squares of the chessboard, find the expected number of steps until the knight returns to its starting square.

Markov Chain Monte Carlo

11.14 See the MCMC toy example. Find the transition matrix for the Markov chain constructed by the Metropolis–Hastings algorithm. Show that $\pi = (0.1, 0.2, 0.3, 0.4)$ is the stationary distribution and that the detailed-balance condition is satisfied.

11.15 Use the Metropolis–Hastings algorithm to simulate a Poisson random variable with parameter λ. Let T be the (infinite) matrix that describes a simple random walk on the integers. From an integer i, the walks moves to $i-1$ or $i+1$ with probability 1/2 each. Show how to simulate a Poisson random variable. Then implement your algorithm and give evidence that it works.

11.16 Modify the simulation code for a bivariate standard normal distribution to simulate a bivariate normal distribution with parameters $\mu_X = 20$, $\sigma_X^2 = 100$, $\mu_Y = -14$, $\sigma_Y^2 = 4$, and $\rho = -0.8$ using the Gibbs sampler.

11.17 Research project: Implement the MCMC cryptography algorithm as described in Example 11.9. Study ways to adjust parameters in order to improve the algorithm for more efficient decoding.

Chapter Review

Chapter review exercises are available through the text website. The URL is `www.wiley.com/go/wagaman/probability2e`.

APPENDIX A

PROBABILITY DISTRIBUTIONS IN R

There are four commands for working with probability distributions in **R**. The commands take the root name of the probability distribution (see Table A.1) and prefix the root with `d`, `p`, `q`, or `r`. These give continuous density or discrete probability mass function (`d`), cumulative distribution function (`p`), quantile (`q`), and random variable simulation (`r`).

TABLE A.1. Probability distributions in R.

Distribution	Root	Distribution	Root
Beta	beta	Log-normal	lnorm
Binomial	binom	Multinomial	multinom
Cauchy	cauchy	Negative binomial	nbinom
Chi-squared	chisq	Normal	norm
Exponential	exp	Poisson	pois
F	f	Student's t	t
Gamma	gamma	Uniform	unif
Geometric	geom	Weibull	weibull
Hypergeometric	hyper		

Probability: With Applications and R, Second Edition. Amy S. Wagaman and Robert P. Dobrow.

Companion Website: www.wiley.com/go/wagaman/probability2e

APPENDIX B

SUMMARY OF PROBABILITY DISTRIBUTIONS

Discrete Distributions

Distribution	Pmf, Expectation, and Variance
Uniform$(1, \ldots, n)$	$P(X = k) = \frac{1}{n}, \quad k = 1, \ldots, n$ $E[X] = \frac{n+1}{2} \qquad V[X] = \frac{n^2 - 1}{12}$
Binomial	$P(X = k) = \binom{n}{k} p^k (1-p)^{n-k}, \quad k = 0, 1, \ldots, n$ $E[X] = np \qquad V[X] = np(1-p)$
Poisson	$P(X = k) = \frac{e^{-\lambda}\lambda^k}{k!}, \quad k = 0, 1, 2, \ldots$ $E[X] = \lambda \qquad V[X] = \lambda$
Geometric	$P(X = k) = (1-p)^{k-1}p, \quad k = 1, 2, \ldots$ $E[X] = \frac{1}{p} \qquad V[X] = \frac{1-p}{p^2}$ $P(X > t) = (1-p)^t$

Probability: With Applications and R, Second Edition. Amy S. Wagaman and Robert P. Dobrow.

Companion Website: www.wiley.com/go/wagaman/probability2e

Discrete Distributions

Distribution	Pmf, Expectation, and Variance
Negative binomial	$P(X=k)=\binom{k-1}{r-1}p^r(1-p)^{k-r}$, $k=r,r+1,\ldots,\quad r=1,2,\ldots$ $E[X]=\frac{r}{p}\qquad V[X]=\frac{r(1-p)}{p^2}$
Hypergeometric	$P(X=k)=\frac{\binom{r}{k}\binom{N-r}{n-k}}{\binom{N}{n}},\quad k=0,1,\ldots,n$ $E[X]=\frac{nr}{N}\qquad V[X]=\frac{n(N-n)r(N-r)}{N^2(N-1)}$

Continuous Distributions

Distribution	Density, cdf, Expectation, and Variance
Uniform(a,b)	$f(x)=\frac{1}{b-a},\quad a<x<b$ $F(x)=\frac{x-a}{b-a},\quad a<x<b$ $E[X]=\frac{a+b}{2}\qquad V[X]=\frac{(b-a)^2}{12}$
Exponential	$f(x)=\lambda e^{-\lambda x},\quad x>0$ $F(x)=1-e^{-\lambda x},\quad x>0$ $E[X]=\frac{1}{\lambda}\qquad V[X]=\frac{1}{\lambda^2}$
Normal	$f(x)=\frac{1}{\sigma\sqrt{2\pi}}\exp\left[\frac{-(x-\mu)^2}{2\sigma^2}\right],-\infty<x<\infty$ $E[X]=\mu\qquad V[X]=\sigma^2$

Continuous Distributions

Distribution	Density, cdf, Expectation, and Variance
Gamma	$f(x) = \frac{\lambda^a (x)^{a-1} e^{-\lambda x}}{\Gamma(a)}, \quad x > 0$ where $\Gamma(a) = \int_0^\infty t^{a-1} e^{-t}\, dt$ $E[X] = \frac{a}{\lambda} \qquad V[X] = \frac{a}{\lambda^2}$
Beta	$f(x) = \frac{\Gamma(\alpha+\beta)}{\Gamma(\alpha)\Gamma(\beta)} x^{\alpha-1}(1-x)^{\beta-1}, \quad 0 < x < 1$ $E[X] = \frac{\alpha}{\alpha+\beta} \qquad V[X] = \frac{\alpha\beta}{(\alpha+\beta+1)(\alpha+\beta)^2}$

APPENDIX C

MATHEMATICAL REMINDERS

Exponents

$$e^{a+b} = e^a e^b$$

$$[e^a]^b = e^{ab}$$

Logarithms

$$\log ab = \log a + \log b$$

$$\log a^r = r \log a$$

$$\log x = y \text{ if and only if } x = e^y$$

Calculus

Limits

$$\lim_{x\to\infty} \left(1 + \frac{a}{x}\right)^x = e^a$$

Derivatives

$$f'(x) = \lim_{h\to 0} \frac{f(x+h) - f(x)}{h}$$

Probability: With Applications and R, Second Edition. Amy S. Wagaman and Robert P. Dobrow.

Companion Website: www.wiley.com/go/wagaman/probability2e

Definite Integral

$$\int_a^b f(x)\,dx = \lim_{n\to\infty} \frac{b-a}{n} \sum_{i=1}^{n} f(x_i),\; a = x_1 < x_2 < \cdots < x_n = b$$

Chain Rule

$$\frac{d}{dx} f(g(x)) = f'(g(x))g'(x)$$

Fundamental Theorem of Calculus

$$\frac{d}{dx} \int_a^x f(t)\,dt = f(x)$$

Taylor Series Expansion for $f(x)$ about $x = a$

$$f(x) = f(a) + f'(a)(x-a) + \frac{f''(a)}{2}(x-a)^2 + \frac{f^{(3)(a)}}{3!}(x-a)^3 + \cdots$$

Integration by Parts for Definite Integrals

$$\int_a^b u\,dv = uv\Big|_a^b - \int_a^b v\,du$$

Series

$$\sum_{k=0}^{\infty} \frac{x^k}{k!} = e^x, \quad \text{for all } x$$

$$\sum_{k=0}^{\infty} x^k = \frac{1}{1-x}, \quad \text{for } |x| < 1$$

$$\sum_{k=0}^{n} x^k = \frac{1-x^{n+1}}{1-x}, \quad \text{for } x \neq 1$$

$$\sum_{k=1}^{n} k = \frac{n(n+1)}{2}$$

$$\sum_{k=1}^{n} k^2 = \frac{n(n+1)(2n+1)}{6}$$

APPENDIX D

WORKING WITH JOINT DISTRIBUTIONS

This appendix is designed to give you practice working with joint distributions, particularly aspects of working with double integrals or partial derivatives that you might not be familiar with. We highly recommend sketching domains for the following examples to help you understand the setup of the integrals.

Example D.1 Suppose X and Y are random variables with joint pdf given by $f(x, y) = x + y$, when $0 < x < 1$ and $0 < y < 1$, and 0, otherwise. We want to find two probabilities: (i) $P(X < 1/2, Y < 1/3)$ and (ii) $P(X < Y)$.

The domain here is a unit square. That makes many problems very easy to setup. Rectangles are similarly easy. Sketch the domain and shade in the region corresponding to the probability. For this example, in (i), this produces a rectangle. Recall that the density curve would be above that domain, and so our idea of a probability being the "area under the curve" is now "volume." The double integral serves to give us this volume.

(i) We want to find $P(X < 1/2, Y < 1/3)$. In your sketch, draw a horizontal line in the desired region, creating a "slice" in the domain of the desired probability. Note that on your horizontal slice, the bounds for X are 0 and 1/2 for the entire region of interest. If we integrated along this slice, we could get the probability

Probability: With Applications and R, Second Edition. Amy S. Wagaman and Robert P. Dobrow.

Companion Website: www.wiley.com/go/wagaman/probability2e

of X being between 0 and 1/2 at some fixed value of Y. But what we want to look at is any value of $Y < 1/3$. So, we imagine moving the slice up and down. Again, note that the bounds on X do not change, and Y goes from 0 to 1/3 on this region. "Slicing" in the horizontal direction sets up double integrals where x is the variable for the inner integral, and y for the outer. Vertical slices reverse this. In this first example though, there are no constraints on the relationship between X and Y so either setup is comparable. Combining what we learned about X and Y's behavior in our desired region, we find that

$$P(X < 1/2, Y < 1/3) = \int_0^{1/3} \int_0^{1/2} (x + y)\, dx\, dy$$

$$= \int_0^{1/3} \left(\frac{x^2}{2} + xy \right) \Big|_0^{1/2} dy = \int_0^{1/3} \left(\frac{1}{8} + \frac{y}{2} \right) dy$$

$$= \frac{y}{8} + \frac{y^2}{4} \Big|_0^{1/3} = \frac{5}{72}.$$

When evaluating the integrals, solve the inner integral first, treating the outer integral's variable as a constant. Thus, in our example, we find the desired probability to be 5/72.

(ii) We want to find $P(X < Y)$. If we sketch this region within our unit square, we find that it corresponds to the triangle above the line $y = x$. The fact that the area is not a rectangle (or square) means that we must be more careful when setting up our double integral to evaluate the desired probability. We could set up the integrals with y in the inner integral (vertical slices) or x in the inner integral (horizontal slices). We start with the vertical slices, and suggest you sketch both as you follow along.

Making a vertical slice within our triangular region shows us how y varies at a fixed x value. In this example, each vertical slice goes from the line $y = x$ to $y = 1$. Thus, y varies from x to 1 along each slice. To accrue the desired volume, we move the vertical slice across all values of x allowed in the region, which we see is from 0 to 1. Thus, we would evaluate the desired probability as

$$P(X < Y) = \int_0^1 \int_x^1 (x + y)\, dy\, dx.$$

We could also approach the problem with horizontal slices. Each horizontal slice shows us that x varies from 0 to y for a fixed y value. Then, to accrue the volume, we would need slices at each value of y from 0 to 1. This gives an equivalent approach to finding the probability which gives

$$P(X < Y) = \int_0^1 \int_0^y (x + y)\, dx\, dy$$

$$= \int_0^1 \left(\frac{x^2}{2} + xy \right) \Big|_0^y \, dy$$

$$= \int_0^1 \frac{3}{2} y^2 \, dy = \frac{1}{2} y^3 \Big|_0^1 = \frac{1}{2}.$$

Note that the constraint imposed by having $x < y$ shows up in the bounds of the inner integral either way we slice. The outer integral bounds will always be numeric (not include either variable).

Triangles are a common shape to encounter in these sorts of problems, and often, either way of "slicing" the region of interest results in a single integral. However, with some triangles and with more complicated shapes, slicing may result in several needed double integrals. At times, slicing one way may result in a single double integral, whereas the other way may result in several double integrals. Thus, it is useful to learn to sketch the region of interest to see if one approach is "easier" than the other in terms of number of integrals needed. In the next examples, we examine more complicated regions. Sketches are strongly encouraged. ■

Example D.2 Suppose X and Y have joint pdf $f(x, y) = 8xy$, for $0 \leq x \leq y \leq 1$, and 0, otherwise. In this setting, we want to find (i) the marginal distribution of X, (ii) $P(X < 1/2)$, and (iii) $P(X + Y < 1)$.

Before doing any computations, sketching the domain shows us that the pdf itself takes positive value in the triangle above the line $y = x$ in the unit square. We encountered this region in the last example when computing a probability, but this time, this is the shape of the domain itself.

(i) For the marginal of X, the idea is to integrate out the other variable, Y. The catch is that the bounds for Y depend on X. Thus, this must be taken into account when finding the marginal of X. Making a vertical slice, because we want to integrate out y, we see that for a fixed x, y ranges from x to 1. This helps us set up our integral and we find the marginal as

$$f_X(x) = \int_x^1 8xy \, dy = 4xy^2|_x^1 = 4x(1 - x^2), \quad 0 < x < 1,$$

and 0, otherwise.

(ii) We want to find $P(X < 1/2)$ from the joint distribution, for the purposes of illustration, not from the marginal we just found. To begin, we sketch the domain, and add this region. Note that you should start with a triangle for the domain, and the constraint that $x < 1/2$ takes off a triangle. There are multiple approaches at this point. You could use the complement rule and find $P(X < 1/2)$ as $1 - P(X > 1/2)$ since $P(X > 1/2)$ looks relatively easy to compute (the region of interest is a triangle). However, we want to demonstrate how differences in slicing may give different numbers of integrals to work with, so we'll actually integrate over the trapezoid.

First, consider vertical slices of the trapezoid. For a fixed x, we can see that y varies over the desired region from x to 1, and x itself goes from 0 to 1/2. This results in a single double integral to find the desired probability. We obtain

$$P(X < 1/2) = \int_0^{1/2} \int_x^1 8xy \, dy \, dx.$$

Next, consider what would happen with horizontal slices. In your sketch, you should see two cases develop. When $0 < y < 1/2$, the x value is dependent on y, and ranges from 0 to y. However, when $y \geq 1/2$, then x just ranges from 0 to 1/2. This means that two double integrals would be required to find the probability. Thus, we would obtain the probability with the following setup:

$$P(X < 1/2) = \int_0^{1/2} \int_0^y 8xy \, dx \, dy + \int_{1/2}^1 \int_0^{1/2} 8xy \, dx \, dy.$$

In both setups, the desired probability is 7/16. We leave the computations to the reader. However, the first setup is arguably easier because there is only one double integral to work with. (Using the marginal we previously found would also have been easier!)

(iii) Finally, we consider finding $P(X + Y < 1)$. First, your domain sketch might look a little messy because of our previous work on this problem. Make a new domain sketch. (A good idea to not get confused!)

When working with probabilities involving a function of both variables, it can be useful to re-express it to focus on one variable, say X in this case. That means we want to find $P(X < 1 - Y)$. In the previous part, we looked at $X < 1/2$, so this is similar, except that rather than adding a vertical constraint at $x = 1/2$ to consider, here, we add a line $x = 1 - y$ to our sketch. (Equivalently, this is $y = 1 - x$, if that helps you for plotting.)

Once you add this constraint, we see that it splits the domain into two triangles. Be sure you know which triangle corresponds to your desired probability. If you test points, it is quickly clear that we want the triangle whose base is the y-axis. The triangles are equal in size, so if the pdf was uniform, we'd know the probability was 1/2. However, that is not the case. So, we need to integrate over the region with our pdf of $8xy$. We can proceed with either horizontal or vertical slices to help set up the double integral(s) required. Take a moment, consider your sketch, and determine which way you want to slice.

Slicing vertically results in only one double integral, whereas horizontal slices will require two double integrals. If we had been considering $P(X + Y > 1)$, this would have been reversed with the other triangle. We present both approaches. Again, both approaches are valid and work, but one may be easier to compute than the other. For slicing vertically, we see that when x is fixed, the values of y vary between the two linear constraints x and $1 - x$. The values of x in the triangle range from 0 to 1/2. Thus, we can solve for the desired probability as

$$P(X+Y<1) = \int_0^{1/2} \int_x^{1-x} 8xy\, dy\, dx$$
$$= \int_0^{1/2} 4xy^2 \Big|_x^{1-x} dx = \int_0^{1/2} 4x(1-2x)\, dx$$
$$= 2x^2 - \frac{8}{3}x^3 \Big|_0^{1/2} = \frac{1}{2} - \frac{1}{3} = \frac{1}{6}.$$

To examine the setup via the horizontal slices, note that the bounds on x change depending on whether y is less than or greater than 1/2. The original boundary constraint matters when y is less than 1/2, and the new constraint $x = 1 - y$ matters when y is greater than 1/2. Thus, the setup for the two double integrals to solve the same probability slicing horizontally yields

$$P(X+Y<1) = \int_0^{1/2} \int_0^{y} 8xy\, dx\, dy + \int_{1/2}^{1} \int_0^{1-y} 8xy\, dx\, dy\,,$$

which evaluates to $1/16 + 5/48 = 1/6$ as expected. ■

We explore one final example to consider some additional shapes for regions of interest, though the pdf is uniform over the domain, and thus, we also encourage the reader to consider this example thinking geometrically.

Example D.3 Suppose X and Y have joint pdf $f(x, y) = 1/8$, for $0 \le y \le 4$, $y \le x \le y + 2$, and 0, otherwise. In this setting, we want to find (i) $P(Y < 2)$ and (ii) $P(X + Y \le 4)$.

We strongly encourage you to sketch the domain before any computations. Here, the domain is a parallelogram. The region could also have been described originally as the parallelogram defined by the points $(0, 0)$, $(2, 0)$, $(6, 4)$, and $(4, 4)$.

(i) For considering $P(Y < 2)$, from your sketch, it should seem that this probability should be 1/2. (Uniform pdfs are nice for geometric arguments!) Indeed, this is the case. Still, we want to practice setting up the appropriate double integral(s). Think carefully about which way you would want to slice here.

In this example, the horizontal slices would result in a single double integral while vertical slices require two double integrals. Try the setup yourself, and then examine the solutions below.

For horizontal slices, you should see that x varies between y and $y + 2$, so the double integral is

$$P(Y<2) = \int_0^{2} \int_y^{y+2} 1/8\, dx\, dy.$$

When slicing vertically, the bounds on y change as x goes past 2. Thus, we get two double integrals as follows

$$P(Y < 2) = \int_0^2 \int_0^x 1/8\, dy\, dx + \int_2^4 \int_{x-2}^2 1/8\, dy\, dx.$$

(ii) For the final probability, $P(X + Y \leq 4)$, we encourage you to resketch the domain, and then add the constraint imposed by the probability, the line $y = 4 - x$. Then determine whether you are interested in the region of the domain above or below the line. You can always try points (such as the origin) quickly to check. The shape we are interested in is a trapezoid.

In this instance, slicing either direction will lead to needing two double integrals to find the desired probability. For problems like this, we advocate for writing down equations of the constraints as notes so you can work with them. For example, the line $y = 4 - x$ came from our desire to find the probability that $x + y \leq 4$. We could also have written the line, less conventionally, as $x = 4 - y$. This is a useful re-expression if we are looking for bounds on x rather than y.

Try the setup of the double integrals yourself, and then examine the solutions below.

When slicing horizontally, we encounter a parallelogram and a triangle as the shapes we need to integrate over. In the parallelogram, the bounds on x are those of the domain. In the triangle, one side comes from the domain and the other from the constraint added by the probability of interest. The change in shapes happens when $y = 1$. Thus, we would obtain

$$P(X + Y \leq 4) = \int_0^1 \int_y^{y+2} 1/8\, dx\, dy + \int_1^2 \int_y^{4-y} 1/8\, dx\, dy = \frac{3}{8}.$$

When slicing vertically, we are integrating over two triangles. The bounds on y change when $x = 2$. The set of double integrals needed for this approach is

$$P(X + Y \leq 4) = \int_0^2 \int_0^x 1/8\, dy\, dx + \int_2^3 \int_{x-2}^{4-x} 1/8\, dy\, dx = \frac{3}{8}.$$

■

If double integrals were new to you, the concept of partial derivatives may also be new. Partial derivatives are used when moving from a joint cdf to a joint pdf. In our final example of the appendix, we demonstrate how to find a joint pdf from a joint cdf, as well as show how the joint cdf was obtained in the first place.

Example D.4 Suppose the joint cdf of X and Y is $F(x, y) = x^2y/2 + xy^2/2$ for $0 < x < 1, 0 < y < 1$. Find (i) the joint pdf of X and Y, and then (ii) validate the cdf from your pdf.

(i) Recall that when working with a single continuous random variable, the derivative of the cdf is the pdf. It is a similar idea when working with jointly distributed continuous random variables, except that you must take the derivative of the cdf in turn for each variable, holding the other variables constant as you do so. These are partial derivatives. As long as you are clear about which variable is being held constant, and which you are currently working with, this is not too difficult. For notation, you will see this written in our context as

$$f(x, y) = \frac{\partial}{\partial x}\frac{\partial}{\partial y}F(x, y) = \frac{\partial}{\partial y}\frac{\partial}{\partial x}F(x, y).$$

In other words, it does not matter in which order you take the derivatives (and sometimes, one way is easier than another). Applying this to our example, we find

$$f(x, y) = \frac{\partial}{\partial x}\frac{\partial}{\partial y}\left(\frac{x^2y}{2} + \frac{xy^2}{2}\right).$$

We will take the partial derivative with respect to y first, and then with respect to x. For example, when working with $x^2y/2$, we treat this as a constant times y, and thus the partial derivative is the constant, even though it contains an x term. Applying this idea yields

$$\begin{aligned} f(x, y) &= \frac{\partial}{\partial x}\left(\frac{x^2}{2} + xy\right) \\ &= x + y, \end{aligned}$$

on the range provided of $0 < x < 1, 0 < y < 1$, and 0, otherwise. This is the same pdf we encountered in our first example in this section.

(ii) Now we want to validate the cdf. In other words, how do we get the cdf from the pdf? The cdf $F(x, y) = P(X \leq x, Y \leq y)$, so we just need to evaluate that probability. We use dummy variables for the integration since we will have x and y in our bounds.

In this example, there are not other constraints on x and y to consider, but you should be careful about this. For example, if $x < y$, you need to be careful about setting up the area you are integrating over. As before, sketches help tremendously! In this example, we are simply integrating over a rectangle out of the domain of the unit square. In more complicated settings, you might need to set up cases and define the cdf for particular regions.

Using dummy variables in our setting, we find that

$$\begin{aligned} F(x, y) &= \int_0^x \int_0^y (s + t)\, dt\, ds \\ &= \int_0^x st + \frac{t^2}{2}\bigg|_0^y ds = \int_0^x ys + \frac{y^2}{2}\, ds \\ &= \frac{y}{2}s^2 + \frac{y^2}{2}s\bigg|_0^x = \frac{x^2y}{2} + \frac{xy^2}{2}, \end{aligned}$$

as desired. ■

SOLUTIONS TO EXERCISES

SOLUTIONS FOR CHAPTER 1

1.3 (i) Choosing a topping; (ii) set of possible toppings. Let n, p, r, denote pineapple, peppers, and pepperoni, respectively. Then

$$\Omega = \{\emptyset, n, p, r, np, nr, pr, npr\}.$$

(iii) $A = \{np, nr, pr\}$.

1.5 (i) Harvesting tomatoes; (ii) set of all possibilities for bad and good tomatoes among the 100; (iii) {At most five tomatoes are bad}.

1.7 (a) $\{R = 0\}$; (b) $\{R = 1, B = 2\}$; (c) $\{R + B = 4\}$; (d) $\{R = 2B\}$; (e) $\{R \geq 2\}$.

1.9 $P(\omega_4) = 1/41$.

1.11 (b) and (c) are valid.

1.13 Show

$$\sum_{\omega} P(\omega) = 1.$$

Probability: With Applications and R, Second Edition. Amy S. Wagaman and Robert P. Dobrow.

Companion Website: www.wiley.com/go/wagaman/probability2e

1.15 $p = 0$ or 1.

1.17 $1/16$.

1.19 (a) $1/6^4$; (b) 0.598; (c) 0.093.

1.21 $1/(n \cdot n-1 \cdots n-k+1)$.

1.23 (a) $P(4432) = 0.2155$. $P(4333) = 0.1054$. (b) $P(5332) = 0.1552; P(5431) = 0.1293; P(5422) = 0.1058$.

1.25 9.10947×10^{-6}.

1.27 (b) Hint: Use the ballot problem.

1.29 Suppose we want to pick an k element subset from $S = \{1, \ldots, n\}$. One way to choose the set is to pick k numbers from S. This can be done in $\binom{n}{k}$ ways. Another way to choose the set is to pick the $n-k$ numbers from S that will not be in the set. This can be done in $\binom{n}{n-k}$ ways.

1.31 (a) 0.4; (b) 0.5; (c) 0.3.

1.33 (a) 0.90; (b) 0.

1.35 0.80.

1.37 (a) 0.2; (b) 0.3; (c) 0.6; (d) 0.9.

1.39 (i) $P(\text{HHHT}, \text{HHTH}) = 2/16$; (ii) $3/8$. (iii) $P(\text{HHHH}, \text{HHTT})$.

1.41 (a)

$$\Omega = \{pp, pn, pd, pq, np, nn, nd, nq, dp, dn, dd, dq, qp, qn, qd, qq\}.$$

The events are: $\{X = 1\} = \{pp\}$, $\{X = 5\} = \{pn, np, nn\}$ $\{X = 10\} = \{pd, nd, dd, dn, dp\}$, $\{X = 25\} = \{pq, nq, dq, qq, qd, qnqp\}$. (c) 3/8.

1.43 (a)

$$\sum_{k=0}^{\infty} \frac{2}{3^{k+1}} = \frac{2}{3} \sum_{k=0}^{\infty} \frac{1}{3^k} = \frac{2}{3} \frac{1}{1-1/3} = 1.$$

(b) 1/27.

1.45 (a) $A \cup B \cup C$; (b) BA^cC^c; (c) $A^cB^cC^c \cup AB^cC^c \cup BA^cC^c \cup CA^cB^c$; (d) ABC; (e) $A^cB^cC^c$.

1.47 1/2.

1.49 0.40.

SOLUTIONS FOR CHAPTER 2

2.3 (a) 0.411; (b) 0.164; (c) 0.547; (d) 0.770.

2.5 $(2p_1 - p_2)/p_1$.

2.7 (a) 0.457; (b) 0.347; (c) 0.758.

2.9 (a) 0; (b) 1; (c) $P(A)/P(B)$; (d) 1.

2.11 (a) False. For instance, consider tossing two coins. Let A be the event that both coins are heads. Let B be the event that the first toss is heads. Then $P(A|B) + P(A|B^c) = 1/2 + 0 = 1/2 \neq 1$. (b) True.

$$P(A|B) = \frac{P(AB)}{P(B)} = \frac{P(B) - P(A^cB)}{P(B)} = 1 - P(A^c|B).$$

2.13 0.00198.

2.15 (a) $p_1 = P(AB|A) = P(AB)/P(A)$; (b) $p_2 = P(AB|A \cup B) = P(AB)/P(A \cup B)$, since AB implies $A \cup B$; (c) Since $A \subseteq A \cup B$, $P(A) \leq P(A \cup B)$ and thus $p_1 \geq p_2$.

2.17

$$\begin{aligned} P(A \cup B|C) &= \frac{P((A \cup B)C)}{P(C)} = \frac{P(AC \cup BC)}{P(C)} \\ &= \frac{P(AC) + P(BC) - P(ABC)}{P(C)} = P(A|C) + P(B|C) - P(AB|C). \end{aligned}$$

2.19 Apply the birthday problem. Approximate probability is 0.63.

2.21 (ii)

$$P(G) = P(G|1)P(1) + P(G|2)P(2) = \frac{3}{5}\left(\frac{1}{2}\right) + \frac{2}{6}\left(\frac{1}{2}\right) = \frac{7}{15}.$$

2.23

$$P(A|B^c) = \frac{P(A) - P(AB)}{1 - P(B)}.$$

2.27 Solve $0.20 = 0.35(0.30) + x(0.70)$ for x.

2.29 (a) 0.32; (b) 76/77 = 98.7% reliable.

2.31 0.10.

2.33 If A and B are independent,

$$\begin{aligned} P(A^cB^c) &= P((A \cup B)^c) = 1 - P(A \cup B) \\ &= 1 - [P(A) + P(B) - P(AB)] = P(A^c) - P(B) + P(A)P(B) \\ &= P(A^c) - P(B)[1 - P(A)] = P(A^c) - P(A^c)P(B) = P(A^c)P(B^c). \end{aligned}$$

2.35 0.25.

2.37 0.04914.

2.39 9.

2.43 The simulation follows:

```
d1 <- function() {
sample(c(3,5,7),1,replace=T)    }
d2 <- function() {
sample(c(2,4,9),1,replace=T)    }
d3 <- function() {
sample(c(1,6,8),1,replace=T)    }
simlist1 <- replicate(10000,d1() > d2())
mean(simlist1)
[1] 0.5638
simlist2 <- replicate(10000,d1() > d2())
mean(simlist2)
[1] 0.562
simlist3 <- replicate(10000,d1() > d2())
mean(simlist3)
[1] 0.5585
```

SOLUTIONS FOR CHAPTER 3

3.3 Let X be the number of toppings. The probability of interest is $P(X = 2)$.

3.5 Let X be the number of bad tomatoes among the 100. The probability of interest is $P(X \leq 5)$.

3.7 (i) $A^c = \{X = 1\} \bigcup \{X \geq 5\}$; (ii) $B^c = \{1 \leq X \leq 3\}$; (iii) $AB = \{X = 4\}$; and (iv) $A \bigcup B = \{X \geq 2\}$.

3.9

$$P(X = 0) = 0.729; P(X = 1) = 0.243; P(X = 2) = 0.027; P(X = 3) = 0.001.$$

3.11

2	3	4	5	6	7	8
0.01	0.04	0.10	0.20	0.25	0.24	0.16

3.13

3	4	5	6	7	8	9
1/27	3/27	6/27	7/27	6/27	3/27	1/27

3.15 (a) 0.126; (b) 0.048; (c) 0.22; (d) 0.999568.

3.17 (a) Yes. $n = 7, p = 0.25$; (b) No fixed number of trials; (c) Sampling without replacement. No independence or fixed probability; (d) No fixed probability; (e) Yes, assuming independence for pages.

3.19 (a) 0.2778; (b) 0.03845; (c) 0.365; (d) 0.058.

3.21 0.0384. It is very unlikely (less than a 4% chance) that the drug is ineffective.

3.23 $\lambda = 4$ and $P(X = 3) = 0.1954$.

3.25 $\lambda = -\ln 0.10$ and the probability of hatching at least two eggs is 0.6697.

3.27 0.1018.

3.29 $(1 - e^{-2\lambda})/2$.

3.31 (a) $P(1 \leq X \leq 3) = 0.7228997$; (b) $P(1 \leq X \leq 3) \approx 0.7228572$.

3.33 Let X have a Poisson distribution with $\lambda = 1$. The series gives the probability that X is even.

3.39 An example simulation follows:

```
lambda <- 2
sim <- rpois(100000,lambda)
sum( sim%%2 ==1)/100000
(1-exp(-2*2))/2
```

SOLUTIONS FOR CHAPTER 4

4.1 1.722.

4.3 (a) $P(W = 0) = 3/6$; $P(W = 10) = 2/6$; $P(W = 24) = 1/6$: (b) $E[W] = 7.33$.

4.5 \$11.

4.9 $E[X!] = e^{-\lambda}/(1-\lambda)$, if $0 < \lambda < 1$.

4.11 (a) $P(T = t, E = z) = e^{-6}4^t 2^z/(t!z!)$; (b) e^{-6}; (c) 0.54; (d) 0.95.

4.13 (a) $c = 6/(n(n+1)(n+2))$; (b) 0.30.

4.15 (a) 1/24. (b) $(e+1)(e+2)(e+3)/24$.

4.17 (a) X is uniform on $\{1, \ldots, n\}$; (b) 1/2.

4.19 (a) $P(X = 1) = 3/8$; $P(X = 2) = 5/8$ and $P(Y = 1) = 1/4$; $P(Y = 2) = 3/4$; (b) No; (c) 1/2; (d) 7/8.

4.21 Let $I_k = 1$ if the kth ball in the sample is red. $E[R] = nr/(r+g)$.

4.23 7.54.

4.25 2.61.

4.27 $V[X] = 1.25n$.

4.29 (a) 61; (b) 75.

4.31 $a(1-a) + b(1-b) + 2(c-ab)$.

4.33 (a) $E[S] = n(2p-1)$; (b) $V[S] = 4np(1-p)$.

4.35 (a) $c = 1/15$; (b) $P(X = 0) = 6/15 = 1 - P(X = 1)$; $P(Y = 1) = 3/15$; $P(Y = 2) = 5/15$; $P(Y = 3) = 7/15$. (c) $\text{Cov}(X, Y) = -2/75$.

4.37 If $X = 12$, then necessarily two $6'$s were rolled and thus $Y = 0$. Thus X and Y are not independent. Show $E[XY] = E[D_1^2 - D_2^2] = 0 = E[X]E[Y]$ to show X and Y are uncorrelated.

4.39 (a) 101; (b) 1410; (c) 3; (d) 1266.

4.41 0.234; 0.1875; 0.03125.

4.45

$$\text{Cov}(X, Y) = \text{Cov}(X, aX + b) = a\,\text{Cov}(X, X) = aV[X].$$

4.49 $P(A|B)$.

4.51 (a) False. (b) True. (c) False.

4.53

$$P(X + Y = k) = \begin{cases} (k-1)/n^2, & \text{for } k = 2, \ldots, n+1 \\ (2n-k+1)/n^2, & \text{for } k = n+2, \ldots, 2n. \end{cases}$$

SOLUTIONS FOR CHAPTER 5

5.1 694.167.

5.3 $V[X] = (p-1)/p^2$.

5.5 (a) 0.078; (b) 0.556; (c) 6.667.

5.7 Send out 11 applications.

5.9 Evaluate $\sum_{k=1}^{n}(k-1)n/(n-(k-1))^2$.

5.11 $E[2^X] = 2p/(1-2(1-p))$, for $1/2 < p < 1$.

5.13

$$E[X^k] = \begin{cases} q-p, & \text{if } k \text{ is odd} \\ q+p, & \text{if } k \text{ is even.} \end{cases}$$

5.17 (a) Bernoulli(0.8).

5.19

$$m_X(t)m_Y(t) = (1-p+pe^t)^m(1-p+pe^t)^n(1-p+pe^t)^{m+n} = m_{X+Y}(t).$$

5.21 (a) 0.082; (b) 0.254; (c) $P(Y=k) = P(X=k+r) = \binom{k+r-1}{r-1}p^r(1-p)^k$, for $k = 0, 1, \ldots$

5.23 $P(X+Y=k) = (k-1)p^2(1-p)^{k-2}, k = 2, 3, \ldots$

5.25 0.183.

5.27 $P(\text{Aidan wins}) = 0.1445$. Aidan gets \$14.45 and Bethany gets \$85.55.

5.29 $X \sim \text{Geom}(p^2)$.

5.31 (a) Hypergeometric. (b) $P(X=k) = \binom{4}{k}\binom{4}{4-k} \Big/ \binom{8}{4}$.

5.33 (a) 0.00395; (b) 0.0179.

5.37 (a) $E[G] = 200$, $V[G] = 160$. (b) $E[G+U] = 300$, $V[G+U] = 160 + 90 - 40 = 210$.

5.39 (a) 0.074 (b) 0.155.

5.41 Apply the multinomial theorem to expand $(1+1+1+1)^4$.

5.47 Let C, D, H, S, be the number of clubs, diamonds, hearts, and spades, respectively, in the drawn cards. Then (C, D, H, S) has a multinomial distribution with parameters $(8, 1/4, 1/4, 1/4, 1/4)$. Desired probability is 0.038.

5.49 0.54381.

SOLUTIONS FOR CHAPTER 6

6.1 (a) $c = 2$; (b) 1/4; (c) 2/3.

6.3 (a) 0; (b) $F(11) - F(3)$; (c) 1; (d) $(1 - F(12))/(1 - F(6))$.

6.5 (a) $c = 1/(e^2 - e^{-2}) = 0.13786$. (b) $P(X < -1) = 0.032$; (c) $E[X] = 1.0746$.

6.7 (b) 0.0988; (c) $(\pi - 2)/2$.

6.9 (a) 1/4; (b) $\frac{\int_{-\infty}^{\infty} x}{(1+x^2)}\, dx$ does not exist.

6.11 $E[X^k] = \frac{(b^{k+1} - a^{k+1})}{((k+1)(b-a))}$.

6.13 $0 \le a \le 1$ and $b = 1 - a$.

6.15

$$
\begin{aligned}
E[aX + b] &= \int_{-\infty}^{\infty} (ax + b) f(x)\, dx = a \int_{-\infty}^{\infty} x f(x)\, dx + b \int_{-\infty}^{\infty} f(x)\, dx \\
&= aE[X] + b.
\end{aligned}
$$

6.17 Not true. $P(X > s + t | X > t) = P(X > s)$, not $P(X > s + t)$.

6.19 $V[X] = \frac{1}{\lambda^2}$.

6.21 0.988.

6.23 0.95.

6.25 (a) 1/16; (b) 8/15; (c) 4/9.

6.27 (a) 5/12; (b) 1/3; (c) 1/48.

6.29 (a) $P(X \le x, Y \le y) = (1 - e^{-x})(1 - e^{-2y}), x > 0, y > 0$; (b) $P(X \le x) = 1 - e^{-x}, x > 0$; (c) 1/3.

6.33 $X \sim \text{Exp}(6)$ and $Y \sim \text{Unif}(1, 4)$.

6.35 3/10.

6.37 $f_X(x) = (2/\pi)\sqrt{1 - x^2}, -1 < x < 1$; Not independent.

SOLUTIONS FOR CHAPTER 7

7.1 (a) About 0.025; (b) 68%; (c) 0.16.

7.3 (a) 87.1; (b) 0.0003; (c) 0.0067.

7.5 G: 0.943; H: 0.868; I: 0.691.

7.7 (a) $E[T] = 1500$; $\text{SD}[T] = 315.18$; (b) 80th percentile is 1765; 90th percentile is 1904; (c) 0.171.

7.9 $x = \mu \pm \sigma$.

7.11 (a) 0.68; (b) 0.997.

7.17 Gamma(3/2, 1/2).

7.23 $V[X] = a/\lambda^2$.

7.25 (a) $E[N_{15}] = V[N_{15}] = 30$; (b) 0.542; (c) 0.476; (d) 0.152; (e) 9:03 and 30 seconds.

7.27 (a) 0.140; (b) 0.026; (c) 0.205.

7.29 $e^{-\lambda(s-t)}(\lambda(s-t))^{k-n}/(k-n)!$

7.31 λs.

7.35 (a) 1320; (b) $\mu = 1/3$ and $\sigma^2 = 2/117$; (c) 0.1133.

7.37 (a) 1/13440; (b) 3/8; (c) 1/105.

7.39 (a) 12; (b) 3/5.

7.41 (a) $E[X] = am/(a-1)$ for $a > 1$; (b) $m^2a/((a-1)^2(a-2))$ for $a > 2$.

7.43 The probabilities for $k = 3, 4, 5, 6$ are, respectively, $0.008, 0.0047, 0.003, 0.0021$.

7.45 $a = 1.0044$.

SOLUTIONS FOR CHAPTER 8

8.1 $Y \sim Exp(\lambda/c)$.

8.3 (a) $3y^2/16, -2 \le y \le 2$; (b) $3/2(w-4)^2, 3 \le w \le 5$.

8.5 $e - 1$.

8.7 For all real x,

$$P(1/X \le t) = P(X \ge 1/t) = 1 - P(X < 1/t).$$

Density of $1/X$ is thus

$$f(t) = f_X(1/t)/t^2 = f_X(t).$$

8.9 For $ma + n < y < mb + n$,

$$P(Y \le y) = P(mX + n \le y) = P(X \le (y-n)/m) = \frac{(y-n)/m - a}{b-a}$$
$$= \frac{y - (ma+n)}{mb - ma},$$

which is the cdf of a uniform distribution on $(ma + n, mb + n)$.

8.11 $Y \sim \text{Geom}(1 - e^{-\lambda})$.

8.15 $F(X) \sim \text{Unif}(0, 1)$.

8.17 $f_{Z^2}(t) = e^{-t/2}/\sqrt{2\pi t}, t > 0$.

8.19

$$f(x) = \left(\frac{1}{t-s}\right)\frac{\Gamma(a+b)}{\Gamma(a)\Gamma(b)}\left(\frac{x-s}{t-s}\right)^{a-1}\left(\frac{t-x}{t-s}\right)^{b-1}, s < x < t.$$

8.21 $1 - X \sim \text{Beta}(b, a)$.

8.23 $F(x) = P(X/Y \le x) = x/(x+1), x > 0$. $P(X/Y < 1) = 1/2$.

8.27 $f(x) = \lambda^3 x^2 e^{-\lambda x}/2, x > 0$.

8.29 $\frac{1}{3}$.

8.31 (a) Exp(2/5); (b) $E[M] = 2.5$; $V[M] = 25/4$.

8.33 (a) $f_M(m) = 2m/\theta^2$; (b) $E[M] = 2/3(\theta)$; (c) 9.

8.35 (a) 0.1314; (b) $\mu = 0.9406$ and $\sigma^2 = 0.00055$; (c) 99/101.

8.37 (a) $\frac{\sqrt{2}}{3}$; (b) $\frac{7}{12}$.

8.39 $\frac{2\sqrt{3}}{3\pi} \approx 0.37$.

8.43 (a) $e^{-4\lambda\pi}$; (b) $f(x) = 2\pi\lambda x e^{-\lambda\pi x^2}, x > 0$.

8.45

$$f_{V,W}(v, w) = \frac{w}{(1+v)^2}\lambda^2 e^{-\lambda w}, \text{ for } v > 0, w > 0.$$

8.47

$$f_{V,W}(v, w) = \frac{4w}{v}, \quad \text{for } 0 < w < v < 1.$$

$$f_W(w) = -4w \log w, \quad \text{for } 0 < w < 1.$$

8.51

$$f(\theta, \phi, r) = \frac{r^2 \sin \phi}{(2\pi)^{3/2}} e^{-r^2/2}, \ 0 \le \theta \le 2\pi, 0 \le \phi \le \pi, r > 0.$$

$$f_R(r) = \sqrt{\frac{2}{\pi}} r^2 e^{-r^2/2}, \ r > 0.$$

SOLUTIONS FOR CHAPTER 9

9.1 (a)

$$f_X(x) = 12x(1-x)^2, 0 < x < 1.$$

$$f_Y(y) = 6y^2 - 4y^3, 0 < y < 1.$$

(b)

$$f_{Y|X}(y|x) = \frac{1}{1-x}, x < y < 1.$$

The conditional distribution is uniform on $(x, 1)$. (c) 0.3125.

9.3 (c) 1/3.

9.5

$$f_{X+Y|X}(t|x) = \lambda e^{-\lambda(t-x)}, t > x.$$

Let $Z \sim \text{Exp}(\lambda)$. Then the desired conditional distribution is the same as the distribution of $Z + x$.

9.11 (a)

$$f_{X|Y=y}(x|y) = (y+1)e^{-x(y+1)}, x > 0,$$

which is an exponential density with parameter $y + 1$.
(b) $E[X|Y = y] = 1/(y + 1)$ and $E[X|Y] = 1/(Y + 1)$.
(c) (1)

$$E[X] = E[E[X|Y]] = E[1/(Y+1)] = \int_0^{e-1} \frac{1}{(y+1)^2}\, dy = \frac{e-1}{e}.$$

(2)

$$E[X] = \int_0^{\infty} x \frac{e^{-x} - e^{-ex}}{x}\, dx = \frac{e-1}{e}.$$

9.13 (a) $3y^2/64, 0 < y < 4$; (b) $(2/5)(x + 2y)/y^2$; (c) 16/15.

9.15

$$E[X|Y=0] = 9/12 \quad \text{and} \quad E[X|Y=1] = 8/12.$$

9.17 (a) As per hint,

$$P(M \leq m, N > n) = 2P(X \leq m, Y > n) = 2m(1-n).$$

$$\begin{aligned} P(M \leq m, N \leq n) &= P(M \leq m) - P(M \leq m, N > n) \\ &= 2m - m^2 - 2m(1-n). \end{aligned}$$

Thus $f(m,n) = 2$, for $0 < m < n < 1$. (b) $f_{N|M}(n|m) = 1/(1-m), m < n < 1$. The conditional distribution is uniform on $(m, 1)$.

9.19 (a) 5/2; (b) $S/2$; (c)

$$E[M|X_1 = x] = \begin{cases} 5/2, & \text{if } x = 1 \\ 11/4, & \text{if } x = 2 \\ 13/4, & \text{if } x = 3 \\ 4, & \text{if } x = 4. \end{cases}$$

(d) $5X_1/2$.

9.21 4.5.

9.23 The probability should be 0.397.

9.25 There are two cases based on the value of X. $P(Y < 4) = 8/9$.

9.27

(a)

$$f_{XY}(t) = \int_0^\infty \frac{1}{y} f_X(t/y) f_Y(y)\, dy,\ t > 0.$$

(c) For real t,

$$f_{X-Y}(t) = \int_0^\infty f_X(t+y) f_Y(y)\, dy.$$

9.29 (a) $E[X] = p\lambda$. (b) $V[X] = p\lambda$.

9.31 $\mu = \$250$, $\sigma = \$25.98$.

9.33 $E[D] = 3/2$; $V[D] = 9/4$.

9.35 (a) $(X_1 + (n-1)\mu)/n$; (b) $(n-1)\sigma^2/n^2$.

9.39 (b) α/β; (c) $\sigma^2 + \alpha/\beta^2$.

9.41 Show $E[(X+Y)(X-Y)] = E[X+Y]E[X-Y]$.

9.43 0.092.

9.45 (a) Bivariate standard normal distribution with $\rho = 1/2$. (b) Normal with (conditional) mean $y/2$ and variance 3/4. (c) 0.193.

SOLUTIONS FOR CHAPTER 10

10.3 (i) $P(S \geq 380) \leq 0.921$; (ii) $P(S \geq 380) \leq 0.324$.

10.5 (a) $P(X \geq 60) \leq 0.833$; (b) $P(X \geq 60) \leq 0.25$.

10.7 $P(\log X \geq c) = P(X \geq e^c) \leq \mu/e^c$.

10.9 $\hat{p} = 10/21$.

10.11

$$np = (X_1 + \cdots + X_n)/n \text{ and } np(1-p) + (np)^2 = (X_1^2 + \cdots + X_n^2)/n.$$

10.13 (a) Let $U_1, \ldots, U_n$ be i.i.d. uniform (0, 1) variables. Let $f(x) = \sin(x)e^{-x^2}$. Then $I \approx (f(U_1) + \cdots + f(U_n))/n$.

(c) Let $X_1, \ldots, X_n$ be i.i.d. normal variables with mean zero and variance 1/2. Let $f(x) = \log(x^2)\sqrt{\pi}$. Then $I \approx (f(X_1) + \cdots + f(X_n))/n$.

(e) Let $X_1, \ldots, X_n$ be i.i.d. Poisson variables with mean one. Let $f(x) = e \cos \cos x$. Then $S \approx (f(X_1) + \cdots + f(X_n))/n$.

10.15 0.508.

10.17 0.9869.

10.19 0.363.

10.21 Let $p = P(\sum_{i=1}^{10} X_i \geq 14)$. (a) $p \leq 0.714$. (b) $p \leq 0.625$. (c) $p \approx 0.103$. (d) $p \approx 0.134$. (e) $p = 0.1355$.

10.23 Consider a sum of a independent exponential random variables with $\mu = \sigma = \lambda$.

10.25 **(a)** $\log Y = \sum_{i=1}^{n} \log X_i$, a sum of i.i.d. random variables. By clt, $\log Y$ is approximately normal.

$$\mu = E[\log X_1] = \int_0^1 \log x \, dx = -1,$$

$$E[(\log X_1)^2] = \int_0^1 (\log x)\, dx = 2$$

and $\sigma^2 = 1$. The result follows.

(b) 0.837.

SOLUTIONS FOR CHAPTER 11

11.1 (a) 0.2688; (b) [0.2308, 0.3846, 0.3846].

11.3 $P(\text{center}) = 1/2$ and $P(\text{leaf}) = 1/(2(k-1))$.

11.5 $P(a) = 3/34$; $P(b) = 5/34$; $P(c) = 8/34$; $P(d) = 8/34$; $P(e) = 7/34$; $P(f) = 2/34$; $P(g) = 1/34$.

11.7 If the transition matrix is symmetric, it is doubly stochastic, and thus the stationary distribution is uniform. Then $\pi_i T_{ij} = T_{ij}/k = T_{ji}/k = \pi_j T_{ji}$.

11.9 Not time-reversible.

11.13 Expected return time for the knight is 168 steps.

REFERENCES

A. Arampatzis and J. Kamps. A study of query length. In *Proceedings of the 31st Annual International ACM SIGIR Conference on Research and Development in Information Retrieval*, SIGIR '08, pp. 811–812. ACM, 2008.

BBC News. Card trick defies the odds. http://news.bbc.co.uk/2/hi/uk_news/50977.stm. Accessed on May 14, 2003.

F. Benford. The law of anomalous numbers. *Proceedings of the American Philosophical Society*, 78:551–572, 1938.

A. Berger and T. P. Hill. *An Introduction to Benford's Law*. Princeton, New Jersey: Princeton University Press, 2015.

G. Blom, L. Holst, and D. Sandell. *Problems and Snapshots from the World of Probability*. New York: Springer-Verlag, 1991.

M. C. Borja and J. Haigh. The birthday problem. *Significance*, 4(3):124–127, 2007.

K. Bryan and T. Leise. The $25,000,000,000 eigenvector. The linear algebra behind Google. *SIAM Review*, 48(3):569–581, 2006.

L. Carroll. *The Mathematical Recreations of Lewis Carroll: Pillow Problems and a Tangled Tale*. New York: Dover Publications, 1958.

G. Casella and E. I. George. Explaining the Gibbs sampler. *American Statistician*, 46(3):167–174, 1992.

A. Cerioli, L. Barabesi, A. Cerasa, M. Menegatti, and D. Perrotta. Newcomb-Benford law and the detection of frauds in international trade. *Proceedings of the National Academy of Sciences of the United States of America*, 116(1):106–115, 2019.

Probability: With Applications and R, Second Edition. Amy S. Wagaman and Robert P. Dobrow.

Companion Website: www.wiley.com/go/wagaman/probability2e

E. Chia and M. F. Hutchinson. The beta distribution as a probability model for daily cloud duration. *Agricultural and Forest Meteorology*, 56(3–4):195–208, 1991.

College Board. 2011 College-Bound Seniors. Total Group Profile Report, 2011.

B. Dawkins. Siobhan's problem: the coupon collector revisited. *American Statistician*, 45:76–82, 1991.

K. Devlin. *The Unfinished Game: Pascal, Fermat, and the Seventeenth-Century Letter that Made the World Modern*. New York: Basic Books, 2008.

P. Diaconis. Dynamical bias in coin tossing. *SIAM Review*, 49(2):211–235, 2007.

P. Diaconis. The Markov chain Monte Carlo revolution. *Bulletin of the American Mathematical Society*, 46(2):179–205, 2008.

P. Diaconis and F. Mosteller. Methods for studying coincidences. *Journal of the American Statistical Association*, 84(408):853–861, 1989.

W. Dorsch, T. Newland, D. Jassone, S. Jymous, and D. Walker. A statistical approach to modeling the temporal patterns in ocean storms. *Journal of Coastal Research*, 24(6):1430–1438, 2008.

C. Durtschi, W. Hillison, and C. Pacini. The effective use of Benford's law to assist in detecting fraud in accounting data. *Journal of Forensic Accounting*, 5(1):17–34, 2004.

S. F. Ebey and J. J. Beauchamp. Larval fish, power plants, and Buffon's needle problem. *American Mathematical Monthly*, 84(7):534–541, 1977.

R. Eckhardt. Stan Ulam, John von Neumann, and the Monte Carlo method. *Los Alamos Science*, 15:131–137, 1987.

A. Ehrenberg. The pattern of consumer purchases. *Applied Statistics*, 8(1):26–41, 1959.

W. Feller. *An Introduction to Probability Theory and Its Applications*. New York: Wiley, 1968.

T. S. Ferguson. Who solved the secretary problem? *Statistical Science*, 4(3):282–296, 1989.

M. Finkelstein, H. G. Tucker, and J. A. Veeh. Confidence intervals for the number of unseen types. *Statistics and Probability Letters*, 37:423–430, 1998.

Sir R. A. Fisher. *The Design of Experiments*. Edinburg, TX: Oliver and Boyd, 1935.

D. Freedman, R. Pisani, and R. Purves. *Statistics*. New York W.W. Norton and Company, 2007.

S. Geman and D. Geman. Stochastic relaxation, Gibbs distributions, and the Bayesian restoration of images. *IEEE Transactions on Pattern Analysis and Machine Intelligence*, 6:721–741, 1984.

A. K. Gupta and S. Nadarajah, editors. *Handbook of Beta Distribution and Its Applications*. New York: Marcel Dekker, 2004.

R. Harris. Reliability applications of a bivariate exponential distribution. *Operations Research*, 16(1):18–27, 1968.

D. Hoffman. Negative binomial control limits for count data with extra-Poisson variation. *Pharmaceutical Statistics*, 2:127–132, 2003.

A. B. Hollowed, T. Amar, S. Barbeau, N. Bond, J. N. Ianelli, P. Spencer, and T. Wilderbuer. Integrating ecosystem aspects and climate change forecasting into stock assessments. *Alaska Fisheries Science Center Quarterly Report Feature Article*, 2011.

M. Huber and A. Glen. Modeling rare baseball events—are they memory less. *Journal of Statistics Education*, 15(1), 2007.

D. Huff. *How to Take a Chance*. New York: W.W. Norton and Company, 1964.

S. Krantz. *Mathematical Apocrypha Redux: More Stories and Anecdotes of Mathematics and Mathematical*. Washington, DC: Mathematical Association of America, 2005.

K. Z. Leder, S. E. Spagniole, and S. M. Wild. Probabilistically optimized airline overbooking strategies, or "Anyone willing to take a later flight?!". *Journal of Undergraduate Mathematics and Its Applications*, 23(3):317–338, 2002.

R. E. Leiter and M. A. Hamdan. Some bivariate probability models applicable to traffic accidents and fatalities. *International Statistical Review*, 41(1):87–100, 1973.

D. Mackenzie. Compressed sensing makes every pixel count. *What's Happening in the Mathematical Sciences*, 7:114–127, 2009. http://www.ams.org/samplings/math-history/hap7-pixel.pdf.

E. B. Mallon and N. R. Franks. Ants estimate area using Buffon's needle. *Proceedings of the Royal Society B: Biological Sciences*, 267(1445):765–770, 2000.

J. S. Marshall and W. M. Palmer. The distribution of raindrops with size. *Journal of Meteorology*, 5:165–166, 1948.

G. P. Masharin, A. N. Langville, and V. A. Naumov. The life and work of A. A. Markov. *Linear Algebra and Its Applications*, 386:3–26, 2004.

N. Mather, S. M. Traves, and S. Y. W. Ho. A practical introduction to sequentially Markovian coalescent methods for estimating demographic history from genomic data. *Ecology and Evolution* 10(1):579–589, 2020.

S. J. Miller. *Benford's Law*. Princeton, New Jersey: Princeton University Press, 2015.

W. Miller, Stephan C. Schuster, Andreanna J. Welch, et al. Polar and brown bear genomes reveal ancient admixture and demographic footprints of past climate change. *Proceedings of the National Academy of Sciences of the United States of America*, 109(36):E2382–E2390, 2012.

D. B. Murray and S. W. Teare. Probability of a tossed coin landing on edge. *Physical Review E (Statistical Physics, Plasmas, Fluids, and Related Interdisciplinary Topics)*, 48(4):2547–2552, 1993.

M. W. Nachman and S. L. Crowell. Estimate of the mutation rate per nucleotide in humans. *Genetics*, 156:297–304, 2000.

M. Newman. Power laws, Pareto distributions, and Zipf's law. *Contemporary Physics*, 46(5):323–351, 2005.

M. Newman. *Networks*. Oxford, England: Oxford University Press, 2018.

M. J. Nigrini. Benford's Law: *Applications for Forensic Accounting, and Fraud Detection*. Hoboken, NJ: Wiley, 2012.

T. Parsons and E. L. Geist. Tsunami probability in the Caribbean region. *Pure and Applied Geophysics*, 165(11–12):2089–2116, 2008.

A. Piovesan, M. C. Pelleri, F. Antonaros, P. Strippoli, M. Caracausi, and L. Vitale. On the length, weight and GC content of the human genome. *BMC Research Notes* 12(1):106, 2019.

Project Gutenberg. http://www.gutenberg.org. Accessed on May 13, 2013.

J. F. Ramaley. Buffon's noodle problem. *American Mathematical Monthly*, 76(8):916–918, 1969.

M. Richey. Evolution of Markov chain Monte Carlo methods. *American Mathematical Monthly*, 117(5):383–413, 2010.

C. Robert and G. Casella. Short history of Markov chain Monte Carlo: subjective recollections from incomplete data. *Statistical Science*, 26(1):102–115, 2011.

K. A. Ross. Benford's law, a growth industry. *American Mathematical Monthly*, 118:571–583, 2011.

S. Ross. *A First Course in Probability*. Upper Saddle River, NJ: Prentice Hall, 2012.

Z. Schechner, J. J. Kaufman, and R. S. Siffert. A Poisson process model for hip fracture risk. *Medical and Biological Engineering and Computing*, 48(8):799–810, 2010.

H. Stern. Shooting darts. *Chance*, 10(3):16–19, 1997.

S. M. Stigler. Isaac Newton as a probabilist. *Statistical Science*, 21(3):400–403, 2006.

L. Tákacs. The problem of coincidences. *Archive for History of Exact Sciences*, 21(3):229–244, 1980.

K. Troyer, T. Gilroy, and B. Koeneman. A nine STR locus match between two apparent unrelated individuals using AmpFISTR Profiler Plus™and COfiler™. *Proceedings of the Promega 12th International Symposium on Human Identification*, 2001.

B. S. Van Der Laan and A. S. Louter. A statistical model for the costs of passenger car traffic accidents. *Journal of the Royal Statistical Society, Series D*, 35(2):163–174, 1986.

M. vos Savant. Game show problem. http://marilynvossavant.com/game-show-problem/. Accessed on May 13, 2013.

T. Watkins. Raindrop size. http://www.sjsu.edu/faculty/watkins/raindrop.htm. Accessed on October 1, 2020.

M. Williams. Can we measure homelessness? A critical evaluation of 'capture-recapture'. *Methodological Innovations*, 5(2):49–59, 2010.

M. Zwahlen, B. E. Neuenschwander, A. Jeannin, F. Dubois-Arber, and D. Vlahov. HIV testing and retesting for men and women in Switzerland. *European Journal of Epidemiology*, 16(2):123–133, 2000.

INDEX

Probability: With Applications and R, Second Edition. Amy S. Wagaman and Robert P. Dobrow.

Companion Website: www.wiley.com/go/wagaman/probability2e

CPSIA information can be obtained
at www.ICGtesting.com
Printed in the USA
JSHW021443260821
18198JS00001B/104